Cutting Edge Technologies for Developing Future Crop Plants

Anita Mann • Naresh Kumar
Ashwani Kumar • Priyanka Chandra
Satish Kumar Sanwal • Parvender Sheoran
Editors

Cutting Edge Technologies for Developing Future Crop Plants

Springer

Editors
Anita Mann
Division of Crop Improvement
ICAR – Central Soil Salinity Research Institute (CSSRI)
Karnal, Haryana, India

Ashwani Kumar
Division of Crop Improvement
ICAR – Central Soil Salinity Research Institute (CSSRI)
Karnal, Haryana, India

Satish Kumar Sanwal
Division of Crop Improvement
ICAR – Central Soil Salinity Research Institute (CSSRI)
Karnal, Haryana, India

Naresh Kumar
Eternal University
Baru Sahib, Himachal Pradesh, India

Priyanka Chandra
Division of Soil and Crop management
ICAR – Central Soil Salinity Research Institute (CSSRI)
Karnal, Haryana, India

Parvender Sheoran
ICAR - Agricultural Technology Application Research Institute, Zone-1
Ludhiana, Punjab, India

ISBN 978-981-96-2507-9   ISBN 978-981-96-2508-6 (eBook)
https://doi.org/10.1007/978-981-96-2508-6

This Springer imprint is published by the registered company Springer Nature Singapore Pte Ltd.
The registered company address is: 152 Beach Road, #21-01/04 Gateway East, Singapore 189721, Singapore

Preface

The major concern of global climate change, one of the sustainable developmental goals, gives rise to numerous environmental cues affecting crop productivity. Both biotic and abiotic stresses are consequences of changing temperature, water scarcity, increasing salinization, insect-pest infestations, etc. Through agriculture, the only futuristic solution seems to be developing climate-resilient crops adapted to changing environment through translation of knowledge into expedited research to achieve the desired outcomes of increasing crop productivity and ensuring long-term global food and nutritional security. In this book, the technological interventions for managing poor soils and water, nutrient management, plant breeding and physiological strategies, microbial diversity, potential of new or under-utilized crops and advanced strategies for crop improvement against various stresses, etc. along with the economics of agricultural investment and production in developing climate-smart crops have been illustrated very precisely. This book will be helpful for the readers in knowing effect of climate at physiological, biochemical, genetic, and molecular level to understand the plant behavior under stress conditions and what interventions/strategies can be used for development of multi-stress tolerant climate resilient plant types. Incorporating several approaches of multi-omics, genome editing, speed breeding, and traditional mechanical strategies would be one of the strategies to develop climate-smart cultivars for targeting the 'zero hunger' goal.

This book contains the current agricultural status and farmer's prospective on climate change and crop productivity and how the research can be expedited from laboratory to field fruitfully. Inevitable effects of climate change and global warming are well recognized nowadays, limiting the natural resources for agricultural activities and threatening the global food security. The world population is expected to reach 10 billion by the end of 2050, enforcing scientific fraternity to think critically about meeting the food demand. With predicted changes in climatic patterns, agricultural science research must now require to be interdisciplinary to formulate solutions while addressing multifaceted problems under the general umbrella complex of plant, soil, and microflora systems.

The information compiled in this book will be a source of different advanced technologies and strategies for development of climate-resilient multi-stress tolerant plants using multi-agricultural approaches. Hence, this book will serve a reference for the early researchers, scientists, and scholars. This book can also serve as reference read for national and international agricultural scientific platforms for defining further policies on mitigation of climate change effects for sustainable crop production.

The first author would like to express deepest gratitude to all the co-authors for their support during the preparation of the whole manuscript of the book. Also, the authors acknowledge the efforts of the senior research fellows, students, and technical staff in the Molecular Plant Physiology Lab for their timeless efforts during the preparation of the whole manuscript.

Above all, the authors sincerely thank the Director, ICAR-Central Soil Salinity Research Institute, Karnal, for all support and cooperation in preparation of this book.

Hearty gratitude for the reviewers of this book, their expertise was instrumental in shaping the research direction of this book. Their insightful feedback enriched the content and strengthened the presented title.

A special thanks to all our authors for their chapter contributions and for believing in this book content and championing it tirelessly.

The editorial team at Springer deserves applause for their insightful feedback and dedication in shaping this manuscript.

Finally, to the readers who embark on this adventure with us, thank you for giving these characters a home in your imagination.

Karnal, Haryana, India	Anita Mann
Baru Sahib, Himachal Pradesh, India	Naresh Kumar
Karnal, Haryana, India	Ashwani Kumar
Karnal, Haryana, India	Priyanka Chandra
Karnal, Haryana, India	Satish Kumar Sanwal
Ludhiana, Punjab, India	Parvender Sheoran

Contents

About the Editors and Contributors

About the Editors

Anita Mann, Principal Scientist in Plant Physiology at ICAR-CSSRI, Karnal, specializes in molecular plant physiology with a focus on enhancing abiotic stress tolerance in crop plants. She is currently leading a genome-editing project on wheat, rice, and mustard for improved salt tolerance using the CRISPR/Cas9 system, funded by the Indian Council of Agricultural Research (ICAR), New Delhi, with a grant of approximately ₹7 crores. Her research emphasizes the physiological, biochemical, and molecular characterization of salt-specific traits in halophytes, wheat and chickpea and the identification and transformation of salt-tolerance genes in crops like rice. She has contributed significantly to the study of salt tolerance in chickpea, pomegranate, okra, and tomato through integrated physiological, biochemical, molecular, and breeding approaches. She has been awarded research fellowships from CSIR and the Department of Science and Technology, New Delhi. With over 80 publications in peer-reviewed journals, book chapters, and conference proceedings, Dr. Mann serves on the editorial and review boards of reputed national and international journals.

Naresh Kumar, Assistant Professor in the Department of Chemistry and Biochemistry at Eternal University, Himachal Pradesh, specializes in plant stress physiology, antioxidative metabolism, and salinity tolerance in crops. He holds a Ph.D. in Biochemistry from CCS Haryana Agricultural University and has made significant contributions to understanding plant adaptation to abiotic stress through biochemical and molecular pathways. Dr. Kumar has published over 30 research articles in high-impact journals, focusing on stress tolerance, gene expression, and antioxidant responses in various plant species. He has also authored 14 book chapters and 6 popular articles in the field of plant biology and stress physiology. His research portfolio includes key projects on enhancing nutrient use efficiency and identifying salt-tolerant genes in grasses. He serves as the Convenor of the Institution's Innovation Council at Eternal University, where he leads research and innovation initiatives. He is a dedicated reviewer for reputed journals and has expertise in molecular biology, bioinformatics, and statistical analysis.

Ashwani Kumar is a Senior Scientist (Plant Physiology) at ICAR-CSSRI, Karnal, where he focuses on crop improvement for salt-affected regions through institutional and externally funded research projects. His research includes the study of nano-fertilizers to enhance crop growth and nutrient use efficiency, as well as genome editing of wheat for improved salt tolerance. He has made significant contributions to understanding the physiological and biochemical mechanisms of different crop plants in varying environments of salinity or drought. His collaborative work spans inter-divisional and inter-institutional research programs, with funding exceeding ₹3 crores from various agencies. With over 50 peer-reviewed research papers, 15 book chapters, and numerous popular articles, technical bulletins, and training manuals to his credit, Dr. Kumar has an extensive publication record. Additionally, he brings over a decade of teaching experience, mentoring students and researchers in plant physiology and crop science.

Priyanka Chandra, Senior Scientist (Soil Microbiology) at ICAR-CSSRI, is actively engaged in research on microbial diversity and dynamics in salt-affected soils. Her work focuses on screening and isolating salt-tolerant microbes and evaluating the impact of mycorrhiza on plant performance under saline conditions. Through physiological, biochemical, and molecular markers, she assesses the plant growth-promoting properties of these microbes. Dr. Chandra has extensively studied plant growth-promoting rhizobacteria (PGPR), particularly their effects on sugarcane growth, and has purified bioactive metabolites from fungal sources. Her research has resulted in over 50 publications, along with technical bulletins and presentations at national and international conferences. Currently, she is also investigating microbial composition and soil processes under conservation agriculture practices in reclaimed salt-affected soils, contributing to sustainable agricultural solutions in challenging environments.

Satish Kumar Sanwal, Principal Scientist (Plant Breeding) and Head of the Division of Crop Improvement at ICAR-CSSRI, specializes in the development of salt-tolerant crop varieties for saline and sodic soils. His work focuses on both traditional and advanced breeding techniques, utilizing molecular markers of marker-assisted selection (MAS). He is developing salt-tolerant lines of chickpea, lentil, quinoa, and linseed, employing breeding, molecular, and genome-wide association studies (GWAS) to identify and map novel genes and QTLs for salt tolerance. He has developed two okra varieties, two garden pea varieties, two lentil varieties, and one variety each of tomato and a tuber crop. Additionally, he has registered three pea genetic stocks, including multiple disease-resistant and triple-podded lines. With over 100 publications, including research papers, book chapters, technical bulletins, and training manuals, he is also a dedicated mentor, guiding Ph.D. students in plant breeding. He is also a Fellow of the Indian Society of Vegetable Science.

Parvender Sheoran, Director of ICAR-Agricultural Technology Application Research Institute (ATARI), Zone-I, Ludhiana, brings over 20 years of expertise in the National Agricultural Research System. A recipient of numerous accolades, he was awarded the ICAR Swami Sahajanand Saraswati Outstanding Extension Scientist Award (2021), Dr. P.S. Deshmukh Young Agronomist Award (2015), CSSRI Best Scientist Award (2017–18), and the Australian Award Fellowship (2017). He is also a Fellow of the Indian Society of Oilseeds Research and the Indian Society of Weed Science. He has developed seven improved varieties and hybrids of oilseed crops at the national and state levels, alongside formulating ten agronomic recommendations tailored for Punjab and Haryana. The evidence-based technological outputs from his research are quite helpful in demonstrating ecosystem-based approaches viewing optimal resource use, improving ecological resilience with socio-economic development of distressed farming community, and developing impact-oriented technological modules. With over 100 publications in national and international journals, Dr. Sheoran's contributions extend to book chapters, technical bulletins, training manuals, and numerous invited lectures.

Contributors

Akanksha Department of Genetics, Plant Breeding and Biotechnology, Dr Khem Singh Gill Akal College of Agriculture, Eternal University, Sirmour, Himachal Pradesh, India

Sunder Singh Arya Department of Botany, MDU, Rohtak, Haryana, India

R. Banyal ICAR- Central Soil Salinity Research Institute, Karnal, Haryana, India

Nirmalendu Basak ICAR–Central Soil Salinity Research Institute, Karnal, Haryana, India

Division of Soil and Crop Management, ICAR-Central Soil Salinity Research Institute, Karnal, Haryana, India

Sandeep Bedwal ICAR–Central Soil Salinity Research Institute, Karnal, Haryana, India

Pragya Bhadauria ICAR-Agricultural Technology Application Research Institute, Patna, Bihar, India

Bhupender ICAR-Central Soil Salinity Research Institute, Karnal, Haryana, India

Seema Bisht Departmrnt of Microbiology, G.B. Pant University of Agriculture and Technology, Pantnagar, Uttarakhand, India

Ashish Kumar Biswas ICAR–Indian Institute of Soil Science, Bhopal, Madhya Pradesh, India

Priyanka Chandra ICAR–Central Soil Salinity Research Institute, Karnal, Haryana, India

Division of Soil and Crop Management, ICAR-Central Soil Salinity Research Institute, Karnal, Haryana, India

Amresh Chaudhary ICAR–Central Soil Salinity Research Institute, Karnal, Haryana, India

Sagar Chaudhary Uttarakhand University of Horticulture and Forestry, Ranichauri, Uttarakhand, India

Mukesh Choudhary ICAR-Indian Institute of Maize Research, Ludhiana, Punjab, India

Priti Choudhary ICAR-Central Soil Salinity Research Institute, Karnal, Haryana, India

Department of Crop Improvement, ICAR-Central Soil Salinity Research Institute, Karnal, Haryana, India

Annu Dahiya Department of Crop Improvement, ICAR-Central Soil Salinity Research Institute, Karnal, Haryana, India

Avni Dahiya ICAR-Central Soil Salinity Research Institute, Karnal, Haryana, India

Hemant Dasila Department of Microbiology, Akal College of Basic Sciences, Eternal University, Baru Sahib, Himachal Pradesh, India

Nidhi Devi Department of Genetics, Plant Breeding and Biotechnology, Dr. Khem Singh Gill Akal College of Agriculture, Eternal University, Sirmour, Himachal Pradesh, India

Sarita Devi Department of Botany, MDU, Rohtak, Haryana, India

Sneha Dobhal VCSGUUHF, College of Forestry, Ranichauri, Tehri Garhwal, Uttarakhand, India

Uttarakhand University of Horticulture and Forestry, Ranichauri, Uttarakhand, India

Suman Dutta Ramakrishna Mission Vivekananda Educational and Research Institute, Kolkata, West Bengal, India

Ram Kishor Fagodiya ICAR-Central Soil Salinity Research Institute, Karnal, Haryana, India

Division of Soil and Crop Management, ICAR-Central Soil Salinity Research Institute, Karnal, Haryana, India

O. P. Gangwar ICAR-IIWBR, Regional Station, Shimla, Himachal Pradesh, India

Kamini Gautam ICAR-Indian Grassland and Fodder Research Institute, Jhansi, India

Sadhana Giri Department of Microbiology, Kanya Gurukul Campus, Gurukula Kangri (Deemed to be University), Haridwar, Uttarakhand, India

Y. S. Jadoun ICAR-Agricultural Technology Application Research Institute, Patna, Bihar, India

Poonam Jasrotia ICAR-Indian Institute of Wheat and Barley Research, Karnal, Haryana, India

Manisha Kabdwal Pantanjali Research Foundation Trust, Haridwar, Uttarakhand, India

Jayanth Kallugudi ICAR-IIWBR, Regional Station, Shimla, Himachal Pradesh, India

Paras Kamboj Department of Agronomy, CCS Haryana Agricultural University, Hisar, Haryana, India

Prem Lal Kashyap ICAR-Indian Institute of Wheat and Barley Research, Karnal, Haryana, India

Noyonika Kaul ICAR-Central Soil Salinity Research Institute, Karnal, Haryana, India

Department of Crop Improvement, ICAR-Central Soil Salinity Research Institute, Karnal, Haryana, India

Gurpreet Kaur ICAR-Central Soil Salinity Research Institute, Karnal, Haryana, India

Gurvarinder Kaur Department of Botany, Punjabi University, Patiala, Punjab, India

Karanveer Kaur Department of Genetics, Plant Breeding and Biotechnology, Dr. Khem Singh Gill Akal College of Agriculture, Eternal University, Sirmour, Himachal Pradesh, India

Ramandeep Kaur Department of Genetics, Plant Breeding and Biotechnology, Dr. Khem Singh Gill Akal College of Agriculture, Eternal University, Sirmour, Himachal Pradesh, India

Suneetha Kota ICAR-Indian Institute of Rice Research, Hyderabad, Telangana, India

Divjot Kour Department of Microbiology, Akal College of Basic Sciences, Eternal University, Baru Sahib, Himachal Pradesh, India

Krishanu ICAR-Central Soil Salinity Research Institute, Karnal, Haryana, India

P. A. Krishnan ICAR- Indian Agricultural Research Institute, Dhemaji, Assam, India

Aman Kumar ICAR–Central Soil Salinity Research Institute, Karnal, Haryana, India

Anjani Kumar ICAR-Agricultural Technology Application Research Institute, Patna, Bihar, India

Arvind Kumar ICAR-Central Soil Salinity Research Institute, Karnal, Haryana, India

Ashwani Kumar ICAR-Central Soil Salinity Research Institute, Karnal, Haryana, India

Manish Kumar ICAR-Central Soil Salinity Research Institute, Karnal, Haryana, India

Mukesh Kumar Department of Microbiology, Chaudhary Charan Singh Haryana Agricultural University, Hisar, Haryana, India

Naresh Kumar Department of Microbiology, Akal College of Basic Sciences, Eternal University, Baru Sahib, Himachal Pradesh, India

Eternal University, Baru Sahib, Himachal Pradesh, India

Parveen Kumar SKUAST-J, Krishi Vigyan Kendra, Ramban, Jammu and Kashmir, India

Rahul Kumar ICAR Research Complex for North-Eastern Region, Lembucherra, Tripura, India

Raj Kumar ICAR-Central Soil Salinity Research Institute, Karnal, Haryana, India

Rohtash Kumar ICAR-Central Soil Salinity Research Institute, Karnal, Haryana, India

Santosh Kumar ICAR-Indian Agricultural Research Institute, Hazaribagh, Jharkhand, India

Satender Kumar ICAR-Central Soil Salinity Research Institute, Karnal, Haryana, India

Suresh Kumar ICAR-Central Soil Salinity Research Institute, Karnal, Haryana, India

Vivek Kumar PAU-Krishi Vigyan Kendra, Sri Muktsar Sahib, Punjab, India

Gayatri Kumari Department of Botany and Plant Physiology, College of Basic Science and Humanities, CCS Haryana Agricultural University, Hisar, Haryana, India

Pritam Kumari ICAR-Indian Institute of Wheat and Barley Research, Karnal, Haryana, India

CCS Haryana Agricultural University, Hisar, Haryana, India

Charu Lata ICAR-IIWBR, Regional Station, Shimla, Himachal Pradesh, India

ICAR-Indian Institute of Wheat and Barley Research, Regional Station, Shimla, Himachal Pradesh, India

B. M. Lokeshkumar ICAR-Central Soil Salinity Research Institute, Karnal, Haryana, India

Isha Madaan Department of Botany, Punjabi University, Patiala, Punjab, India

Government College of Education, Jalandhar, Punjab, India

Damini Maithani Department of Microbiology, School of Allied Sciences, Dev Bhoomi Uttarakhand University, Dehradun, Uttarakhand, India

Subhasis Mandal ICAR-National Dairy Research Institute, Karnal, Haryana, India

Anshul Sharma Manjul ICAR-IIWBR, Regional Station, Shimla, Himachal Pradesh, India

ICAR-Indian Institute of Wheat and Barley Research, Regional Station, Shimla, Himachal Pradesh, India

Anita Mann ICAR-Central Soil Salinity Research Institute, Karnal, Haryana, India

Department of Crop Improvement, ICAR-Central Soil Salinity Research Institute, Karnal, Haryana, India

Rashmi Mishra Department of Botany, Punjabi University, Patiala, Punjab, India

Monika Department of Botany and Plant Physiology, College of Basic Science and Humanities, CCS Haryana Agricultural University, Hisar, Haryana, India

Shubham Pandey Department of Microbiology, University of Delhi, New Delhi, Delhi, India

Pankaj Department of Botany and Plant Physiology, College of Basic Science and Humanities, CCS Haryana Agricultural University, Hisar, Haryana, India

Himani Panwar Department of Botany and Microbiology, Gurukula Kangri (Deemed to be University), Haridwar, Uttarakhand, India

Chandramani Patel Jawaharlal Nehru University, New Delhi, Delhi, India

Kailash Prajapat ICAR-Central Soil Salinity Research Institute, Karnal, Haryana, India

Division of Social Science Research, ICAR-Central Soil Salinity Research Institute, Karnal, Haryana, India

Nitish Ranjan Prakash ICAR-Central Soil Salinity Research Institute, Karnal, Haryana, India

Govind Prasad ICAR-Central Soil Salinity Research Institute, Karnal, Haryana, India

Pramod Prasad ICAR-IIWBR, Regional Station, Shimla, Himachal Pradesh, India

K. V. Praveen ICAR-Indian Agriculture Research Institute, New Delhi, Delhi, India

Priti K. L. Mehta Dayanand College for Women, Faridabad, Haryana, India

Arvind Kumar Rai ICAR–Central Soil Salinity Research Institute, Karnal, Haryana, India

Division of Soil and Crop Management, ICAR-Central Soil Salinity Research Institute, Karnal, Haryana, India

Rajkumar ICAR-Central Soil Salinity Research Institute, Karnal, Haryana, India

Poonam Ranga ICAR-Central Soil Salinity Research Institute, Karnal, Haryana, India

Department of Crop Improvement, ICAR-Central Soil Salinity Research Institute, Karnal, Haryana, India

K. T. Ravikiran ICAR-CSSRI, Regional Research Station, Lucknow, Uttar Pradesh, India

ICAR-Central Soil Salinity Research Institute, RRS, Lucknow, Uttar Pradesh, India

Satish Kumar Sanwal ICAR-Central Soil Salinity Research Institute, Karnal, Haryana, India

S. Sarvade Department of Forestry, Jawaharlal Nehru Krishi VishwaVidhyalaya, Jabalpur, Madhya Pradesh, India

Anish Sehrawat Department of Botany, Maharshi Dayanand University, Rohtak, Haryana, India

Anita Rani Sehrawat Department of Botany, Maharshi Dayanand University, Rohtak, Haryana, India

Deepti Rani Sehrawat Department of Anthropology, Panjab University, Chandigarh, Punjab, India

Gargi Sharma Division of Soil and Crop Management, ICAR-Central Soil Salinity Research Institute, Karnal, Haryana, India

Imran Sheikh Department of Genetics, Plant Breeding and Biotechnology, Dr. Khem Singh Gill Akal College of Agriculture, Eternal University, Sirmour, Himachal Pradesh, India

Parvender Sheoran Ludhiana, Punjab, India

ICAR-Agricultural Technology Application Research Institute, Ludhiana, Punjab, India

Awtar Singh Division of Soil and Crop Management, ICAR-Central Soil Salinity Research Institute, Karnal, Haryana, India

Kajal Singh Department of Botany and Microbiology, Gurukula Kangri (Deemed to be University), Haridwar, Uttarakhand, India

Palampreet Singh Department of Botany, Punjabi University, Patiala, Punjab, India

Geetika Sirhindi Department of Botany, Punjabi University, Patiala, Punjab, India

Sanchika Snehi Institute of Agricultural Sciences, BHU, Varanasi, Uttar Pradesh, India

Pragati Srivastava ICAR National Institute of Plant Biotechnology, IARI PUSA, New Delhi, Delhi, India

Sukriti Krishi Vigyan Kendra, CCS Haryana Agricultural University, Rohtak, Haryana, India

Parul Sundha ICAR–Central Soil Salinity Research Institute, Karnal, Haryana, India

Division of Soil and Crop Management, ICAR-Central Soil Salinity Research Institute, Karnal, Haryana, India

Akriti Thakur ICAR-IIWBR, Regional Station, Shimla, Himachal Pradesh, India

R. Thribhuvan ICAR – Central Research Institute for Jute and Allied Fibres, Nilganj, West Bengal, India

Vikrant Tyagi Department of Genetics, Plant Breeding and Biotechnology, Dr. Khem Singh Gill Akal College of Agriculture, Eternal University, Sirmour, Himachal Pradesh, India

Viabhav Kumar Upadhayay Departmrnt of Microbiology, G.B. Pant University of Agriculture and Technology, Pantnagar, Uttarakhand, India

Kamlesh Verma Division of Soil and Crop Management, ICAR-Central Soil Salinity Research Institute, Karnal, Haryana, India

ICAR-Central Soil Salinity Research Institute, Karnal, Haryana, India

Gajender Yadav ICAR-Central Soil Salinity Research Institute, Karnal, Haryana, India

Manish Yadav ICAR-Central Soil Salinity Research Institute, Karnal, Haryana, India

Rajender Kumar Yadav ICAR-Central Soil Salinity Research Institute, Karnal, Haryana, India

Division of Soil and Crop Management, ICAR-Central Soil Salinity Research Institute, Karnal, Haryana, India

Sujata Yadav ICAR-Central Soil Salinity Research Institute, Karnal, Haryana, India

Department of Crop Improvement, ICAR-Central Soil Salinity Research Institute, Karnal, Haryana, India

Abbreviations

ABA	Abscisic Acid
ACC	1-Aminocyclopropane-1-Carboxylate
AFLP	Amplified Fragment Length Polymorphism
AMF	Arbuscular Mycorrhizal Fungi
APX	Ascorbate Peroxidase
AR5	Fifth Assessment Report
ATP	Adenosine Triphosphate
AUM	Animal Unit Month
BRs	Brassinosteroids
Bt	*Bacillus thuringiensis*
bZIP	Basic Leucine Zipper Transcription Factor
CAIT	Climate Analysis Indicators Tool
CAM	Crassulacean Acid Metabolism
CAS	Climate Action Strategy
Cas9	CRISPR Associated Protein 9
CAT	Catalase
CBA	Cost-Benefit Analysis
CBB	Calvin-Benson-Bassham Cycle
CBP	Capacity Building Program
CCS	Carbon Capture and Storage
cDNA	Complementary DNA
CDPK	Calcium-Dependent Protein Kinase
CH_4	Methane
CNG	Compressed Natural Gas
CO_2	Carbon Dioxide
COP	Conference of the Parties
CRISPR	Clustered Regularly Interspaced Short Palindromic Repeats
CRP	Comprehensive Risk Profiling
CSA	Climate-Smart Agriculture
CSIR	Council of Scientific and Industrial Research
DHAR	Dehydroascorbate Reductase
DREB	Dehydration-Responsive Element-Binding Protein
EEZ	Exclusive Economic Zone
EMI	Environmental Management Indicator

ET	Ethylene
ETS	Emissions Trading System
$FADH_2$	Flavin Adenine Dinucleotide
FAO	Food and Agriculture Organization
Fd	Ferredoxin
F-gases	Fluorinated Gases
FGDG	Flue Gas Desulfurization Gypsum
FNR	Ferredoxin-NADP+ Reductase
Fv/Fm	Maximum Quantum Efficiency of Photosystem II
GA_3	Gibberellic Acid
GABA	Gamma-Aminobutyric Acid
GAD	Glutamate Decarboxylase
GAE	Gallic Acid Equivalent
GAs	Gibberellins
GCF	Green Climate Fund
GEF	Global Environment Facility
GHE	Greenhouse Effect
GHG	Greenhouse Gases
GIS	Geographic Information System
GM	Genetically Modified
GMD	Global Monitoring Division
GMO	Genetically Modified Organism
GPX	Glutathione Peroxidase
GR	Glutathione Reductase
GSH	Glutathione
GST	Glutathione S-Transferase
GWP	Global Warming Potential
HFCs	Hydrofluorocarbons
HIRA	Hazard Identification and Risk Assessment
HR	Hypersensitive Response
HSPs	Heat Shock Proteins
IAA	Indole Acetic Acid
ICT	Information and Communications Technology
IDB	Inter-American Development Bank
IEA	International Energy Agency
IFAD	International Fund for Agricultural Development
ILUC	Indirect Land Use Change
IP	Inositol Phosphate
IPCC	Intergovernmental Panel on Climate Change
ISP	International Soil Partnership
JRC	Joint Research Centre
LDN	Land Degradation Neutrality
LEA	Late Embryogenesis Abundant
LHC	Light Harvesting Complex
LPG	Liquefied Petroleum Gas

MAPK	Mitogen-Activated Protein Kinase
MDGs	Millennium Development Goals
MDHAR	Monodehydroascorbate Reductase
MEP	Methylerythritol Phosphate Pathway
miRNA	MicroRNA
MOEFCC	Ministry of Environment, Forest and Climate Change
mRNA	Messenger RNA
MSWC	Municipal Solid Waste Compost
MYB	MYB Transcription Factors
N_2	Nitrogen Gas
N_2O	Nitrous Oxide
NADH	Nicotinamide Adenine Dinucleotide
NADPH	Nicotinamide Adenine Dinucleotide Phosphate
NAPCC	National Action Plan on Climate Change
NB-LRR	Nucleotide-Binding Leucine-Rich Repeat
NGO	Non-Governmental Organization
NGS	Next-Generation Sequencing
NMHS	National Mission on Himalayan Studies
NMNF	National Mission on Natural Farming
NO_2	Nitrogen Dioxide
NOx	Nitrogen Oxides
NRF	Natural Resource Framework
NRM	Natural Resource Management
NUE	Nutrient Use Efficiency
OEC	Oxygen-Evolving Complex
OPEC	Organization of the Petroleum Exporting Countries
ORF	Open Reading Frame
PA	Paris Agreement
PCR	Polymerase Chain Reaction
PDP	Policy Development Process
PEP	Phosphoenolpyruvate
PEPC	Phosphoenolpyruvate Carboxylase
PFCs	Perfluorocarbons
PGPR	Plant Growth-Promoting Rhizobacteria
PGR	Plant Growth Regulator
PLC	Phospholipase C
PLD	Phospholipase D
PM H^+-ATPase	Plasma Membrane H^+-ATPase
POD	Peroxidase
PPC	Pyruvate Carboxylase
PQ	Plastoquinone
PR	Pathogenesis-Related Proteins
QTL	Quantitative Trait Locus
RAPD	Random Amplified Polymorphic DNA
REE	Renewable Energy Expansion

RES	Renewable Energy Source
RFLP	Restriction Fragment Length Polymorphism
RNAi	RNA Interference
ROS	Reactive Oxygen Species
RT-PCR	Reverse Transcription Polymerase Chain Reaction
RUBISCO	Ribulose-1,5-bisphosphate Carboxylase/Oxygenase
RuBP	Ribulose-1,5-bisphosphate
RuDP	Ribulose Diphosphate
SA	Salicylic Acid
SAS	Salt-Affected Soils
SCS	Soil Carbon Sequestration
SDG	Sustainable Development Goals
SDI	Sustainable Development Indicator
siRNA	Small Interfering RNA
SMs	Secondary Metabolites
SNP	Single Nucleotide Polymorphism
SOC	Soil Organic Carbon
SOD	Superoxide Dismutase
SRI	System of Rice Intensification
SSR	Simple Sequence Repeat
SWC	Soil and Water Conservation
TCA	Tricarboxylic Acid Cycle
TFs	Transcription Factors
UHI	Urban Heat Island
UN	United Nations
UNDP	United Nations Development Programme
UNEP	United Nations Environment Programme
UNFCCC	United Nations Framework Convention on Climate Change
USEPA	United States Environmental Protection Agency
UV	Ultraviolet
VIGS	Virus-Induced Gene Silencing
WBCSD	World Business Council for Sustainable Development
WEF	Water-Energy-Food Nexus
WFP	World Food Programme
WHO	World Health Organization
WMO	World Meteorological Organization
WRI	World Resources Institute
WRKY	WRKY Transcription Factor Family
WUE	Water Use Efficiency

Part I

Agricultural Preview, Sustainable Practices and Policy Perceptions

Farmer's Perspective on Climate Change and Agricultural Production

1

Parvender Sheoran, Parveen Kumar, Vivek Kumar, Ashwani Kumar, and Anita Mann

Abstract

The agriculture sector has the unique distinction of being a contributor to climate change as well as a victim of this phenomenon suffering from its adverse impacts. The sector impacts climate change through the release of greenhouse gases (GHGs), including carbon dioxide (CO_2), methane (CH_4), and nitrous oxide (N_2O). The impact of this has been evident from the various reports of the Intergovernmental Panel on Climate Change (IPCC). One point emphasizes that human activities have driven all global warming over the past 200 years, causing a current temperature increase of 1.1 °C above pre-industrial levels. This increase in temperature has contributed to more frequent and severe weather events, causing growing harm to both people and the environment. Reports indicate that around 3.3–3.6 billion people are living in situations that are highly susceptible to the impacts of climate change. The increased temperatures have put our food and nutritional security under threat. Climate change is projected to lead to around 250,000 additional deaths annually between 2030 and 2050, primarily due to undernutrition, diarrhea, malaria, and heat stress. Fortunately, this sector also provides us with various adaptation as well as mitigation strategies. The adoption of various climate-resilient and resource-conserving technologies can

P. Sheoran (✉)
Ludhiana, Punjab, India
e-mail: Parvender.Sheoran@icar.gov.in

P. Kumar
SKUAST-J, Krishi Vigyan Kendra, Ramban, Jammu and Kashmir, India

V. Kumar
PAU-Krishi Vigyan Kendra, Sri Muktsar Sahib, Punjab, India

A. Kumar · A. Mann
ICAR-Central Soil Salinity Research Institute, Karnal, Haryana, India

A. Mann et al. (eds.), *Cutting Edge Technologies for Developing Future Crop Plants*,
https://doi.org/10.1007/978-981-96-2508-6_1

help us to reduce the emission of GHGs and mitigate the adverse effects of rising temperatures. At the same time, a shift toward nonconventional practices such as a transition to renewable energy sources, increasing energy use efficiency, and protecting and restoring forests and other critical ecosystems have the potential to restore this planet to its original state.

Keywords
GHGs · IPCC · Climate change · Adaptation · Mitigation

1.1 Introduction

It is now well established that the rise in global temperature and the consequent climate change as a result of it has emerged as a potent threat to life on this planet. The rising temperature, largely due to various anthropogenic activities, has already taken a huge toll on human lives and our rich biodiversity (Shivanna 2022). Natural resources like land, water, and air are also witnessing the harmful effects of rising temperatures and it is manifested in the form of pollution, degradation, and desertification. A research finding of the World Meteorological Organization (WMO) also authenticates that the planet is now approximately one degree warmer than it was before the onset of widespread industrialization (BBC 2020). The World Meteorological Organization (WMO) has confirmed that 2023 stands as the warmest year on record, with the global average near-surface temperature reaching 1.45 ± 0.12 °C above the pre-industrial average of 1850–1900. This marks 2023 as the hottest year in the 174-year observational record, surpassing previous records set in 2016, which was 1.29 ± 0.12 °C above the 1850–1900 average, and in 2020, which recorded 1.27 ± 0.13 °C above the same baseline (WMO 2024a, b). The Intergovernmental Panel on Climate Change (IPCC) has also reported that each of the past three decades has shown a warming trend, with the 2000s being the hottest decade recorded (Alexander 2016). In a study of global climate models and projected development scenarios, Agarwal and Singh (2010) reported that global warming could increase by 1.4–5.8 °C over the coming century. The latest IPCC report emphasizes that human activities have driven all global warming over the past 200 years, leading to a current temperature increase of 1.1 °C above pre-industrial levels. This warming has intensified the frequency and severity of hazardous weather events, causing escalating damage to both people and the environment (UN Environment Programme 2023). The situation is expected to worsen, and the report issues a strong warning. There is over a 50% likelihood that the global temperature increase will reach or exceed 1.5 °C between 2021 and 2040 across various scenarios. In a high-emissions scenario, this threshold could be reached even earlier, potentially between 2018 and 2037 (Locci and Jammoul 2023).

It is widely understood that the primary driver of global warming is the increased concentration of greenhouse gases (GHGs) in the atmosphere. The most common GHGs are carbon dioxide (CO_2), methane (CH_4), and nitrous oxide (N_2O), largely generated by human activities like fossil fuel combustion and changes in land-use patterns (Yoro and Daramola 2020). Of these gases, CO_2 is the most abundant and significant. Over the past 50 years, atmospheric CO_2 levels have seen one of the most notable increases, rising sharply from 280 ppm in the pre-industrial era to 416 ppm. By 2100, this concentration could potentially double (GMD 2011).

The regions that are more vulnerable to adverse effects of rising temperatures are home to an estimated 3.3 to 3.6 billion people (IPCC 2023a). The vulnerabilities include their proneness to food and nutritional insecurity, and various types of ailments like malaria, diarrhea, and heat stress. Additionally, research indicates that climate change could result in approximately 250,000 additional deaths annually between 2030 and 2050 due to various vulnerabilities. All this will have a bearing on the finances of the sufferer in the form of increased spending on health. A study conducted by the World Health Organization (WHO) estimated that the direct healthcare costs, excluding expenses in health-determining sectors like agriculture, water, and sanitation, could range from US$ 2 to 4 billion annually by 2030 (WHO 2023). The developing and undeveloped third world countries which have weak health infrastructure and inadequate human resources will be the worst sufferers. This is due to their limited ability to prepare for and respond to challenges without external support.

In terms of the potential impact of climate change on food security, data from 2021 indicate that approximately 720 million to 811 million people were undernourished, with around 200,000 facing catastrophic levels of food insecurity (FAO, IFAD, UNICEF, WFP, and WHO 2021). By 2050, climate change is expected to expose an additional 8–80 million people to the risk of hunger. In a 2006 study, Shreshta also highlighted that rising temperatures, climate extremes, increased CO_2 and other GHGs, and changing rainfall patterns will severely threaten global food production. The extent of food insecurity will depend on the severity of future warming and the efficiency of adaptation strategies.

The State of Climate in Asia 2023 report highlights significant regional variations in the impacts of climate change. It revealed that Asia was the most disaster-prone region globally in 2023, facing the highest number of weather, climate, and water-related hazards (WMO 2024a, b). The highest number of reported causalities and economic losses was caused by floods and storms as the impact of heat waves became more severe. The report also underscored the rising trends in critical climate change indicators, including surface temperature, glacial retreat, and sea level rise. These changes are expected to have significant impacts on communities, economies, and ecosystems prevalent across the whole Asian region.

1.2 Farmers' Perception About Climate Change

Climate change is now a reality and has become an all-pervasive phenomenon. Farmers, in particular, are increasingly experiencing its impacts and dealing with the consequences. A study by Jha and Gupta (2021) in Bihar explored farmers' perceptions of climate change. Findings showed that the majority of farmers believed temperatures had risen over the past 20 years. Specifically, 87% noticed more hot days, 4% observed fewer, and 9% reported no change. Regarding cold days, 44% reported an increase, 47% observed a decrease, and 9% noted no change. Additionally, 86% of farmers observed shifts in precipitation; of these, 47% noted a decline, 39% saw an increase, and 14% reported no change. The study also highlighted that 80% of farmers were aware of climate trends and sought adaptive measures. Key socio-economic factors—including age, gender, household size, education, off-farm income, and farm size—significantly influenced farmers' decisions to adapt. A similar study conducted in the marshlands of South Kivu, Democratic Republic of Congo, examined farmers' perceptions of climate change and adaptation strategies. Balasha et al. (2023) found that both men (77%) and women (73%) reported experiencing climate change, noting shifts in temperature and precipitation patterns. Farmers' perceptions aligned with local climate data, which indicated a slight increase in temperature and a decrease in rainfall over the past decade, particularly between 2013 and 2019. The study documented common impacts observed by farmers, including an increase in pests (90%), reduced soil fertility (75%), and crop failures due to flooding. In response, farmers adopted various adaptation methods, such as crop diversification, planting low-maintenance crops, and using mulch and manure. These practices were influenced by factors like experience, information exchange among farmers, livestock ownership, and awareness of climate-related threats to crops.

1.3 Climate Change and Agriculture

Agriculture is both a contributor to and a sector impacted by climate change. It primarily contributes to climate change through GHG emissions, including CH_4, CO_2, and N_2O. Methane emissions in agriculture arise from the microbial breakdown of organic material in oxygen-poor (anaerobic) environments (Chauhan et al. 2014). For instance, paddy fields, when kept flooded, create conditions conducive to CH_4 production (Mahajan et al. 2017). Additionally, livestock contribute significantly to CH_4 emissions through enteric fermentation in ruminant animals. The burning of crop residues, common in areas like Punjab and Haryana, not only pollutes the air but also releases GHGs into the atmosphere (Sharma et al. 2024; Kumar et al. 2023a). Similarly, CO_2 is also emitted during agricultural activities, particularly from fuel combustion in machinery, which contributes to the trapping of solar radiation and raises surface temperatures. Nitrous oxide, another potent GHG, is released into the atmosphere primarily due to the application of nitrogen-based fertilizers.

India's contribution to global GHG emissions is around 5% of the global total of 55 billion tonnes of CO_2 equivalent. Agriculture accounts for approximately 11–12% of this global emission, with Indian agriculture contributing about 1% (Pathak et al. 2014). In India, the energy sector is the largest source of GHG emissions (65%), followed by agriculture (18%) and industry (16%). In 2010, the agricultural sector, encompassing both crop production and livestock, emitted around 420 million tonnes of CO_2 equivalents (Pathak 2015). Between 1970 and 2010,

GHG emissions from Indian agriculture rose by approximately 75%t, largely due to the indiscriminate use of inorganic fertilizers and other agricultural inputs (Pathak et al. 2014).

By 2030, the emissions of GHGs from the agriculture sector are projected to be nearly 60% higher as compared to 1990 levels, with N_2O emissions from soils expected to rise by 75% due to the increased use of chemical fertilizers. Methane emissions from various agricultural sources are also anticipated to rise significantly. Research indicates that, if current trends continue, GHG emissions from India's agriculture sector could increase by 17% by 2030 compared to 2010 levels (Pathak et al. 2014).

The impact of rising temperatures on crops varies depending on the crop type and the region where it is grown. Increased atmospheric CO_2 from anthropogenic activities, especially fossil fuel combustion, results in a "CO_2 fertilization effect." While this effect can partially counteract some negative impacts of climate change on agriculture, it also leads to lower concentrations of essential micronutrients in crops (IPCC 2023). It is important to note that CO_2 fertilization has a minimal impact on crops such as maize, which utilize the Hatch–Slack (C_4) photosynthetic pathway (Ainsworth and Long 2020). The rise in sea levels will result in the loss of some agricultural land, and the melting glaciers due to high temperatures will also impact the availability of water needed for irrigation (Biemans et al. 2019). Other related impacts include soil erosion, shifts in soil fertility, and alterations in the length of growing seasons. On a positive note, the thawing of previously frozen land could create more arable land. However, rising temperatures also accelerate the growth of bacteria like *Salmonella* and mycotoxin-producing fungi, which harm food safety, increase food loss, and drive up prices (IPCC 2023).

1.3.1 Effect of Climate Change on Field Crops

The increase in temperature and the consequent changes in weather patterns come with the risk of adversely affecting the yield of cultivated crops and in extreme cases even there is the possibility of simultaneous crop failures (Schmitt et al. 2022). This will ultimately have a bearing on the food and nutritional security of the populace all across the globe. The climate-induced crop yield reduction due to rising temperatures has already been reported worldwide (Wing et al. 2021; Liu et al. 2022). Research studies done in West Africa have reported that the increased intensification of extreme weather events has reduced millet and sorghum productivity by 10–20 and 5–15%, respectively (Sultan et al. 2019). Likewise, in Europe, heat extremes from 1950 to 2019 have become more frequent and increasingly occur consecutively, while cold extremes have become less common (Pradhan et al. 2022). The loss in crop production due to the severity of heat waves and drought has also been reported in Europe. In Europe, yield declines have intensified over time, increasing threefold over a 50-year period from 2.2% between 1964 and 1990 to 7.3% from 1991 to 2015 (Brás et al. 2021). The 2021 floods in Europe, which severely affected Belgium, caused significant agricultural damage and led to long-term issues like soil erosion (Alan 2021). In China, extreme rainfall events over the past two decades have reduced rice production by approximately 8% (Jin et al. 2023). Climate change impacts on key staple crops—rice, wheat, maize, and soybean—have been extensively studied, as these four crops provide roughly two-thirds of the calories consumed by humans directly and indirectly as animal feed (Chuang et al. 2017). Soybean growth and leaf area are notably reduced under temperatures beyond their historical range (Burroughs et al. 2023). Projections by the U.S. National Research Council (2011) indicate that maize will fail to reproduce at temperatures above 35 °C (95 °F), and soybean at temperatures above 38.8 °C (101.8 °F) (US Global Change Research Program 2013).

1.3.2 Emergence of New Insect Pests

Another impact of the rising temperatures will be the emergence of new insect pests and plant diseases. Climate change and extreme weather events have a major impact on agricultural pests. Temperature is the primary environmental factor influencing insect population dynamics, so it is expected that an increase in temperature could lead to an increase in the geographic range of insects/pests along with increased overwintering survival. It will also result in more generations, a higher risk of invasions by these insect species, and an increased likelihood of plant diseases spread by insects. Their interactions with host plants and natural enemies are also likely to shift (Skendžić et al. 2021). Sharma and colleagues observed that global warming could promote the spread of insect-transmitted plant diseases, driven by geographic expansion and growing populations of insect vectors (Sharma et al. 2005; Sharma 2014). The primary insect order responsible for transmitting plant viruses is the sap-feeding Hemiptera (Skendžić et al. 2021). Among these, aphids represent the major group of vectors, carrying over 275 virus species, with many aphid species capable of transmitting plant viruses (Skendžić et al. 2021). Aphids play a key role as virus carriers in temperate zones worldwide, while whiteflies are more common in warmer climates, especially thriving in temperate regions where crops are cultivated in greenhouse environments (Canto et al. 2009). The high reproductive rates and short development cycles of these sucking insects make them particularly responsive to climate change (Hance et al. 2007).

Climate change can impact the migration and long-distance dispersal abilities of virus vectors. For instance, aphids can travel considerable distances when favorable thermal conditions lift them into the air, where atmospheric currents enable horizontal movement (Fereres et al. 2017). In North America, this form of long-distance transport has been associated with severe viral outbreaks. Persistent low-pressure winds from the southern Great Plains have carried aphids northward to corn-growing regions in Minnesota, leading to significant epidemics (Zeyen et al. 1987). All these studies thus reveal that there exists an intricate relationship between weather conditions and insect pest occurrence and there is a need to understand this intricate relationship for adopting timely and effective pest management strategies (Kumar et al. 2021).

1.3.3 Effect of Climate Change on Livestock

The livestock sector is a substantial source of GHG emissions, accounting for approximately 14.5% of emissions of GHGs at the global level (Gerber et al. 2013). Climate change poses challenges to livestock worldwide, leading to issues such as increased heat stress, feed shortages, and heightened risks from parasites and vector-borne diseases (IPCC 2023). Temperature changes affect many essential aspects of livestock production, including water supply, growth, reproduction, and overall health. Livestock farming also uses around 8% of global human water use, and rising temperatures could double or triple animals' water needs (Nardone et al. 2010). Additionally, rising sea levels may increase saltwater intrusion into coastal freshwater aquifers, adding salinity and raising concentrations of biological, chemical, and heavy metal contaminants, which could further impact livestock productivity (Karl et al. 2009; Nardone et al. 2010).

1.3.4 Effect of Climate Change on Fisheries

Fisheries are another sector that provides a crucial source of livelihood for many impoverished and particularly vulnerable communities. Approximately 43.5 million people worldwide are directly employed in the fisheries sector, with the majority based in developing countries (Mohammed and Uraguch 2013). As in the case of crops and livestock, fisheries are significantly impacted by climate change, though the effects vary by each fishery (Weatherdon et al. 2016). Marine aquatic ecosystems are reported to be affected by ocean acidification (Doney 2006), increasing ocean temperatures (Bindoff et al. 2007), and ocean deoxygenation. In freshwater ecosystems, changing water temperatures contribute to habitat loss for fish (USEPA2017). Increased ocean temperatures and acidity make it harder for marine species like corals, oysters, and shrimp to build shells, a process called calcification (Anonymous 2020). Climate change is also altering fish distribution patterns. The IPCC warns that if ocean temperatures continues to rise, up to 90% of coral reefs could be destroyed. A 5 °C increase in global temperature could result in the loss of nearly two-thirds of global fish species by 2100 (Robitzski 2020). The greatest reduction in fishery potential is expected in tropical areas, especially in the South Pacific (Barange et al. 2018).

The economic and geopolitical effects will be profound, particularly for countries heavily reliant on this sector. The intersection of agriculture, climate change, and food security is increasingly complex and severe for developing nations. Among these, India has been identified as particularly vulnerable to climate-related risks. The frequency and intensity of extreme weather events will negatively impact food security by reducing availability, accessibility, stability, and utilization.

1.4 Social Ramifications of Climate Change

Climate change will also have profound effects on social structures. The social ramifications of climate change include worsening health for populations more vulnerable to rising temperatures, increased climate migration and displacement, as well as higher rates of hunger and malnutrition in regions where food production becomes insufficient (Moon 2023). While climate change affects everyone, its impacts vary across regions, gender, age, and socio-economic status. In this regard, it is also important to consider findings from the IPCC. The report indicated that marginalized and vulnerable populations will be the most affected by rising temperatures. The resource poor who do not have access to even the basic resources are more at risk from the threats of climatic change. This is due to of their reduced ability to adapt to severe events like droughts or floods (IPCC 2022). They are unable to cope with these extreme events, which directly impact their livelihoods and ability to produce food. Studies have revealed that both men and women involved in natural resource sectors like agriculture are likely to face the effects of climate change (Osman-Elasha 2009). However, these impacts will differ by gender. Women are often more vulnerable to climate change than men, largely because they make up a significant portion of the world's poor and are more reliant on threatened natural resources (Osman-Elasha 2009).

1.5 Climate Change and Impact on Women

According to the United Nations, approximately 80% of most affected by climate change are the "women" (Waheed 2023). This fact has been further substantiated by recent studies, which have established a clear link between gender inequality and climate change. Let us clear it by some case studies. When Bangladesh was hit in the aftermath of Cyclone Gorky, women made up 91% of the fatalities (IUCN 2021). In 2003, heat waves that erupted across Europe, three-fourths of the total fatalities were found to be in women. These numbers are severe, considering the projections that people dying from climate change will outnumber those dying from all infectious diseases. Already, there exist many disparities between two genders in terms of equality and accessibility. Our society also does not provide equal access to girls in terms of education, health, and so many other opportunities. When women lack access to education and fair employment opportunities, they are less likely to acquire the information and support necessary to effectively manage the effects of climate change (UNFCC 2016).

The female folk face specific, life-threatening challenges following climate-related tragedies. The risk of gender-based violence increases, as families may resort to practices like child marriage of girls for survival during severe droughts (IUCN 2021). After Cyclone Idai hit Zimbabwe, Malawi, and Mozambique in 2019, approximately 75,000 pregnant women lacked access to reproductive healthcare, sanitation, and clean water facilities, and faced heightened risks of sexual assault and harassment in camps (IUCN 2021). In Fiji, following Cyclone Winston in 2016, gender minorities were denied access to shelters, increasing their vulnerability to injury and death (IUCN 2021). Differences between men and women are evident in their roles, responsibilities, decision-making power, and access to resources, including land, natural resources, and opportunities (Osman-Elasha 2008). Globally, women have less access than men to land, credit, agricultural resources, technology, training, and extension services, which are essential for building resilience to climate change (Aguilar 2008). It can thus be concluded that climate change further intensifies existing gender inequalities.

1.6 Climatic Displacement and Refugee

The term "climate refugees" has been in use since 1985, when United Nations Environment Programme (UNEP) expert Essam El-Hinnawi defined climate refugees, also known as climate or environmental migrants, as individuals "forced to leave their traditional habitat, temporarily or permanently, due to significant environmental disruptions" (Bates 2002). In 2023, McAllister reported that extreme weather, rising temperatures, and ecosystem damage are endangering millions of climate refugees. For these individuals, climate migration is not a distant issue but an immediate reality, as they experience the severe impacts of the climate crisis firsthand. The number of climate migrants is projected to grow substantially as global climate threats intensify (McAllister 2023).

According to the UN Refugee Agency (UNHCR), between 2008 and 2016, an average of 21.5 million people were forcibly displaced each year due to severe weather events, such as extreme temperatures, floods, wildfires, and storms (McAllister 2023). This figure is expected to rise sharply in the coming decades. The Institute for Economics and Peace (IEP), an Australian think tank, estimates that by 2050, as many as 1.2 billion people could be displaced globally due to climate change and natural disasters.

1.7 Mitigation and Adaptation Strategies to Combat Climate Change

As discussed in depth, rising temperatures are adversely impacting both human and natural systems, leading to reduced crop productivity, loss of biodiversity, and posing a threat to food and nutritional security. The decline in crop productivity has a bearing on the life and livelihoods of farmers, particularly in India, where a large segment of the population earns its livelihood from this sector. As a climate-sensitive sector, agriculture is especially vulnerable to the adverse effects of climate change. This thus underlines an urgent need and a focused dual strategy of mitigation and adaptation to at least halt the adverse impacts of climate change and restore this planet to its original form. Our strategies and action plans must be integrated with innovative technologies that balance adaptation and mitigation, reducing trade-offs while maximizing synergies. There are numerous feasible, effective, and low-cost options available in agriculture, the only need is to replicate them and scale them up in the near term (IPCC 2023). As agriculture contributes to climate change through the emission of various GHGs, the solution also lies in promoting various adaptation and mitigation strategies.

1.7.1 Adaptation Strategies

As far as adaptation is concerned, it means the state or process of changing to suit a new situation. We have to change the way we do farming; we have to shift our farming from a chemical intensive to one that is based on the use of local resources. For an agriculture that depends on external inputs, we have to do it in a way where the use of external inputs is zero. Agronomic practices like minimum or no-till agriculture, which do not disturb the soil, result in the conservation of water, soil surface coverage with crop residues, and the adoption crop rotations. In addition to reducing the water requirements, these practices also ensure that the top fertile layer does not get eroded by fast-moving winds or intense rains (Arshad et al. 2023; Fagodiya et al. 2024). Similarly, sowing across the slope helps reduce soil erosion and the loss of nutrients, which supports soil fertility (Hilger et al. 2013). Planting in rows with a seed-cum-fertilizer drill or maize planter lowers the seed rate per unit area, maintains an ideal plant population, and decreases the risk of lodging at maturity, ultimately providing farmers with better yields. Farmers do not go for the recommended package of practices out of ignorance and often indulge in the use of more than the recommended dose of chemical fertilizers (Sheoran et al. 2023) which ultimately results in increased emission of GHGs. The farmers need to be aware of the appropriate doses and the right combinations of fertilizers. Local resources such as plant litter, cow dung and urine, and other organic waste materials can be effectively used to make compost or vermicompost, which can substitute for inorganic fertilizers, helping to lower input costs and improve soil health (Manna et al. 2018). Bunding is a technique used to control soil erosion and retain soil moisture, with various methods including compartmental bunding, block bunding, and contour bunding (Kumar and Kumar 2016). Sloped areas can be planted with grasses and trees for additional stability. Employing climate-resilient and region-specific crop varieties is essential for agriculture to effectively tackle the variations in climate (Sheoran et al. 2022). Developing and selecting varieties that can withstand heat, drought, flooding, chilling, and salinity stresses is crucial for maintaining and enhancing crop yields in the face of climate change challenges (Sheoran et al. 2021). Recently, 109 high-yielding, climate-resilient, and bio-fortified varieties of 61 crops including 37 food crops and 24 horticultural crops were released by the Hon'ble Prime Minister of India Sh. Narendra Modi at the Indian Agricultural Research Institute (IARI), New Delhi, to cope with the anticipated climatic vulnerabilities and ensure the nutritional security of citizens (Anonymous 2024).

Mulching is another resource conservation technology that provides many benefits such as soil moisture conservation (Kader et al. 2019), alteration of soil temperature (Kader et al. 2017), and reduction of compaction after heavy rains (Oliveira and Merwin 2001; Iqbal et al. 2020). Mulching involves covering the soil using different materials, such as plastic sheets, straws, dust, and peat, or by keeping a vegetative cover in place (Patil et al. 2013). In wheat, mulching has been reported to be beneficial. It increases tolerance to terminal heat stress by lowering canopy temperature and maintaining optimal water potential in plant leaves (Ram 2020). Several technologies, including the happy seeder, smart seeder, and zero-till drill, have been developed by agricultural scientists along with mulching techniques (Sharma et al. 2024; Kumar et al. 2023a; Fagodiya et al. 2023). These interventions and innovations can help mitigate the adverse effects of high temperatures on wheat (Sidhu et al. 2015). Water-efficient technologies such as Direct Seeded Rice (DSR) (Kaur et al. 2023) and the System of Rice Intensification (SRI) (Tong et al. 2022) should be widely promoted and adopted on a large scale. In addition to water saving, methods like DSR help in saving ground water, fuel, labour, and a reduction of incidence of insect pests (Singh et al. 2022; Singh et al. 2023a). Therefore, it is also necessary that central as well as state governments should come up with appropriate policies and incentives for the farming community to motivate them to adopt these resource conservation techniques.

Given the water scarcity in rainfed or Kandi regions, which are expected to be severely impacted by rising temperatures, it is crucial to implement appropriate water harvesting systems (Kumar et al. 2008). One effective approach is to collect water in pond-like structures. This stored water can then be used for emergency irrigation of rainfed crops such as maize, wheat, and other fodder crops. Rooftop harvesting is another viable option, where rainwater can be collected directly from the rooftops (Anchan and Prasad 2021). Sprinkler and drip irrigation are also water-efficient irrigation technologies (Singh et al. 2023b). Drip technology ensures that water directly gets into the root zone of the plants through a network of economically designed plastic pipes and low-discharge emitters (Mansingh et al. 2023). The technology maximizes crop productivity through an increase in crop yields, and also the area under cultivation can be increased by using the same amount of water that would have been consumed while flooding the fields. In a way, it also results in the income of the farmers. In highly cold regions that are covered by snow for almost half of a year such as Ladakh and Lahul Sapiti in Himachal Pradesh, protected cultivation in polyhouses, low tunnels, and trenches can be successfully used in high-altitude areas for year-round vegetable cultivation. This can prove beneficial especially, for smallholders, wherein multi-storied cropping is quite possible (Reddy 2016). Diversification through the use of region-specific Integrating Farming System Models can be a boon for farming community generating regular income through different enterprises (Bhagat et al. 2024). It also optimizes the use of local resources, eliminating the need for external inputs and significantly reducing the cost of cultivation (Paramesh et al. 2022).

For livestock, studies suggest diversifying animal breeds (within species) and a shift to mixed crop-livestock systems as effective adaptation measures. By diversifying animal and crop varieties, the tolerance to climate variability and to diseases and insect pest outbreaks will also be improved (Rojas-Downing et al. 2017). Additionally, improving animal nutrition and genetics is crucial. It is important to mention that enteric fermentation is a significant source of greenhouse gases in livestock (Rojas-Downing et al. 2017).

1.7.2 Mitigation Strategies

Mitigation strategies involve actions aimed at reducing or preventing GHG emissions from anthropogenic activities. Key strategies include transitioning to renewable energy sources, improving energy efficiency, adopting regenerative agricultural practices, and protecting and restoring forests and vital ecosystems (UNDP 2024). Currently, fossil fuels are the primary source of GHG emissions, making the shift to renewable energy sources such as solar, wind, and geothermal power critical. Regenerative agricultural practices, such as natural farming that enhances soil health, reducing emissions from livestock, employing direct seeding methods, and using cover crops, not only contribute to mitigation efforts but also improve resilience and reduce the financial burden on farmers (UNDP 2024). The excessive use of fertilizers in conventional farming greatly contributes to GHG emissions and climate change. A 2022 NITI Aayog report says that for every metric ton of fertilizer produced approximately 1.1 metric tons of CO_2 is released, making a substantial contribution to GHG emissions. An alternative agricultural approach to mitigate the negative effects of chemical farming is natural farming (Kumar et al. 2023a, b; Rana et al. 2023). The government of India is also focusing hard on promoting natural frming in niche areas. The Agricultural and Farmers' Welfare Ministry has also fixed a target of roping in one crore farmers under natural farming in the next 2 years. The method of natural farming works in harmony with natural biodiversity, fostering the complexity of living organisms, both plants and animals (NMNF 2024). Another important component of our ecosystem is the forests. The forest acts as a carbon sink, absorbing CO_2 and reducing the overall impact of GHGs in the atmosphere (Popo-Ola et al. 2012). Therefore, deforestation should be discouraged and new plantations should be done on waste and community lands so as to clean the environment and restore the degraded soils. Besides, various crop diversification practices, like agro-forestry and agri-horticultural systems being a key component in combating climate change, have the potential to sustain soil organic matter and improve soil physical properties and nutrient cycling (Kumar et al. 2023a, b). Additionally, ecosystems like grasslands, peatlands, and wetlands, along with coastal biomes such as mangrove forests, play a crucial role in carbon sequestration. These ecosystems also support biodiversity and help enhance climate resilience (UNDP 2024).

Various adaptation and mitigation strategies are one side of the coin, there is another face to this issue, and that concerns the formulation of appropriate policies with a bottom-up approach. The participation of the farming community in describing any line of action will ensure that the work at the grassroots level will be done effectively. An enabling ecosystem with the participation of all relevant stakeholders is crucial to achieving the net zero emissions target. At the same time, it is also essential to focus on strengthening human resources and institutional infrastructure, expansion and a strong connection of basic and applied research, and enhancement of inter-disciplinary and inter-institutional collaborations along with policy support (Dhillon and Sohu 2024; Singh et al. 2019). The government has to recognize that increased investment in research and development is needed to enable the scientific community to create, improve, integrate, and promote climate-smart innovations, technologies, and practices. For this, the necessary budgetry support should be available. Despite a rise in global funding since the AR5 report (IPCC 2014), current support remains inadequate, particularly for developing countries, and hampers progress in mitigation, adaptation, and climate-resilient development (Dhillon and Sohu 2024). Financial support both from the public as well as private sector by way of Corporate Social Responsibility (CSR) or other means has to be there. It must also be noted that small efforts at our own individual level by way of conserving our natural resources and various practices as mentioned in this chapter can make a big impact. This can be our contribution to ensuring that the development that is taking place today will not compromise with the needs of our coming generations.

References

Aggarwal PK, Singh AK (2010) Implications of global climatic change on water and food security. In: Global change: impacts on water and food security, pp 49–63

Aguilar L (2008) Is there a connection between gender and climate change? International Union for Conservation of Nature (IUCN), Office of the Senior Gender Adviser. Paper for presentation at the 3rd Global Congress of Women in Politics and Governance, Manila, Philippines, 19–22 October, 2008

Ainsworth EA, Long SP (2020) 30 years of free-air carbon dioxide enrichment (FACE): what have we learned about future crop productivity and its potential for adaptation? Glob Chang Biol 27(1):27–49

Alan H (2021) Heavy rainfall and flash floods: another catastrophe for farmers? The Brussels Times. Brussels. Archived from the original on 16 July 2021. Retrieved on 16 August, 2024

Alexander LV (2016) Global observed long-term changes in temperature and precipitation extremes: a review of progress and limitations in IPCC assessments and beyond. Weather Clim Extrem 11:4–16

Anchan SS, Prasad HCS (2021) Feasibility of roof top rainwater harvesting potential – a case study of south Indian university. Clean Eng Technol 4:100206. https://doi.org/10.1016/j.clet.2021.100206

Anonymous (2020) Effects of climate change on fisheries sector. BASU, Patna. https://www.basu.org.in/wp-content/uploads/2020/03/Effects-of-Climate-Change-on-Fisheries-Sector.pdf. Assessed on 16 August, 2024

Anonymous (2024, August 11) PM Modi unveils 109 high-yielding, climate-resilient crop varieties. The Economic Times. https://economictimes.indiatimes.com/news/economy/agriculture/. Assessed on 12 Aug 2024

Arshad T, Naqve M, Mukhtiar A, Javaid MM, Mahmood A, Nadeem MA, Khan BA (2023) Conservation tillage for sustainable agriculture. In: Hasanuzzaman M (ed) Climate-resilient agriculture, vol 1. Springer, Cham. https://doi.org/10.1007/978-3-031-37424-1_15

Balasha AM, Munyahali W, Kulumbu JT, Okwe AN, Fyma JNM, Lenge EK, Tambwa AN (2023) Understanding farmers perception of climate change and adaptation practices in the marshlands of south Kivu, Democratic Republic of Congo. Clim Risk Manag 39:100469

Barange M, Bahri T, Baveridge MCM, Cochrane KL, FungeSmith S, Poulain F (2018) Impacts of climate change on fisheries and aquaculture; synthesis of current knowledge, adaptation and mitigation options. FAO, Rome

Bates DC (2002) Environmental refugee? Classifying human migration caused by environmental change. Popul Environ 23(5):465–477

BBC (2020) Climate change: Where we are in seven charts and what you can do to help. https://www.bbc.com/news/science-environment-46384067. Assessed on 16 Aug 2024

Bhagat R, Walia SS, Sharma K, Singh R, Singh G, Hossain A (2024) The integrated farming system is an environmentally friendly and cost-effective approach to the sustainability of agri-food systems in the modern era of the changing climate: a comprehensive review. Food Energy Secur 13:e534. https://doi.org/10.1002/fes3.534

Biemans H, Siderius C, Lutz AF, Nepal S, Ahmad B, Hassan T, von Bloh W, Wijngaard RR, Wester P, Shrestha AB, Immerzeel WW (2019) Importance of snow and glacier meltwater for agriculture on the indo-Gangetic plain. Nat Sustain 2(7):594–601

Bindoff NL, Willebrand J, Artale V, Cazenave A, Gregory J, Gulev S, Hanawa K, Le Quéré C, Levitus S, Nojiri Y, Shum CK, Talley LD, Unnikrishnan A (2007) Observations: oceanic climate change and sea level. In: Solomon S, Qin D, Manning M, Chen Z, Marquis M, Averyt KB, Tignor M, Miller HL (eds) Climate change 2007: the physical science basis. Contribution of working group I to the fourth assessment report of the intergovernmental panel on climate change. Cambridge University Press, Cambridge

Brás T, Seixas J, Carvalhais N, Jägermeyr J (2021) Severity of drought and heatwave crop losses tripled over the last five decades in Europe. Environ Res Lett 16(6):065012. https://doi.org/10.1088/1748-9326/abf004

Burroughs CH, Montes CM, Moller CA, Mitchell NG, Michael AM, Peng B, Kimm H, Pederson TL, Lipka AE, Bernacchi CJ, Guan K, Ainsworth EA (2023) Reductions in leaf area index, pod production, seed size, and harvest index drive yield loss to high temperatures in soybean. J Exp Bot 74(5):1629–1641. https://doi.org/10.1093/jxb/erac503

Canto T, Aranda MA, Fereres A (2009) Climate change effects on physiology and population processes of hosts and vectors that influence the spread of hemipteran-borne plant viruses. Glob Chang Biol 15:1884–1894. https://doi.org/10.1111/j.1365-2486.2008.01820.x

Chauhan BS, Kaur P, Mahajan G, Randhawa RK, Singh S, Kang MS (2014) Global warming and its possible impact on agriculture in India. Adv Agron 123:65–121. https://doi.org/10.1016/B978-0-12-420225-2.00002-9

Chuang Z, Bing L, Shilong P, Xuhui W, David LB, Yao H, Mengtian H, Yitong Y, Simona B, Philippe C, Jean-Louis D, Joshua E, Frank E, Ivan JA, Tao L, Erda L, Qiang L, Pierre M, Christoph M, Shushi P, Josep P, Alex RC, Daniel W, Tao W, Donghai W, Zhuo L, Yan Z, Zaichun Z, Senthold A (2017) Temperature increase reduces global yields of major crops in four independent estimates. Proc Natl Acad Sci USA 114(35):9326–9331. https://doi.org/10.1073/pnas.1701762114

Dhillon BS, Sohu VS (2024) Climate change shocks and crop production: the foodgrain bowl of India. Indian J Agron 69(1):1–10

Doney SC (2006) The dangers of ocean acidification. Sci Am 294(3):58–65

Fagodiya RK, Singh A, Singh R, Rani S, Kumar S, Rai AK, Sheoran P, Chandra P, Yadav RK, Sharma PC, Biswas AK (2023) The food-energy-water-carbon nexus of the rice-wheat production system in the western indo-Gangetic plain of India: an impact of irrigation system, conservational tillage and residue management. Sci Total Environ 860:160428

Fagodiya RK, Sharma G, Verma K, Rai AK, Prajapat K, Singh R, Chandra P, Sheoran P, Yadav RK, Biswas AK (2024) Computation of soil quality index after fifteen years of long-term tillage and residue management experiment (LT&RE) under rice wheat system. Agric Syst 219:104039

FAO, IFAD, UNICEF, WFP and WHO (2021) The State of Food Security and Nutrition in the World 2021. Transforming food systems for food security, improved nutrition and affordable healthy diets for all. Rome, FAO. https://doi.org/10.4060/cb4474en

Fereres A, Irwin ME, Kampmeier GE (2017) Aphid movement: process and consequences. In: van Emden HF, Harrington R (eds) Aphids as crop pests, 2nd edn. CABI Publishing, Wallingford, pp 196–224

Gerber HS, Henderson B, Mottet A, Opio C, Dijkman J, Falcucci A, Tempio G (2013) Tackling climate change through livestock: a global assessment of emission and mitigation opporunities. FAO, Rome

Hance T, Van Baaren J, Vernon P, Boivin G (2007) Impact of extreme temperatures on parasitoids in a climate change perspective. Ann Rev Entomol 52:107–126

Hilger T, Keil A, Lippe M, Panomtaranichagul M, Saint-Macary C, Zeller M, Pansak W, Dinh TV, Cadisch G (2013) Soil conservation on sloping land: technical options and adoption constraints. In: Fröhlich HL, Schreinemachers P, Stahr K, Clemens G (eds) Sustainable land use and rural development in Southeast Asia: innovations and policies for mountainous areas. Springer environmental science and engineering. Springer, Berlin, Heidelberg. https://doi.org/10.1007/978-3-642-33377-4_7

IPCC (2014) Climate change 2014: mitigation of climate change. Contribution of working group III to the 5th assessment report of the intergovernmental panel on climate change. In: Edenhofer OR, Pichs-Madruga Y, Sokona E, Farahani S, Kadner K, Seyboth A, Adler I, Baum S, Brunner P, Eickemeier B, Kriemann J, Savolainen S, Schlomer C, von Stechow T, Minx JC (eds) . Cambridge University Press, Cambridge

IPCC (2022) Climate change 2022: impacts, adaptation and vulnerability. In: Pörtner H-O, Roberts DC, Tignor M, Poloczanska ES, Mintenbeck K, Alegría A, Craig M, Langsdorf S, Löschke S, Möller V, Okem A, Rama B (eds) Contribution of working group II to the sixth assessment report of the intergovernmental panel on climate change. Cambridge University Press, Cambridge, p 3056. https://doi.org/10.1017/9781009325844

IPCC (2023) Food, fibre and other ecosystem products. In: Climate change 2022—impacts, adaptation and vulnerability: working group II contribution to the sixth assessment report of the intergovernmental panel on climate change. Cambridge University Press, Cambridge, pp 713–906. https://doi.org/10.1017/9781009325844.007

IPCC (2023a) Summary for policymakers. In: Core Writing Team, Lee H, Romero J (eds) Climate change 2023: synthesis report. contribution of working groups I, II and III to the sixth assessment report of the intergovernmental panel on climate change. IPCC, Geneva, pp 1–34. https://doi.org/10.59327/IPCC/AR6-9789291691647.001

Iqbal R, Raza MAS, Valipour M, Saleem MF, Zaheer MS, Ahmad S, Toleikiene M, Haider I, Aslam MU, Nazar MA (2020) Potential agricultural and environmental benefits of mulches – a review. Bull Natl Res Cent 44:75. https://doi.org/10.1186/s42269-020-00290-3

IUCN. 2021. Climate change is not gender neutral. International Union for Conservation of Nature. https://iucn.org/news/gender/202106/climate-change-not-gender-neutral. Assessed on 16 Aug 2024

Jha CK, Gupta V (2021) Farmers perception and factors determining the adaptation decisions to cope with climate change: an evidence from rural India. Environ Sustain Indic 10(22):100112

Jin F, Yiwei J, Xuhui W, Laurent L, Philippe C, Jakob Z, Yin W, Yanhong T, Christoph M, Heidi W, Bo Y, Yali W, Qihui W, Xiaoqing C, Weichen H, Yongqiang L, Pengjun Z, Shilong P, Feng Z (2023) Extreme rainfall reduces one-twelfth of China's rice yield over the last two decades. Nat Food 4(5):416–426. https://doi.org/10.1038/s43016-023-00753-6

Kader MA, Senge M, Mojid MA, Nakamura K (2017) Mulching type-induced soil moisture and temperature regimes and water use efficiency of soybean under rain-fed condition in Central Japan. Int Soil Water Conserv Res 5:302–308. https://doi.org/10.1016/j.iswcr.2017.08.001

Kader MA, Singha A, Begum MA, Jewel A, Khan FH, Khan NI (2019) Mulching as water-saving technique in dryland agriculture: review article. Bull Natl Res Cent 43:147. https://doi.org/10.1186/s42269-019-0186-7

Karl TR, Melillo JM, Peterson TS (2009) Global climate change impacts in the United States. US global change research programme. Cambridge University Press

Kaur J, Kumar V, Singh G, Sabhikhi HS, Gill GS, Rampal VK (2023) Comparative analysis of different paddy establishment methods in Punjab. J Agric Dev Policy 33(1):122–126

Kumar P, Kumar A (2016) Climate change and agriculture. Daily Excelsior. https://www.dailyexcelsior.com/climate-change-and-agriculture/. Assessed on 16 Aug 2024

Kumar MD, Patel A, Singh OP (2008) Rainwater harvesting in the water-scarce regions of India: potential and pitfalls. In: Amarasinghe, Upali A; Sharma, Bharat R (eds) Strategic analyses of the National River Linking Project (NRLP) of India, Series 2. Proceedings of the workshop on analyses of hydrological, social and ecological issues of the NRLP, New Delhi, India, 9–10 October 2007. Colombo, Sri Lanka: International Water Management Institute (IWMI), pp 289–314

Kumar V, Aulakh CS, Kooner R, Kaur J (2021) Nutrition and weather effects on population dynamics of insect-pests in potato (*Solanum tuberosum*). Indian J Agric Sci 91(11):1690–1694

Kumar A, Singh VK, Suvana S (2023a) Climate-resilient agriculture for sustainable production systems. XXII biennial national symposium climate smart agronomy for resilient production systems and livelihood security ICAR-CCARI, Goa, India 22 24 November, 2023, pp 3–9

Kumar V, Singh G, Sharma K (2023b). Productivity and economics of wheat under different sowing methods in South-western Punjab. In: XXII Biennial national symposium of indian society of agronomy, 22–24 November, 2023 at ICAR-CCARI, Ela, Goa, pp 418–419

Liu L, Hao L, Zhang Y, Zhou H, Ma B, Cheng Y, Tian Y, Chang Z, Zheng Y (2022) The CO_2 fertilization effect on leaf photosynthesis of maize (*Zea mays* L.) depends on growth temperatures with changes in leaf anatomy and soluble sugars. Front Plant Sci 13:890928. https://doi.org/10.3389/fpls.2022.890928

Locci E, Jammoul H (2023) Climate diplomacy: international and national responses. Dar Abaad, Beirut—Liban, pp 15–67

Mahajan G, Kumar V, Chauhan BS (2017) Rice production in India. In: Chauhan BS, Khawar J, Mahajan G (eds) Rice production worldwide. Springer International Publishing AG, pp 53–91

Manna MC, Rahman MM, Naidu R, Sahu A, Bhattacharjya S, Wanjari RH, Patra AK, Chaudhari SK, Majumdar K, Khanna SS (2018) Bio-waste management in subtropical soils of India: future challenges and opportunities in agriculture. Adv Agron 152:87–148. https://doi.org/10.1016/bs.agron.2018.07.002

Mansingh B, Banik R, Bag S, Sunaratiya A, Das T (2023) Drip irrigation system: an effective approach for increasing water use efficiency. In: Green technologies for sustainable agriculture. Biotech Books, New Delhi, pp 247–262. https://doi.org/10.5281/zenodo.7985739

McAllister S (2023) There could be 1.2 billion climate refugees by 2050. Here's what you need to know. Zurich Magazine. https://www.zurich.com/media/magazine/2022/there-could-be-1-2-billion-climate-refugees-by-2050-here-s-what-you-need-to-know. Assessed on 16 Aug 2024

Mohammed EY, Uraguch ZB (2013) Impacts of climate change on fisheries: implications for food security in sub-Saharan Africa. In: Hanjra MA (ed) Global food security: emerging issues and economic implications. Nova Science Publishers, Inc., pp 113–135

Moon MP (2023) Food and health security impact of climate change in Bangladesh: a review. J Water Climate Change 14(10):3484–3495. https://doi.org/10.2166/wcc.2023.131

Nardone A, Ronchi B, Lacetera MS, Ranieri MS, Bernabucci U (2010) Effects of climate change on animal production and sustainability of livestock systems. Livest Sci 130:57–69

NITI Aayog Annual Report (2022). https://www.niti.gov.in/sites/default/files/2023-02/Annual-Report-2022-2023-English_1.pdf

NMNF (2024) Prospects and challenges of Natural Farming. National Mission on Natural Farming Management and Knowledge Portal. https://naturalfarming.dac.gov.in/NaturalFarming/Concept. Assessed on 16 Aug 2024

Oliveira MT, Merwin IA (2001) Soil physical conditions in a New York orchard after eight years under different groundcover management systems. Plant Soil 234:233–237

Osman-Elasha B (2008) Gender and climate change in the Arab Region. In: Proceedings of the second conference entitled "Women in the concept and issues of human security- Arab and international perspectives" by the Arab Women Organization, Abu Dhabi, United Arab Emirates on 11–13 November 2008, pp 335–363

Osman-Elasha B (2009) Women...In the shadow of climate change. UN Chronicle 3&4(46) https://www.un.org/en/chronicle/article/womenin-shadow-climate-change. Assessed on 16 Aug 2024

Paramesh V, Ravisankar N, Behera U, Arunachalam V, Kumar P, Rajkumar RS, Misra SR, Kumar RM, Prusty AK, Jacob D, Panwar AS, Mayenkar T, Reddy VK, Rajkumar S (2022) Integrated farming system approaches to achieve food and nutritional security for enhancing profitability, employment, and climate resilience in India. Food Energy Secur 11:e321. https://doi.org/10.1002/fes3.321

Pathak H (2015) Greenhouse gas emission from Indian agriculture: trends, drivers and mitigation strategies. Proc Indian Natl Sci Acad 81(5):1133–1149

Pathak H, Bhatia A, Jain N (2014) Greenhouse gas emission from Indian agriculture: trends, mitigation and policy needs. Indian Agricultural Research Institute, New Delhi

Patil SS, Kelkar TS, Bhalerao SA (2013) Mulching: a soil and water conservation practice. Res J Agric Forestry Sci 1(3):26–29

Popo-Ola FS, Aiyeloja AA, Adedeji GA (2012) Sustaining carbon sink potentials in tropical forest. J Agric Soc Res 12(1):64–73

Pradhan P, Seydewitz T, Zhou B, Lüdeke MKB, Kropp JP (2022) Climate extremes are becoming more frequent, co-occurring, and persistent in Europe. Anthr Sci 1:264–277. https://doi.org/10.1007/s44177-022-00022-4

Ram H (2020) Mulching, farmyard manure and exogenous application of osmo-protectants confer terminal heat tolerance in wheat (*Triticum aestivum*). Indian J Agron 65:271–277

Rana RK, Sheoran P, Keshava, Singh R, Singh RK, Gautam US (2023) Training manual for uniform out-scaling of natural farming through KVKs. ICAR ATARI, Zone-I, p 30

Reddy PP (2016) Protected cultivation. In: Sustainable crop protection under protected cultivation. Springer, Singapore. https://doi.org/10.1007/978-981-287-952-3_1

Robitzski R (2020) 60% of the world's fish species at risk of extinction due to climate change. World Economic Forum. Assessed on 16 Aug 2024

Rojas-Downing MM, Nejadhashemi AP, Harrigan T, Woznicki SA (2017) Climate change and livestock: impacts, adaptation, and mitigation. Clim Risk Manag 16:145–163

Schmitt J, Offermann F, Söder M, Frühauf C, Finger R (2022) Extreme weather events cause significant crop yield losses at the farm level in German agriculture. Food Policy 112:102359

Sharma HC (2014) Climate change effects on insects: implications for crop protection and food security. J Crop Improv 28:229–259. https://doi.org/10.1080/15427528.2014.881205

Sharma HC, Dhillon MK, Kibuka J, Mukuru SZ (2005) Plant defense responses to sorghum spotted stem borer, *Chilo partellus* under irrigated and drought conditions. Int Sorghum Millets Newsletter 46:49–52

Sharma K, Kumar V, Singh G (2024) Extent of adoption of paddy residue management techniques in Sri Muktsar sahib district. Indian Res J Ext Educ 24(3):126–132

Sheoran P, Kumar A, Sharma R, Prajapat K, Kumar A, Barman A, Raju R, Kumar S, Dar YJ, Singh RK, Sanwal SK (2021) Quantitative dissection of salt tolerance for sustainable wheat production in sodic agro-ecosystems through farmers' participatory approach: an Indian experience. Sustain For 13(6):3378

Sheoran P, Sharma R, Kumar A, Singh RK, Barman A, Prajapat K, Kumar S, Sharma PC (2022) Climate resilient integrated soil–crop management (CRISCM) for salt affected wheat agri–food production systems. Sci Total Environ 837:155843

Sheoran P, Kamboj P, Kumar A, Kumar A, Singh RK, Barman A, Prajapat K, Mandal S, Yousuf DJ, Narjary B, Kumar S (2023) Matching N supply for yield maximization in salt–affected wheat agri–food systems: on-farm participatory assessment and validation. Sci Total Environ 875:162573

Shivanna KR (2022) Climate change and its impact on biodiversity and human welfare. Proc Indian Natl Sci Acad 88(2):160–171

Sidhu HS, Singh M, Singh Y, Blackwell J, Lohan SK, Humphreys E, Jat ML, Singh V, Singh S (2015) Development and evaluation of the turbo happy seeder for sowing wheat into heavy rice residues in NW India. Field Crop Res 184:201–212

Singh RK, Singh A, Sheoran P, Sharma DK, Sharma PC (2019) 'Mera Gaon, Mera Gaurav' policy: a source of hypothesis for the co-production of adaptive agricultural knowledge. Indian J Agric Sci 89(9):1410–1414

Singh R, Singh A, Sheoran P, Fagodiya RK, Rai AK, Chandra P, Rani S, Yadav RK, Sharma PC (2022) Energy efficiency and carbon footprints of rice-wheat system under long-term tillage and residue management practices in western indo-Gangetic Plains in India. Energy 244:122655

Singh G, Kumar V, Kaur J (2023a) Adoption status and perception of farmers regarding different rice establishment technologies in Punjab. J Commu Mobil Sustain Dev 18(3):871–876

Singh R, Singh A, Kumar S, Fagodiya RK, Sheoran P, Rai AK, Rani S, Chandra P (2023b) Effect of mini-sprinkler irrigation, tillage and residue on productivity, profitability and resource saving in rice–wheat system in western Indo-Gangetic Plains of India. Paddy Water Environ 21(4):479–495

Skendžić S, Zovko M, Živković IP, Lešić V, Lemić D (2021) The impact of climate change on agricultural insect pests. Insects 12(5):440. https://doi.org/10.3390/insects12050440

Sultan B, Defrance D, Iizumi T (2019) Evidence of crop production losses in West Africa due to historical global warming in two crop models. Sci Rep 9(1):12834. https://doi.org/10.1038/s41598-019-49167-0

Tong LA, Ulubaşoğlu MA, Guven C (2022) Growing more Rice with less water: the system of Rice intensification and water productivity in Vietnam. Aust J Agric Resour Econ 66(3):581–611

UN Environment Programme (2023) Climate change 2023: synthesis report. https://www.unep.org/resources/report/climate-change-2023-synthesis-report. Assessed on 16 Aug 2024

UNDP (2024) What is climate change mitigation and why is it urgent? United Nations Development Programme. https://climatepromise.undp.org/news-and-stories/what-climate-change-mitigation-and-why-it-urgent. Assessed on 16 Aug 2024

UNFCC (2016) Why climate change is not gender neutral. United Nations Framework Convention on Climate Change. https://unfccc.int/news/climate-action-needs-gender-action. Assessed on 16 Aug 2024

US Global Change Research Program (2013) Corn and soybean temperature response. https://web.archive.org/web/20130512133320/http://nca2009.globalchange.gov/corn-and-soybean-temperature-response

USEPA (2017) Climate action benefits report. Climate action benefits: freshwater fish. https://19january2017snapshot.epa.gov/cira/climate-action-benefits-freshwater-fish_.html. Assessed on 16 Aug 2024

Waheed Z (2023) Climate change's greatest victims are women and girls. UNICEF. https://www.unicef.org/rosa/blog/climate-changes-greatest-victims-are-women-and-girls. Assessed on 16 Aug 2024

Weatherdon LV, Magnan AK, Rogers AD, Sumulia UR, Cheung WWL (2016) Observed and projected impacts of climate change on marine fisheries, aquaculture, coastal tourism and human health: an update. Front Mar Sci 3:48. https://doi.org/10.3389/fmars.2016.00048

WHO (2023) World Health Organization. Climate change. https://www.who.int/news-room/fact-sheets/detail/climate-change-and-health. Assessed on 16 Aug 2024

Wing IS, Cian ED, Mistry MN (2021) Global vulnerability of crop yields to climate change. J Environ Econ Manag 109:102462. https://doi.org/10.1016/j.jeem.2021.102462

WMO (2024a) Climate change indicators reached record levels in 2023: WMO. https://wmo.int/news/media-centre. Assessed on 16 Aug 2024

WMO (2024b) State of the Climate in Asia 2023. https://wmo.int/publication-series/state-of-climate-asia-2023. Assessed on 16 Aug 2024

Yoro KO, Daramola MO (2020) CO_2 emission sources, greenhouse gases, and the global warming effect. In: Rahimpour MR, Farsi M, Makarem MA (eds) Advances in carbon capture, 1st edn. Woodhead Publishing, pp 3–28

Zeyen RJ, Stromberg EL, Kuehnast EL (1987) Long-range aphid transport hypothesis for maize dwarf mosaic virus: history and distribution in Minnesota, USA. Ann Appl Biol 111:325–336

2 Greenhouse Gas Emission: Problems, Global Reality, and Future Perspectives

Paras Kamboj, Sukriti, Ram Kishor Fagodiya, and Parvender Sheoran

Abstract

Greenhouse gas (GHG) emissions are a critical driver of global climate change, posing significant environmental, economic, and social challenges. This chapter explores the multifaceted problems of GHG emissions, including rising global temperatures, extreme weather events, ocean acidification, sea level rise, biodiversity loss, and adverse health impacts. It provides a comprehensive analysis of the current global reality of emissions, highlighting the major sources such as fossil fuel combustion, agriculture, deforestation, and waste—and the uneven geographical distribution of both emissions and climate impacts. While developed nations have historically contributed the most to GHG emissions, emerging economies are now playing an increasing role, although disparities in per capita emissions and climate vulnerabilities persist. In response to these challenges, global efforts like the Paris Agreement aim to limit temperature rise and promote carbon neutrality, yet significant hurdles remain. The chapter examines future perspectives on reducing GHG emissions, including technological innovations such as carbon capture and renewable energy advancements, sustainable practices such as regenerative agriculture and circular economies, and policy mechanisms such as carbon pricing. It also addresses the importance of behavioral

P. Kamboj
Department of Agronomy, CCS Haryana Agricultural University, Hisar, Haryana, India

Sukriti
Krishi Vigyan Kendra, CCS Haryana Agricultural University, Rohtak, Haryana, India

R. K. Fagodiya
ICAR-Central Soil Salinity Research Institute, Karnal, Haryana, India

P. Sheoran (✉)
ICAR-Agricultural Technology Application Research Institute, Ludhiana, Punjab, India
e-mail: Parvender.Sheoran@icar.gov.in

A. Mann et al. (eds.), *Cutting Edge Technologies for Developing Future Crop Plants*,
https://doi.org/10.1007/978-981-96-2508-6_2

changes and international cooperation in ensuring a just transition to a low-carbon future. Despite progress, challenges related to political resistance, economic costs, and equity between nations must be addressed to mitigate the worst effects of climate change and achieve long-term sustainability.

Keywords

GHGs · IPCC · Climate change · Adaptation · Mitigation

2.1 Introduction

Greenhouse effect, global warming, and climate change are some of the words we regularly keep listening to in our daily lives; on every international platform this topic is highlighted as the most important one, and we also feel climate change is a reality not a hoax anymore and it traces its origin in eighteenth and nineteenth centuries after the spread of industrial revolution in Europe and other countries. During this industrial revolution, the world has witnessed that greenhouse gas emissions have increased and since that period till now, we are continuously observing a rising global average temperature. Now the question is, what are the major causes of this? The biggest cause is the burning of fossil fuels, let us understand with an example of coal-based thermal power plant which is a necessity for electricity generation but the same thermal power plant is responsible for the generation of greenhouse gases. Greenhouse gases are known to trap the long-wave terrestrial radiation. When all these terrestrial radiations get trapped in the lower atmosphere they increase heat levels and become the main reason for the rise in temperature. Similarly, another important reason can be changes in the land use pattern as forest area is decreasing and being converted to nonforest use purposes, for example, infrastructure and agricultural practices. Land which used to function as a carbon sink now acts as a carbon emitter. Rising consumption tendencies are also directly impacting climate changes as a direct result of continuous resource extraction. Every aspect of life is affected by these interrelated phenomena of global warming, greenhouse gases, and climate change, for example, the recent droughts in Amazonian rainforests, the increase in intensity and frequency of tropical cyclones in Indian coastal lines, and the increase in monsoon erraticism, arrival of heat waves in Europe and cold waves in North America, weakening of the polar vortex, these all are outcomes of climate change. The year 2023 has been recorded as the hottest year on record as the global average temperature had increased by 1.48 °C compared to the pre-industrial level.

In today's world, there are various challenges, among which increasing population and climate change due to greenhouse gas (GHGs) emissions, such as CO_2 (carbon dioxide), CH_4(methane), and N_2O (nitrous oxide), are the topmost challenges. To meet the increasing population's energy requirement, we have to rely on fossil fuels, which adversely affect the environment and social and economic conditions.

Greenhouse gases are defined by their radiative forcing, which changes the Earth's atmospheric energy balance; typically, expressed as watts per square meter (W m^{-2}) (IPCC 1996).

It is represented by positive and negative values, where a positive value indicates an increase or addition of energy level to the atmosphere, whereas a negative value indicates an increase in the level returning to space or a decrease in the energy level in the atmosphere. On the basis of the source of occurrence, GHGs can be classified into naturally occurring (such as CO_2, CH_4, O_3 (ozone), N_2O, and also includes water vapor) and anthropogenic activities (such as chlorofluorocarbon, hydro fluorocarbon, perfluorocarbon, and sulfur hexa fluorides). These gases can trap heat resulting in an increase in the earth's temperature. CO_2, CH_4, and N_2O are long lived in the atmosphere, and are the major contributors to positive increases in radiative forces (IPCC 1996).

In the United States, energy contributes 87%, agriculture 6%, industries (4.5%), waste (2.7%), and others (< 1%) to the GHG emission (USEPA 2007). India's GHG emissions from the energy sector accounted for 68.7% of the total emissions followed by the agriculture sector which accounted for 19.6% of total emissions (WRI CAIT 4.0 2017). Within the energy sector, electricity and heat generation account for 49% of emissions, and construction and manufacturing account for 24% of emissions (Fig. 2.1). Globally, industrialization and automobiles account for the drastic increase in GHG emissions. About 20% of CO_2 emissions are from transportation, of which 75% emissions are from road transportation (Siddique et al. 2021).

The emission of greenhouse gases is the major source of climate change worldwide and sharp changes in global temperatures create a serious threat to living beings (Lacis et al. 2010). Climate change will lead to about 0.25 million deaths

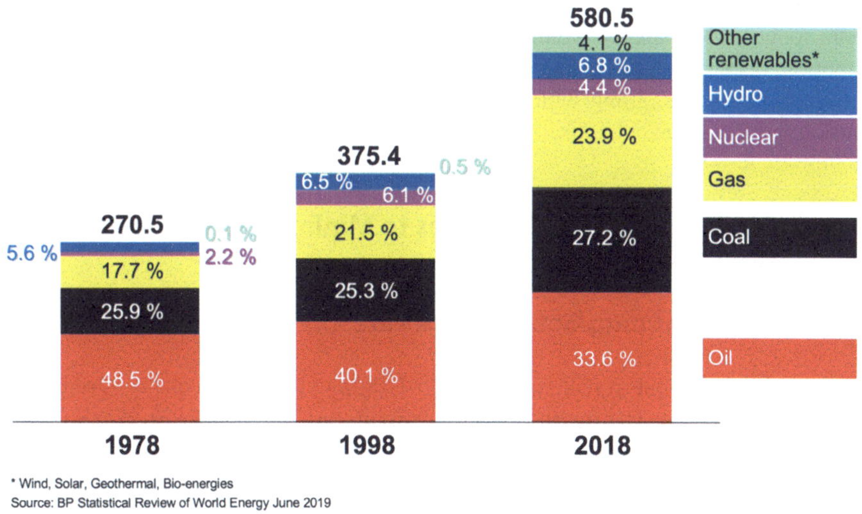

Fig. 2.1 Global primary energy consumption 1978–2018 (Exajoules). (Source: Kober et al. 2020a, b, CC BY 4.0)

annually from higher temperatures, malaria, diarrhea, and malnutrition; and 7 million deaths from heart disease, cancer, lung disease, and stroke from 2030 to 2050 (Helliwell et al. 2020). As per WHO, air pollution and climate change ranked topmost among the top ten serious threats to global health threats. Global climate change caused fluctuating sea levels, which further resulted in the extinction of numerous species (Songqi et al. 2021).

Over the past 70 years, the growth rate of CO_2 concentration has experienced a remarkable increase. In the 1960s, the average growth rate of CO_2 concentration was measured at 0.7×10^{-6}, which escalated to 2×10^{-6} between the years 2005 and 2019. There has been 100 times increase in the growth rate of CO_2 concentration over the last 70 years. In the 1960s, the average growth rate of CO_2 concentration was 0.7×10^{-6} and further increased to 2×10^{-6} from 2005 to 2019 (WMO 2019).

Over the past few years (196–1990), there has been a continuous rise in the global temperature, with an increase averaging 1.1 °C. The rise in temperature might seem small and insignificant but can substantially impact climate change and can cause global warming (Kumar et al. 2022). There is an increase in the temperature of land and ocean by 1.59 °C (1.34–1.83 °C) and 0.88 °C (0.68–1.01 °C), respectively (IPCC 2021).

Among all kinds of anthropogenic GHG emissions, agriculture accounts for nearly 12% of global emission (Linquist et al. 2012). Methane production from agricultural soils is limited to anaerobic soil conditions and depends upon various factors such as carbon content, temperature, and bulk density/soil texture (Mitra et al. 2002). Nitrous oxide production from soils depends on two complementary processes, viz. nitrification and denitrification. These two processes depend on wet and dry transition in the soil (Abao et al. 2000). Drying promotes nitrification while denitrification is boosted when the soils are rewetted. Agriculture is a potential source as well as a sink for greenhouse gases (Zaman et al. 2021). Among the anthropogenic NO_2 and NH_3 emissions, agriculture accounts for 80% and 70% contribution, respectively. These are mainly caused by the use of livestock manure and inorganic fertilizers, along with this 40% of the anthropogenic CH_4 is caused by enteric fermentation.

2.2 Good Agricultural Practices to Minimize Greenhouse Gas Emission

2.2.1 Smart Farming/Smart Agriculture

With the advancement of technology, smart farming vis-a-vis precision agriculture came into existence. The motto of smart agriculture is to enhance productivity by improving yield and profitability along with reducing the environmental footprint in the form of greenhouse gas emissions while maintaining sustainability (Fagodiya et al. 2023), with the help of different techniques such as judicious irrigation, need-based fertilizers and pesticides for crops as well as real-time estimation of weeds or diseases, and different weather parameters and soil conditions, which ultimately leads to adequate and judicious use of agrochemicals (Alam et al. 2020). The use of smart technology in agriculture may include field monitoring which helps in accurate data collection of farm situation (Prathibha et al. 2017) and judicious use of agrochemicals and water (Singh et al. 2023). The use of greenhouses helps to maximize production while controlling the micro-climate condition. With the control of temperature and moisture, compost management can be done to prevent fungal and microbial contaminants (Villa-Henriksen et al. 2020). Livestock farming has become a boon with the advancement of technology, which helps monitor grazing in open pasture. Smart farming also helps in maintaining air quality and ventilation while reducing greenhouse gas emissions from livestock farms. A smart agriculture technique helps in the detection of climate change, greenhouse gas emissions, and carbon footprint. All these come under climate-smart agriculture.

2.2.2 Climate-Smart Agriculture (CSA)

Climate-smart agriculture combines traditional farming practices with modern technology to aim for high productivity while maintaining sustainability and keeping check on climate constraints while reducing greenhouse gas emissions (Pinto et al. 2020; McNunn et al. 2020). Different field studies presented in the literature proved that climate-smart agriculture improves the quality and quantity of the produce, makes efficient use of farm input, helps to generate more revenue, and ultimately reduces greenhouse gas emissions (Fagodiya et al. 2023; Pinto et al. 2020; McNunn et al. 2020; Sapkota et al. 2014). Different climate-smart practices (Fig. 2.2) are compiled in this review.

Integrated pest management has been demonstrated to be effective in reducing chemical use. Focus is placed on carbon footprint reduction and sustainable land management in agro-silviculture and fodder management. Zero tillage and minimum tillage are energy-efficient methods that promote water infiltration and organic matter retention while requiring less energy to prepare the soil (Pittelkow et al. 2015). Weather smart practices use technology based on crop agro advisory to warn farmers of weather uncertainties and provide them with recommendations (Gangopadhyay et al. 2019). Legumes improve soil quality and nitrogen content when intercropped (with other primary crops in different rows).

Nutrient-smart practices can increase the effectiveness of nutrient use. These practices may include split dose application in crop fields and site-specific nutrient management, which helps to optimize the supply of nutrients according to the needs of crops (Banayo et al. 2018). Leaf color charts can also be used to quantify the amount of nitrogen that is needed in crops like wheat, rice, and maize. Green manuring (the growing of legumes in cropping systems) and intercropping (the planting of other major crops in alternate rows) with legumes enhance soil quality and nitrogen content.

Utilizing local knowledge and technology: Knowledge-smart practices increase productivity while lowering environmental impact. These practices include improved crop varieties (tolerant to abiotic and biotic stresses), contingent crop

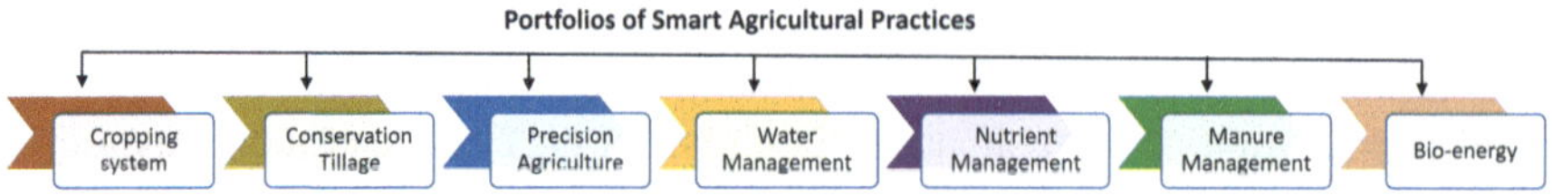

Fig. 2.2 Smart agricultural practices for the mitigation of GHG emissions

plans, and risk management strategies to be ready for unfavorable weather conditions (Pagliacci et al. 2020). There is plenty of literature supporting CSA technology. A study was conducted by McNunn et al. (2020) in the Indo-Gangetic Plains of India to monitor the potential economic benefits of CSA. Their study found that farmers preferred laser land leveling, improved crop varieties with crop diversification, and zero tillage practices over other CSA practices like site-specific nutrient management and residue management. The use of enhanced crop types, laser leveling, and zero tillage each boost the farmer's net profit, but combining all three of these practices increases crop production and, ultimately, productivity. Another study by Pinto et al. (2020) and Mann et al., (2018) found that the use of CSA technology in the agriculture of maize, wheat, and rice was successful in lowering GHG emissions. This study demonstrated that even in unfavorable climatic conditions, strategies with the potential to boost food production can improve food security. CSA must be used in all facets of agriculture if we want to reduce greenhouse gas emissions significantly.

The author analyzed two different cases with two different farmers. For the first farmer, crop yield was the primary concern than carbon footprint, but in the case of the second farmer he wanted to optimize his farm production as well as minimize the GHG emissions. For the first farmer, the reduction in GHG emission was 44 Mt. CO_2 year^{-1}. However, the reduction in GHG emission in the case of the second farmer was 101 Mt. CO_2 year^{-1}. The higher amount of GHG emission reduction undoubtedly was at the cost of production. In the case of the second farmer where GHG reduction was a priority on a large scale, there was a yield reduction of 21 Mt. year^{-1}. Smith et al. (2007) identified some prominent CSA practices that are highly effective in reducing greenhouse gas emissions for example, technologies helpful in the management of cropland and grazing land, restoration of degraded lands, and need-based and judicious use of manures, fertilizers, pesticides, and water. McNunn et al. (2020) used a process-based biochemistry modeling framework in 11 states of the United States for the estimation of regional-specific GHG emission reduction under the influence of some CSA practices in soybean and maize farms. They followed different models based on the organic carbon content (up to a depth of 50 cm), CH_4, and N_2O emissions. The results were converted to CO_2 equivalents, and to estimate regional variability the results were weighted spatially to a country scale. The potential of a shift from conventional to zero tillage on the country scale was estimated and found a mean reduction of 1477 kg CO_2-eq ha^{-1} year^{-1} and included CH_4, N_2O, and SOC flux reductions of −17 kg, 549 kg, and 945 kg, CO_2 -eq ha^{-1} year^{-1}, respectively, with a standard deviation of 605 kg CO_2- eq ha^{-1} year^{-1}. Negative reduction indicated an increase in emissions. By using the cover crop method further reduction in CO2 -eq was found possible. By improving the timing of nitrogenous fertilizer, more than 413 kg CO_2 -eq can be reduced. McNunn et al. (2020) claimed that multiple CSA practices can reduce greenhouse gas emissions of 2861 kg CO_2-eq. The author also suggested that the use of maize stover can satiate the need for fossil fuel as a feedstock for bio-energy or other bio-products generation.

2.3 Carbon Sequestration and Its Role in Greenhouse Gas Mitigation

The main drivers of anthropogenic greenhouse gas emission are carbon dioxide (CO_2), methane (CH_4), and nitrogen oxides (NO_x) which are mainly released through the burning of fossil fuels and biomass, i.e., surface or subsurface decomposition of organic matter. Increased atmospheric CO_2 leads to increased terrestrial carbon storage mainly in the form of photosynthesis, vegetation, and soil responses to continuous warming and change in the water cycle (IPCC 2021). The main sinks of CO_2 are oceans, plants, and soils (IPCC 2021). According to Solomon et al. (2007), soil serves as a major sink for carbon reserve, it can store more than two-thirds of terrestrial carbon. Soil plays an important role in mitigating greenhouse gas emissions (Eurostat 2021). Various strategies discussed in the literature reveal the potential of soil to reduce carbon emissions and suggest other sequestration opportunities (Lehmann et al. 2006, Lal 2004; Fagodiya et al.; 2024b). Carbon sequestration supports different United Nations Sustainable Development Goals (SDGs), such as SDG 2 "End Hunger," SDG 13 "combat climate change," and SDG 15 "Terrestrial Ecosystem and Biodiversity."

Terrestrial carbon sequestration has numerous benefits while compensating human-mediated emissions, helping in reducing net increase CO_2 concentration, increasing ecosystem function, and services of natural resources by enhancing their quality, decreasing nutrient losses, erosional risk, restoring degraded lands, improving habitats, water retention, and land use efficiency (Lal 2007; Lal et al. 2007). Being the largest terrestrial reservoir, soil plays an important role in climate change (Dorrepaal et al. 2009). Globally soil harbors 1526 Pg C of soil organic carbon and 940 PgC of inorganic carbon (Lal et al. 2021). One meter depth of soil accommodates 2500 Pg C and vegetation contains 620 Pg C which is jointly thrice the amount of atmospheric carbon, i.e., 880 Pg C (Lal et al. 2021).

The total amount of carbon held within soil and vegetation plays an important role in modulating the global carbon cycle as well as the earth's climatic balance (Elbasiouny et al. 2022). By storing carbon in the form of organic matter, soil acts as a carbon sink (Fageria 2012). This organic matter can be derived from plant and animal residues, microbes, and carbonates (Berner and Lasaga 1989), with time this organic matter formed into more stable forms, i.e., humus (Kwiatkowska-Malina 2018). Carbon storing capacity of soil not only improves soil fertility but also mitigates the level of CO_2 in the atmosphere which is a potent GHG (Elbasiouny et al.

2022). The carbonation process involves the reaction of carbon dioxide with basic metal oxides and hydroxides present in the soil, as noted by Manning et al. (2013). Additionally, carbonates constitute a substantial portion of the global carbon reservoir (Alongi et al. 2016).

Soil inorganic carbon plays an important role in carbon sequestration in the soil while the carbonation process is slow and depends upon various factors such as the availability of basic cation, soil moisture, and temperature (Barman and Dash 2022). Vegetation is crucial in the photosynthesis process, as it absorbs atmospheric CO_2 and transforms it into organic compounds, thereby functioning as a carbon sink. Following decomposition, this plant material contributes to the soil organic carbon (SOC) pool (Condron et al. 2010). Mangroves and peatland ecosystems sequester large amounts of carbon compared to other ecosystems (Palit et al. 2022).

Total carbon stored in the soil and vegetation helps in reducing the quantity present in the atmosphere (Swingland et al. 2002), this indicates the role of soil and vegetation as a tool for climate change mitigation mainly by reducing atmospheric carbon dioxide levels. Hence, adopting good land management strategies helps in the enhancement of carbon sequestration in soil as well as in vegetation such as afforestation, and sustainable agricultural practices can be good instruments for combating climate change (Lal et al. 2015; Fagodiya et al. 2024a, b).

As discussed in the previous section of this review good soil management practices can enhance the CO_2 sequestration potential of every type of soil. Different sustainable practices like organic farming, cover crops, and minimum or no tillage help to enhance CO_2 sequestration while excessive tillage or land conversion can deteriorate the carbon storage potential of soils (Zerssa et al. 2021). The sequestration of atmospheric CO_2 by the soil is relatively small, while maximum sequestration occurs through the process of photosynthesis. Hence, the net sequestration in an ecosystem is the result (Kell 2011) of net primary productivity, which is equal to the difference between photosynthesis and respiration (Luo et al. 2004) (von Lützow and Kögel-Knabner 2009; Herold et al. 2014).

The level of carbon accumulation in soil depends upon the difference between input through the plant residue, root exudates, and output through the losses of this added organic matter by the process of respiration, decomposition, erosion, leaching, nutrient depletion, etc. (Fuss et al. 2018; Poeplau and Don 2015) under original condition (moisture, productivity, and temperature). Intensively managed ecosystems always store higher carbon levels than original conditions (Six et al. 2002). The accumulation of SOC mostly depends on its stability against microbial decomposition than other environmental factors (Atkinson et al. 2010). According to Chaplot et al. (2010), the potential of carbon sequestration is much lower than estimated in any system, which is mainly due to its faster decomposition through soil microflora. Soil degradation and higher CO_2 emissions from soil decrease its SOM reserves (Atkinson et al. 2010).

Anthropogenic activity such as changes in land use patterns where the natural ecosystem is converted to an agricultural ecosystem also decreases SOC drastically by exposing the surface soil to higher temperatures, which reduces soil moisture content in the root zones, making the soil drained agricultural soil (Guo and Gifford

2002; Pulleman et al. 2000). Strategies that allow an increase in soil organic carbon storage can be an integral part of integrated sustainability policies (Smith 2004). SOC serves as a source of major plant nutrients, the process of ion exchange regulates charge density, improves soil aggregation, reduces susceptibility to erosion, increases water retention in low moisture potential sites, serves as source of substrate for soil biota to improve soil biodiversity, and reduces surface runoff by promoting the water infiltration capacity of the soil. SOM acts as a buffer against abrupt changes in pH and soil temperature. Significant factors concerning economics and the environment that can be tackled by SOC encompass the decrease in sediment load within surface water bodies, which facilitates the biodegradation and filtration of pollutants, in addition to the mitigation of greenhouse gas emissions from soils into the atmosphere.

Strategies that improve water quality, biomass productivity, and reduction in GHG emission mainly involve restoring degraded soil and ecosystem, these strategies should address enhancing soil carbon reserves, optimizing productivity, reduce decomposition rates such as selection of varieties with good root biomass (Kell 2011), intercropping, cover cropping, crop rotation, better use of soil amendments (Bronick and Lal 2005), and eliminating burning (Batjes 1998).

2.3.1 Factors Affecting Soil Organic Carbon Status

2.3.1.1 Climate

Climate conditions such as temperature and precipitation influence the distribution and storage of soil organic carbon. Low temperature and waterlogging reduce the decomposition and mineralization of soil organic carbon which leads to the accumulation of organic carbon (Viscarra Rossel et al. 2014). Hence, cold and humid region soils contain more organic carbon than soils of hot or dry climates (Lal 2004; Hobbie et al. 2000). In terrestrial ecosystems higher precipitation indicates higher plant primary productivity and therefore more carbon into the soil (Wiesmeier et al. 2019). Wet conditions promote more weathering of parent rock which often acidifies leading to reduced decomposition of organic matter change in moisture, and temperature also modulates the microbial decomposition of organic matter (Intergovernmental Panel On Climate Change 2023; von Lützow and Kögel-Knabner 2009; Bronick and Lal 2005; Conant et al. 2011; Jobbágy and Jackson 2000).

2.3.1.2 Topography

Topography can be defined by three major aspects; geographic location, altitude, and slope. These three characteristics influence vegetation and the erosive nature of a location and both these influence soil carbon storage (Chaplot et al. 2010; Grimm et al. 2008). The geographic location of a place determines the type of vegetation that grows at that place and the decomposition of different vegetative species reacts differently toward the decomposition process. Altitude and slope are also important as higher altitudes have cooler climatic conditions which may limit the growth of certain vegetation. Steeper slopes experience more rapid erosional loss of soil, consequently erosion may enhance mineralization rates leading to higher carbon emission and less storage capacity of soil (Quinton et al. 2010).

Topographical characteristics also influence precipitation and water flow dynamics which influence soil moisture content. Higher or lower soil moisture content both negatively affect the microbial population present in a soil as a result accumulation of organic matter will be more. Higher soil erosion at the site of disturbance causes loss of organic carbon but the same erosion leads to buried and storage of carbon below the surface thus potentially reducing emission from the soils.

2.3.1.3 Parent Material

The parent material plays a crucial role in defining the intrinsic characteristics of soil, as it directly influences the soil's mineral composition. This mineralogy governs nutrient and water retention, as well as their availability to the plant. For example, soil rich in clay content is often derived from Basaltic parent material and has high nutrient and water retention capacity (Sollins et al. 1988), while sandy soils that usually originate from granite or sandstone parent material these soils drain faster and have less nutrient holding capacity. The soils derived from limestone often remain at near-neutral pH, which governs high microbial growth and availability of nutrients (Hamati 2022; Mineo and Majumdar 1996). Albeit this correlation occurs, but still the relation between parent material and SOC is still less direct. Soil type which is more dependent on climate and vegetation is more associated with SOC storage.

2.3.1.4 Organism

It mainly covers natural vegetation and soil biota which influence SOM storage in soil. The more profound the soil profile, the stored SOC will be more older (Stockmann et al. 2013). This also supports the finding that a greater amount of SOC will be present near the surface and decreases with deeper soil depth (Lal et al. 2007; Grimm et al. 2008; Vasques et al. 2010). The types of vegetation influence the carbon storage in the ecosystem. Shallow-rooted conifer species accumulate less organic matter than deep-rooted deciduous trees and they also slow litter transformation rate (Jandl et al. 2007). This indicates the different kinds of contributions of the different species to the process of SOC storage. Soil microorganisms, especially heterotrophs play an important role in the cycling and transformation of SOM. Heterotrophs such as bacteria, fungi, and protozoa are involved in the process of decomposition to fulfill their nutritional needs, and as a byproduct of this decomposition process nutrients are released into the soil for absorption. The byproducts produced by the microbes like microbial residues and exudates make up to 80% of the carbon in the stable fraction of organic carbon. This indicates that soil microbes address the SOM both ways. Through mineralization and stabilization of organic matter in mineralization, microbes convert organic matter into inorganic nutrients, during stabilization, decomposed organic matter transforms into humus a stable form of organic matter.

Soil microorganisms form complexes with soil particles and influence various soil properties and processes such as porosity, infiltration, and aggregate stability.

By improving soil structure and fertility, microbes enhance soil capacity to sequester carbon, hence helping in mitigating greenhouse gas emissions.

2.3.2 Soil Properties

Soil texture influences the capacity of soil to store carbon. Well-structured soil facilitates good aeration and moisture retention. Loamy soils provide a conducive environment for microbial growth thereby facilitating organic matter decomposition and helping in carbon storage. Clay soil having high cation exchange capacity can store vast amounts of organic matter promoting high carbon sequestration. Clay soil having poor aeration can hinder microbial activity. Sandy soil has lower cation exchange capacity and hence has less storing capacity of organic matter. Clay content influences soil aggregation, specific surface area and has an expanding nature. All these characters affect the rate of soil organic carbon decomposition.

2.4 Conclusion

The role of soil in mitigating climate change is important as it is the largest terrestrial carbon reservoir. Improved soil carbon reserves have many benefits, including a reduction in greenhouse gas emissions, improved soil health and fertility, and better agricultural production. Soil organic carbon can be increased with good agricultural practices which may include the incorporation of organic matter, crop rotation, and cover crops. However, soil does not have the infinite capacity to sequester carbon, it always needs continuous sustainable management. The implementation of sustainable management practices amid alterations in land use constitutes fundamental strategies for the sequestration of terrestrial organic carbon. The application of sustainable management practices during changes in land use forms cornerstone strategies for terrestrial organic carbon sequestration. Development of novel approaches and strengthening collaborations may serve as the backbone to carbon reserves in the soils, which is also essential to fulfilling the United Nations Agenda 2030 and Sustainable Development Goals.

References

Abao EB, Bronson KF, Wassmann R, Singh U (2000) Simultaneous records of methane and nitrous oxide emissions in rice-based cropping systems under rainfed conditions. In: Wassmann R, Lantin RS, Neue H-U (eds) Methane emissions from major rice ecosystems in Asia. Springer, Netherlands, pp 131–139. https://doi.org/10.1007/978-94-010-0898-3_12

Alam M, Alam MS, Roman M, Tufail M, Khan MU, Khan MT (2020) Real-time machine-learning based crop/weed detection and classification for variable-rate spraying in precision agriculture. In: 2020 7th international conference on electrical and electronics engineering (ICEEE), pp 273–280. https://doi.org/10.1109/ICEEE49618.2020.9102505

Alongi DM, Murdiyarso D, Fourqurean JW, Kauffman JB, Hutahaean A, Crooks S, Lovelock CE, Howard J, Herr D, Fortes M, Pidgeon E, Wagey T (2016) Indonesia's blue carbon: a globally significant and vulnerable sink for seagrass and mangrove carbon. Wetl Ecol Manag 24(1):3–13. https://doi.org/10.1007/s11273-015-9446-y

Atkinson CJ, Fitzgerald JD, Hipps NA (2010) Potential mechanisms for achieving agricultural benefits from biochar application to temperate soils: a review. Plant Soil 337(1):1–18. https://doi.org/10.1007/s11104-010-0464-5

Banayo NPMC, Haefele SM, Desamero NV, Kato Y (2018) On-farm assessment of site-specific nutrient management for rainfed lowland rice in The Philippines. Field Crop Res 220:88–96. https://doi.org/10.1016/j.fcr.2017.09.011

Barman D, Dash SK (2022) Stabilization of expansive soils using chemical additives: a review. J Rock Mech Geotech Eng 14(4):1319–1342. https://doi.org/10.1016/j.jrmge.2022.02.011

Batjes NH (1998) Mitigation of atmospheric CO2 concentrations by increased carbon sequestration in the soil. Biol Fertil Soils 27(3):230–235. https://doi.org/10.1007/s003740050425

Berner RA, Lasaga AC (1989) Modeling the geochemical carbon cycle. Sci Am 260(3):74–81

Bronick CJ, Lal R (2005) Soil structure and management: a review. Geoderma 124(1):3–22. https://doi.org/10.1016/j.geoderma.2004.03.005

Chaplot V, Bouahom B, Valentin C (2010) Soil organic carbon stocks in Laos: spatial variations and controlling factors. Glob Chang Biol 16(4):1380–1393. https://doi.org/10.1111/j.1365-2486.2009.02013.x

Conant RT, Ryan MG, Ågren GI, Birge HE, Davidson EA, Eliasson PE, Evans SE, Frey SD, Giardina CP, Hopkins FM, Hyvönen R, Kirschbaum MUF, Lavallee JM, Leifeld J, Parton WJ, Megan Steinweg J, Wallenstein MD, Martin Wetterstedt JÅ, Bradford MA (2011) Temperature and soil organic matter decomposition rates – synthesis of current knowledge and a way forward. Glob Chang Biol 17(11):3392–3404. https://doi.org/10.1111/j.1365-2486.2011.02496.x

Condron L, Stark C, O'callaghan M, Clinton P, Huang Z (2010) The role of microbial communities in the formation and decomposition of soil organic matter. Springer, Berlin/Heidelberg, pp 81–118

Dorrepaal E, Toet S, van Logtestijn RSP, Swart E, van de Weg MJ, Callaghan TV, Aerts R (2009) Carbon respiration from subsurface peat accelerated by climate warming in the subarctic. Nature 460:7255. https://doi.org/10.1038/nature08216

Elbasiouny H, El-Ramady H, Elbehiry F, Rajput VD, Minkina T, Mandzhieva S (2022) Plant nutrition under climate change and soil carbon sequestration. Sustain For 14:2. https://doi.org/10.3390/su14020914

Eurostat (2021) In: Kotzeva M, Brandmüller T, Fohgrub B, Önnerfors Å (eds) Eurostat regional yearbook. Publications Office of the European Union

Fageria NK (2012) Role of soil organic matter in maintaining sustainability of cropping systems. Commun Soil Sci Plant Anal 43(16):2063–2113. https://doi.org/10.1080/00103624.2012.697234

Fagodiya RK, Singh A, Singh R, Rani S, Kumar S, Rai AK, Sheoran P, Chandra P, Yadav RK, Sharma PC, Biswas AK (2023) The food-energy-water-carbon nexus of the rice-wheat production system in the western indo-Gangetic plain of India: an impact of irrigation system, conservational tillage and residue management. Sci Total Environ 860:160428

Fagodiya RK, Sharma G, Verma K, Rai AK, Prajapat K, Singh R, Chandra P, Sheoran P, Yadav RK, Biswas AK (2024a) Computation of soil quality index after fifteen years of long-term tillage and residue management experiment (LT&RE) under rice wheat system. Agric Syst 219:104039

Fagodiya RK, Sharma G, Verma K, Singh A, Singh R, Sheoran P, Rai AK, Prajapat K, Kumar S, Chandra P, Rani S (2024b) Fourteen-years impact of crop establishment, tillage and residue management on carbon input, soil carbon sequestration, crop productivity and profitability of rice-wheat system. Eur J Agron 161:127324

Fuss S, Lamb WF, Callaghan MW, Hilaire J, Creutzig F, Amann T, Beringer T et al (2018) Negative emissions – part 2: costs, potentials and side effects. Environ Res Lett 13(6):063002. https://doi.org/10.1088/1748-9326/aabf9f

Gangopadhyay PK, Khatri-Chhetri A, Shirsath PB, Aggarwal PK (2019) Spatial targeting of ICT-based weather and agro-advisory services for climate risk management in agriculture. Clim Chang 154(1):241–256. https://doi.org/10.1007/s10584-019-02426-5

Grimm R, Behrens T, Märker M, Elsenbeer H (2008) Soil organic carbon concentrations and stocks on Barro Colorado Island – digital soil mapping using random forests analysis. Geoderma 146(1):102–113. https://doi.org/10.1016/j.geoderma.2008.05.008

Guo LB, Gifford RM (2002) Soil carbon stocks and land use change: a meta analysis. Glob Chang Biol 8(4):345–360. https://doi.org/10.1046/j.1354-1013.2002.00486.x

Hamati S (2022) Ecophysiology of Juniperus virginiana encroachment in Ohio. Kent State University. https://etd.ohiolink.edu/acprod/odb_etd/etd/r/1501/10?clear=10&p10_accession_num=kent1651146580975921

Helliwell JF, Layard R, Sachs JD et al (2020) World happiness report 2020. Sustainable Development Solutions Network, New York

Herold N, Schöning I, Michalzik B, Trumbore S, Schrumpf M (2014) Controls on soil carbon storage and turnover in German landscapes. Biogeochemistry 119(1):435–451. https://doi.org/10.1007/s10533-014-9978-x

Hobbie SE, Schimel JP, Trumbore SE, Randerson JR (2000) Controls over carbon storage and turnover in high-latitude soils. Glob Chang Biol 6(S1):196–210. https://doi.org/10.1046/j.1365-2486.2000.06021.x

Intergovernmental Panel On Climate Change (2023) Climate change 2021 – the physical science basis: working group I contribution to the sixth assessment report of the intergovernmental panel on climate change, 1st edn. Cambridge University Press. https://doi.org/10.1017/9781009157896

Intergovernmental Panel on Climate Change (IPCC) (1996) Climate change1995. The science of climate change. The contribution of Working Group I to the second assessment report of the intergovernmental Panelon climate change. Cambridge University Press, New York

IPCC (2021) Climate change 2021: the physical science basis. In: Masson-Delmotte V, Zhai P, Pirani A, Connors SL, Péan C, Chen Y, Goldfarb L, Gomis MI, Matthews JBR, Berger S et al (eds) Contribution of working group I to the sixth assessment report of the intergovernmental panel on climate change. Cambridge University Press, Cambridge

Jandl R, Lindner M, Vesterdal L, Bauwens B, Baritz R, Hagedorn F, Johnson DW, Minkkinen K, Byrne KA (2007) How strongly can forest management influence soil carbon sequestration? Geoderma 137(3):253–268. https://doi.org/10.1016/j.geoderma.2006.09.003

Jobbágy EG, Jackson RB (2000) The vertical distribution of soil organic carbon and its relation to climate and vegetation. Ecol Appl 10(2):423–436

Kell DB (2011) Breeding crop plants with deep roots: their role in sustainable carbon, nutrient and water sequestration. Ann Bot 108(3):407–418. https://doi.org/10.1093/aob/mcr175

Kober T, Schiffer HW, Densing M, Panos E (2020a) Global energy perspectives to 2060 – WEC's world energy scenarios 2019. Energ Strat Rev 31:100523

Kober T, Schiffer HW, Densing M, Panos E (2020b) Global energy perspectives to 2060 – WEC's world energy scenarios 2019. Energ Strat Rev 31:100523. https://doi.org/10.1016/j.esr.2020.100523

Kumar A, Singh P, Raizada P, Hussain CM (2022) Impact of COVID-19 on greenhouse gases emissions: a critical review. Sci Total Environ 806:150349

Kwiatkowska-Malina J (2018) Qualitative and quantitative soil organic matter estimation for sustainable soil management. J Soils Sediments 18(8):2801–2812. https://doi.org/10.1007/s11368-017-1891-1

Lacis AA, Schmidt GA, Rind D, Ruedy RA (2010) Atmospheric CO2: principal control knob governing earth's temperature. Science 330:356–359

Lal R (2004) Soil carbon sequestration to mitigate climate change. Geoderma 123:1–22

Lal R (2007) Carbon sequestration. Philos Trans R Soc B: Biol Sci 363(1492):815–830. https://doi.org/10.1098/rstb.2007.2185

Lal R, Follett RF, Stewart BA, Kimble JM (2007) Soil carbon sequestration to mitigate climate change and advance food security. Soil Sci 172(12):943. https://doi.org/10.1097/ss.0b013e31815cc498
Lal R, Negassa W, Lorenz K (2015) Carbon sequestration in soil. Curr Opin Environ Sustain 15:79–86. https://doi.org/10.1016/j.cosust.2015.09.002
Lal R, Monger C, Nave L, Smith P (2021) The role of soil in regulation of climate. Philos Trans R Soc B: Biol Sci 376(1834):20210084. https://doi.org/10.1098/rstb.2021.0084
Lehmann J, Gaunt J, Rondon M (2006) Bio-char sequestration in terrestrial ecosystems-a review. Mitig Adapt Strateg Glob Chang 11:403–427
Linquist B, van Groenigen KJ, Adviento-Borbe MA, Pittelkow C, van Kessel C (2012) An agronomic assessment of greenhouse gas emissions from major cereal crops. Glob Chang Biol 18(1):194–209. https://doi.org/10.1111/j.1365-2486.2011.02502.x
Luo Y, Su B, Currie WS, Dukes JS, Finzi A, Hartwig U, Hungate B, McMurtrie RE, Oren R, Parton WJ, Pataki DE, Shaw RM, Zak DR, Field CB (2004) Progressive nitrogen limitation of ecosystem responses to rising atmospheric carbon dioxide. Bioscience 54(8):731–739
Mann A, Meena BL, Kumar A, Sheoran P, Yadav RK, Sharma PC (2018) Mruda swasthay card apnayen santulit urvark prabhandhan dwara fasal utpadan badhayen. CSSRI/Karnal/Extension bulletin/2018/02pg58
Manning DAC, Renforth P, Lopez-Capel E, Robertson S, Ghazireh N (2013) Carbonate precipitation in artificial soils produced from basaltic quarry fines and composts: an opportunity for passive carbon sequestration. Int J Greenhouse Gas Control 17:309–317. https://doi.org/10.1016/j.ijggc.2013.05.012
McNunn G, Karlen DL, Salas W, Rice CW, Mueller S, Muth D, Seale JW (2020) Climate smart agriculture opportunities for mitigating soil greenhouse gas emissions across the U.S. Corn-Belt. J Clean Prod 268:122240. https://doi.org/10.1016/j.jclepro.2020.122240
Mineo L, Majumdar SK (1996) Ectomycorrhizae in oaks (Quercus alba, Q. Rubra) in northeastern Pennsylvania woodlands: morphology, frequency and implied physiology and ecology. In: Mukerji KG (ed) Concepts in mycorrhizal research. Springer, Netherlands, pp 315–331. https://doi.org/10.1007/978-94-017-1124-1_11
Mitra S, Wassmann R, Jain MC, Pathak H (2002) Properties of rice soils affecting methane production potentials: 2. Differences in topsoil and subsoil. Nutr Cycl Agroecosyst 64(1):183–191. https://doi.org/10.1023/A:1021175404418
Pagliacci F, Defrancesco E, Mozzato D, Bortolini L, Pezzuolo A, Pirotti F, Pisani E, Gatto P (2020) Drivers of farmers' adoption and continuation of climate-smart agricultural practices. A study from northeastern Italy. Sci Total Environ 710:136345. https://doi.org/10.1016/j.scitotenv.2019.136345
Palit K, Rath S, Chatterjee S, Das S (2022) Microbial diversity and ecological interactions of microorganisms in the mangrove ecosystem: threats, vulnerability, and adaptations. Environ Sci Pollut Res 29(22):32467–32512. https://doi.org/10.1007/s11356-022-19048-7
Pinto AD, Cenacchi N, Kwon H-Y, Koo J, Dunston S (2020) Climate smart agriculture and global food-crop production. PLoS One 15(4):e0231764. https://doi.org/10.1371/journal.pone.0231764
Pittelkow CM, Liang X, Linquist BA, van Groenigen KJ, Lee J, Lundy ME, van Gestel N, Six J, Venterea RT, van Kessel C (2015) Productivity limits and potentials of the principles of conservation agriculture. Nature 517(7534):7534. https://doi.org/10.1038/nature13809
Poeplau C, Don A (2015) Carbon sequestration in agricultural soils via cultivation of cover crops – a meta-analysis. Agric Ecosyst Environ 200:33–41. https://doi.org/10.1016/j.agee.2014.10.024
Prathibha SR, Hongal A, Jyothi MP (2017) IOT based monitoring system in smart agriculture. In: 2017 international conference on recent advances in electronics and communication technology (ICRAECT), pp 81–84. https://doi.org/10.1109/ICRAECT.2017.52
Pulleman MM, Bouma J, van Essen EA, Meijles EW (2000) Soil organic matter content as a function of different land use history. Soil Sci Soc Am J 64(2):689–693. https://doi.org/10.2136/sssaj2000.642689x

Quinton JN, Govers G, Van Oost K, Bardgett RD (2010) The impact of agricultural soil erosion on biogeochemical cycling. Nat Geosci 3(5). https://doi.org/10.1038/ngeo838

Sapkota TB, Majumdar K, Jat ML, Kumar A, Bishnoi DK, McDonald AJ, Pampolino M (2014) Precision nutrient management in conservation agriculture based wheat production of Northwest India: profitability, nutrient use efficiency and environmental footprint. Field Crop Res 155:233–244. https://doi.org/10.1016/j.fcr.2013.09.001

Siddique A, Shahzad A, Lawler J, Mahmoud KA, Lee DS, Ali N et al (2021) Unprecedented environmental and energy impacts and challenges of COVID-19 pandemic. Environ Res 193:110443

Singh R, Singh A, Kumar S, Fagodiya RK, Sheoran P, Rai AK, Rani S, Chandra P (2023) Effect of mini-sprinkler irrigation, tillage and residue on productivity, profitability and resource saving in rice–wheat system in western indo-Gangetic Plains of India. Paddy Water Environ 21(4):479–495

Six J, Conant RT, Paul EA, Paustian K (2002) Stabilization mechanisms of soil organic matter: implications for C-saturation of soils. Plant Soil 241(2):155–176. https://doi.org/10.1023/A:1016125726789

Smith P (2004) Carbon sequestration in croplands: the potential in Europe and the global context. Eur J Agron 20(3):229–236. https://doi.org/10.1016/j.eja.2003.08.002

Smith P, Martino D, Cai Z, Gwary D, Janzen H, Kumar P, McCarl B, Ogle S, O'Mara F, Rice C, Scholes B, Sirotenko O, Howden M, McAllister T, Pan G, Romanenkov V, Schneider U, Towprayoon S, Wattenbach M, Smith J (2007) Greenhouse gas mitigation in agriculture. Philos Trans R Soc B: Biol Sci 363(1492):789–813. https://doi.org/10.1098/rstb.2007.2184

Sollins P, Robertson GP, Uehara G (1988) Nutrient mobility in variable- and permanent-charge soils. Biogeochemistry 6(3):181–199. https://doi.org/10.1007/BF02182995

Solomon D, Lehmann J, Kinyangi J, Amelung W, Lobe I, Pell A, Riha S, Ngoze S, Verchot L, Mbugua D et al (2007) Long-term impacts of anthropogenic perturbations on dynamics and speciation of organic carbon in tropical forest and subtropical grassland ecosystems. Glob Chang Biol 13:511–530

Songqi P, Caineng Z, Yong L et al (2021) Major biological events andfossil energy formation: on the development of energy science under theearth system framework. Petrol Explor Dev 48(3):498e509

Stockmann U, Adams MA, Crawford JW, Field DJ, Henakaarchchi N, Jenkins M, Minasny B, McBratney AB et al (2013) The knowns, known unknowns and unknowns of sequestration of soil organic carbon. Agric Ecosyst Environ 164:80–99. https://doi.org/10.1016/j.agee.2012.10.001

Swingland IR, Bettelheim EC, Grace J, Prance GT, Saunders LS, Malhi Y, Meir P, Brown S (2002) Forests, carbon and global climate. Philos Trans R Soc London, Ser A 360(1797):1567–1591. https://doi.org/10.1098/rsta.2002.1020

US EPA (2007) 2007 Draft U.S. Greenhouse Gas Inventory Report: DRAFTInventory of U.S. Greenhouse Gas Emissions and Sinks: 1990e2005.United States Environmental Protection Agency, Office of AtmosphericPrograms (6207J), Washington, DC. Available from: http://epa.gov/climatechange/emissions/usinventoryreport07.html.

Vasques GM, Grunwald S, Comerford NB, Sickman JO (2010) Regional modelling of soil carbon at multiple depths within a subtropical watershed. Geoderma 156(3):326–336. https://doi.org/10.1016/j.geoderma.2010.03.002

Villa-Henriksen A, Edwards GTC, Pesonen LA, Green O, Sørensen CAG (2020) Internet of things in arable farming: implementation, applications, challenges and potential. Biosyst Eng 191:60–84. https://doi.org/10.1016/j.biosystemseng.2019.12.013

Viscarra Rossel RA, Webster R, Bui EN, Baldock JA (2014) Baseline map of organic carbon in Australian soil to support national carbon accounting and monitoring under climate change. Glob Chang Biol 20(9):2953–2970. https://doi.org/10.1111/gcb.12569

von Lützow M, Kögel-Knabner I (2009) Temperature sensitivity of soil organic matter decomposition – what do we know? Biol Fertil Soils 46(1):1–15. https://doi.org/10.1007/s00374-009-0413-8

Wiesmeier M, Urbanski L, Hobley E, Lang B, von Lützow M, Marin-Spiotta E, van Wesemael B, Rabot E, Ließ M, Garcia-Franco N, Wollschläger U, Vogel H-J, Kögel-Knabner I (2019) Soil organic carbon storage as a key function of soils – a review of drivers and indicators at various scales. Geoderma 333:149–162. https://doi.org/10.1016/j.geoderma.2018.07.026

World Meteorological Organization, Global Atmosphere Watch (2019). WMO greenhouse gas bulletin. https://library.wmo.int/doc_num.php?explnum_id¼4022

World Resources Institute Climate Analysis Indicators Tool WRI CAIT 4.0 (2017) Emissions including land-use change and forestry. Global Warming Potentials are from the Intergovernmental Panel on Climate Change Second Assessment Report

Zaman M, Heng L, Müller C (2021) Measuring emission of agricultural greenhouse gases and developing mitigation options using nuclear and related techniques: applications of nuclear techniques for *GHGs*. Springer Nature, p 337

Zerssa G, Feyssa D, Kim D-G, Eichler-Löbermann B (2021) Challenges of smallholder farming in Ethiopia and opportunities by adopting climate-smart agriculture. Agriculture 11(3). https://doi.org/10.3390/agriculture11030192

Improving Soil Properties Through Various Soil Amendments Under Changing Climate Scenario

3

Parul Sundha, Nirmalendu Basak, Arvind Kumar Rai, Priyanka Chandra, Amresh Chaudhary, Sandeep Bedwal, Aman Kumar, and Rajender Kumar Yadav

Abstract

Worldwide ~831 million hectares of land are salt-affected (SAS). A considerable area of SAS is sodic in nature and carries a considerable quantity of Na and carbonate/bicarbonate, with Na-saturated clay/mineral, which have the potential to supply OH^- than Ca ions. The physico-chemical properties and microbial activities of this soil are deleteriously affected by the disproportional availability of CO_3^{2-}/HCO_3^- of Na and have serious problems with crop productivity. The projection on climate change rising temperature, elevated CO_2 concentration, and evapotranspiration may affect soil sodification and salt dynamics. For restoration of sodic soil, an application of external Ca^{2+} or acid former is recommended to neutralize carbonate and bicarbonate. Gypsum is the cheapest and most readily available amendment for sodicity reclamation. With limited availability and questionable purity of gypsum, alternate amendments are being searched from time to time. Different approaches have been suggested to apply inorganic in conjunction with organic materials to reclaim such soils. This chapter focuses on selected studies that have identified several materials like polymers, nanomaterials, acidified biochars, elemental sulfur, municipal solid waste compost (MSWC), and flue gas desulfurization gypsum (FGDG) and have the potential to replace gypsum as an amendment in future. Practical issues associated with these materials warrant scientific solutions are also highlighted. Under precise management strategies, climate change mitigation and adaptation strategies are described for pedogenic processes and soil properties. This generated information can be utilized for developing soil sodicity reclamation strategies for land degradation neutrality under changing climatic conditions.

P. Sundha · N. Basak (✉) · A. K. Rai · P. Chandra · A. Chaudhary · S. Bedwal · A. Kumar
R. K. Yadav
ICAR–Central Soil Salinity Research Institute, Karnal, Haryana, India
e-mail: nirmalendu.basak@icar.gov.in

A. Mann et al. (eds.), *Cutting Edge Technologies for Developing Future Crop Plants*,
https://doi.org/10.1007/978-981-96-2508-6_3

Keywords

Salinity · Sodic soils · Phosphogypsum · Climate change · Soil degradation

3.1 Introduction

It is estimated around 50% of the arable cropland of the world is affected by salinity and poor-quality water irrigation. India contains 6.73 Mha salt-affected soils, with 3.77 Mha of sodic soils carrying a disproportionately larger quantity of Na and its carbonate and bicarbonate salts (Sharma et al. 2015). The higher concentration of the salts deteriorates soil structure, water permeability, and air transmission, thereby affecting soil chemical and biological properties and constraining agricultural production (Basak et al. 2022a, b, 2023). The management and reclamation of such soils are dependent on the presence of salt type and quantity, soil matrix, crop growth patterns, climatic conditions (precipitation, temperature, relative humidity), and quality of irrigation water (Chhabra 2004). Different physical, chemical, and biological adaptive measures are being followed to bring degraded soils under cultivation (Ahmad et al. 2013; Singh et al. 2015). During the reclamation process, the Na^+ ions are replaced by bivalent Ca^{2+} ions in the root zone. The most basic reclamation agent is gypsum (Kanwar et al. 1965; CSSRI 2006; Abrol and Bhumbla 1979), including dilute sulfuric acid, ferrous sulfate or elemental sulfur, and organic amendments like city waste compost, sewage sludge, farmyard manure, chicken manure, and biochar (Sundha et al. 2020, 2023a, b).

3.2 Climate Change and Soil Sodicity

Climate change causes a global rise in temperature, rainfall aberration, and higher greenhouse gas emissions (GHGs), leading to deteriorated ecosystems and biodiversity losses (Shivanna 2022). Therefore, climate change can disrupt current efforts to meet the prime obligations and responsibilities of livelihood security and a sustainable food system. Changing precipitation, temperature patterns, and GHG concentrations impact agro-ecosystems at local to national scales in space and time.

Salt-affected soils (SAS) are loaded with undesirable electrolytes, and sodic soil is inherently saturated with exchangeable Na and shows alkali-to-alkaline reactions (Rai et al. 2021a). Therefore, rising temperature, elevated CO_2 concentration, and evapotranspiration demand-associated environmental factors regulate the pedogenic processes and soil properties (Basak et al. 2022b; Eswar et al. 2021). SASs coexist with poor-quality underground water. Therefore, areas under poor quality irrigation will be more vulnerable to the impacts of climate change due to the deterioration in the scenario of water resources.

Reclamation and management of sodic soil is the prime agenda of the Union Govt. for greening the sodic soil. Several materials, including gypsum, elemental sulfur, acidified biochars, flue gas desulfurization gypsum (FGDG), city waste compost, pressmud, sugarcane distillery spent wash, nanomaterials like polymers, have been prescribed for the reclamation of sodic soil (Rai et al. 2021b). These amendments are the sources of Ca^{2+} or mobilizers of native calcium carbonate or supply H_3O^+ on chemical or microbial oxidation (Basak et al. 2022a). The neutralization of soluble alkalinity (CO_3^{2-}/HCO_3^-) by supplying H_3O^+/Ca^{2+}_{aq} and exchange reaction [$2Na^+$–(Clay) + $Ca^{2+}_{aq} \rightarrow Ca^{2+}$–(Clay) + Na^+] are the primary mechanisms for neutralization of alkalinity and lower down the Na^+ saturation. Further, the best available water is prescribed for irrigation and desalinating the SASs and is suitable for cultivation. The reclamation potential of the S° amendment with the introduction of *Sphingomonas olei* is reported by Rai et al. (2024).

The areas of salt-affected soils are likely to increase because of faulty irrigation practices or the use of poor-quality irrigation and climate change. Soil sodification is a dynamic process mainly governed by changes in temperature, moisture, and partial pressure of the CO_2 in the soil. The tentative area under SASs will reach nearly 16.2 million ha by 2050 if the current rate of reclamation of sodification is discontinued (CSSRI 2015).

3.3 Availability of Soil Reclamation Sources

Different chemical ameliorants for saline-sodic soil reclamation are gypsum, pyrites, aluminum chloride, elemental sulfur, phosphogypsum, FGD gypsum, silicious chalk, etc. (Abrol and Bhumbla 1979; Sharma and Swarup 1997; CSSRI 2006; Sundha et al. 2023a, b). The amendments are broadly categorized into inorganic/chemical and organic agents based on their availability and neutralization potential. Besides, organic sources such as farmyard manure (FYM), pressmud, corn stalks (Li and Keren 2009), city waste compost (Meena et al. 2016; Sundha et al. 2018; Singh et al. 2017), sewage sludge (ICAR–CSSRI Annual Report 2018), and crop residue (Choudhary et al. 2011) are also used in conjunction with gypsum to amend degraded soils. Soil organic matter improves the soil structure and aggregation (Oades 1993), increases hydraulic conductivity (Hussain et al. 2001), improves the infiltration rate and cation exchange capacity, more vital enzymatic activities for enhanced nutrient uptake by the plants (Von Lutzow et al. 2002; Jalali and Ranjbar 2009). Industrial by-products obtained include phosphogypsum, fluorogypsum, borogypsum, and FGD (flue gas desulfurization) gypsum, from fertilizer industries and national power plants.

3.4 Present Status of Use of Reclamation Agents in Sodic Soils

3.4.1 Gypsum

chemically $CaSO_4$, $2H_2O$, has been the most commonly used amendment in sodic soils for years. The gypsum requirement (GR) is calculated based on the degree of salts present and factors like soil structure, water quality for irrigation, choice of a crop variety, and desired extent of overall improvement of soil. The reaction activity of the gypsum starts straightaway on its application in the soil via the replacement of Na^+ from the exchange complex by Ca^{2+} ions through leaching. Gypsum is made available to farmers through different agencies across the country like the World Bank, European Union, Haryana Land Reclamation and Development Corporation and other developmental agencies. The success of India in this field lies in the reclamation of 2.10 M ha of alkali lands with the largest area reclaimed in Punjab (0.79 M ha), then Uttar Pradesh (0.73 M ha), followed by Haryana (0.35 M ha), and to an extent of 0.70 M ha in other states. ICAR-CSSRI is continuously contributing in these states for the betterment of the soil and livelihood status through the addition of food grain per annum. By now reclaimed alkali soils in Uttar Pradesh added around 4.30 Mt. of additional food grains per annum, approximately 2% of total food grain production. The use of alkali water (high RSC) in sodic soils raises the problem of secondary salinization. The reclaimed soils again exhibit a rise in pH leading to loss of crop productivity. There is a huge yield penalty in major food grain crops like rice and wheat. High RSC water (> 2.5 me L^{-1}) irrigation often leads to spikelet sterility (unfilled grains), more popularly called chaffy grain (Sheoran et al. 2021). Such situations are usually accompanied by changes in climate, erratic rainfalls, and higher evapotranspiration (ET) causing increased salt load in the soils. Such conditions are constantly being taken over under the mandate of the institute for minimized losses. Many studies on high RSC (5–10 meL^{-1}) neutralization through gypsum application waters improved rice and wheat yields (Choudhary et al. 2011).

3.4.2 Phosphogypsum (PG)

Phosphogypsum (PG) is the by-product of the phosphoric acid plants. It is more soluble compared to mined gypsum. Presently, 70 million tonnes of phosphogypsum have been stock piled, and the Department for Promotion of Industry and Internal Trade (DPIIT) is focused on reducing the quantity of unutilized phosphogypsum. Around 9.16 million tonnes (MT) of phosphogypsum was generated from the fertilizer industries during 2021–22 in India and stockpiled to estimate 70 MT phosphogypsum in the vicinity of fertilizer plants. Phosphogypsum addition significantly decreased the values of pH, EC, exchangeable sodium percentage, sodium absorption ratio, and exchangeable Na^+ in bauxite residue (pH 10.43) during the production of alumina from bauxite ore (Tian et al. 2021). Phosphogypsum application at 2–20 t ha^{-1} along with Azolla and cyanobacteria also decreased the total CH_4 emission by 9–32% in paddy soil (Ali et al. 2007). Phosphogypsum could be used as an alternative amendment in acidic soils by thermal pyrolysis treatments (300–700 °C) of phosphogypsum waste mixed with banana peduncle to nutrient-rich alkaline material having 25.8% carbon, 10% potassium, and 12% sulfur (Karim et al. 2021). Production of biochar through coplasma processing of phosphogypsum and nutrient-rich biomass had potassium in the range of 4.2–12.7% and sulfur ranging from 13.3 to 17.8% and toxic metals below the regulatory limit of land application (Kumar et al. 2018).

The application of phosphorgypsum (at 10 Mg ha^{-1}), followed by Karnal grass as the first crop, resulted in the reduction of soil pH and exchangeable sodium percentage (ESP). The application of PG at 10 Mg ha^{-1} resulted in greater yields of both rice and wheat than other treatments (Nayak et al. 2013). The use of PG is associated with serious chemical and radioactive contamination of the surrounding soil, water, and air (Tayibia et al. 2009). The environmental issues of relatively high levels of natural uranium-series radionuclides and other impurities such as fluoride sulfate and trace elements restrict its application in agriculture. Recently, the Government in India has focused on the use of PG through circular economy in the agriculture sector as an ameliorant for sodic soil.

3.4.3 Iron Pyrite (FeS_2)

Iron Pyrite (FeS_2) could be used as a reclamation agent for saline-sodic soils. Generally, pyrites of low commercial value (15–30% total sulfur) are preferred (Tozsin et al. 2014). It generates sulfuric acid through sequential biochemical oxidation, which aids in the dissolution of native calcium carbonate ($CaCO_3$). The free calcium ions facilitate Na displacement from the exchange phase. The sulfuric acid produced during the chemical oxidation reaction $2FeS_2 + 7O_2 + 2H_2O = 2FeSO_4 + 2H_2SO_4$ also causes $CaCO_3$ dissolution.

Further, the soil microorganism *Thiobacillus ferrooxidans* takes part in bacterial oxidation of Fe (II).

$4FeSO_4 + O_2 + 2H_2SO_4 = 2Fe_2(SO_4)_3 + 2H_2O$. Subsequently $Fe_2(SO_4)_3$ is reduced and pyrite is oxidized by a chemical reaction as follows:

$$Fe_2\left(SO_4\right)_3 + FeS_2 = 3FeSO_4 + 2S$$

The generated elemental sulfur (S) was then oxidized by *Thiobacillus thiooxidans* and liberated sulfuric acid $2S + 3O_2 + 2H_2O = 2H_2SO_4$

$$Na_2CO_3 + H_2SO_4 = CO_2 + H_2O + Na_2SO_4\left(\text{leachable}\right)$$

$$CaCO_3 + H_2SO_4 = CaSO_4 + H_2O + CO_2$$

$$Na\left[\text{Clay micelle}\right]Na + CaSO_4 = \left[\text{Clay micelle}\right]Ca + Na_2SO_4\left(\text{leachable}\right)$$

3.4.4 Industrial Waste by-Products

Industrial Waste By-products are used for the amelioration of degraded soils. Fly ash is one such product from thermal power plants generated during the combustion of coal. In India, the coal used for thermal power generation has an ash content of around 40%. There is a generation of 120–150 million tons of coal fly ash from the existing 120 coal-based thermal power plants in India (Lokeshappa and Dikshit

2011). The expected increase of fly ash to 175 million tonnes (Mt) would require 40,000 hectares of additional land for storage (Lokeshappa and Dikshit 2011) creating environmental concerns. Their application in agricultural lands would be a method of recycling waste and a source of Ca^{2+} in the soil and increasing yield in rice and wheat crops (Kumar and Singh 2003). Conjunctive use of organic amendments such as FYM, press mud, sewage sludge, crop residues, and city waste compost improved degraded/marginal soil (Lokeshappa and Dikshit 2011; Li et al. 2005).

The process of biomethanation of raw spent wash to obtain the methane gas generates the by-product biomethanated spent wash (PBSW). This organic by-product is rich in organic carbon (OC), Ca, and K beneficial to the soil for supplying nutrients. The application of gypsum and spent wash (at 5 LL ha^{-1}) increased the seed yield of soybean by 85% over control. Raw spent wash at 5.0 lakh L ha^{-1} reduced the ESP to significant levels (Biennial Reports, CSSRI 2012–2014). There was an improvement in the *Sodic Haplusterts* with the application of primary biomethanated spent wash of 2.5 cm (PBSW), followed by leaching with two pore volumes of irrigation water.

Pressmud is an industrial waste by-product of the sugar mills. It is estimated that every 100 tonnes of sugarcane crushed produce about 3 tonnes of pressmud cake rich in organic sources. Utilization of pressmud in agriculture is well reported for improving organic matter content, structure, water infiltration, and decreasing pH (Partha and Sivasubramanian 2006; Ibrahim et al. 2012). Calcareous saline-sodic soils showed a decrease in the EC and SAR with the application of sulfidated pressmud (Yaduvanshi and Yadav 1990; Sheoran et al. 2021).

3.4.5 Manure

Manure application to agricultural lands is an economical and environmentally sustainable mechanism for enhancing the concentration of water-soluble salts, improving soil nutritional content, and increasing crop production. It increases the CEC and soluble and exchangeable K, limiting the entry of Na^+ into the exchange complex and promoting crop growth in saline (Walker and Bernal 2008). A considerable decrease in ESP (about 50%) and electrical conductivity of the soil from 9 dS m^{-1} to as low as 1 dS m^{-1} after 5 years was reported with the application of cotton gin crushed compost and poultry manure at rates of 5 and 10 t ha^{-1} (Tejada et al. 2006).

3.5 Futuristic Amendments to Combat Soil Sodicity

Sodic soil requires high Ca sources for reclamation. The most common amendment adopted is gypsum. Regarding the issue of availability and purity grade of gypsum, researchers are now keen to explore the industrial by-products, organic sources, and conjunctive application of organic and chemical products for fast and economically viable alternatives. Some of the sources recently applied in sodic soil reclamation are discussed below.

3.5.1 Acid and Acid-Forming Substances

The reaction of sulfuric acid with soil calcium carbonate releases calcium sulfate and neutralizes the free sodium carbonate in soils in replacing the adsorbed sodium. Sulfuric acid is being used in many parts of the world for the reclamation of calcareous sodic soils and decreasing the alkalinity of irrigation water (Griffen and Silvertooth 1999; Amrhein 2000; Ziskin et al. 2024). The constraint in the large-scale use of sulfuric acid for improving sodic soils is the handling (Ghafoor and Muhammed 1981; Ghafoor et al. 1986). Another alternate technology to rehabilitate sodic soils is a sulfurous acid generator (SAG). The burning of the sulfur produces SO_2 which is dissolved in a fraction (10–15%) of tube well water to form sulfurous acid (H_2SO_3). H_2SO_3 neutralizes the carbonate (CO_3^{2-}) and bicarbonate ($HCO_3^{\ -}$) ions of water and reduces the RSC of treated water. Elemental S is also used for amending the calcareous-sodic soil (Overstreet et al. 1951). Elemental S is oxidized to H_2SO_4 by microbial activities and converts sodium carbonate and bicarbonate to sodium sulfate. Elemental S is insoluble in water and works upon oxidation under moist conditions (Rai et al. 2024). The oxidation of elemental S is mediated by *Thiobacillus thiooxidans*, which requires warm, well-aerated, and moist soil with low pH conditions

$$2S + 3O_2 = 2SO_3\left(\text{microbiological oxidation}\right)$$
$$SO_3 + H_2O = H_2SO_4$$

The elemental S works in sodic soils by forming H_2SO_4, which changes the $NaHCO_3$ to the less harmful and leachable sodium sulfate (Na_2SO_4) and decreases the pH

$$NaHCO_{3+}\ H_2SO_4 = Na_2SO_4\left(\text{Leachable}\right) + H_2O + CO_2$$
$$Na_2CO_{3+}\ H_2SO_4 = Na_2SO_4\left(\text{Leachable}\right) + H_2O + CO_2$$
$$Na^+ - [\text{Soil}] - Na^+ + H_2SO_4 = H^+ - [\text{Soil}] - H^+ + Na_2SO_4\left(\text{Leachable}\right)$$

Elemental S furnishes Ca^{2+} (soluble) indirectly with reacting native soil lime ($CaCO_3$). The released Ca^{2+} performs the exchange reaction with Na^+ on the clay complex

$$CaCO_3 + H_2SO_4 = CaSO_4 + H_2O + CO_2$$
$$Na^+ - [\text{Soil}] - Na^+ + CaSO_4 = Ca^{2+} - [\text{Soil}] + Na_2SO_4\left(\text{Leachable}\right)$$

The advantages of these reactions are that the sodium carbonate and bicarbonate change to sodium sulfate, a mild neutral salt, and both the carbonate and bicarbonate anions are removed from the system. When gypsum is used, a portion of carbonate may remain as a calcium compound ($CaCO_3$) (Brady and Weil 2014). Microbial oxidation converts the S to sulfate (Lawrence and Germida 1988). Microbial spices *Thiobacillus* produces sulfuric acid when elemental S is applied to soil, which reduces soil pH (Garcia Junior 1992). The dissociated hydrogen can displace

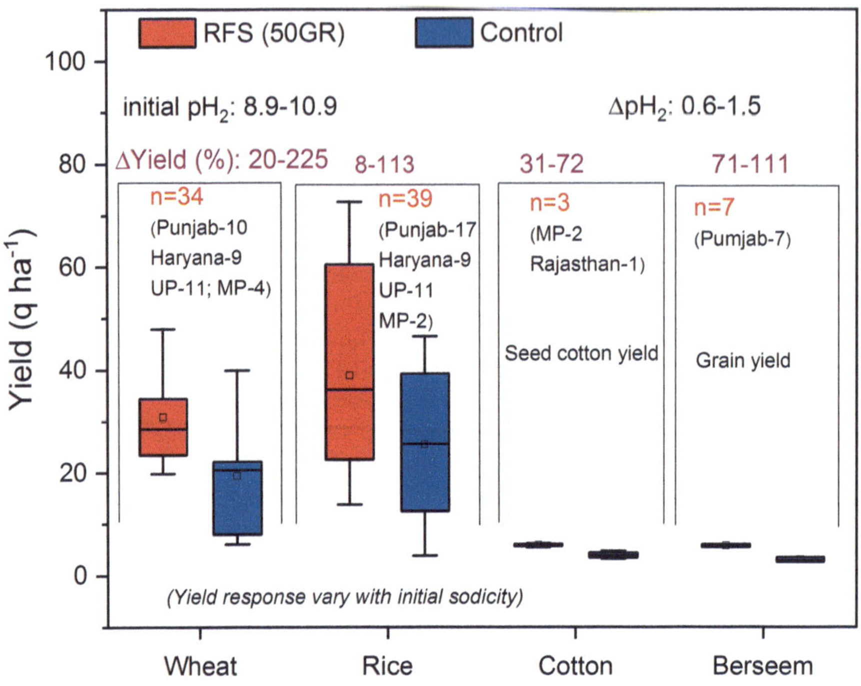

Fig. 3.1 Effect of RFS application on the crop yield and soil pH in different agroecologies

sodium from the exchange complex (Rupela and Tauro 1973). Expanding petroleum industries in the country have led to an increased availability of the by-product elemental S, offering an option for use in sodic soil reclamation. Depending on the severity of sodicity, the application of elemental sulfur-based formulation (Reliance formulated S) increased crop yield by 8–225% compared to unreclaimed sodic soils. The application of RFS also caused a 0.6–1.5 unit decrease in soil pH after one season of the amendment application (Fig. 3.1). The assessment conducted by CSSRI, Karnal, India, that the economic feasibility analysis assuming 12% opportunity cost of the capital revealed that the benefit-cost ratio varied from 1.22 to 1.70 in different agroecologies with the payback period in the range of 2–3 years for Reliance formulated S addition to reclaim sodic soil.

3.5.2 Sea Extracted Gypsum (Marine Gypsum)

The production of common salt by solar evaporation produces marine gypsum as a by-product in the coastal areas of Gujarat and Tamil Nadu. The Salt Commissioner, Jaipur, accounts for the production of 142,289 tonnes in 2010–11 and 166,665 tonnes in 2011–12 of marine gypsum from Gujarat and Tamil Nadu. The marine gypsum carries 89.7–92.6% of $CaSO_4$, $2H_2O$; additionally, it bears 0.5–2.1, 0.6, 3.4, and 3.5–7.7% of NaCl, $MgCl_2$, $MgSO_4$, and insoluble constituents, respectively

(Basak et al. 2023). Marine gypsum is produced at a rate of 30–50 kg per tonne of salt production. This is one emerging source of sodic soil reclamation with increased solar salt production in the nation. It contains NaCl, $MgCl_2$, and $MgSO_4$ as impurities which may increase the ionic strength of the aqueous solution by decreasing its activity coefficient and enhancing reclamation efficiency (Jha et al. 2022).

3.5.3 Flue Gas Desulfurization Gypsum (FGD Gypsum)

Over the past few decades, there has been wide interest in developing strategies for the utilization of flue gas desulfurization gypsum (FGDG), a by-product of scrubbing sulfur from combustion gases in coal-fired power generation plants. It is produced when coal is burned to produce electricity, heat, or other forms of energy. The applications of FGDG in sodic soils supply Ca^{2+} ions to replace the Na^+ in the soil matrix and improve physico-chemical properties, control soil and nutrient loss, supplement micronutrients for soil, and increase crop yield; thereby increasing productivity (Chen and Dick 2011; Mandal 2022; Zhao et al. 2021). Co-utilization of FGDs with organic materials (manures, composts, and bio-solids) may also provide many benefits when used on land. Soils with added FGDGs have less surface crusting and compaction, greater water infiltration and water-holding capacity, greater aggregate stability, and less water runoff and erosion (Norton and Zhang 1998). The application of wet and semidry FGD gypsum as an alkali soil resulted in decreased soil pH, ESP, and increased corn and alfalfa yield (Sakai et al. 2004; Wang et al. 2008). Besides beneficial effects, some reports also showed an increase in the concentrations of trace elements (e.g., As, Se, and B) in the plant tissue after land application of FGDG (Punshon et al. 2001). The FGD gypsum can be utilized as an alternative to mined gypsum in the Indian context provided its efficacy is established and issues related to by-product utilization are addressed through planned research (Sundha et al. 2023a, b). The treatment of barren saline-sodic soils with FGD gypsum would be of considerable benefit to both agricultural development as crop production improvement and reclamation of sodic soil and the improvement of local ecosystems (Table 3.1; Fig. 3.2).

3.5.4 City Waste Compost and Sewage Sludge

Environmental waste management is based on the recycling, reuse, and composting of waste organic sources (Raut et al. 2008). This could be used in the reclamation

Table 3.1 Crop performance (Mg ha^{-1}) after reclamation of sodic soil with FGD gypsum

Treatment	Rice	Wheat	Barseem
Control	1.35 ± 0.24	1.74 ± 0.55	0.25 ± 0.03
FGDG 50GR	3.85 ± 1.38	3.53 ± 0.10	0.68 ± 0.03
Gypsum 50 GR	3.46 ± 0.93	3.44 ± 0.08	0.66 ± 0.03

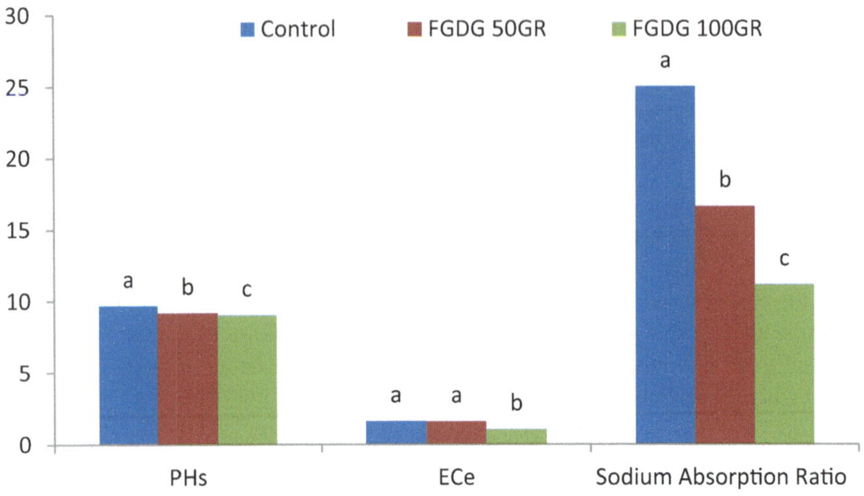

Fig. 3.2 Application of FGD gypsum reduces pH_s, EC_e (dS m^{-1}), and sodium absorption ratio ($mmol^{1/2}$ $L^{-1/2}$) of sodic soil

and rehabilitation of degraded soils. It helps meet the crop nitrogen requirements and other plant essential nutrients. The use of MSWC in reclaiming sodic soils by improving soil structure and aggregation (Oades 1993) increased hydraulic conductivity (Kosmas and Moustakas 1990; Hussain et al. 2001) and greater cation exchange capacity (Von Lutzow et al. 2002; Jalali and Ranjbar 2009). The addition of gypsum and MSWC declined soil pH (Sundha 2023a, b, Vance et al. 1998); accelerated NaCl leaching, decreased exchangeable sodium percentage; and increased water infiltration, water-holding capacity, and aggregate stability in saline soils (Tejada et al. 2006). It increases beneficial soil organisms and reduces plant pathogens (Abawi and Widmer 2000). The application of 10 t ha^{-1} compost prepared from MSW + farm wastes+ earthworms + waste decomposer in combination with gypsum reduced the soil pH from 9.80 to 8.84 and reduced the exchangeable sodium percent (ESP) from 78.3 to 28.0 under the rice–wheat cropping system in the sodic soils (Singh et al. 2017). High SAR water application in saline-sodic soil amendment with gypsum and MSWC showed higher reclamation efficiency and greater leaching of Ca with the application of 20 t ha^{-1} MSWC (Sundha et al. 2018). Gypsum application in sodic soil reduces N and P availability; MSWC application in conjunction with gypsum compensates for the N and P shortage and cuts short the demand for gypsum for sodic soil reclamation (Sundha et al. 2017). The interactive effect of organic (compost) and inorganic amendments and the quality of the irrigation water influence P dynamics and leaching losses during sodic-soil reclamation (Sundha et al. 2023a, b). The rapid acidulation of the compost with elemental S (S°) and S°– oxidizers declined pH with a substantial amount of equivalent alkali neutralization potential (EANP) in the different wastes within 28 days of incubation

(Bedwal et al. 2023). MSW has a wide scope for usage in degraded soils through managed scientific interventions and government-institutional support.

3.5.5 Polymer-Based Nanomaterials

Polymers and nanomaterials are found effective in improving degraded soils. It activates the rapid flocculation and sedimentation of colloidal particles in water bodies. The negatively charged colloids are neutralized via cationic hydrolysis and impurities are incorporated into the amorphous hydroxide precipitate ("sweep flocculation") are the mechanisms for sodic water reclamation (Duan and Gregory 2003). The hard degraded soils are dispersed and aggregated to improve permeability, porosity, and water-holding capacity. The inorganic polymers reduce the soil pH supplying the H^+ ions by hydrolysis and facilitating the dissolution of $CaCO_3$ to provide Ca^{2+} ions for ameliorating sodic soils (Zouboulis et al. 2007). The application of PAM enhances the Na^+ leaching through improved water movement and also reduces the soil electrical conductivity and SAR and increases the pecan nut yield by 34% over control (Ganjegunte et al. 2011).

3.5.6 Acid-modified Biochar

Biochar is a solid carbonaceous residue, produced at high temperatures (300–1000 °C) and low oxygen conditions and is a potential soil amendment source. Biochar application in saline environments reported a reduction in pH and ESP of soils (Chaganti and Crohn 2015; Kim et al. 2016; Liu et al. 2017). It also improves the soil water-holding capacity and nutrients in soil solution (Akhtar et al. 2015). Co-application of compost and biochar and acidified biochars significantly works in the reclamation of sodic soil (Lashari et al. 2013; Chaganti et al. 2015; Zadeh et al. 2018). The long-term application effects are further areas of research in the application of biochar as an organic amendment in sodic soils.

3.5.7 Dried Biogas Slurry

The biogas slurry (BGS) is the by-product from biogas plants as a source of organic amendment for the nutrient-deficient degraded sodic soils. It is reported that 1 m^3 of BGS contains approximately 0.16–1.05 kg of N depending on the type of feedstock used during fermentation (Mukhtiar et al. 2024). The application of biogas slurry improves the soil C pool and N, P, and K percentage in the soil by approximately 8.78, 11.01, and 14.33%, respectively, and reduced N_2O emissions (Yadav et al. 2023). The microbial communities present in the slurry also mitigate the nitrogen leaching loss through soil microbial nitrate transformations (Sun et al. 2023) and stimulate the bacteria and fungi communities in SOC cycling (Tang et al. 2021). Research on co-composting of biogas slurry with pressmud, and rice straw to produce vermimanure improved the shoot, root, and plant yield (Srivastava et al. 2024). The conjoint application of slurry (25% N requirement) and sub-optimal doses (75%) of NPK fertilizers resulted in long-term sustainability and increased the productivity of the diversified rice-based cropping systems (Mohanty et al. 2024), and BGS is reported to reduce the synthetic fertilizer requirement by 20% (Mukhtiar et al. 2024). The holistic approach toward integration of agri-waste management, biogas digestate utilization, and policy support for soil environment stabilization needs to focus on the successful implementation of the bio-circular economy to ensure food safety and sustainable development.

3.6 Response of Reclamation Sources Under Changing Climate

The changes in the climate adversely affect the productivity of soils in salt-affected areas across the world. The rise in the air temperature and abrupt changes in the rainfall regimes exacerbate the areas under sodicity with very low crop yields (Khamidov et al. 2022). The use of low-cost organic amendments like city waste compost, sewage sludge, biochar, and dried biogas slurry improves the soil's physico-chemical status. The conjunctive use of organic and inorganic amendments improves the soil aggregate stability promoting better water infiltration and retention, and improved water use efficiency for enhanced plant growth and crop productivity. Moisture retention in soils alters the salt distribution in the soil layers, increases organic matter mineralization and nutrient availability in soil (Saudy et al. 2020). The application of chemical fertilizers and humic acid (HA) increased the maize grain yield by 18.32% in a saline-alkali soil in Northwest China (Wang et al. 2023). Application of composted poultry manure increased the tiller number, plant height, shoot dry weight, root dry weight, and thousand kernel weight in the rice plants and significantly increased the relative abundance of plant growth-promoting bacteria such as *Thiobacillius, Methylobacter, Rhodoferax, Lysobacter*, and *Flavobacterium* saline-alkali soil (Zhang et al. 2022). The use of organic soil amendments ensures not only the agronomic benefits but also the climate mitigation potential and protection of the environment. It is reported that the partial replacement of nitrogen (N) fertilizers with organic fertilizers can potentially reduce N_2O emissions with no yield losses (Rubin et al. 2023; Xie et al. 2022).

3.7 Conclusion

Soil sodicity deleteriously affects crop growth and productivity in arid and semi-arid regions and several irrigated commands. The nature and properties of these sodic soils are physico-chemically disturbed, and it requires specific approaches of rehabilitation to grow the crop. Therefore, specific chemical, biological, and technical measures are required to reclaim such types of problems. Gypsum application is the most adopted method for the amendment of sodic soils. But, gypsum consumption in other sectors leads to issues of low quality and agricultural grade gypsum countrywide. Researchers are working on reduced gypsum application in conjunction with organics. Other available sources like elemental sulfur, pressmud, flue gas desulfurization gypsum (FGDG), and MSWC are merging alternatives for the reclamation of sodic soils with larger stock availability. Other formulations polymers, nanomaterials, and acidified biochars need more intense research on their reclamation potential. However, environmental sustainability and associated risk assessment are important before wide applicability. There is a need for large-scale pilot studies to assess the efficacy and efficiency of emerging sources in refurbishing sodic soils and increasing agricultural productivity. Soil sodicity is a serious environmental threat leading to land degradation and affecting crop production and livelihood of farmers in India, covering 3.77 Mha of sodic soil.

References

Abawi GS, Widmer TL (2000) Impact of soil health management practices on soil borne pathogens, nematodes and root diseases of vegetable crops. Appl Soil Ecol 15:37–47

Abrol IP, Bhumbla DR (1979) Crop response to differential gypsum application in a highly sodic soil and the tolerance of several crops to exchangeable sodium under field conditions. Soil Sci 127:79–85

Ahmad S, Ghafoor A, Akhtar ME, Khan MZ (2013) Ionic displacement and reclamation of saline–sodic soils using chemical amendments and crop rotation. Land Degrad Dev 24:170–178. https://doi.org/10.1002/ldr.1117

Akhtar SS, Andersen MN, Liu F (2015) Residual effects of biochar on improving growth, physiology and yield of wheat under salt stress. Agric Water Manag 158:61–68. https://doi.org/10.1016/j.agwat.2015.04.010

Ali M, Lee C-H, Kim P (2007) Effect of Phospho-gypsum on reduction of methane emission from rice paddy soil. Korean J Environ Agric 26:131. https://doi.org/10.5338/KJEA.2007.26.2.131

Amrhein C (2000) Revisiting acid injection into irrigation water. Golf Course Management

Basak N, Rai AK, Barman A, Mandal S, Sundha P, Bedwal S, Kumar S, Yadav RK, Sharma PC (2022a) Salt affected soils: global perspectives. In: Shit PK, Adhikary PP, Bhunia GS, Sengupta D (eds) Soil health and environmental sustainability: application of geospatial technology. Springer International Publishing, Cham, pp 107–129. https://doi.org/10.1007/978-3-031-09270-1_6

Basak N, Rai AK, Sundha P, Meena RL, Bedwal S, Yadav RK, Sharma PC (2022b) Assessing soil quality for rehabilitation of salt-affected agroecosystem: a comprehensive review. Front Environ Sci 10. https://doi.org/10.3389/fenvs.2022.935785

Basak N, Rai AK, Sundha P, Chandra P, Bedwal S, Patel S, Yadav RK, Chander Sharma P (2023) 4 – soil management for salt-affected soil. In: Meena SK, Ferreira ADO, Meena VS, Rakshit A, Shrestha RP, Rao CS, Siddique KHMBT-ASS (eds) Agricultural soil sustainability and carbon management. Academic Press, pp 99–128. https://doi.org/10.1016/B978-0-323-95911-7.00011-6

Bedwal S, Basak N, Rai AK, Sundha P, Patel S, Raj D, Mann A, Kumar S, Chandra P, Yadav RK (2023) Converting waste to resource: utilisation of elemental S and consortia of halo–alkaliphilic S–oxidisers for preparation of acidified compost from organic wastes. Bioresour Technol Rep 24:101620. https://doi.org/10.1016/j.biteb.2023.101620

Biennial Reports (2012) 14ICAR-CSSRI, pp 152–153

Brady NC, Weil RR (2014) The nature and properties of soils, 14th edn. Pearson Publication

Chaganti VN, Crohn DM (2015) Evaluating the relative contribution of physicochemical and biological factors in ameliorating a saline-sodic soil amended with composts and biochar and leached with reclaimed water. Geoderma 259-260:45–55. https://doi.org/10.1016/j.geoderma.2015.05.005

Chaganti VN, Crohn DM, Simunek J (2015) Leaching and reclamation of a biochar and compost amended saline-sodic soil with moderate SAR reclaimed water. Agric Water Manag 158:255–265. https://doi.org/10.1016/j.agwat.2015.05.016

Chen L, Dick WA (2011) Gypsum as an agricultural amendment: general use guidelines. Ohio State University Extention

Chhabra R (2004) Classification of salt-affected soils. Arid Land Res Manag 19(1):61–79. https://doi.org/10.1080/15324980590887344

Choudhary OP, Ghuman BS, Bijay-Singh TN, Buresh RJ (2011) Effects of long-term use of sodic water irrigation, amendments and crop residues on soil properties and crop yields in rice–wheat cropping system in a calcareous soil. Field Crop Res 121:363–372. https://doi.org/10.1016/j.fcr.2011.01.004

CSSRI (2006) CSSRI: a journey to excellence (1969–2006), Central Soil Salinity Research Institute, Karnal, India p. 156

CSSRI (2015) Vision 2050. ICAR-Central Soil Salinity Research Institute, Karnal, India, pp 31. www.cssri.res.in

CSSRI Annual Report (2017–2018) pp 21–24. www.cssri.res.in

Duan J, Gregory J (2003) Coagulation by hydrolysing metal salts. Adv Colloid Interf Sci 100:475–502

Eswar D, Karuppusamy R, Chellamuthu S (2021) Drivers of soil salinity and their correlation with climate change. Curr Opin Env Sust. https://doi.org/10.1016/j.cosust.2020.10.015

Ganjegunte GK, Sheng Z, Braun RJ (2011) Salinity management using an anionic polymer. J Environ Qual 40:1314–1321

Garcia junior O. (1992) O enxofre e suas transformacoes microbianas. In: Cardoso EJB, Tsai SM, Neves MCP (eds) Microbiplpgia do solo. Sociedade Brasileira de Ciencia do Solo, Campinas, pp 1–15

Ghafoor A, Muhammed S (1981) Comparison of H_2SO_4, HCl, HNO_3, and gypsum for reclaiming calcareous saline-sodic soil and for plant growth. Irrigation Drainage Flood Control Res Council Pakistan 11:69–75

Ghafoor A, Muhammed S, Mujtaba G (1986) Comparison of gypsum, H_2SO_4, HCl and $CaCl_2$ for reclaiming subsoiled calcareous saline-sodic Khurrianwala soil series. J Agric Res 24:179–183

Griffen J R and Silverttoth J C. 1999. Evaluation of an acid soil conditioner in an irrigated cotton production system. Arizona Cotton Report Series, pp 260–274

Hussain N, Hassan G, Arshadullah M, Mujeeb F (2001) Evaluation of amendments for the improvement of physical properties of sodic soil. Int J Agric Biol 3(3):319–322

Ibrahim M, Han KH, Ha SK, Zhang YS, Hur SO (2012) Physico-chemical characteristics of disturbed soils affected by accumulate of different texture in South Korea. Sains Malaysiana 41:285–291

Jalali M, Ranjbar F (2009) Effects of sodic water on soil sodicity and nutrient leaching in poultry and sheep manure amended soils. Geoderma 153(1–2):194–204. https://doi.org/10.1016/j.geoderma.2009.08.004

Jha SK, Mishra VK, Damodaran T, Singh YP, Sharma PC (2022) Utilization of marine gypsum as an alternative to mineral gypsum in the reclamation of degraded sodic soil and as a means of enhancing crop productivity. Land Degrad Dev 33(13):2276–2288. https://doi.org/10.1002/ldr.4282

Kanwar JS, Bhumbla DR, Singh NT (1965) Studies on the reclamation of saline and sodic soils in the Punjab. Indian J Agric Sci 35:43–57

Karim A, Kumar M, Ray A, Hariprasad D, Dhal N (2021) Biomass mediated conversion of acidic Phosphogypsum into alkaline material through thermal treatments. J Sci Ind Res 80:924–928

Khamidov M, Ishchanov J, Hamidov A, Donmez C, Djumaboev K (2022) Assessment of soil salinity changes under the climate change in the Khorezm region, Uzbekistan. Int J Environ Res Public Health 19. https://doi.org/10.3390/ijerph19148794

Kim HS, Kim KR, Yang JE, Ok YS, Owens G, Nehls T, Wessolek G, Kim KH (2016) Effect of biochar on reclaimed tidal land soil properties and maize (*Zea mays* L.) response. Chemosphere 142:153–159

Kosmas C, Moustakas N (1990) Hydraulic conductivity and leaching of an organic saline-sodic soil. Geoderma 46(4):363–370

Kumar D, Singh B (2003) Land Degrad Dev 14:285–299
Kumar M, Karim A, Mohapatra S, Panda C, Singh S (2018) Co-plasma processing of banana peduncle with phosphogypsum waste for production of lesser toxic potassium–sulfur rich biochar. J Mater Cycles Waste Manag 21:107–115. https://doi.org/10.1007/s1016
Lashari MS, Liu Y, Li L, Pan W, Fu J, Pan G, Yu X (2013) Effects of amendment of biochar-manure compost in conjunction with pyroligneous solution on soil quality and wheat yield of a salt stressed cropland from Central China great plain. Field Crop Res 144:113–118. https://doi.org/10.1016/j.fcr.2012.11.015
Lawrence JR, Germida JJ (1988) Relationship between microbial biomass and elemental Sulphur oxidation in agricultural soils. Soil Sci Soc Am 52:672–677
Li F, Keren R (2009) Calcareous sodic soil reclamation as affected by corn stalk application and incubation: a laboratory study. Pedosphere 9:465–475
Li Z, Li SR, Shen JF, Ho XJ, Cao XG (2005) Effect on water hold capacity of desertificated soil by adding coal ash and sludge mixtures. Environ Earth Sci 33(2):74–78
Liu S, Meng J, Jiang L, Yang X, Lan Y, Cheng X, Chen W (2017) Rice husk biochar impacts soil phosphorous availability, phosphatase activities and bacterial community characteristics in three different soil types. Appl Soil Ecol 116:12–22. https://doi.org/10.1016/j.apsoil.2017.03.020
Lokeshappa B, Dikshit AK (2011) Disposal and management of coal fly ash, Proc. ICLST, 11–14
Mandal A (2022) Necessity for quantified measurement of soil sodicity and selection of suitable gypsum amendment for proper reclamation of sodic soil. Pedosphere 33:231. https://doi.org/10.1016/j.pedsph.2022.06.038
Meena MD, Joshi PK, Narjary B, Sheoran P, Jat HS, Chinchmalatpure AR, Yadav RK, Sharma DK (2016) Effects of municipal solid waste compost, rice-straw compost and mineral fertilisers on biological and chemical properties of a saline soil and yields in a mustard–pearl millet cropping system. Soil Research 54(8):958–969. https://doi.org/10.1071/SR15342
Mohanty S, Saha S, Saha B, Asif SM, Poddar R, Ray M, Mukhopadhyay SK, Hazra G (2024) Substitution of fertilizer-N with biogas slurry in diversified rice-based cropping systems: effect on productivity, carbon footprints, nutrients and energy balance. Field Crop Res 307:109242. https://doi.org/10.1016/j.fcr.2023.109242
Mukhtiar A, Mahmood A, Zia MA, Ameen M, Dong R, Shoujun Y, Javaid MM, Khan BA, Nadeem MA (2024) Role of biogas slurry to reclaim soil properties providing an eco-friendly approach for crop productivity. Bioresour Technol Rep 25:101716. https://doi.org/10.1016/j.biteb.2023.101716
Nayak AK, Mishra VK, Sharma DK, Jha SK, Singh CS, Shahabuddin M, Shahid M (2013) Efficiency of Phosphogypsum and mined gypsum in reclamation and productivity of rice–wheat cropping system in sodic soil. Commun Soil Sci Plant Anal 44(5):909–992. https://doi.org/10.1080/00103624.2012.747601
Norton LD, Zhang XC (1998) In: Wallace A, Terry RE (eds) Handbook of soil conditioners. Substances that enhance the physical properties of soil. Marcel Dekker, New York
Oades JM (1993) The role of biology in the formation, stabilization and degradation of soil structure. Geoderma 56:377–400
Overstreet R, Martin JC, King HM (1951) Gypsum, sulfur, sulfuric acid for reclaiming an alkali soil of the Fresno series. Hilgardia 21:113–126
Partha SN, Sivasubramanian V (2006) Recovery of chemicals from pressmud: a sugarcane industry waste. Indian Chem Eng 48:161–163
Punshon T, Adriano DC, Weber JT (2001) Effect of flue gas desulfurization residue on plant establishment and soil and leachate quality. J Env Quality 30:1071–1080. https://doi.org/10.2134/jeq2001.3031071x
Rai AK, Basak N, Sundha P (2021a) Chemistry of salt-affected soils. In: Minhas PS, Yadav RK, Sharma PC (eds) Managing salt-affected soils for sustainable agriculture. Indian Council of Agricultural Research (ICAR), New Delhi, pp 128–149
Rai AK, Basak N, Sundha P (2021b) Alternate amendments for reclamation of alkali soils. In: Minhas PS, Yadav RK, Sharma PC (eds) Managing salt-affected soil for sustainable agriculture. Indian Council of Agricultural Research (ICAR), New Delhi, pp 332–345

Rai AK, Basak N, Chandra P, Sundha P, Kaur H, Bedwal S, Patel S, Kumar S, Choubey K, Meena RL, Sharma PC, Yadav RK (2024) Occurrence of Sphingomonas olei with elemental S oxidation capability in sodic soil: potential role in sodicity reclamation and plant growth promotion. Ecotoxicol Environ Saf 278:116396. https://doi.org/10.1016/j.ecoenv.2024.116396
Raut MP, William SPMP, Bhattacharyya JK, Chakrabarti T, Devotta S (2008) Microbial dynamics and enzyme activities during rapid composting of municipal solid waste – a compost maturity analysis perspective. Bioresour Technol 99(14):6512–6519
Rubin R, Oldfield E, Lavallee J, Griffin T, Mayers B, Sanderman J (2023) Climate mitigation through soil amendments: quantification, evidence, and uncertainty. Carbon Manag 14:2217785. https://doi.org/10.1080/17583004.2023.2217785
Rupela OP, Tauro P (1973) Isolation and characterization of *Thiobacillus* from alkali soils. Soil Biol Biochem 5:891–897
Sakai Y, Matsumoto S, Sadakata M (2004) Alkali soil reclamation with flue gas desulfurization gypsum in China and assessment of metal content in corn grains. Soil Sediment Contam 13:65–80. https://doi.org/10.1080/10588330490269840
Saudy HS, Hamed MF, Abd El-Momen WR, Hussein H (2020) Nitrogen use rationalization and boosting wheat productivity by applying packages of humic, amino acids, and microorganisms. Commun Soil Sci Plant Anal 51:1036–1047. https://doi.org/10.1080/00103624.2020.1744631
Sharma P, Swarup A (1997) Comparison of pyrites varing in water-solblesulfr with gypsm for the reclamation of alkani soil under a rice-wheat rotation. Biol Fertil Soils 24:96–101
Sharma DK, Thimmppa K, Chinchmalatpure AR, Mandal AK, Yadav RK, Choudhari SK, Kumar S, Sikka AK (2015) Assessment of production and monetary losses from salt-affected soils in India. Technical Bulletin: ICAR-CSSRI/ Karnal/Tech.Bulletin/2015/. ICAR-Central Soil Salinity Research Institute, Karnal
Sheoran P, Basak N, Kumar A et al (2021) Ameliorants and salt tolerant varieties improve rice-wheat production in soils undergoing sodification with alkali water irrigation in Indo–Gangetic Plains of India. Agric Water Manag 243:106492. https://doi.org/10.1016/j.agwat.2020.106492
Shivanna KR (2022) Climate change and its impact on biodiversity and human welfare. Proc Indian Natl Sci Acad 88:160–171. https://doi.org/10.1007/s43538-022-00073-6
Singh K, Mishra AK, Singh B, Singh RP, Patra DD (2015) Tillage effects on crop yield and physicochemical properties of sodic soils. Land Degrad Dev 27(2):223–230. https://doi.org/10.1002/ldr.2266
Singh YP, Arora S, Mishra VK, Dixit H, Gupta RK (2017) Composting of municipal solid waste and farm wastes for its use as amendment in sodic soil. J Soil Water Conserv 16(2):172–177
Srivastava PK, Tiwari GN, Sinha ASK (2024) Enhanced vermicomposting of rice straw and pressmud with biogas slurry employing Eisenia fetida: production, characterization, growth, and toxicological risk assessment. J Environ Manag 352:120032. https://doi.org/10.1016/j.jenvman.2024.120032
Sun R, Ding J, Li H, Wang X, Li W, Li K, Ye X, Sun S (2023) Mitigating nitrate leaching in cropland by enhancing microbial nitrate transformation through the addition of liquid biogas slurry. Agric Ecosyst Environ 345:108324. https://doi.org/10.1016/j.agee.2022.108324
Sundha P, Basak N, Rai A K, Yadav R, Sharma D, Sharma P (2017) N and P release pattern in saline-sodic soil amended with gypsum and municipal solid waste compost. J Soil Salinity Water Quality 9:145–155
Sundha P, Basak N, Rai AK, Yadav RK, Sharma DK (2018) Utilization of municipal solid waste compost in reclamation of saline-sodic soil irrigated with poor quality water. J Indian Soc Soil Sci 66(1):28–39. https://doi.org/10.5958/0974-0228.2018.00004.X
Sundha P, Basak N, Rai AK, Yadav RK, Sharma PC, Sharma DK (2020) Can conjunctive use of gypsum, city waste composts and marginal quality water rehabilitate saline-sodic soils? Soil Tillage Res 200:104608. https://doi.org/10.1016/j.still.2020.104608
Sundha P, Mukhopadhyay R, Basak N et al (2023a) Characterization of flue gas desulphurized (FGD) gypsum of a coal-fired plant and its relevant risk of associated potential toxic elements in sodic soil reclamation. Sci Rep 13:19787. https://doi.org/10.1038/s41598-023-45706-y

Sundha P, Basak N, Rai AK, Yadav RK, Sharma PC (2023b) Irrigation water quality, gypsum, and city waste compost addition affect P dynamics in saline-sodic soils. Environ Res 216(2):1819–1833. https://doi.org/10.1016/j.envres.2022.114559

Tang Y, Luo L, Carswell A, Misselbrook T, Shen J, Han J (2021) Changes in soil organic carbon status and microbial community structure following biogas slurry application in a wheat-rice rotation. Sci Total Environ 757:143786. https://doi.org/10.1016/j.scitotenv.2020.143786

Tayibia H, Chourab M, Lopeza FA, Alguacila FJ, Lopez-Delgadoa A (2009) Environmental impact and management of phosphogypsum. J Environ Manag 90:2377–2386

Tejada M, Garcia C, Gonzalez JL, Hernandez MT (2006) Use of organic amendment as a strategy for saline soil remediation: influence on the physical, chemical and biological properties of soil. *Soil Biol Biochem* 38(6):1413–1421

Tian T, Zhang C, Zhu F, Yuan S, Guo Y, Xue S (2021) Effect of phosphogypsum on saline-alkalinity and aggregate stability of bauxite residue. Trans Nonferrous Metals Soc China 31:1484–1495. https://doi.org/10.1016/S1003-6326(21)65592-9

Tozsin G, Arol A, Cayci G (2014) Use of waste pyrite as an alternative to gypsum for alkaline soil amelioration. Int J Mining Reclamation Environ 29:1–11. https://doi.org/10.1080/17480930.2013.873257

Vance WH, Tisdell JM, McKenzie BM (1998) Residual effects of surface application of organic matter and calcium salts on the sub-soil of a red-brown earth. Aust J Exp Agric 38:595–600

Von Lutzow M, Leifeld J, Kainz M, Kogel-Knabner I, Munch JC (2002) Indications for soil organic matter quality in soils under different management. Geoderma 105:243–258

Walker DJ, Bernal MP (2008) The effects of olive mill waste compost and poultry manure on the availability and plant uptake of nutrients in a highly saline soil. Bioresour Technol 99:396–403

Wang SJ, Chen CH, Xu XC, Li YJ (2008) Amelioration of alkali soil using flue gas desulfurization byproducts: productivity and environmental quality. Environ Pollut 151:200–204

Wang Y, Gao M, Chen H, Chen Y, Wang L, Wang R (2023) Organic amendments promote saline-alkali soil desalinization and enhance maize growth. Front Plant Sci 14. https://doi.org/10.3389/fpls.2023.1177209

Xie L, Li L, Xie J, Wang J, Anwar S, Du C, Zhou Y (2022) Substituting inorganic fertilizers with organic amendment reduced nitrous oxide emissions by affecting Nitrifiers' microbial community. Land 11. https://doi.org/10.3390/land11101702

Yadav R, Sudhishri S, Khanna M, Lal K, Dass A, Kushwaha HL, Bandyopadhyay K, Dey A, Kushwah A, Nag RH (2023) Temporal characterization of biogas slurry: a pre-requisite for sustainable nutrigation in crop production. Front Sustain Food Syst 7. https://doi.org/10.3389/fsufs.2023.1234472

Yaduvanshi NPS, Yadav DV (1990) Effect of Sulphidation Pressmud and nitrogen fertilizer on biomass nitrogen economy, plant composition in sugarcane and on soil chemical properties. J Agric Sci 114:259–263

Zadeh FM, Jalili B, Bahmanyar MA (2018) Rehabilitation of calcareous saline-sodic soil by means of biochars and acidified biochars. Land Degrad Dev 29(10):3262–3271

Zhang Z, Liu H, Liu X, Chen Y, Lu Y, Shen M, Dang K, Zhao Y, Dong Y, Li Q, Li J (2022) Organic fertilizer enhances rice growth in severe saline–alkali soil by increasing soil bacterial diversity. Soil Use Manag 38:964–977. https://doi.org/10.1111/sum.12711

Zhao Y, Wang S-J, Jia L, Zhuo Y-Q, Li Y, Zhang W (2021) Fertility and biochemical activity in a sodic soil 17 years after reclamation with flue gas desulfurization gypsum. J Integr Agric 20:3312. https://doi.org/10.1016/S2095-3119(20)63446-3

Ziskin R, Dag A, Yermiyahu U, Levy GJ (2024) Different amendments for combating soil sodicity in an olive orchard. Agric Water Manag 299:108837. https://doi.org/10.1016/j.agwat.2024.108837

Zouboulis AI, Tzoupanos ND, Moussas PA (2007) Inorganic pre-polymerisedcoagulants: current status and future trends. In: Proceedings of the 3rd IASME/WSEAS international conference on energy, environment, ecosystems and sustainable development, pp 292–300

4 Paradigm Shift from Traditional to Innovative Extension Approaches in Changing Agricultural Scenario for Better Crop Productivity

Pragya Bhadauria, Y. S. Jadoun, Anjani Kumar, and Parvender Sheoran

Abstract

Agriculture plays a critical role in India's economy, with most farmers owning less than 1 hectare of land. The Green Revolution made India self-reliant in food grain production through high-yielding varieties (HYVs) of seeds, improved irrigation, and better use of fertilizers and pesticides. However, challenges remain in making agricultural technology accessible, especially to poor and risk-prone farmers. There has been a shift from a traditional top-down approach in agricultural extension services to people-centred methods, where farmers are partners in the innovation process. Extension services have expanded beyond transferring technology to farmers, now focusing on linking farmers to markets, reducing vulnerability and fostering micro-enterprises. The goal is to improve farm incomes and rural livelihoods by providing technical, managerial and entrepreneurial skills. Agricultural extension services face challenges, such as the need to modernize and address the changing demands of farmers. New approaches emphasize participatory research, the use of ICT and mass media, precision agriculture tools, market-driven approaches and group farming. There is also a focus on including women and youth in agricultural processes. Collaboration between government, private sectors, NGOs and farmers is vital for advancing agricultural development. Overall, extension services must evolve to help farmers become more competitive in a globalized, sustainable agriculture system. They need to expand their role beyond just increasing productivity to improving liveli-

P. Bhadauria (✉) · Y. S. Jadoun · A. Kumar
ICAR-Agricultural Technology Application Research Institute, Patna, Bihar, India
e-mail: pragya.bhadauria@icar.gov.in

P. Sheoran
ICAR-Agricultural Technology Application Research Institute, Ludhiana, Punjab, India

A. Mann et al. (eds.), *Cutting Edge Technologies for Developing Future Crop Plants*,
https://doi.org/10.1007/978-981-96-2508-6_4

hoods through market integration, financial support and innovative technologies.

Keywords

Extension approaches · Farmers · Technology · Sustainable

4.1 Introduction

In developing countries like India, the importance of agriculture as a viable driver for economic development and sustainable livelihood is widely acknowledged. Indian agriculture is essentially small farm agriculture with the majority of farmers owning less than 1 hectare (ha) of land which constitutes over 80 percent of farming households in the country. However, the Green Revolution made the country self-reliant in the agricultural research production of food grains. This was achieved through systematic and planned interventions during the 1960s by, for example, the introduction of seeds of high-yielding varieties (HYVs), the efficient use of chemical-based pesticides and fertilizers, improved irrigation and institutionalized credit facilities. During this period, national and international public sector institutions played a crucial role in agricultural research, developing and releasing many new initiatives. However, to date making agricultural technology generation and dissemination relevant to the needs of farmers particularly the poorest and those in complex, risk-prone environments is a huge challenge to the agricultural system. Over the past two decades, shifts have begun in extension approaches from technology transfer towards more people-centred agricultural extension. The paradigm shifts from traditional to innovative extension approaches in agriculture reflect the evolving needs of farmers and the challenges they face in a dynamic agricultural environment.

A paradigm is defined as "the set of general theoretical assumptions, laws, and methods of application adopted by the members of a specific scientific community" (Willis 2007). For a period, a paradigm offers model problems and solutions for a community of practitioners. This definition highlights one of the key characteristics of a paradigm: its capacity to evolve and be replaced by a new one. As new schools of thought emerge, scientific knowledge advances, the needs of stakeholders change, paradigms shift and transform over time. Traditionally, extension services were viewed as channels for transferring technologies developed by research institutions to farmers. However, this system has faced criticism for failing to achieve the desired developmental impacts in developing countries. With the evolving landscape of agriculture, marked by globalization, commercialization and a growing emphasis on sustainability, the role of extension services is expanding to encompass a broader range of objectives. These objectives include better linking of farmers to input and output markets, reducing vulnerability, amplifying the voices of the rural poor and fostering the development of micro-enterprises. Presently, the extension systems are being transformed, moving from top-down instruction towards farmer-to-farmer exchange and joint learning. The use of new information technologies is expanding, allowing information sharing between farmers. As a result, farmers are increasingly being seen as partners in the innovation process, rather than merely recipients of national and international research and extension. But still, we have to go long way as the progress has been slow and fragmented.

4.2 Definition of Agricultural Extension

Extension, in general terms, is a function that can be applied to various areas of society. It operates in the industrial, health and education sectors, as well as agricultural and rural development. Originally derived from university extension (Mosher 1976), the term "extension" is therefore applicable to various areas of development. Agricultural extension operates within a broader knowledge system that includes research and agricultural education. In a more precise way, it is educating the farmers about the advanced technologies of agricultural practices and the application to meet their needs. Although one should be very clear that there is a closer relationship between agricultural extension and agricultural research because the knowledge that agricultural extension transfers is usually generated by agricultural research through applied and adaptive agricultural research development. In principle, agricultural extension receives relevant information from the agricultural education system and feeds back field observations to this system.

In the broadest interpretation, agricultural extension provides non-formal agriculturally related continuing adult education—for multiple audiences: farmers, spouses, youth and community and for various purposes (including agricultural development, community resource development, group promotion and cooperative organizational development). Albeit, the only purpose of agricultural extension is to disseminate information to raise the production and profitability of the farmers (agricultural production performance). However, when agricultural extension is combined with rural extension goals, the extension function ranges even more widely in its purposes. Rural extension, for instance, includes non-agricultural activities such as micro-enterprise development (Echeverría 1998). In a broader interpretation, the purpose of agricultural extension is to advance not alone production knowledge but the whole range of agricultural development tasks, such as credit, supplies, marketing and markets (agricultural process development).

4.3 Challenges of Current Agricultural Extension

The extension services in India are highly pluralized. Numerous agencies are now providing extension services in varied nature through different approaches. Broadly, extension service providers can be categorized into three systems: public extension system, private extension system and individual-based (farmer to farmer or lead farmer to farmer) extension system. While the first two systems are more or less organized in nature, the latter system is largely unorganized. In the public extension system, Agricultural Research and Education has been the primary domain of ICAR Institutes and Agricultural Universities (SAUs, CAU, DU). These research and educational institutions also have their own extension initiatives to reach farmers. In the public sector, extension services were traditionally assumed to be the conduits for transferring technologies developed by the research system to the farmers. The system, however, has been under severe criticism for not being able to contribute to desired developmental impacts in the country. With changing circumstances of agriculture and increasing trends of globalization, commercialization and drive towards sustainability, extension is being looked upon to play an expanded role with a diverse set of objectives, which include better linking of farmers to input and output markets, reducing the vulnerability, and enhancing the voice of the rural poor and developing micro-enterprises. Presently, the biggest challenge with extension is how to transform and strengthen pluralistic agricultural extension and advisory systems in moving towards the broader goal of increasing farm income and improving rural livelihoods. The focus is primarily on the technical knowledge, management skills, entrepreneurial skills and information services that small-scale farm households will need to improve their livelihoods in the rapidly changing global economy.

Therefore, in the current scenario agricultural extension has to go beyond production, with a focus on making farming and associated activities profit-oriented just like any other business organizations. Empowering farmers to become active players in an increasingly globalized system means moving beyond the traditional focus on farmers and technologies to farmer relations with other actors through markets. This calls for changes in the prevailing agriculture scenario so as to meet the new demands that can only be addressed by enhancing the capacities of the extension functionaries. Farmers have to be convinced that extension systems and the information they communicate are valuable for income generation in particular and for improving their living standards in general. Assisting resource-poor farmers with appropriate technology may provide the opportunity for rural households to increase their productivity and incomes. Consequently, there is a dire need for a shift from a single commodity, mono-disciplinary base to a farming system and multi-disciplinary-based approach together with a change from a top-down extension model to a participatory approach to technology assessment and adoption (Singh et al. 2019).

4.4 Opportunities in Agricultural Extension

Though the agricultural extension system has taken many pro active measures to help farmers, there is a need for more involvement and formulation of innovative practices to enable them to address different challenges. Innovative extension approaches and services can play an important role in addressing many of these challenges. Perhaps, there is no agency at the ground level, other than agricultural extension services that can provide knowledge support to farmers and other intermediaries who are serving rural communities at the same time support programme implementation. Although considering the changing nature of agriculture and the evolving challenges, producers currently need a wider range of support, including organizational, marketing, technological, financial and entrepreneurial.

But to play this role effectively, the extension should expand its mandate beyond disseminating information on technologies so that it can better respond to the evolving demands for support and services of the farmers. Extension providers need to be properly equipped so as to address changes in the development scenario, as well as to meet the emerging demands and needs of farmers and FPOs, especially on agribusiness, value addition and marketing. Extensionists need to be trained in next-generation extension tools and media. This includes organizing user/producer groups, linking farmers to markets, engaging in research planning and technology selection, enabling changes in policies and linking producers to a range of other support and service networks. Moreover, extension services have to integrate financial products such as microloans, weather-indexed insurance and digital financial tools to enhance farmers' capacity to adopt new practices and manage risks.

4.5 Factors Shaping Extension Paradigms

There are factors that collectively influence the evolution of agricultural extension paradigms, guiding how extension services are conceptualized, implemented and evaluated to meet the needs of diverse farming communities. It mainly includes:

1. *Socio-Economic Factors:* Demographics such as age, gender, education level and socioeconomic status of farmers affect how extension services are tailored. Local customs, traditions and social norms can shape the acceptability and adoption of agricultural practices promoted by extension services. The role of community leaders, farmer groups, cooperatives and informal networks is critical in disseminating information and facilitating collective action among farmers.
2. *Technological Factors:* New technologies such as precision farming, biotechnology, climate-resilient crops and integrated pest management systems require extension services to update and adapt their methodologies to incorporate these advancements. The rise of digital tools, mobile phones, internet access and social media has transformed the way extension services are delivered, enabling more personalized, timely and widespread dissemination of information.
3. *Environmental Factors:* Increasingly, erratic weather patterns, extreme events and shifting growing seasons force extension paradigms to include climate-smart agricultural practices, sustainable land management and disaster preparedness. The condition and availability of natural resources (such as water, soil quality and biodiversity) influence extension programmes to focus on sustainable practices, conservation agriculture and ecosystem management.
4. *Natural Resource Management:* The condition and availability of natural resources (such as water, soil quality and biodiversity) influence extension programmes to focus on sustainable practices, conservation agriculture and ecosystem management.

5. *Political and Institutional Factors:* National or regional development policies and goals can influence extension paradigms. Priorities like poverty reduction, food security or income diversification may direct extension efforts towards certain crops, farming practices or income-generating activities. Policies regarding agriculture, rural development, education and research shape extension paradigms by determining funding, resource allocation and priority areas for extension services. Local governance can lead to more tailored and context-specific extension activities. Collaboration between government agencies, NGOs, private sector companies, research institutions and farmer organizations influences the design, delivery and reach of extension programmes.
6. *Educational and Research Factors:* The presence of agricultural education institutions, vocational training centres and continuous learning opportunities influence the skills and knowledge base of extension workers and farmers. Research institutions and universities play a key role in generating new agricultural knowledge and technologies, which in turn shape extension paradigms by providing evidence-based practices and innovations.
7. *Global Trends and International Organizations:* Global Policy Frameworks and International agreements, such as the Sustainable Development Goals (SDGs), influence national agricultural extension priorities and paradigms. International Organizations: Organizations like the Food and Agriculture Organization (FAO), World Bank and others provide frameworks, funding and technical assistance that shape extension practices. Privatization has caused "commodification" of agricultural knowledge (Rivera and Cary 1997). Extension services are viewed as tools to generate income. Countries particularly the low-income countries promote services to provide practical, income-generating agricultural information.

4.6 New Approaches of Agricultural Extension

1. *Farming System Perspective:* The problem of the farm/farmer cannot be understood or solved by looking at single elements alone but by involving the interaction among different components. Empowering farmers to be active players in agricultural research programmes involves supporting farmers and their organizations through innovative funding, guaranteed stakes in R&D organizations and platforms for dialogue. Research organizations must simultaneously establish avenues for engagement with farmers, develop innovative learning methods, and offer incentives and recognition for those who are champions of change.
2. *Participatory Research Methods:* Methods of deficiencies in the technology and the process that generated it occur due to inadequate participation in different stages of the process by those intended to benefit. Participation eliminates weaknesses of the traditional "top-down" approach to R&D. People learn best, and more willingly apply what they have learned, when they do it themselves. People themselves have the capacity to solve their own problems and bring

about a change. It also involves group-based learning and problem-solving activities. Farmers collaborate with researchers to experiment with new varieties or practices and provide feedback, ensuring that new technologies are contextually relevant. These approaches are also being used to diagnose market chain challenges and opportunities, and facilitate change in market systems. Models like farmer's school emphasizes experiential learning where farmers learn by doing in their own fields.

3. *Mass Media Support and Use of Information and Communication Technologies (ICTs):* Extensionists should equip themselves with latest and effective ICT and mass media tools at free or at a nominal cost to increase the timely and transparent flow of information among different stakeholders. Further, masses through ICTs can also revolutionize the interaction through Information Kiosks, Telecentres, toll-free Call Centres, websites, mobile phones software applications, etc. Apps and SMS services provide real-time information on weather forecasts, pest and disease alerts, market prices and best practices. They enable farmers to make timely and informed decisions. Online courses and platforms like MOOCs (Massive Open Online Courses) offer farmers access to educational resources and training. Social media and messaging platforms like WhatsApp, Facebook and YouTube are increasingly used to share videos, success stories, tutorials and expert advice.
4. *Use of Data Analytics and Precision Agriculture Tools:* Satellite imagery, drones and Geographic Information Systems (GIS) provide detailed data on soil health, crop growth, water usage and more, allowing for targeted interventions. AI-based models predict weather patterns, disease outbreaks and pest infestations, providing actionable insights for farmers. New advanced instruments like personal digital assistants may be provided to the extension agents for technical information, communicating, field recording, database maintenance and scheduling.
5. *Holistic Approach:* New extension approaches focus not just on production but also on sustainability, climate adaptation and resilience. Practices such as climate-smart agriculture, conservation tillage, integrated pest management and organic farming are being promoted.
6. *Market-Driven Approach:* The present extension services need to be more market focused and adopt a systems approach to work in ways that strengthen the roles and opportunities of other factors in a value chain, such as farm workers, input suppliers, warehouse managers, transport companies and financial services so that they can all work towards a common business goal. Production and marketing of agricultural and allied sector products through the creation of basic market facilities and market information for the farmers are very essential. There should be massive scaling up of federal e-commerce/digital platforms like e-NAM for farmers and traders, which must be given the much-needed impetus in the present scenario.
7. *Role of Group Farming:* The extension approach needs to be changed from an individual to a group or association approach to have effective decision making. The Interest Groups, SHGs or Cooperatives have been very successful models

for effective production or marketing. Farmer groups are primarily organizations of farmers, typically informal in nature with a membership of 15–50 farmers or even more that may be resource-orientated organizations and market-orientated organizations. Besides, skills in mobilizing farmers and facilitating interaction are very much needed to secure the coordination of different agencies to broker gains for farmers. Input-output transactional costs can be reduced and social participation can be increased as a result of farmer's organization into commodity groups.

8. *Gender and Youth Inclusion:* There is a greater focus on inclusive approaches that engage women and youth, recognizing their unique roles and potential contributions to agriculture. Programmes are tailored to address their specific needs, preferences and constraints. It often involves youth-based agri-tech start-ups that offer new technologies like precision agriculture tools, blockchain for supply chain management and digital advisory services.
9. *Networking of Institutions:* Agricultural development is best promoted when all components involved in a common goal will interact effectively. Since the extension efforts by Indian Council of Agricultural Research (ICAR) institutes have very limited reach, efforts should be made to collaborate with different agencies or actors involved in agricultural extension. In this context, different universities their associated KVKs and public/private institutions can play a significant role. In addition to this, multi-stakeholder collaboration between government agencies, private sector players, NGOs and farmer organizations is becoming common, allowing for greater resource mobilization, innovation and wider outreach.
10. *Integrating Agricultural Knowledge and Information System (AKIS):* Agricultural research, extension and education operate in one system that generates knowledge and information for farmers. Research and extension are not separate institutions. Farmers, agricultural educators, researchers and extension personnel must be integrated to harness knowledge and information from various sources for better farming and improved livelihoods.
11. *Policy Measures:* The agricultural sector not only proved to be a good source of income generation but also provides employment opportunities to the unemployed. So, appropriate policy measures should be undertaken to strengthen government schemes and programmes. The agriculture-based policy should be based on the principles of social acceptability, economic feasibility, technical viability and resource conservation ability.

4.7 Emerging Challenges in Extension

- *Competitive and Quality-Conscious Agriculture:* Recent developments that present challenges to agricultural research and innovation in developing countries are confronting new priorities in a rapidly changing world (e.g. stronger demand for competitive and quality-conscious agriculture) and adapting to changes within a more complex innovation systems framework where there are a greater number of actors and linkages to consider.

- *Redefining the Role of Government in Agricultural Research and Service Provision:* Government has to play a critical role in defining the role of the private sector, civil society and end users. Strengthening the demand side of agricultural research and services to ensure that these programmes are more responsive and accountable to end users.
- *Developing a Clear Understanding of the Institutional Structures.* It is needed at the national, regional and sub-regional levels for agricultural research and service provision and whether, and how, this understanding would imply changes in the current structures present at national, regional and global levels. Similarly, developing a clear understanding of the institutional structures needed at every level for agricultural education within the emerging food and agricultural innovation systems. Ensuring stakeholder participation and developing local, regional and global partnerships and alliances is the need of the system.
- *Facilitating Development of Innovative Funding Instruments*: It make public institutions more sustainable, reduce donor dependence and enhance co-financing by end users and others. Facilitating strategies to improve coordination of internal and external support for food and agricultural innovation in developing nations, while enhancing collaboration and integrating agricultural research with broader science and technology policies.

4.8 Conclusion

The development of sustainable agriculture in a country will depend largely on the effectiveness of its agricultural extension strategies, approaches, service delivery, methodology and processes. Through the years, extension shifts its focus from one function to another as a response to the changing goals of the government leading to the emerging issues in the agricultural sector. The role of agricultural extension in the next decade should be quite different from what it was 10 years ago or even now. Its role as a facilitator of the agricultural knowledge system would only increase by using bottom-up or participatory approach giving importance to the farmers in planning, implementing and making decisions about extension works. These new approaches broaden the role of farmers by being not just recipients of extension services but as co-implementers of these services. The paradigm shifts from traditional to innovative extension approaches reflecting a more holistic, inclusive and technology-driven strategy to address the complex and dynamic challenges faced by farmers today. These approaches are crucial for achieving sustainable agricultural development, enhancing food security and improving rural livelihoods in an ever-changing global landscape.

References

Echeverria RG (1998) Agricultural research policy issues in Latin America: an overview. World Dev 26(6):1103–1111

Mosher AT (1976) Thinking about rural development. Agricultural Development Council, Inc, p 350

Rivera WM, Cary JW (1997) Privatizing agricultural extension. In: Swanson BE, Bentz RP, Sofranko AJ (eds) Improving agricultural extension, a reference manual. FAO, Rome

Singh RK, Singh A, Sheoran P, Sharma DK, Sharma PC (2019) 'Mera Gaon, Mera Gaurav' policy: a source of hypothesis for the co-production of adaptive agricultural knowledge. Indian J Agric Sci 89(9):1410–1414

Willis J (2007) Foundations of qualitative research: interpretive and critical approaches. SAGE Publications, Inc, Thousand Oaks

Impact of Agri-Based Technological Interventions for Managing Poor Soils and Policy Insights

5

Suresh Kumar, Ram Kishor Fagodiya, Subhasis Mandal, Kailash Prajapat, Rajkumar, Rohtash Kumar, Bhupender, and K. V. Praveen

Abstract

Abiotic stresses are caused by various environmental factors, including land degradation is a significant environmental challenge on a global scale, leading to serious socio-economic repercussions. In India, the current assessment indicates that around 6.73 million hectares of land are impacted by salinity and sodicity. This chapter offers a comprehensive overview of the technologies proposed for the management of salt-affected soils. These interventions have shown a positive influence on agricultural productivity and yield, thus contributing to the improvement of the socio-economic status of farmers in affected areas. Additionally, the chapter addresses critical technical, economic, and policy-related challenges that impede the progress of implementation and scaling of these technologies. In light of future demands and challenges, several key recommendations are also presented.

Keywords

Salt-affected soils · Economic impact · Technologies and constraints

S. Kumar (✉) · R. K. Fagodiya · K. Prajapat · Rajkumar · R. Kumar · Bhupender
ICAR-Central Soil Salinity Research Institute, Karnal, Haryana, India
e-mail: suresh.kumar10@icar.gov.in; Ram.Fagodiya@icar.gov.in; Kailash.Prajapat@icar.gov.in; rajkumar2@icar.gov.in

S. Mandal
ICAR-National Dairy Research Institute, Karnal, Haryana, India
e-mail: S.Mandal@icar.gov.in

K. V. Praveen
ICAR-Indian Agriculture Research Institute, New Delhi, Delhi, India
e-mail: praveen.velayudhan@icar.gov.in

A. Mann et al. (eds.), *Cutting Edge Technologies for Developing Future Crop Plants*,
https://doi.org/10.1007/978-981-96-2508-6_5

5.1 Introduction

Land degradation is a widespread global environmental concern, resulting in profound negative economic and social repercussions. Currently, ~50% of the global population is experiencing the detrimental effects of land degradation (Verburg et al. 2019). Among the different processes of land degradation, soil salinity is considered one of the most harmful types of land degradation. Salinity-affected soils are characterized by an excess of neutral soluble salts, whereas sodic soils are dominated by an excess of carbonates and bicarbonates. This leads to excessive alkalinity upon hydrolysis and poor physical properties (Ismail 2009). The global extent of salt-affected soils (SAS) is estimated to range from approximately 932 to 1125 million hectares (Sparks 2003; Paz et al. 2023). This widespread issue of soil salinity presents a formidable challenge to agricultural production, significantly jeopardizing global efforts to achieve food security. Notably, Szabolcs (1989) was the pioneer in creating the first world map of salt-affected soils, estimating the area of SAS at 955 million hectares. Subsequently, Wicke et al. (2011), utilizing the Harmonized World Soil Database, revised this figure to approximately 1125 million hectares. Additionally, Ivushkin et al. (2019) employed remote sensing techniques, data from the World Soil Information System, and modelling approaches to arrive at an estimate of 1050 million hectares. The most current assessment is provided by the FAO Global Map of Salt-affected soils, which categorizes the findings into 424 million hectares of topsoil (0–30 cm) and 833 million hectares of subsoil (30–100 cm). Moreover, the alarming pace of land degradation can be judged by the fact that every second, an equivalent of four football fields of healthy land becomes degraded, adding up to a total of 100 million hectares each year (UNCCD 2024a).

In India, the current situation reveals that 6.73 million hectares of land is affected by salinity and sodicity, and 32–84% of groundwater in different states is of poor quality. Approximately 251.71 million people are exposed to land degradation, which is 18.39% of the total population (UNCCD 2024b). Furthermore, it is predicted that the area under salt-affected soils (SAS) will expand to ~16 million hectares by 2050, primarily due to climate change, the use of unsustainable practices, and other anthropogenic activities (CSSRI 2015). The estimates suggest that India incurs soil salinity losses to an extent equivalent to 17 million tons of food grains annually, or an annual economic loss of $US 3.0 billion, which is expected to increase to $US 5.6 billion by the end of the next decade (Sharma et al. 2015). In order to prevent, lessen, or reverse land degradation, it is essential to implement sustainable land management (SLM) practices. Over the years a host of SLM technologies and practices were developed, demonstrated, and out-scaled in the field. These SLM technologies include salt-tolerant crop varieties, the application of gypsum for reclaiming sodic soils, guidelines for using poor-quality waters, and the rehabilitation of saline lands using alternative land use or agroforestry models. For example, in India, the utilization of gypsum and sub-surface drainage (SSD) methods has resulted in the reclamation of approximately 2.07 million hectares of alkali soils and ~ 71,000 hectares of waterlogged saline soils (Kumar et al. 2022). These technological solutions have improved land productivity, production, and thereby socioeconomic conditions of the communities depending on the salinity-affected regions (Mandal et al. 2018a; Sheoran et al. 2021b). However, with the increasing effects of climate change and the use of poor quality groundwater, strategies must be formulated to tackle the new challenges associated with SASs. For instance, there is a need for the advancement of suitable technologies for the diagnosis and treatment of land affected by secondary salinization and re-sodification in various regions of the country (CSSRI 2020). In this chapter, we have synthesised the impact of key SAS managing technologies, their status of adoption, and barriers in their out-scaling.

5.2 Extent of Salt-Affected Soils in India

The salt-affected soils (SAS) cover a total area of 6.73 mha, with saline and sodic soils accounting for 2.95 mha and 3.77 mha, respectively (Fig. 5.1). Saline soils make up 43.95% of the total SAS area, while sodic soils make up 56.05%. Within the 2.95 mha of salinity-affected soils, 1.75 mha are inland salinity and 1.20 mha are distributed along the east and west coasts of India. The majority of salinity-affected soils, approximately 82.4% (2.44 million hectares), are located in states such as Gujarat, Andhra Pradesh, West Bengal, Orissa, Tamil Nadu, and the Andaman and Nicobar Islands. Inland plains have saline soils predominantly in Gujarat (17%), Rajasthan (6.6%), Maharashtra (5.9%), Karnataka (0.06%), Haryana (1.6%), and Bihar (1.6%). Sodic soils are mainly found in the Indo-Gangetic Plain, arid and semiarid regions of Western and Central India, and the Peninsular Region in Southern India. Moreover, more than half of the SAS area, specifically 53.4% (3.59 million hectares), is concentrated in Gujarat and Uttar Pradesh. When considering other states such as Maharashtra, West Bengal, Rajasthan, and Tamil Nadu, these

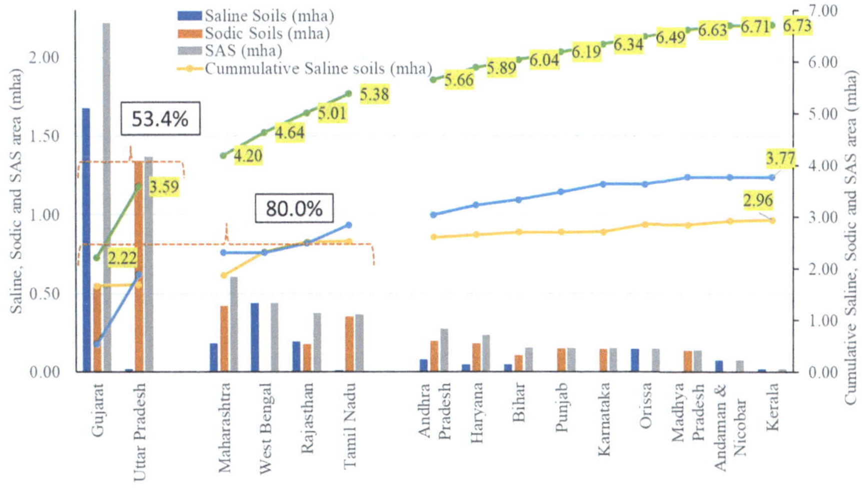

Fig. 5.1 Distribution of salt-affected soils in India

six states collectively account for 80% of the total SAS area. In the Indo-Gangetic Plain, sodic soils cover 47.4% (1.79 million hectares) of the total area, spanning across Uttar Pradesh, Haryana, Punjab, and Bihar. Similarly, 34% (1.28 million hectares) of sodic soils are present in the arid and semiarid regions of Gujarat, Rajasthan, Madhya Pradesh, and Maharashtra (Mandal et al. 2010).

5.3 Projection on the Extent of Salt-Affected Soils in India

Projections show that rising sea levels coupled with intensified weather patterns due to climate change will contribute to the infiltration of salt into terrestrial and aquatic systems. Consequently, this is anticipated to lead to a marked expansion in the area of salt-affected soils in the future (Hassani et al. 2021). Furthermore, the growing variability in the temperature and rainfall distributions is expected to lead to substantial declines in crop yields, particularly in arid and semiarid areas (Enfors and Gordon 2007). The rise in sea level, along with the resulting saltwater intrusion and more frequent cyclonic storms, could affect the productivity potential of coastal agroecosystems. Rising temperatures are likely to enhance evapotranspiration losses, which in turn would result in a greater accumulation of salts (Yeo 1998). Climate change can also aggravate and speed up the pace of secondary salinization in the SASs (Bhardwaj et al. 2019; Ashwin et al. 2020; Mukhopadhyay et al. 2021). In the coastal areas, seawater ingress into the freshwater aquifers and croplands due to sea level rise and cyclonic events is a challenge that gets accentuated by climate change (Mandal et al. 2018b). By 2050, the area of salt-affected soils (SASs) is forecasted to grow to roughly 16 million hectares, attributed to climate change and/or human activities (CSSRI 2015). Estimates indicate that India suffers losses from soil salinity equivalent to 17 million tons of food grains annually, leading to an economic loss of approximately $US 3.0 billion per year, a figure that is expected to climb to $US 5.6 billion by the end of the next decade (Sharma et al. 2015).

5.4 Workable Technological Options for Managing the Salt-Affected Soils

Within the timescale of several human generations, soils are recognized as non-renewable resources. Therefore, there is a need to develop management strategies that avert irreversible degradation of soil resources. This endeavour seeks to promote agricultural systems that are sustainable, emphasizing the importance of fulfilling present needs without undermining the ability of future generations to satisfy their own needs (Powlson et al. 2014). The widespread development of secondary salinity in irrigated areas and the incessantly building-up of the problem of resodification (Gharaibeh et al. 2014; Tripathi and Singh 2010) and resalinization (Amin 2004; Valipour 2014) of the ameliorated soils have garnered considerable interest of the researcher, policy planner, and natural resource managers. For sustainable management of SAS, a mix of appropriate technologies and practices is to be

Table 5.1 Technological options for the management of salt-affected soils and institutions involved in their out-scaling

Technology/ practice	Options	Institutions involved in the process of out-scaling and implementations
Amendments	Gypsum, Sulphur-based formulations, flue gas desulphurization gypsum, organic amendments	State agencies, central and state reclination scheme
Drainage management	Subsurface drainage, cut-soiler technology,	It is spread through various state government agencies, RKVY (reclamation of problem soil), NGOs, private players, co-operatives, farmers, etc.
Salt-tolerant crop varieties	Varieties of wheat, rice, and mustard	Breeder seeds are multiplied by various state agricultural departments or private seed firms and then supplied to the farmers
Alternative land-use options	Tree species *Prosopis juliflora, Acacia nilotica,* and *Tamarix articulate Eucalyptus tereticornis,* halophytes for phytoremediation	State departments, forestry, and horticulture department, farmer
Groundwater management	Groundwater recharge, rainwater harvesting	State groundwater departments, NGOs, farmers
Irrigation water treatment	Gypsum bed, pyrite bed	Farmers
Controlled-irrigation options	Micro-irrigation system, sub-surface irrigation	Central and state scheme, state agriculture, horticulture, and groundwater departments
Land use modification		State departments, NGOs, etc.
Salinity management	Leaching to reduce the solutes from the top layers of the soil	

Source: Kumar et al. (2022), Mandal and Kumar (2024), Shaygan and Baumgartl (2022), Singh et al. (2022)

implemented along with investment support of the institutions involved in the management and reclamation of the problematic soils (Barman et al. 2021; Basak et al. 2021; Datta and Dayal 2000; Lambin and Meyfroidt 2011; Sharma and Singh 2017; Sheoran et al. 2021a). The relevant technological alternatives and the institutions engaged in their expansion are outlined in Table 5.1. Furthermore, the subsequent section discusses the effects of essential SAS management technologies.

5.4.1 Impact of Salt-Tolerant Crop Varieties of Mustard (STMVs)

Several STMVs, namely CS 52, CS 54, CS 56, CS 58, CS 60, CS 61, CS 62, and CS 64, have been developed. For the economic impact assessment, CS 52, CS 54, CS 56, CS 58, and CS 60 were considered. These varieties can tolerate sodicity up to a

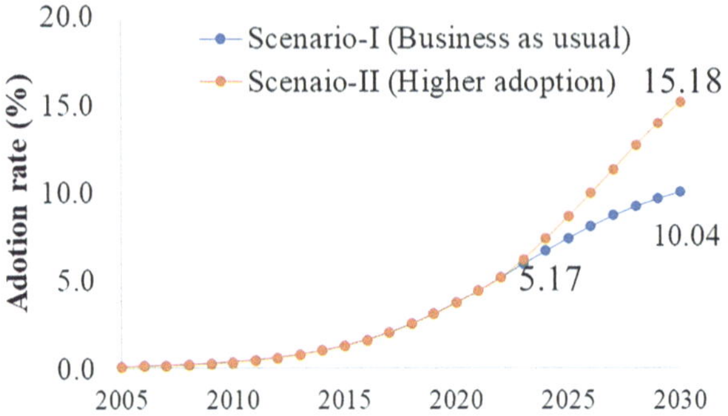

Fig. 5.2 Adoption rate of STMVs. (Adopted from: Mandal et al. 2024)

pH of 9.4 and salinity up to 12 dSm^{-1}, producing grain yields between 2.0 and 2.9 t ha^{-1} in normal soil and 1.6 to 2.2 t ha-1 under salt-stressed conditions (Mandal et al. 2024). The adoption rate of STMVs was estimated at 5.17% in 2022 (Fig. 5.2), with potential growth to 10.04% by 2030 under the Business-as-Usual (BAU) scenario, considering the expansion of salt-affected areas and exogenous growth in mustard cultivation. Additionally, the projected economic surplus was estimated under two alternative scenarios for the period 2005–2030. In Scenario I, with a maximum adoption rate of 10.04% by 2030, and in Scenario II, with a maximum adoption rate of 15.18% by 2030, the latter scenario considers the likely increase in salt-affected areas and other environmental stresses (such as low temperature) in key regions, leading to higher demand and adoption of STMVs. The estimated consumer and producer surplus up to 2022 was INR 1872 crores and INR 2677 crores, respectively. By 2030, these figures are expected to rise to INR 7998 crores and INR 11434 crores, respectively, with a maximum adoption rate of 10.04%. The total economic surplus was INR 4549 crores up to 2022 and is projected to reach INR 19432 crores by 2030 (Fig. 5.2). Furthermore, under the scenario where the adoption rate reaches 15.18% in 2030, driven by increased demand for STMV seeds and higher awareness of STMVs, the total economic surplus is estimated to be INR 24075 crores with INR 9909 and 14,166 crores for consumer and producer surplus (Fig. 5.3), respectively (Mandal et al. 2024).

5.4.2 Impact of CSR-30

The ICAR-Central Soil Salinity Research Institute in Karnal introduced CSR 30, the first sodicity-tolerant Basmati rice variety. This variety is highly favoured by farmers in Haryana. It was officially recognized and approved by the Central Variety Release Committee (CVRC) in 2001 for cultivation in salt-affected regions of Haryana, Uttar Pradesh, and Punjab. In 2012, it was re-affirmed as Basmati CSR 30 by the CVRC. This variety can endure soil sodicity up to a pH of approximately 9.5 and salinity levels up to an EC of around 7.0 dSm^{-1}. Basmati CSR 30 can yield up

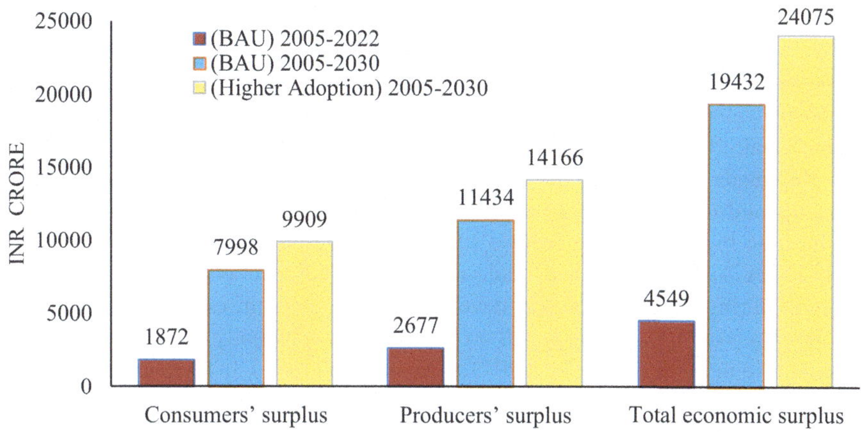

Fig. 5.3 Estimated value of the economic surplus from the STMVs. (Adopted from: Mandal et al. 2024)

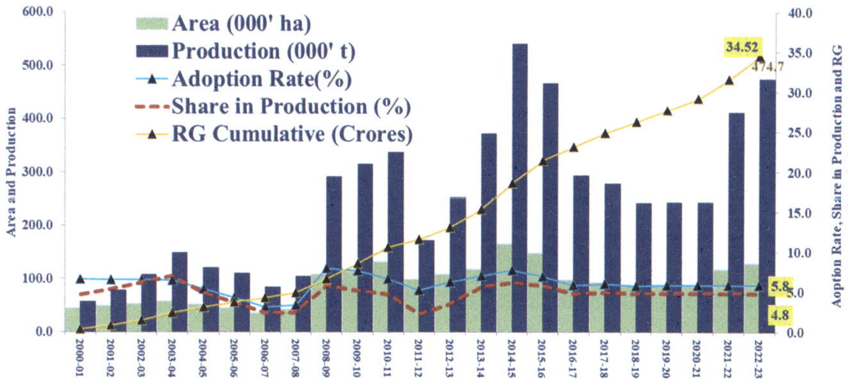

Fig. 5.4 Share in area and production of CSR-30 in Basmati rice production

to 2.8 t ha^{-1} in saline soils. It features long, slender, and highly aromatic grains, with excellent head rice recovery and an impressive elongation ratio when cooked (ICAR-NIAP 2020).

The total area under Basmati rice is around 2.2 mha, of which nearly 0.13 mha is under CSR-30. This implies that CSR-30 accounts for 5.8% of the total area under Basmati rice (Fig. 5.4) and contributes 4.8% of the total Basmati production (9.98 mt). Moreover, in Haryana, the share of CSR-30 is nearly 11.3% of the total area under Basmati. The annual revenue generated by CSR-30 is around INR 1501 crores.

5.4.3 Land Modification: Waterlogged Sodic

In the Indo-Gangetic plain of India, due to canal irrigation, a dual predicament, involving waterlogging and the development of sodic soils, has been observed (Dagar et al. 2022). Insufficient drainage systems, along with persistent seepage from the canals, have caused the water table to elevate (less than one meter), leading to the upward movement of salts onto the topsoil. Consequently, the affected land has become highly alkaline (with a pH exceeding 9) and unproductive (Mandal et al. 2022). Additionally, inadequate attention to drainage systems worsened the situation during irrigation project development. As a result, even though benefits may be observed initially, the negative consequences eventually surpass the positive impacts (Singh 2009; Joshi 1987, 2011; Joshi and Jha 1991). In the *Sharda Sahayak Canal Command area*, approximately 0.50 million hectares of sodic lands are currently impacted by shallow water table conditions, rendering them unsuitable for economically viable cultivation. Despite efforts to reclaim these lands using conventional gypsum-based methods, a portion of these lands (0.18 million hectares) continue to face the dual challenge of shallow groundwater tables (less than one meter) and high sodic conditions (soil pH exceeding 9), as noted by Singh et al. (2008) and Bhardwaj et al. (2019). Consequently, there has been a decrease in land-water productivity and a loss of livelihoods for farming families in the affected region of the canal area. ICAR-Central Soil Salinity Research Institute (ICAR-CSSRI), Regional Research Station (RRS) Lucknow, Uttar Pradesh, conducted a slew of experiments with different types of land modifications to collect and utilize seepage water from the canal. The successful demonstration of harvesting and controlling canal seepage water through these land modifications in farmers' fields highlighted a promising approach to cultivating multiple crops on land that was previously considered unproductive.

Financial feasibility analysis focused on the two most successful models (which were entirely managed by the farmers themselves) with areas of 0.60 ha (crop-fish based) and 0.46 ha (fish-crop based). The initial investment was computed taking into account the existing rate of INR 90 per cubic meter of soil for excavation as per the prevailing norms. Initial investment was calculated as INR 282,000 and INR 212,000 for the respective areas, using the prices from the year 2018–19.

The financial feasibility analysis (Table 5.2) demonstrated that both model types were highly attractive for future investment, with positive internal rate of return (IRR) values of 37% and 56%, as well as net present value (NPV) amounts of INR 249,741 and INR 368,752, and benefit-cost ratios (BCR) of 1.37 and 1.22. The pond-based (fish-crop) system exhibited a rapid return on the initial investment, with a payback period of 1.75 years, while the crop-fish-based system had a longer payback period of 2.6 years. The pond-based system was noted for its intensive utilization, characterized by high costs and high returns due to the dense fish-rearing activities.

Table 5.2 Impact indicators of land modification models

Particular	Land modification as fish-based (0.60 ha)	Land modification as fish-based (0.46 ha)
Area under intervention (ha)	0.60	0.46
Initial investment (INR)	2,82,000	2,12,040
Economic life (years)	10	10
Average annual O&M cost (INR)	91,884	336,410
Average annual gross return (INR)	200,529	487,653
Average annual net return (INR)	108,645	151,243
Output–input ratio	2.18	1.45
NPV (INR per ha) and IRR (%)	210,350 and 34%	338,549 and 53%
BCR and PBP (years)	1.34 (2.82)	1.20 (1.85)
Break-even size (ha)	0.45	0.38

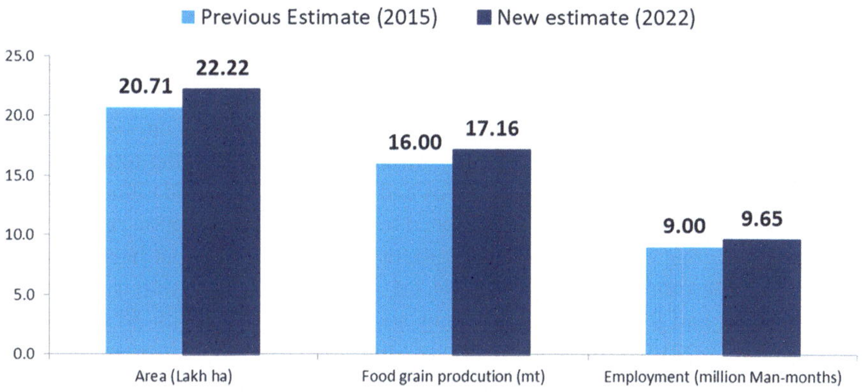

Fig. 5.5 Extent of area reclaimed by gypsum application

5.4.4 Gypsum Technology for Reclamation of Sodic Soils

Gypsum is extensively applied as an amendment for the reclamation of sodic soils that face structural challenges and obstructed water flow. The introduction of gypsum markedly enhances the physical characteristics of the soil, as indicated by improved flocculation, enhanced stability of aggregates, and a more efficient infiltration rate (Mandal and Kumar 2024). Up to 2015, approximately 20.71 lakh ha (10 lakh ha is equal to 1 million ha) or 2.07 million hectares of sodic land have been restored through the use of gypsum (10–15 t ha^{-1}), resulting in an annual contribution of around 16.60 million tonnes of incremental food grains to the national food supply (Fig. 5.5). The investment in gypsum technology has been shown to be financially viable, as indicated by a bnefit cost ratio (BCR) of 2.47, an internal rate of rReturn (IRR) of 67%, and a payback period of 3 years (Tripathi 2011). The revenue obtained from agriculture in the reclaimed sodic region amounts to INR 253200 million (Chinchmalatpure et al. 2015). As per the latest estimate, the reclaimed area has increased from 20.71 lakh ha (2015) to 22.22 lakh ha in the year 2022, indicating the pace of the reclamation of the sodic soils is around 18,809 ha per year. With increased reclaimed area, the estimated contribution in food grain production is around 17.16 mt in the national food basket (Fig. 5.4).

5.4.5 Sub-surface Drainage for Reclamation of Waterlogged Saline Soils

The sub-surface drainage (SSD) system is designed to facilitate the removal of excess water from the root zone in saline soils affected by waterlogging, which in turn helps to lower soil salinity. This method has had a profound effect on agricultural productivity in India, particularly in restoring the viability of waterlogged saline soils (Gupta 2002).

Extensive efforts have been directed towards the improvement, refinement, and promotion of the SSD system to address the interconnected issues of salinity and waterlogging (Chinchmalatpure et al. 2020). Developed in 1980, SSD technology has been successfully implemented in ~75,000 ha in states like Haryana, Rajasthan, Gujarat, Punjab, Andhra Pradesh, Maharashtra, and Karnataka through various schemes and programmes in the country, which is mere ~3% of the total affected area. To treat the remaining affected areas and achieve land degradation neutrality (LDN) goal of India, there is an urgent need to accelerate the adoption rate of SSD technology (Kumar et al. 2024). At the current prices (2023), the per hectare adoption cost was estimated at INR 1,21,794 for medium to light textured soils with 67 m spacing, and INR 1,40,345 for heavy textured black soils with 30 m spacing (Kumar et al. 2024). The implementation of SSD led to a threefold increase in farmers' income post-installation. Financial analysis indicated a payback period of 2 years, NPV of Rs. 113,000 per hectare, BCR of 2.71, and IRR of 40% (Mandal and Kumar 2024).

5.4.6 Impact of KRL-210

The area covered is 22,279 ha, which constitutes 1.27% of the total salt-affected land. The estimated additional yield is 5 qha^{-1}, representing a 14% increase compared to the HD 2967 variety in soils with high salinity (pH levels >9.3 and RSC >5.0). The economic surplus generated during the period from 2010–11 to 2018–19 amounts to INR 4040 million. In comparison to its counterparts, KRL-210 shows a reduced incidence of lodging and is recognized for its exceptional quality in the making of *chapattis* (Sheoran et al. 2021c).

5.5 Constraints in Dissemination and Out-Scaling of SAS Management Technologies

Previous initiatives have demonstrated promising outcomes in the management of salt-affected regions, yielding positive benefits. The developed strategies are both technologically feasible and economically viable; however, the instances of success tend to be inconsistent and primarily confined to regions where research institutions are actively engaged (Mandal and Raju 2021). Technologists frequently fail to recognize the reasons for farmers' reluctance to adopt financially viable technologies. Consequently, the adoption of SAS managing technologies is slow (Qadir et al. 2007; Mann et al. 2020; Mandal and Raju 2021). Therefore, it is crucial to delve into the issues and constraints faced by the farmers. Recognizing these issues is crucial for tackling challenges in technology dissemination in saline regions and for converting research into practices that provide significant societal benefits (Mandal and Raju 2021; Van den Burg et al. 2024). All the constraints in the dissemination of the SAS management technologies are categories into three major groups, namely technical, economic, and policy constraints (Table 5.3). Key technical constraints include the non-availability of micro-scale maps and data for location-specific problems; small and fragmented land holdings; non-availability of quality amendments, etc. Major economic constraints are high initial costs for resource-poor farmers and poor access to credit facilities.

Similarly, policy and institution-related issues are lack of training and technical guidance and few dedicated schemes and programme for managing the SAS. Effective management of the SAS necessitates a precise identification of the underlying problem, specifically whether it is related to sodicity, salinity, or waterlogged salinity since the reclamation strategies are tailored to the sites, and based on physical and socio-economic settings (Shaygan and Baumgartl 2022). In this context, the utilization of digital soil maps is becoming increasingly significant for scientists and land managers as they enhance the ability to identify issues and apply tailored solutions that are specific to the location (Khaledian and Miller 2020; Barman et al. 2021).

Detailed inventories of affected soils with specific characteristics, distribution patterns, and potential for utilization will be highly useful for policy planners and project implementation agencies (Yarazari 2022). For instance, the situation seems to have become critical in shallow aquifers situated close to the coastline, however,

Table 5.3 Key constraints in scaling-up of the SAS managing technologies

Constraint	Description
Technical	Non-availability of micro-scale map of the location-specific problems Fragmentation of land Non-availability quality amendments
Economic	High initial cost for resource-poor farmers of the affected area Poor access to credit facilities
Policy and institutional	Lack of training and technical guidance A few dedicate schemes and programme

accurate quantitative data are largely absent, which is crucial for formulating the necessary prevention and management measures to address this challenge. Therefore, there is an urgent need to map the SAS at the micro-scale preferably at village and block levels. Dayal et al. (1999) noted a host of farm-level technical challenges hindering the under-utilization of low-quality water.

These challenges encompass the inadequate and delayed access to canal water, the unavailability of organic materials and amendments, and such issues aggravate with fragmentation of land holdings as users are not able to take advantages of bulk buying. Mandal and Raju (2021) noted that in case of gypsum application major bottlenecks are high cost (approximately INR 75000 per ha) and impurities Raju et al. (2022) found that in adoption of SSD technology for reclamation of waterlogged saline soils in Maharashtra, the high initial investment requirement was also perceived as a key barrier for resource-poor farmers. The implementation of the SSD becomes even more difficult when there is insufficient institutional support in the form of credit and subsidies to such resource-poor farmers having limited financial capacity.

5.6 Policy Insights

The design of effective and efficient policies to encourage farmers to adopt SAS managing technologies and practices (Prager et al. 2011) is a crucial step to achieving the desired level of adoption. These policies are expected to augment the productivity of existing agricultural lands as well as to bring the abandoned lands under crop production in a socially acceptable and economically viable manner. This has become indispensable for ensuring the food and nutritional security of the burgeoning world population, witnessing a number of formidable challenges such as land degradation, freshwater scarcity, and climate change. Successful adoption of appropriate technological options requires enabling of socio-economic setting, including adequate policy support. Some specific policy recommendations are: (a) quick mapping of the database on salt-affected soils and poor-quality water; (b) to overcome the issues of impurity and depleting supply of gypsum there is a need to develop other cost-effective alternatives for reclamation of sodic soils; (c) development and refinement of technologies for the management of sub-surface sodicity; (d) increasing the institutional support for faster adoption of the sub-surface drainage technologies; (e) development and dissemination of location-specific agroforestry and horticultural model; (f) development of salt-tolerant crop varieties for multiple stress tolerance that coexisted with water scarcity and other abiotic-stresses; (g) strengthening of forward and backward linkages (for instance, formulation of farmer producers' organization) particularly for resource-poor smallholders for purchasing quality soil amendments. Finally, sustainable management soils require appropriate technologies to be adopted at the desired scale for better social outcomes. Therefore, actual field-level scaling-up of SAS managing technologies must be ensured by efficient institutions with adequate financial support.

References

Amin AA (2004) The extent of desertification on Saudi Arabia. Environ Geol 46:22–31

Ashwin HS, Rekha MV, Champa BV, Alur AS, Nagaraja MS, Bhardwaj AK (2020) Secondary salinization risks associated with irrigation water from borewell, lift and canal in the typical black soils of southern India. J Nat Resour Conserv Manag 1(2):96–103

Barman A, Sheoran P, Yadav RK, Abhishek R, Sharma R, Prajapat K, Singh RK, Kumar S (2021) Soil spatial variability characterization: delineating index-based management zones in salt-affected agroecosystem of India. J Environ Manag 296:113243

Basak N, Sheoran P, Sharma R, Yadav RK, Singh RK, Kumar S, Krishnamurthy T, Sharma PC (2021) Gypsum and pressmud amelioration improve soil organic carbon storage and stability in sodic agroecosystems. Land Degrad Dev 32(15):4430–4444

Bhardwaj AK, Mishra VK, Singh AK, Arora S, Srivastava S, Singh YP, Sharma DK (2019) Soil salinity and land use-land cover interactions with soil carbon in a salt affected irrigation canal command of Indo-Gangetic Plain. Catena 180:392–400

Chinchmalatpure AR, Ali S, Kulshrestha N, Singh RK, Bundela DS, Kumar P, Sharma DK (2015) Intellectual property management and commercialisation of ICAR-CSSRI technologies for management of salt affected and waterlogged soils of India. Technical Bulletin, ICAR-Central Soil Salinity Research Institute, Karnal, Haryana, India, p 62

Chinchmalatpure AR, Vibhute SD, Kaledhonkar MJ, Kad SV, Kumar S, Camus D et al (2020) Performance evaluation of subsurface drainage system under waterlogged saline vertisols for sugarcane crop in Ukai Kakrapar Canal command, Gujarat. J Agri Eng 57(3):248–258

CSSRI (2015) Vision-2050. Karnal, India: ICAR-Central Soil Salinity Research Institute, p 31. https://cssri.res.in/vision-2050/

CSSRI (2020) Annual report-2020. Karnal, India: ICAR-Central Soil Salinity Research Institute, p 120. https://cssri.res.in/download/annualrepart-2019-20/

Dagar JC, Singh G, Gupta SR (2022) Sustainable land-use Systems for Rehabilitation of highly sodic lands in the Indo-Gangetic Plains of North-Western India: synthesis of long-term field experiments. J Soil Salinity Water Quality 14(1):1–14

Datta KK, Dayal B (2000) Irrigation with poor quality water: an empirical study of input use, economic loss, and coping strategies. Indian J Agric Econ 55(1):26–37

Dayal B, Datta KK, Chauhan CPC, Minhas PS (1999) Operational research project on saline irrigation in Karanpur, Mathura (UP): socio-economic considerations and impact analysis, p 35, Bulletin No. 2/99, Central soil salinity Research Institute, Karnal

Enfors EI, Gordon LJ (2007) Analysing resilience in dryland agro-ecosystems: a case study of the Makanya catchment in Tanzania over the past 50 years. Land Degrad Dev 18(6):680–696

Gharaibeh MA et al (2014) Reclamation of highly calcareous saline-sodic soil using low quality water and phosphogypsum. Appl Water Sci 4:223–230

Gupta SK (2002) A century of subsurface drainage research in India. Irrig Drain Syst 16:69–84

Hassani A, Azapagic A, Shokri N (2021) Global predictions of primary soil salinization under changing climate in the 21st century. Nat Commun 12:6663. https://doi.org/10.1038/s41467-021-26907-3

ICAR-NIAP (2020) "Economic Impact of ICAR Research Some Recent Evidence" ICAR-National Institute of Agricultural Economics and Policy Research Indian Council of Agricultural Research, New Delhi, pp 35

Ismail AM (2009) Development of technologies to harness the productivity potential of salt-affected areas of the Indo-Gangetic, Mekong, and Nile River basins. Colombo, Sri Lanka: CGIAR Challenge Program on Water and Food. https://hdl.handle.net/10568/3785

Ivushkin K, Harm B, Bregt AK, Pulatov A, Kempen B, de Sousa L (2019) Global mapping of soil salinity change. Remote Sens Environ 231:111260

Joshi PK (1987) Effect of surface irrigation on land degradation – problems and strategies. Indian J Agric Econ 42(3):416–423

Joshi PK (2011) Conservation agriculture: an overview. Indian J Agric Econ 66(1)

Joshi PK, Jha D (1991) Farm-level effects of soil degradation in Sharda Sahayak irrigation project. Washington, D. C: International Food Policy Research Institute. Working Paper on Future Growth in Indian Agriculture 1

Khaledian Y, Miller BA (2020) Selecting appropriate machine learning methods for digital soil mapping. Appl Math Model 81:401–418

Kumar R, Singh A, Bhardwaj AK, Kumar A, Yadav RK, Sharma PC (2022) Reclamation of salt-affected soils in India: progress, emerging challenges, and future strategies. Land Degrad Dev 33(13):2169–2180

Kumar S, Kumar S, Narjary B, Lal M, Bundela DS, Yadav RK, Chaudhari SK (2024) Financial viability of sub-surface drainage technology for reclamation of waterlogged saline soils. Technical folder, ICAR-CSSRI, Karnal, p 4

Lambin EF, Meyfroidt P (2011) Global land use change, economic globalization, and the looming land scarcity. Proc Natl Acad Sci 108(9):3465–3472

Mandal and Kumar (2024) Rejuvenating salt affected ecologies for land degradation neutrality under changing climate. In: International Salinity conference (ISC-2024), p 5

Mandal S, Raju R (2021) Reclamation of salt-affected soils: microeconomic issues and policies. In: Managing salt-affected soils for sustainable agriculture, pp 617–631

Mandal AK, Sharma RC, Singh G, Dagar JC (2010) Computerized database on salt affected soil in India. CSSRI Publ. No. 2/2010, Karnal, p 15 https://krishi.icar.gov.in/jspui/bitstream/123456789/25582/1/CSSRI_database_2010.pdf

Mandal S, Raju R, Kumar A, Kumar P, Sharma PC (2018a) Current status of research, technology response and policy needs of salt-affected soils in India–a review. J Indian Soc Coast Agric Res 36(2)

Mandal UK, Nayak DB, Samui ARPAN, Jana AK, Mullick SOURAV, Lama TD et al (2018b) Trend of sea-level-rise in West Bengal coast. J Indian Soc Coastal Agric Res 36(2):64–73

Mandal UK, Nayak DB, Samui A, Jana AK, Mullick S, Lama TD, Raut S, Burman D (2018c) Trend of sea-level-rise in West Bengal coast. J Indian Soc Coastal Agric Res 36(2):64–73

Mandal S, Mishra VK, Verma CL, Sharma PC (2022) land modification models for restoring waterlogged sodic soils in Uttar Pradesh – the innovations, economics and financial viability. *ICAR-CSSRI/Working Paper/47/2021*, ICAR-Central Soil Salinity Research Institute, p 2

Mandal S, Kumar S, Singh J, Jain R, Kandpal A (2024) Economic impact of salt-tolerant mustard varieties. Policy Brief. https://niap.icar.gov.in/pdf/policy-briefs/PolicyBrief56.pdf

Mann A, Kumar A, Sanwal SK, Sharma PC (2020) Sustainable production of pulses under saline lands in India. In: Legume crops-prospects, production and uses. IntechOpen. https://doi.org/10.5772/intechopen.90304

Mukhopadhyay R, Sarkar B, Jat HS, Sharma PC, Bolan NS (2021) Soil salinity under climate change: challenges for sustainable agriculture and food security. J Environ Manag 280:1–12

Paz AM, Amezketa E, Canfora L, Castanheira N, Falsone G, Gonçalves MC et al (2023) Salt-affected soils: field-scale strategies for prevention, mitigation, and adaptation to salt accumulation. Ital J Agron 18(2)

Powlson DS, Gregory PJ, Whalley WR, Quinton JN (2014) Soil management in relation to sustainable agriculture and ecosystem services. Food Policy 36:572–587

Prager K, Schuler J, Helming K, Zander P, Ratinger T, Hagedorn K (2011) Soil degradation, farming practices, institutions and policy responses: an analytical framework. Land Degrad Dev 22:32–46

Qadir M et al (2007) Phytoremediation of sodic and saline-sodic soils. Adv Agron 96:197–247

Raju R, Vibhute SD, Bundela DS, Kumar A (2022) Socio-economic impact and constraints in adoption of technology for reclamation of waterlogged saline soils in Maharashtra, India. Indian J Soil Conserv 50(3):240–248

Sharma DK, Singh A (2017) Current trends and emerging challenges in sustainable management of salt-affected soils: a critical appraisal. In: Bioremediation of salt affected soils: an Indian perspective, pp 1–40

Sharma DK, Thimmappa K, Chinchmalatpure A, Mandal AK, Yadav RK, Chaudhury SK, Satyendra K, Sikka A (2015) Assessment of production and monetary losses from salt-affected soils in India. Technical Bulletin 5/2015, p 99. Karnal, India: ICAR-Central Soil Salinity Research Institute

Shaygan M, Baumgartl T (2022) Reclamation of salt-affected land: a review. Soil Systems 6(3):61

Sheoran P, Basak N, Kumar A, Yadav RK, Singh R, Sharma R, Kumar S, Singh RK, Sharma PC (2021a) Ameliorants and salt tolerant varieties improve rice-wheat production in soils undergoing sodification with alkali water irrigation in Indo–Gangetic Plains of India. Agric Water Manag 243:106492

Sheoran P, Kumar A, Singh A, Kumar A, Parjapat K, Sharma R, Singh RK, Yadav RK, Dang YP, Sharma PC (2021b) Pressmud alleviates soil sodicity stress in a rice–wheat rotation: effects on soil properties, physiological adaptation and yield-related traits. Land Degrad Dev 32(9):2735–2748

Sheoran P, Kumar A, Sharma R, Prajapat K, Kumar A, Barman A, Raju R, Kumar S, Dar YJ, Singh RK, Sanwal SK (2021c) Quantitative dissection of salt tolerance for sustainable wheat production in sodic agro-ecosystems through farmers' participatory approach: an Indian experience. Sustain For 13(6):3378

Singh G (2009) Salinity-related desertification and management strategies: Indian experience. Land Degrad Dev 20:367–385

Singh GP, Sharma DK Mishra VK, Singh YP, Nayak AK, Singh Ranbir, Kaledhonkar MJ, Varshney PK (2008) Integrated farming system model for waterlogged sodic soils. Technical Bulletin No. 1/2008, ICAR-Central soil Salinity Research Institute, Regional Research Station, Lucknow –226005 (India), p 12

Singh A, Kumar A, Kumar R, Sheoran P, Yadav RK, Sharma PC (2022) Multivariate analyses discern shared and contrasting eco-physiological responses to salinity stress of Ziziphus rootstocks and budded trees. S Afr J Bot 146:573–584

Sparks DL (2003) Environmental soil chemistry: an overview. Environ Soil Chem 2:1–42

Szabolcs I (1989) Salt-affected soils. CRC Press, Boca Raton

Tanji K, Kielen N (2002) Agricultural drainage water management in arid and semi-arid areas

Tripathi RS (2011) Socio-economic impact of reclaiming salt affected lands in India. J Soil Salinity Water Quality 3(2):110–126

Tripathi KP, Singh B (2010) Impact of different forest plantations and cropping systems on reclamation of sodic soils. Indian J For 33:135–142

UNCCD (2024a) Our legacy, our future. https://www.unccd.int/sites/default/files/2024-04/DDD24%20key%20messages%20EN.pdf

UNCCD (2024b) What proportion of the population is exposed to land degradation and drought? https://data.unccd.int/country-overview?country=IND&layer=LAND_DROUGHT_E5#country-profile-map

Valipour M (2014) Drainage, waterlogging, and salinity. Arch Agron Soil Sci 60:1625–1640

Van den Burg S, Deolu-Ajayi AO, Nauta R, Cervi WR, van der Werf A, Poelman M, van der Meer IM (2024) Knowledge gaps on how to adapt crop production under changing saline circumstances in The Netherlands. Sci Total Environ:170118

Verburg PH, Metternicht G, Allen C, Debonne N, Akhtar-Schuster M, Inácio da Cunha M, Karim Z, Pilon A, Raja O, Sánchez Santivañez M, Şenyaz A (2019) Creating an enabling environment for land degradation neutrality and its potential contribution to enhancing well-being, livelihoods and the environment. A report of the science-policy interface. United Nations Convention to Combat Desertification (UNCCD), Bonn, Germany

Wicke B, Smeets E, Dornburg V, Vashev B, Gaiser T, Turkenburg W, Faaij A (2011) The global technical and economic potential of bioenergy from salt-affected soils. Energy Environ Sci 4(8):2669–2681

Yarazari SP (2022) Constraints in adoption of saline soil management practices by the farmers of belagavi district. Mysore J Agric Sci 56(1):320–326

Yeo A (1998) Predicting the interaction between the effects of salinity and climate change on crop plants. Sci Hortic 78(1–4):159–174

Part II

Frontier Advancements in Crop Breeding and Improvement

Advanced Strategies for Crop Improvement Against Abiotic Stresses: An Integrated View from Breeding to Genomics

6

Jayanth Kallugudi, Charu Lata, R. Thribhuvan, K. T. Ravikiran, Suman Dutta, P. A. Krishnan, Chandramani Patel, Akriti Thakur, Anshul Sharma Manjul, Pramod Prasad, and O. P. Gangwar

Abstract

Abiotic stressors comprising salinity, drought, high temperatures, flooding, low nutrient availability, and heavy metal toxicity are reducing the productivity of agriculture throughout the globe. Global climate change intensifies these stresses and interferes with vital plant functions such as photosynthesis, respiration, and nutrient uptake, which lowers crop yields and increases food insecurity. Conventional breeding approaches have contributed to developing stress-resilient crop plants by selecting desirable traits within existing genetic variations or through controlled crossing. Traditional techniques like pure line and mass selection, pedigree breeding, and backcross breeding have successfully introduced traits such as drought and salinity tolerance in various crops. Nevertheless,

J. Kallugudi · C. Lata (✉) · A. Thakur · A. S. Manjul · P. Prasad · O. P. Gangwar
ICAR-IIWBR, Regional Station, Shimla, Himachal Pradesh, India
e-mail: jayanth.kallugudi@icar.gov.in; Charu.Sharma@icar.gov.in

R. Thribhuvan
ICAR – Central Research Institute for Jute and Allied Fibres, Nilganj, West Bengal, India

K. T. Ravikiran
ICAR-CSSRI, Regional Research Station, Lucknow, Uttar Pradesh, India

S. Dutta
Ramakrishna Mission Vivekananda Educational and Research Institute, Kolkata, West Bengal, India

P. A. Krishnan
ICAR- Indian Agricultural Research Institute, Dhemaji, Assam, India

C. Patel
Jawaharlal Nehru University, New Delhi, Delhi, India

A. Mann et al. (eds.), *Cutting Edge Technologies for Developing Future Crop Plants*,
https://doi.org/10.1007/978-981-96-2508-6_6

genetic variation highly dependent on environmental conditions makes these methods time-consuming and restricted. Advanced molecular breeding techniques, including marker-assisted selection (MAS), genomics, genome editing, and genetic engineering techniques, offer more precise and efficient solutions. MAS leverages molecular markers to improve and augment the breeding process, enabling the accumulation of favorable genes for multiple abiotic stresses and making plants tolerant to multiple stresses. This amalgamation of conventional and modern breeding strategies is indispensable for developing crops that can withstand the increasing frequency and intensity of abiotic stresses. Thereby, these modern practices ensure sustainable agricultural productivity and global food security in changing climate scenarios.

Keywords

Abiotic stress · Crop improvement · MAS · Genomics · Genome editing

6.1 Introduction

6.1.1 Overview of Abiotic Stresses in Agriculture

The population on the globe is very intensively increasing every day; it is projected that by 2050 it will increase to 9.7 billion. The intensive increase in abiotic stresses due to climate changes and other factors can pose a severe threat to plants and crop production (Alexandratos and Bruinsma 2012). Abiotic stresses refer to the nonliving environmental factors that can adversely affect crop growth and productivity. These stresses pose a substantial impact on agricultural productivity worldwide and mainly arise from nonliving environmental factors such as salinity, drought, extreme temperatures (both heat and cold), nutrient deficiencies, water logging, and heavy metal toxicity. Each type of abiotic stress can significantly impact plant physiology and metabolism, leading to reduced crop yields and compromised food security by disrupting essential plant processes such as photosynthesis, respiration, and nutrient uptake, resulting in decreased growth, yield losses, and even crop failure. Under the present scenario climate change conditions increase the occurrence and intensity of these stresses, demanding novel approaches to sustainable agriculture and food security. Crop plants are continuously exposed to multiple stressors and are clearly attested by multi-billion-dollar losses in yield across the globe (Gonzalez Guzman et al. 2022). Different stress targets specific physiological functions and biochemical pathways in plants, leading to a range of impacts on growth, development, and productivity (Krasensky and Jonak 2012; Mann et al. 2019). For example, salinity impacts osmotic balance and nutrient uptake, whereas nutrient imbalances can lead to deficiencies or toxicities affecting overall plant health. Temperature stresses can damage cellular structures and disrupt metabolic processes, while water deficiency/drought stress primarily affects water uptake and turgor pressure. On the other hand, water excess/logging stress can lead to root oxygen deprivation and increased disease susceptibility.

Globally, drought affects over 70 million hectares of rice-growing land, making it one of the most severe abiotic stresses. It leads to water scarcity, which hampers plant growth and reduces yield. On the other hand, high salt concentrations that interfere with plant water uptake and nutrient balance cause salinity to affect over 100 million hectares of agricultural land, making it uncultivable (Mann et al. 2023a, b). Extreme temperatures, including both heat and cold stress, can cause cellular damage and affect enzymatic activities in cells, further impairing plant development (Athar and Ashraf 2009). The effects of these abiotic stresses on crop yield are profound; they not only lower the amount but also lower the quality of the produce, resulting in financial losses and threatening the security of the world's food supply. Addressing these challenges is essential for ensuring a stable food supply in the face of growing environmental uncertainties (Taunk et al. 2019).

6.1.2 Importance of Developing Stress-Resilient Crops

Abiotic stress can interfere with a number of physiological and biochemical functions that occur within plant cells, which can limit the growth, development, and productivity of the plant. Along with this, stresses can also make plants more susceptible to diseases, pests, and other biotic stresses. Hence, improving agricultural productivity and sustainability is of great global importance. There is an urgent need for crop improvement to ensure food security for the ever-increasing population. The objective of crop improvement is to develop improved plant varieties that exhibit desirable traits such as increased yield, tolerance to environmental stresses, disease resistance improved nutritional content, and better post-harvest characteristics (Gao 2021).

6.1.3 Current Challenges in Agriculture Due to Climate Change

Climate change exacerbates agriculture's problems by increasing the frequency and intensity of abiotic stresses. Rising temperatures, unpredictable rainfall patterns, and increasing soil salinity are a few of the immediate effects of climate change that adversely affect crop production (Mann et al. 2020; Dietz and Vogelsang 2023). These challenges necessitate the development of stress-tolerant crops that can withstand harsh conditions while maintaining productivity. Crop improvement plays an important role in sustainable agriculture by enhancing crops' resilience to abiotic stresses. Conventional breeding methods, along with modern biotechnological approaches such as genome editing and genetic engineering, offer promising solutions for developing stress-tolerant crop varieties. Techniques like CRISPR/Cas9 have revolutionized crop improvement by enabling precise modifications at the DNA level, thereby enhancing stress tolerance and yield (Kamali and Singh 2023).

Climate change is creating unprecedented challenges for agriculture. Droughts, floods, and heat waves are becoming more often and devastating for agricultural output and food security throughout the world. For example, prolonged droughts can result in substantial water shortages, limiting the growth and productivity of crops. Similarly, severe rainfall and flooding can cause soil erosion and nutrient leaching, which further reduce agricultural production (EPAU 2023).

6.1.4 Role of Crop Improvement in Sustainable Agriculture

To address these challenges, it is critical to develop stress-resilient crops. Crop improvement through conventional breeding and current biotechnology technologies plays a crucial role in increasing the resistance of food crops to diverse abiotic and biotic stresses. Through the incorporation of traits such as drought tolerance, salt tolerance, heat resistance, and resistance to biotic stresses, researchers can develop crop varieties that can withstand adverse environmental conditions and maintain crop yields.

Traditional breeding methods are an important aspect of crop improvement to develop stress-resilient crop verities. By selecting superior germplasm and cross-breeding plants with desirable characteristics, breeders can develop new varieties that can tolerate diverse stresses and are better adaptable to changing environmental conditions. Additionally, modern techniques are very useful tools that take less time less effort, and more productive. Nowadays the key strategies in crop improvement are the use of genomics and transcriptomics. These advanced techniques allow researchers to identify and manipulate genes associated with stress tolerance (Mann et al. 2021). These techniques work at the genomic level and pinpoint the exact role of genes with in silico modes. For example, genomics can help in targeting genetic variations linked to drought resistance, while transcriptomics can reveal gene expression patterns under both normal and stress conditions. This knowledge can be used to develop crops with enhanced stress resilience through gene editing techniques such as CRISPR-Cas9 (Yang et al. 2021). Crop enhancement includes not just genetic techniques, but also the application of sustainable agriculture practices. Crop resilience can be enhanced by strategies such as integrated pest management (IPM) and conservation agriculture practices. IPM combines biological, cultural, and chemical management strategies to manage pests and diseases, lowering dependency on chemical pesticides and minimizing environmental effects (Kalogiannidis et al. 2022). Conservation agriculture, which includes techniques like low tillage and crop rotation, helps to preserve soil health and improve water retention, making crops more resistant to climate change (Mann et al. 2018; Joshi et al. 2021). In this chapter, we focus on the need to minimize the negative effects of abiotic stressors and guarantee food security. Along with this covering the advanced strategies for crop improvement against abiotic stresses including advanced breeding techniques with genomics, which can give a comprehensive approach to handle the problems posed by climate change and maintain agricultural productivity.

6.2 Traditional Breeding Approaches

Plant breeding is a co-evolutionary process between plants and human beings. It involves altering the plants genetically, physiologically, and morphologically to make them more beneficial for humans. The basis of plant breeding is the selection of desired types among the variants in terms of quality, yield, and resistance to biotic and abiotic stresses.

6.2.1 Conventional Breeding Approaches

The choice of breeding method for the development of stress-tolerant varieties depends on several factors such as screening techniques, mechanism of tolerance, sources of tolerance gees, mode of action, and heritability. Drought tolerance in crops such as groundnut, common bean, chickpea, wheat, and soybean has been incorporated using traditional breeding approaches (Meena et al. 2016). The salt tolerance traits in many self and cross-pollinated crops have also been incorporated through conventional breeding methods. Dewey (1962) gave a strategy of recurrent selection for salt tolerance breeding. A number of abiotic stress-tolerant crop varieties are developed using the conventional breeding approaches as presented in Table 6.1. The conventional breeding approaches can be identified based on different criteria: (i) selection based on existing variance and (ii) selection based on variance created by controlled mating/crossing.

Table 6.1 Varieties developed against abiotic stress using conventional breeding approaches

Stress	Crop	Variety	Breeding approach
Salinity	Rice	CSR 60, CSR 56, CSR 46, CSR-43, CSR-23, CSR-30, CSR-36, CSR-27, CSR-13, and CSR-10	Hybridization
		CSR-43	Farmers' participatory variety selection
	Mustard	CS-52, CS-56, CS-54, CS-58, CS-60, and CS-64	Hybridization
	Chickpea	Karnal Chana-1	Hybridization
	Wheat	KRL 1–4, KRL-19, KRL-210, and KRL-213	Hybridization
	Sugarcane	CO-453 and CO-62175	Hybridization
Drought	Rice	Sahbhagidhan (IR74371-70-1-1)	Pedigree selection with shuttle breeding
		PTB10, Nagina 22, BR19	Pureline selection
		Bala, Annada, Annapurna, Parijat, Kanchan, and Rasi	Pedigree selection
		MR219-9, MR219-4, MK-D-2, MK-D-3, and TaromMahalli	Mutation breeding
	Wheat	Katya, Berkut, SeriM82, Dharwad dry, Aragon 03, Excalibur, and Borlaug 100	Hybridization

6.2.1.1 Selection Based on Existing Variance

This is the most primitive form of plant breeding where the genetic variation already present in the population forms the basis of selection. This early phase of breeding extends from the origin of agriculture until the first hybridization in the 1760s by Koelreuter (Roberts 1929). Late in the nineteenth and the twentieth centuries, hybridization gained more importance among plant breeders.

Domestication: Nature has its own way of creating genetic variation. Mutation was one such event, even though the frequency of mutation for a particular gene is very low, considering the large population in the field, plant mutations are quite frequent. Some of these mutations were selected and conserved by the farmers across many generations. This led to the conversion of many plant species from their wild variants to more beneficial cultivated types. The accumulation of these mutations is a major base of domestication syndrome. As a consequence of selection, the genetic diversity among the cultivated types was reduced compared to the wild type which in turn has led to the founder effect. Many domesticated lines are now cultivated, having been selected over many generations for their yield and resistance to biotic and abiotic stress, and are known as landraces. Until recently, landraces are the most valuable genetic resources for plant breeding operations. The crop wild relatives (CWR) and landraces act as major reservoirs of several resistance genes to abiotic stresses such as drought, mineral toxicity, salt, and heat tolerance. In maize, the drought and salinity tolerance genes are found in Eastern gamagrass and waterlogging tolerance in *Z. nicarguensis*. In rice, genes for both drought and heat tolerance are found in *O. glaberrima*, *O. barthii,* and *O. longistaminata*; salinity tolerance genes in *Oryza coarcata;* and cold tolerance genes in *O. rufipogon* (Mammadov et al. 2018).

Pureline and Mass selection: The earliest method of breeding based on the law of inheritance has been the selection of plants among the landraces with the assumption that the best individuals will produce superior progenies compared to the progeny of a random sample from the population. In self-pollinating crops such as rice, wheat, and tomato, selecting a few lines (Pureline selection) and producing inbred progenies give some lines with superior performance compared to the original landraces. Whereas in out-crossing crops such as maize, the mass selection of superior lines over a few generations increases the desirable allele frequency in the population. The major drawback is that the purelines are normally less stable than diverse populations in facing stress and have limited capacity for long-term adaptation. Few varieties have been developed using pureline selection such as salt-tolerant rice varieties CSR-1, CSR-2, and CSR-3 and deep water tolerance varieties Jaladhi, Neeraja, and Hangeshwari (Meena et al. 2016).

6.2.1.2 Selection Based on Variance Created by Controlled Mating/Crossing

The selection process in the landraces erodes the genetic diversity and eventually it becomes limiting. Recombination through crossing produces a combination of different alleles in many loci resulting in a virtually infinite number of genotypes. This genetic variability is significant to develop varieties having resilience to emerging biotic and abiotic stresses.

Pedigree Breeding: Pedigree breeding involves crossing of parents to generate a segregating population followed by selection in every generation along with keeping the pedigree record until a set of derived lines having beneficial traits from both the parents are obtained. It allows breeders to identify the best line with considerable uniformity. In rice, pedigree selection has been used to incorporate salt and drought tolerance into varieties such as CSR-10, CSR-13, CSR-23, and CR Dhan-402 (Meena et al. 2016).

Shuttle Breeding: This has been highly utilized for improving rice by IRRI along with CRRI, Cuttack. Since abiotic stresses are location-specific, shuttle breeding is a highly suitable approach. In shuttle breeding, the advanced breeding materials are evaluated at two or more locations to evaluate their adaptability, and the best-suited lines will be again crossed and evaluated at a different location. The Eastern Indian Rainfed Lowland (EIRL) shuttle breeding network developed new varieties such as Bhudev, NDR-8002, and CR-2003–2 with submergence tolerance. Under the ICAR-IRRI collaborative project, salt tolerance varieties CSR-23 and CSR-27 were developed through shuttle breeding (Mishra 1994).

Backcross Breeding: This method allows breeders to transfer desirable traits from a donor to a popular acceptor parent without transferring any undesirable traits. However, the abiotic stresses regulated by multiple minor QTLs cannot be easily transferred using backcross breeding. The backcross is advantageous if one parent has more loci with favorable alleles than the other trait under consideration is governed by major genes (Dudley 1984). In wheat, Al tolerance genes have been transferred from the Carazinho variety to the Egret variety through backcross breeding (Fisher and Scott 1987). Drought tolerance has been transferred from wild rice *O. meridionalis* and *O. nivara* to cultivated *O. sativa* (Pham et al. 2006). Several backcross recombinant inbred lines (BILs) have been developed such as BILs between *O. sativa Nipponbare* and *O. rufipogon* which are used for QTL detection on drought tolerance (Thanh et al. 2011).

Pedigree Breeding: The majority of the varieties developed in self-pollinating species have been developed through the pedigree method. The pedigree method consists of crossing two parents and producing segregating generations through self-pollination and selection until a set of derived lines with superior characteristics from both parents is obtained. This method is well suited for the improvement of qualitative traits with high heritability values for which selection can be made in the early segregating generations, but it is constrained by the fact that yield is evaluated only at the end on inbred lines when the seed is available for replicated trials; hence there is little room for improvement in yield potential.

6.2.2 Limitations of Conventional Plant Breeding Techniques

- Time-Consuming Process: Traditional plant breeding is a slow process, often taking many years (sometimes decades) to develop a new variety with desired traits. On average, it takes about 12 years from the year of crossing to final release from a varietal release committee.

- Limited Genetic Variation: Conventional breeding relies on existing genetic variation within a species or closely related species. This limits the pool of traits that breeders can work with, potentially excluding beneficial traits that exist in more distantly related species or wild relatives. For certain traits such as salt tolerance, it is very difficult to find donors from primary gene pool.
- Polygenic Traits: Several key agricultural traits, such as yield, heat tolerance, drought tolerance, and disease resistance, are controlled by multiple genes (polygenic). Identifying and combining these traits through traditional breeding is complex and often inefficient.
- Unintended Linkage: When selecting for specific traits, other undesirable traits can be inadvertently selected due to genetic linkage. This is particularly observed when the donor of the target trait is a wild relative, which makes it challenging to achieve the perfect blend of traits without introducing unwanted characteristics.
- Environmental Dependency: The expression of many traits is influenced by environmental conditions, making it difficult to consistently select desired traits across different environments. Breeders need to test new varieties in multiple locations and over several years to ensure trait estimates with high confidence and select genotypes with stability and promising performance.
- Low Precision: Conventional breeding methods, such as crossbreeding and backcrossing, are less precise compared to modern molecular techniques. This can result in a large proportion of offspring that do not possess the desired traits, requiring extensive selection and backcrossing to achieve the target genotype.
- Limited by Sexual Compatibility: Conventional breeding can only occur between sexually compatible plants. This limits the introduction of beneficial traits from unrelated species or more distantly related organisms.
- Pest and Disease Resistance: As pests and pathogens evolve, conventional breeding may struggle to keep pace with the need for new resistant varieties. The time required to breed and release new resistant varieties can lag the speed of evolving threats.

These limitations have propelled the development and adoption of more advanced techniques such as marker-assisted selection (MAS), genetic modification (GM), and genome editing (e.g., CRISPR/Cas9), which can offer more precision, speed, and flexibility in plant breeding.

6.3 Molecular Breeding Techniques

6.3.1 Marker-Assisted Selection (MAS)

Marker-assisted selection holds significant importance in developing abiotic stress tolerance in crop species by enhancing the breeding process and precision in trait introgression. Through molecular markers, MAS facilitates precise selection and transfer of favorable genes and alleles from donor/wild/related species into elite crop varieties leading to a reduced number of breeding cycles, enhanced genetic gains, and expediting the release of new varieties. Along with broadening the genetic base of the improved varieties, MAS also enables the accumulation of favorable target genes/QTLs for multiple abiotic stresses. Recent advances in sequencing technologies have helped to improve the efficiency and precision of MAS in identifying loci and genetic variants associated with abiotic stress tolerance and transferring them into popular varieties.

6.3.1.1 Molecular Markers

A molecular marker is a specific sequence of DNA or RNA that can be identified and used to distinguish between different individuals, species, or even specific locations within a genome. These markers are used based on their association/linkage with the traits or the genes governing them. Some of the molecular markers are listed in Fig. 6.1.

6.3.1.2 Marker-Assisted Pedigree Selection (MAPS)

This method combines the MAS with the pedigree breeding approach. Marker-assisted pedigree selection (MAPS) facilitates the discovery of transgressive segregants along with enhancing both high heritable and low heritable traits. The utilization of polymorphic markers from F2 generation till F6 or F8 increases the precision of selection. Moreover, the polymorphic markers can be used to identify homozygous desirable genotypes among the parents early in the breeding. Consequently, the application of MAPS will reduce the number of breeding generations to develop a desired variety and number of families/lines handled per generation.

6.3.1.3 Marker-Assisted Backcross Breeding (MABB)

MABB is the most sought method for the transfer of genes governing biotic and abiotic stress tolerance along with nutritional traits. It often involves the transfer of traits from a donor variety which can be a wild relative or landrace or a related species into a well-adapted and popular between the donor and the recurrent parent (RP) to ensure maximum retention of the recurrent parent genome and recombinant selection using two markers flanking the gene of interest to ensure minimum donor parent genome in the selected lines and the concomitant linkage drag. The selected lines will then be evaluated alongside donor and recipient parents for target trait expression and the best lines will be released as new varieties.

6.3.1.4 Marker-Assisted Recurrent Selection (MARS)

Recurrent selection is a population improvement technique that uses cyclical selection and intermating of selected individuals to increase the frequency of favorable alleles in the population, allowing for recombination. Alongside evaluating breeding value, simultaneous selfing and test cross progeny testing are also performed. The complexity is reduced by using molecular markers to select genotypes. F2 or F3 families are genotypes and their respective families derived are evaluated for the desirable traits.

Classification of molecular markers

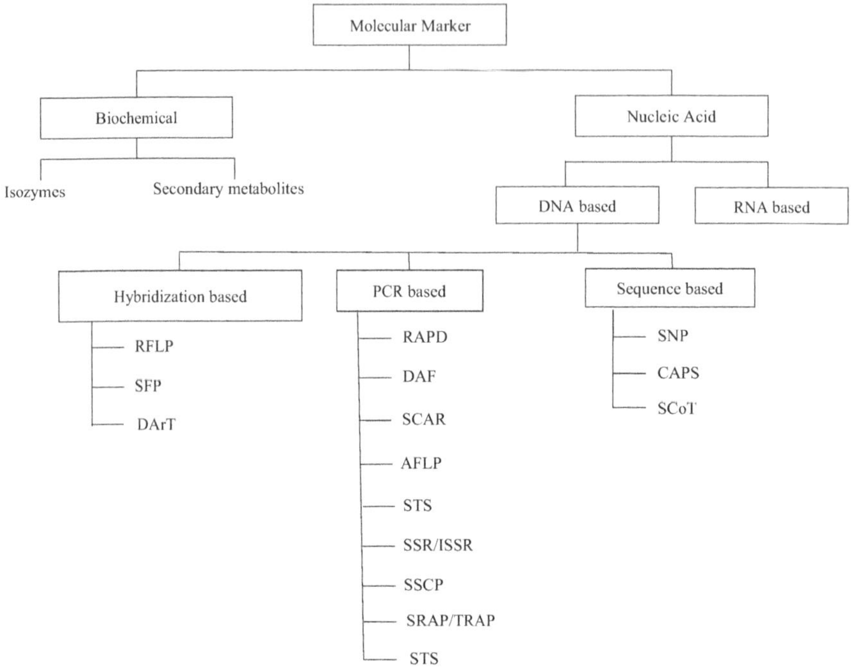

Fig. 6.1 *RFLP* restriction fragment length polymorphism, *SFP* single feature polymorphisms, *DArT* diversity arrays technology, *RAPD* randomly amplified polymorphic DNA, *DAF* DNA amplification fingerprinting, *SCAR* sequence characterized amplified region, *AFLP* amplified fragment length polymorphism, *STS* sequence-tagged site, *SSR* simple sequence repeats, *ISSR* Inter SSR, *SSCP* single-stranded conformation polymorphism, *SRAP* sequence-related amplified polymorphism, *TRAP* Target region amplified polymorphism, *SNP* single nucleotide polymorphism, *STS* sequence tagged sites, *CAPS* cleaved amplified polymorphic sequences, *ScoT* start codon targeted (SCoT) polymorphism

In MABB, the variety is developed using the backcross method. MABB incorporates the use of molecular markers at every backcross generation to select desirable lines with the trait of interest along with a limited background genome from the recipient's parent. It involves foreground selection using molecular markers that show close linkage with the gene of interest to select the lines possessing the gene under transfer, and background selection using polymorphic markers distributed across the genome.

Wheareas in MARS, F2 or F3 families are genotyped and their respective families derived are evaluated for the desirable traits. The marker effects are estimated which are then used to select progenies in the next generation which will be intermated among themselves. The progenies derived will again be genotyped and their progeny families are evaluated phenotypically and the cycle repeats (Gokidi and Bhanu 2016). MARS has been modified to include markers linked to QTLs along with other genomic markers to enhance the genetic gains (Bernardo and Charcosset 2006).

6.3.2 Case Studies of MAS in Developing Stress-Tolerant Crops

6.3.2.1 MAS for Drought

Several QTLs have been found for drought stress tolerance, and these QTLs have been exploited for marker-assisted backcrossing (MABB) to generate drought-tolerant cultivars in a few crops such as rice, wheat, chickpea, and sorghum. More than 13 drought-tolerant rice cultivars have been developed utilizing MABB (Vinod et al. 2019). QTLs like qDTY1.1, qDTY3.1, qDTY32, and qDTY12.1 are frequently used for MABB in rice. Several studies have stacked these QTLs into mega varieties of respective countries to improve their drought tolerance, such as Lalat (qDTY1.1, qDTY3.1, and qDTY12.1; Singh et al. 2022), Naveen (qDTY1.1, qDTY2.2, and qDTY4.1; Ramayya et al. 2021), Pusa Basmati 1 (qDTY1.1; Dhawan et al. 2021), and Pusa 44 (qDTY12.1, qDTY2.1, and qDTY3.1; Dwivedi et al. 2021). In sorghum, MABB has been used to transfer QTLs (Stg1, Stg2, Stg3a, Stg3b, and Stg4) for stay green types which contribute to drought tolerance in many varieties such as R16, RTx700, and Tabat (Kassahun et al. 2010; Kamal et al. 2021). In chickpea, QTLs for drought tolerance have been introgressed into Pusa 372, Pusa362, and DCP 92–3 (Bharadwaj et al. 2021) using MABB.

6.3.2.2 MAS for Salinity

A significant QTL, Saltol governing salt tolerance at the seedling stage, has been identified in rice. Accordingly *Saltol* has been introgressed into several popular indica rice varieties, including Pusa Basmati 1 (PB1), Pusa Basmati 1121 (PB1121), Pusa Basmati 1509 (PB1509), Pusa 44, Sarjoo 52, Aishwarya, and Improved Samba Mashuri from India; as well as into varieties AS996, Q5DB, and Bac Thom 7 from Vietnam; and into Binadhan 7 from Bangladesh. Molecular-assisted breeding (MABB) techniques have also been employed to introgress *Saltol* into the Russian japonica rice cultivar Novator (Usatov et al. 2015) and MS11 (Han et al. 2020).

6.3.2.3 MAS for Heat Stress

Many numbers of QTLs have been identified to be governing heat tolerance in crops, but only a handful of them have been used for MABB to develop heat-tolerant varieties. MABB was employed to harness the heat stress tolerance in wheat. Two QTLs were transferred from WH730 into elite wheat line HD2733 for anthesis and yield under stress (Amasiddha et al. 2016). Likewise, MABB has been employed to introduce heat shock proteins (*Hsp70* and *sHsp*) from the heat stress-tolerant chili genotype AVPP0702 into the sensitive variety Kulai (Usman et al. 2018). Similar introgressions have been made in rice such as QTLs (*qHTSF1.1* and *qHTSF4.1*) into Improved White Ponni which resulted in superior spikelet fertility under heat stress (Vivitha et al. 2017).

6.3.2.4 MAS for Submergence

Submergence tolerance is one of the important traits in rice. Approximately 7 million hectares of rice fields in eastern India endure medium (25–50 cm) to deep (50–100 cm) stagnant water flooding (Kuanar et al. 2017) leading to significant yield losses. The groundbreaking discovery came with the identification of Sub1 QTL from the FR13A cultivar of rice. Several PCR-based markers were also developed to introgress this QTL into elite varieties using MABB. Subsequently using MAS, many famous submergence tolerance varieties were developed such as the popular variety Swarna-Sub1, Samba-Sub1 (India and Nepal), IR64-Sub1 (India and Bangladesh), and Bahuguni Dhan-1 (Nepal), establishing submergence tolerance (*Sub1*) as a paramount example of successful marker-assisted selection.

6.3.2.5 MAS for Heavy Metal Toxicity

Heavy metals are those with density exceeding 5 g/cm^3, including metals such as Cu, Pb, Cd, Zn, Hg, Cr, Ni, among others. Modern biotechnological methods play a crucial role in enhancing plant tolerance to heavy metals. Rice is a major crop that serves as a source of cadmium (Cd). MAS has been employed to develop Cd-safe rice by establishing allelic divergence of quantitative trait loci (QTLs) affecting Cd accumulation. Several QTLs have also been identified such as *OsHMA3* (Ueno et al. 2010). Recently MAS has been used to create low-grain Cd indica lines such as *Lushansimiao* with Cd levels of <0.2 mg/kg (Sun et al. 2022).

6.3.3 Genomic Selection

Genomic selection (GS), a Marker-Assisted Selection (MAS) strategy, enables genotype selection based on the cumulative effects of all markers, irrespective of their individual magnitudes. It is a predictive selection approach in which the progeny's phenotype or breeding value is predicted using marker effects. As a result, a genomic prediction model is created utilizing regression techniques such as Bayesian regression, ridge regression, or machine learning methods such as random forest and K-Nearest Neighbors (KNN) algorithms. Phenotypic and genotypic data from a training population (TP) (usually a breeding population is used) are used to create models. Genomic Estimated Breeding Values (GEBV) are calculated using solely genotype data or marker effects from the validation set. The predicted GEBVs based on phenotype are then compared to the actual phenotype to determine the model's prediction accuracy. A prediction model with high accuracy is deemed suitable for genomic selection in the breeding population. The same markers that were utilized to genotype the training set for the prediction model are also used to genotype the BP. The genotype data are then utilized to estimate the GEBVs of the individuals in the BP. The BP undergoes genomic selection based on GEBV-derived characteristics. Genomic selection allows for a large reduction in the number of generations necessary, saving time and money while attaining higher genetic improvements.

6.3.3.1 Genomic Selection for Abiotic Stress Tolerance

This method allows breeders to identify superior candidates based on genetic prowess rather than observable features. Unlike Marker-Assisted Selection (MAS), which targets individual traits with known genetic markers, GS provides a comprehensive and efficient technique for simultaneously improving numerous quantitative traits by combining genome-wide marker data and powerful statistical models. Most stress tolerance qualities are governed by multiple polygenes with little cumulative effects and low trait heritability values, making phenotypic selection alone or in a combination with MAS ineffective in addressing these traits. In contrast, GS appears as a critical answer in this area, arguing that markers scattered across the genome have both positive and negative effects on the target phenotype. Numerous studies have integrated genomic selection into breeding programs for enhancing abiotic stress tolerance across various crops. For example, Hassen et al. (2018) demonstrated rice breeding for adaptation to alternate wetting and drying utilizing genomic prediction approaches that take into account genotype-by-environment interactions, providing a solid framework for breeding under abiotic stress conditions. Shikha et al. (2017) evaluated the performance of various prediction models in forecasting the breeding values of maize hybrids in drought-prone regions, and Bayes B outperformed the other models studied. Similarly, Bartholome et al. (2023) tested a heterogeneous rice reference panel at different salinity levels, finding substantial genotype-by-condition interactions and reporting predictive capacities ranging from 0.25 to 0.64 for morphological features and 0.05 to 0.40 for stress response indices. These findings emphasize the huge potential for genomic selection to be easily integrated into breeding programs aimed at enhancing abiotic stress tolerance in the near future.

6.4 Genetic Engineering and Biotechnology

Though MAS has been successfully utilized for the development of stress-resilient varieties, there are only QTLs that are frequently targeted. There are several stresses such as cold stress and mineral nutrient stress where MAS is yet to be utilized. There are several reasons for this, one of which is the lack of robust QTLs, such as Saltol for salt tolerance in rice, for these traits. Furthermore, the incorporation of certain genetic segments (QTLs) related to stress tolerance sometimes introduces unfavorable agronomic traits from the original parents (Ahmad et al. 2018). Added to this is the lack of sufficient recovery of the target phenotype in the MAS-derived variety compared to the donor parent. These problems can be offset to a certain extent through genetic engineering which offers a more efficient method for incorporating advantageous genes compared to traditional or molecular breeding techniques. Additionally, it would be the exclusive choice in cases where the desired genes come from species that are genetically distinct, distantly related, or from sources outside the plant kingdom like microorganisms. Genetic engineering is the process of isolating a specific gene, combining it with a suitable vector to form a recombinant DNA molecule, and then introducing that gene into the plant genome to achieve the desired function.

6.4.1 Transgenic Approach

The process of enhancing crops through the transgenic method relies on recombinant DNA technology. The initial step involves isolating the desired gene, which is then cloned into a specific vector along with necessary selectable markers. The gene construct is introduced into the plant cell, which is then subjected to specific selection criteria. Afterward, the plant cell undergoes regeneration, resulting in the development of a complete plant with the transgene integrated into its genome. Agrobacterium-mediated transformation is widely used across the world, the process of Agrobacterium-mediated plant genetic transformation necessitates the existence of two genetic components situated on the bacterial Ti-plasmid. The primary crucial element is the T-DNA, characterized by conserved 25-base pair imperfect repeats located at the ends of the T-region, known as border sequences. The second region is the virulence (vir) region, which consists of the bacterial protein machinery responsible for T-DNA processing, and transfer is encoded by at least seven major loci, namely virA, virB, virC, virD, virE, virF, and virG. The VirA and VirG proteins function as two-component regulators that stimulate the transcription of additional vir genes located on the Ti-plasmid. The VirB, VirC, VirD, VirE, and potentially VirF proteins play a role in the processing, transfer, and integration of the T-DNA from *A. tumefaciens* into a plant cell (Hoekema et al. 1983). Other methods such as microinjection, electroporation, Biolistic (gene gun) method, and chemical-based methods of gene transfer like PEG-mediated gene transfer are also being used for the transformation and development of transgenic lines. The genes governing tolerance to various abiotic stresses can be transferred through this transgenic approach.

Abiotic stress-inducible genes can be categorized into two groups:

1. Osmo-protectants, antifreeze proteins, chaperones, and detoxifying enzymes are examples of genes that produce products that directly battle stress.
2. Stress response-regulating genes, gene expression, and signal transduction pathways, such as transcription factors and protein kinases.

An Agrobacterium-mediated transformation system can be used to modify, transfer, and activate stress-inducible and regulatory genes in diverse species under stress circumstances. This mechanism causes molecular, biochemical, and physiological changes that result in increased growth, development, and yield of plants under stressful conditions (Bhatnagar-Mathur et al. 2008).

Plants employ a universal physiological survival mechanism in response to various stimuli. The process involves the buildup of low-molecular-weight organic molecules that have great solubility in water and are not toxic at large concentrations. These molecules, sometimes referred to as compatible solutes, prevent plants from stress by aiding in cellular osmotic adjustment. They help to protect cellular structures from damage and denaturation by acting as reactive oxygen species (ROS) scavengers and/or low-molecular-weight chaperones. Additionally, they play a key role in maintaining the integrity of cell membranes and stabilizing proteins. Another method for creating stress-resistant plants involves manipulating the expression of genes that encode regulatory proteins such as kinase protein or transcription factor involved in signal transduction pathways. This manipulation aims to imitate or strengthen the plant's natural adaption processes.

6.4.2 Target Molecules for Transgenic Approach in Developing Abiotic Stress Tolerance

Prior knowledge of the biosynthetic path and the modulation of key regulatory enzymes involved in the production and degradation of suitable solutes is required for genetically engineering plants to improve abiotic stress tolerance.

6.4.2.1 Proline

Osmotic adjustment is the term used to describe the regulation of solute potential (Ψs) by plants in response to salinity or drought stress. It involves increasing the number of solute particles in the plant cell and is a crucial strategy for drought tolerance in plants, and plants achieve this by accumulating compatible solutes or compatible osmolytes which are a chemically diverse and highly soluble group of organic compounds that do not disrupt cellular metabolism, even when present in high concentrations. The tobacco plants that had the moth bean (*Vigna aconitifolia*) Δ1-pyrroline-5-carboxylate synthase (*p5cs*) gene overexpressed produced 10- to 18 times more proline than the control plants. Additionally, these plants exhibited higher biomass under stressful conditions (Kishor et al. 1995). When Arabidopsis plants were genetically modified by introducing an antisense proline dehydrogenase cDNA, they showed increased levels of proline and improved ability to tolerate salt stress (up to 600 mM NaCl) and freezing temperatures (down to −27 °C) (Nanjo et al. 1999). A gene LcSAIN2 (*Leymus chinensis* salt-induced 2) from sheep grass (*Leymus chinensis*) is known to be an activator for several stress-responsive genes and is induced by various abiotic stresses, especially salt stress. The Arabidopsis plant that was genetically modified to have an increased expression of LcSAIN2 exhibited increased tolerance to salinity stress. The overexpression of the *LcSAIN2* gene led to an enhancement of four stress-responsive genes MYB2, RD26, RD29B, and RAB18 along with a greater accumulation of soluble sugars and free proline. These plants also exhibited an elevated level of osmolytes, particularly proline, and enhanced expression of additional stress-inducible genes such as *RD29B*, *RAB18*, and transcription factors *MYB2* and *RD26*. These factors are believed to be responsible for the increased ability of the plants to tolerate salt (Li et al. 2013).

6.4.2.2 LEA Proteins

Low molecular weight proteins known as "late-embryogenesis-abundant" (LEA) proteins are created during embryo maturation and seed desiccation. LEA proteins are divided into six separate groups based on sequence similarity (Wise 2003). Several abiotic stress such as salinity, drought, and low temperatures induce their transcription (Ingram and Bartels 1996). Overexpressing BnLEA4-1 cDNA resulted in drought, salt, and temperature tolerance, demonstrating its significance in abiotic stress (Dalal et al. 2009). The introduction of a novel *LEA* gene (DQ663481) derived from *Tamarix androssowii* into tobacco plants led to a decrease in the number of wilted leaves and an increase in plant height under drought conditions (Wang et al. 2006). The study conducted by Cheng et al. (2002) demonstrated that two genes from wheat, PMA80 (which encodes for the LEA group I protein) and PMA1959 (which encodes for the LEA group 2 protein), were found to enhance drought tolerance in transgenic rice. RoyChoudhury et al. (2007) found that overexpressing Rab16A, a gene from the salt-tolerant indica rice genotype Pokkali, increased salt stress tolerance in tobacco. This was achieved by increasing the production of osmolytes, delaying the onset of damage symptoms, and maintaining mineral balance. The transgenic maize plant, which had the maize Rab28 LEA gene overexpressed using the constitutive maize ubiquitin promoter, exhibited enhanced desiccation tolerance (Amara et al. 2013).

6.4.2.3 Transporter Genes

A crucial approach to enhancing tolerance to abiotic stress, particularly salt stress involves assisting plants in re-establishing homeostasis in challenging environments, ultimately restoring both ionic and osmotic balance. By controlling the production of H+ pumps, namely the vacuolar type H + -ATPase and the vacuolar pyrophosphatase (V-PPase), plants are able to maintain high concentrations of K+ in the cytosol and exclude Na + ions. By upregulating these transporters, it was possible to induce salt tolerance in plants, thereby enhancing their capacity to survive in saline environments. The tobacco seedling with increased expression of the apple vacuolar *H + -ATPase* subunit A (*MdVHA-A*) exhibited improved tolerance to drought. This was seen through the development of longer lateral roots and osmotic balance adjustments (Dong et al. 2013). Overexpressing the Na+/H+ antiporter gene *AtNHX1*, which codes for a tonoplast antiporter, in plants enhances their ability to tolerate salt in transgenic Arabidopsis plants (Apse et al. 1999). A study on the vacuolar Na1/H1 antiporter (*PgNHX1*) found in *Pennisetum glaucum*, a glycophyte known for its capacity to tolerate high levels of drought, salinity, and heat stress. The researchers also introduced this vacuolar *Na+/H+* antiporter into rice through overexpression, which led to a significant increase in salt tolerance. The transgenic rice plants exhibited enhanced root system development, initiated flowering, and produced seeds when subjected to a concentration of 150 mM NaCl (Verma et al. 2007).

6.4.2.4 Antioxidants

Abiotic stresses usually cause the production of reactive oxygen species (*ROS*), which are eliminated through two main mechanisms: direct detoxification by non-enzymatic antioxidants such as ascorbate (*ASH*), reduced glutathione (*GSH*), carotenoids, and tocopherols or through the action of antioxidative enzymes including glutathione-S-transferase (*GST*), ascorbate peroxidase (*APX*), superoxide dismutase, and glutathione peroxidase (*GPX*) (Mann et al. 2024). Several anti-oxidant enzymes have been expressed in crops like tobacco and alfaalfa against various stresses. The expression of genes encoding ascorbate peroxidase (*APX*), dehydro-ascorbate (*DHA*) reductase (*DHAR*), and copper-zinc superoxide dismutase (*CuZnSOD*) in the chloroplasts of tobacco has shown increased tolerance to oxidative stress such as paraquat and salt. The plants expressing all three genes had approximately two times higher expression of DHAR, high ratio of oxidized glutathione (*GSSG*) to reduced glutathione (*GSH*), and reduced ascorbate (*AsA*) to DHA making them more tolerant to NaCl(100 mM) compared to negative control (Lee et al. 2007). Glutathione S-transferases (GST, EC 2.5.1.18) are involved in abiotic stress response, however limited information is available regarding gene function in tomatoes. The *LeGSTU2* coding for glutathione S-transferases has been introduced into Arabidopsis plants via *Agrobacterium tumefaciens*-mediated transformation. The expression pattern revealed that *LeGTSU2* was more expressive in roots and glowers and was induced by salt, osmotic, and heat stress. Increased resistance to osmotic stress and salt was shown by the transgenic plants, indicating the beneficial role of LeGTSU2 in salt and drought tolerance.

6.4.2.5 Transcription Factor

The dehydration-responsive element/C-repeat (*DRE/CRT*) is a conserved DNA sequence of 9 base pairs, TACCGACAT that functions as a cis-acting element. It has a crucial role in conferring tolerance to both dehydration and cold conditions. Two DRE binding proteins, namely *DREB1* and *DREB2*, were identified through the yeast one-hybrid screening technique (Liu et al. 1998). Cold stress causes an increase in DREB1 expression, whereas drought and salinity cause an increase in DREB2 expression (Liu et al. 1998). The overexpression of *OsDREB1* in *Oryzae sativa* and *ZmDREB1A* in *Zea mays* has been found to enhance tolerance to salt, drought, and cold stress (Ito et al. 2006). Stress causes plants to release NAC transcription factors, which have been shown to have a significant influence on development. The NAC protein is distinguished by the existence of a remarkably conserved domain known as the NAC domain (Kikuchi et al. 2000; Nakashima et al. 2012). By overexpressing SNAC1 in rice, the drought resistance of the plant is significantly improved during the reproductive stage, without any alterations in its physical characteristics or yield. The transgenic rice exhibited heightened responsiveness to abscisic acid and decreased transpiration rate due to stomatal closure (Hu et al. 2006). In the plant kingdom, MYB transcription factors are a significant class that is essential for controlling the expression of genes that react to stress. MYB proteins are categorized into three subfamilies, namely *MYBR1R2R3*, *MYBR2R3*, and MYB related, based on the occurrence of repetitions in their DNA-binding domain (Stracke et al. 2001). Arabidopsis plants overexpressing *TaMYB73* exhibited an enhanced ability to tolerate high levels of salinity caused by NaCl, LiCl, and KCl. Furthermore, compared to the wild-type plants, the transgenic plants showed a greater degree of germination capability (He et al. 2012).

6.4.3 Genome Editing Technologies

Genome editing facilitates precise and desirable modifications in an organism's genome by introducing point mutations, including insertions or deletions (indels) of specific nucleotides, and modifying native sequences or replacing them by recruiting specific nucleases. Over the years, several tools of genome editing have been developed to achieve these modifications, such as meganucleases, zinc-finger nucleases (ZFNs), transcription activator-like effector nucleases (TALENs), and CRISPR–Cas9, with CRISPR/Cas9 being the most recent and popular technique. This popularity is attributable to CRISPR/Cas9's ability to induce desired mutations efficiently, cheaply, quickly, and accurately. This technique is highly versatile, as it can also be used for applications such as the repression of target genes and their epigenetic modifications. Furthermore, by using multiple guide RNA cassettes within a single CRISPR/Cas9 construct, several genes can be targeted simultaneously. This tolerance capability is particularly relevant for abiotic stress, where resilience is controlled by multiple genes involved in different stress signaling pathways. There are several reviews that provide insightful information on CRISPR/Cas9, its mechanism, and its modifications. The succeeding sections elaborate on the successful application of this technique for inciting resilience for abiotic stresses in crop plants.

6.4.3.1 Drought Stress

CRISPR/Cas9-mediated frameshift mutation in *OsERA1* (enhanced response to ABA1) resulted in oseral mutants showing enhanced response to ABA as well as drought stress through increased stomatal regulation and primary root growth (Ogata et al. 2020). The role of OsSAPK2 (Osmotic stress/ABA-activated protein kinase 2) has been demonstrated by the use of CRISPR/Cas9. The *SAPK2* mutants so generated showed insensitive phenotype toward ABA during germination and post germination stages and increased drought and ROS stress sensitivity (Lou et al. 2017). Thus, SAPK2 was found to confer tolerance to drought by promoting stomatal closure and osmotic adjustment, inducing the expression of stress-responsive genes and increasing ROS scavenging. CRISPR/Cas9-induced deletion mutants of *OsDST* gene, encoding a zinc finger transcription factor, resulted in increased leaf width and reduced stomatal density. This was attributed to the downregulation of several genes involved in stomatal development such as SPCH1, MUTE, and ICE1 (Kumar et al. 2020). CRISPR/Cas9-mediated overexpression of *OsNAC14* in rice resulted in a greater number of panicles and grain filling rate and drought tolerance. This was mostly due to the overexpression of defense-related genes and DNA repair genes in the mutants. Further, the drought tolerance of these mutants was observed at both vegetative and reproductive stages (Shim et al. 2018). Overexpression of *OsERF83*, an AP2/ERF transcription factor through CRISPR/Cas9, has significantly enhanced the drought tolerance of mutants by providing protection to photosystems I and II from drought stress. Through CSRISPR/Cas9, it was elucidated that *OsGPX*1 positively regulates osmotic tolerance in rice by mediating the oxidation of bZIP68, a transcription factor (Zhou et al. 2022). *OsSAP*

(Senescence-associated protein) was demonstrated to be a positive regulator of drought tolerance in rice where CRISPR/Cas9-edited mutants showed reduced growth and survival under drought stress conditions than the wild counterparts (Park et al. 2022). Though two drought-related genes *TaDREB2* and *TaERF3* were targeted for CRISPR/Cas9, the objective of the study was mainly to demonstrate the applicability of technology in polyploids such as wheat using protoplasts. Through CRISPR/Cas9 knockout, it was found that the 108 bp deletion in the promoter region of TaNAC071-A alters its expression and the so obtained mutants are drought sensitive. This deletion contains two cis-regulator elements interacting with *TaMYBL1*, which enhances the expression of TaNAC071-A and confers drought tolerance in wheat (Mao et al. 2022). Using CRISPR/Cas9 a new allele was created for *ARGOS8*, a negative regulator of ethylene responses in maize and associated with drought tolerance in maize. This is achieved by either inserting or replacing the native promoter of *RGOS8* with the *GOS2* promoter to elevate the expression of the gene compared to the native alleles (Shi et al. 2017). Through the genome-wide association study (GWAS) of 627 RNA-Seq data sets (drought stress) of 224 maize genotypes, a gene, abh2, encoding an abscisic acid 8′-hydroxylase (*ABAox*), associated with the oxidative degradation of ABA was identified. The maize mutants obtained through CRISPR/Cas9 knockout of this gene showed significantly higher survival under drought stress than the wild types validating its role in drought tolerance.

Through CRISPR/Cas9, two mutants, gain of function and loss of function (quadruple mutant) mutants, were created for BnaA6.RGA, which codes for DELLA proteins, negative regulator of GA singalong (Wu et al. 2020). The gain of function mutants showed increased tolerance to drought and stomatal hypersensitivity to ABA, whereas loss of function mutants showed reduced drought tolerance and ABA sensitivity. The relevance of GmMYB118 in drought responses of soybean was elucidated through CRIPSR/Cas9 editing. The mutants showed lesser survival rates under drought stress attributed to lesser production and accumulation of osmolytes such as proline, chlorophyll degradation, and poor ROS scavenging. In a seminal study to demonstrate the application of CRISPR/Cas9 genome editing in chickpea, two drought-responsive genes 4-coumarate ligase (4CL) and Reveille 7 (RVE7), the former associated with lignin synthesis and the latter being a transcription factor regulating circadian rhythm, were successfully targeted for editing. A photoperiod-associated transcription factor CYCLING DOF FACTOR 1 (*StCDF1*) in potatoes regulates tuber formation. This is in turn under the regulation of long noncoding RNA, *StFLORE*. Both of these also affect the drought response in potatoes. CRISPR/Cas9-edited plants for *StFLORE* showed increased sensitivity to moisture stress. Conversely, its overexpression enhanced drought tolerance. Thus, the role of *StCDF1* and *StFLORE* in drought and tuber development regulation in potatoes could be elucidated through genome editing. The role of another MAPK transcription factor, *SlMAPK3* in drought stress in tomatoes was elaborated using CRISPR/Cas9. The mutant slmpk3 showed drought sensitivity in terms of high ROS accumulation and membrane damage. The mutants exhibited altered expression of several genes susceptible to drought, including SlLOX, SlDREB, and SlGST. Similar to this, CRIPSR/Cas9 was used to show how the nonexpressor of pathogenesis-related gene 1 (NPR1), a regulator of plant defense responses, functions in tomato drought responses. The knockout mutants slnpr1 showed reduced drought tolerance due to increased membrane injury, ROS production, and induction of various drought-related genes as in the previous study. Reducing the activity of gibberellin in tomatoes can inhibit leaf expansion and reduce stomatal conductance and thus can contribute to drought tolerance. This was demonstrated by editing *GID1* (gibberellin-insensitive dwarf1), a GA receptor. In 2020, Illouz-Eliaz found that the single mutants exhibited drought tolerance, xylem vessel proliferation, and whole-plant transpiration. Lateral organ boundaries domain (LBD) containing gene, LBD40 was shown to be a negative regulator of drought tolerance, in addition to regulating lateral organ development. The CRISPR/Cas9 knockout mutants retained water under drought, thus showing only slight wilting symptoms compared to overexpressed lines.

In addition to drought tolerance, by targeting genes such as Semi-rolled leaf1 and 2 (SRL1 and SRL2), drought avoidance mechanisms such as leaf rolling were induced in the plants. For instance, rice mutants for these genes showed decreased stomatal number conductance with the concomitant decrease in transpiration and drought tolerance. Moreover, the mutant plants displayed low lipid peroxidation and higher ABA content and ROS scavenging enzymes. At the proteomic level, a total of 377 proteins showed differential accumulation in mutant and wild-type plants.

6.4.3.2 Salinity Stress

One of the key tolerance mechanisms is the maintenance of ion homeostasis in the presence of salinity, and many ion transporters help with this. Several crops have used CRISPR/Cas9 to determine their role in this respect. Salt sensitivity in barley was caused by the CRISPR/Cas9 knockout of HVP10, vacuolar H + -pumping pyrophosphatase, which led to decreased growth, elevated shoot Na + content, xylem Na + loading, and transportation. The loss of function in ZmHKT1 caused by CRISPR/Cas9 resulted in increased Na + buildup in the xylem sap, demonstrating its role in Na + exclusion from the xylem sap in maize (Zhang et al. 2018). The role of SlSOS1 in salt tolerance in tomatoes was investigated using CSRIPSR/Cas9 editing. Two knockout mutants, slsos1–1 and slsos1–2, displayed increased Na + buildup and lower K+ in roots, indicating the importance of SlSOS1 in ion homeostasis in tomatoes under salt stress (Wang et al. 2021). NHX (Na+/H+ exchanger) are membrane-localized cation antiporters that regulate ion homeostasis under salt stress. In soybean, one such gene, NHX5, which is located in the Golgi complex, was knocked out, and the resulting mutants had shorter roots and less fresh weight than wild types (Sun et al. 2021). Similarly, using CRISPR/Cas9 to knock out SlHAK20, a clade IV HAK/KUP/KT transporter, resulted in enhanced Na + buildup in the mutants' roots, which was also found in the shoot after 1 week of salinity treatment. However, the K+ content was similar across mutant and wild types. This demonstrates that SlHAK20 regulates Na + levels in roots and inhibits its transit from the root to the shoot.

Auxin, cytokinin, ABA, ethylene, and other plant growth regulators play important roles in modifying plant responses to a variety of abiotic stressors, including salinity. CRISPR/Cas9 was used to show how these growth regulators affect different components of stress signal transduction pathways. The knockout of OsDOF15, which interacts directly with the promotor of OsACS1 and suppresses ethylene production, resulted in decreased primary root elongation under salt stress. This demonstrates that ethylene-induced root inhibition under salt stress is mediated by OsDOF 15. SAUR (small auxin up RNA) genes are among the least well-studied auxin-responsive gene families in plants. Of these genes, the SAUR41 subfamily is strongly stimulated by ABA. A quadruple mutant saur41/40/71/72, an auxin-responsive family member triggered by ABA, created in Arabidopsis via CRISPR/Cas9 editing, was hypersensitive to salt during seedling establishment, which was observed to decrease when driven by an ABA-responsive promote. CRISPR/Cas9-targeted alteration of OsRR2, a B-type response regulator transcription factor involved in cytokinin signaling, led to improved seedling salt tolerance in rice. Mutants of this gene were also created in the background of a GMS (genic male sterility) and restorer line using CRISPR/Cas9 to create a salt-tolerant rice hybrid, known as the "third-generation hybrid rice system." Two additional response regulator genes, OsRR9 and OsRR10, have also been shown to have a deleterious effect on salt stress in rice. The double mutant osrr9/osrr10 generated using CRISPR/Cas9 showed higher salt tolerance than the wild type, owing principally to the overexpression of several high-affinity potassium transporters and, as a result, optimal K+ homeostasis. In soybean, knocking down ABA-inducted transcription repressors (AITRs), which are feedback regulators of ABA signaling, resulted in improved salt tolerance. The gmaitr double (gmaitr36) and quintuple (gmaitr23456) mutants outperformed wild types in terms of germination, seedling growth, plant height, pods per plant, and test weight under salt. Editing OsVDE, violaxanthin de-epoxidase, an enzyme that plays an important part in abscisic acid production, revealed that it is a negative regulator of salt tolerance in rice. The mutant lines accumulated more ABA, had enhanced stomatal closure, and survived better under salt stress. Similar results were reported when an auxin response factor (ARF) was altered in tomato SlARF4 with CRIPSR/Cas9 (Bouzroudet al. 2020).

Transcription factors are critical to plant abiotic stress responses, and CRISPR/Cas9 was successfully used to determine the involvement of several transcription factors in plant salt tolerance. In rice, OsRAV2, an AP2/ERF domain-containing RAV (related to ABI3/VP1) transcription factor, was found to be salinity sensitive. GT-1, a regulator region inside this gene situated −664 from the transcription start site, was found to be salt responsive using CRIPSR/Cas9. The role of two of these in rice, SAPK1 and SAPK2, was discovered using CRIPSR/Cas9-based genome editing. Under salt stress, mutant plants with SAPK1 and SAPK2 alterations displayed lower germination, growth, chlorophyll content, chlorotic symptoms, and overall survival (Lou et al. 2018). A mutant with a 366 bp deletion was produced by targeting OsDST, a zinc finger transcription factor, in rice variety MTU1010 using CRISPR/Cas9, resulting in high seedling stage salinity tolerance as well as moderate tolerance to osmotic stress (Kumar et al. 2020). OsGTγ-2, a trihelix transcription factor, interacts with GT-1 and promotes salt tolerance in rice. Knocking down OsGTγ-2 with CRISPR/Cas9 provided evidence. The resulting mutants (osgtγ-2) demonstrated salt hypersensitivity. This TF even interacts with multiple K transporters, helping to maintain ion homeostasis. NAC is a transcription factor that has been linked to salt stress signaling on numerous occasions. Bo et al. (2019) altered OsNAC041, and the resulting mutants demonstrated increased plant height relative to wild types, but with enhanced salt sensitivity. The mutants showed a considerable increase in salt tolerance at the germination and vegetative phases after selectively editing the negative response region of the hybrid proline-rich protein1 (HyPRP1) expressing genes in tomato SlHyPRP1 using CRISPR/Cas9. CRISPR/Cas9-based deletion of SlABIG1, a transcription factor from the HD-ZIP II subfamily, resulted in improved tomato salt tolerance by maintaining optimal chlorophyll content, photosynthesis, proline, and lowered ROS and Na + buildup (Ding et al. 2022). Another negative regulator of salt stress has been identified using CRISPR/Cas9, namely OsbHLH024, a basic helix-loop-helix (bHLH) transcription factor. A deletion mutant (osbhlh024) demonstrated higher growth, chlorophyll content, effective antioxidant activity, and improved ion homeostasis (Alam et al. 2022). This was also attributable to the upregulation of several ion transporter genes, including OsSOS1, OsHKT1, 3, and OsHAK7.

6.4.3.3 Temperature Stress

Global climate change is resulting in temperature extremes and the consequent abiotic stress—heat stress and cold stress. Several genes contribute to plant tolerance to temperature stresses. The CRISPR/Cas9 technique has been utilized to either elucidate the role of genes in heat/cold tolerance or the generation of tolerant genotypes. CRISPR/Cas9 genome editing of OsNAC006, a NAC transcription factor, resulted in mutants that are drought and heat sensitive. Two other NAC transcription factors, ONAC127 and ONAC129, were characterized using CRISPR/Cas9. The resultant knockout mutants displayed shriveled grains which were exacerbated further under heat stress (Ren et al. 2021). This was due to their regulation of assimilate transport particularly under heat stress conditions. Loss of function mutants of another NAC transcription factor, OsNTL3, ntl3-1 and ntl3-2 generated through CRISPR/Cas9 showed heat sensitivity. However, mutants with a truncated form of OsNLT3 (OsNTL3ΔC), excluding the transmembrane domain, showed heat stress tolerance at the seedling stage. CRISPR/Cas9 editing of a negative regulator of salt tolerance, hybrid proline-rich protein 1 (HyPRP1), resulted in mutants with a lack of functional domains, proline-rich domain (PRD), and eight cysteine-motif (8CM). The 8CM mutants of two tomato cultivars, HK and 15 T01, showed moderate heat stress tolerance in addition to tolerance to other stresses (Tran et al. 2023). Genome editing was utilized to establish the role of a calcium-dependent protein kinase (CPKs), CPK28 in heat stress responses of tomato. CRISPR/Cas9-edited mutant cpk28 showed heat sensitivity due to increased ROS accumulation and scavenging (Hu et al. 2021). It was further found that this is due to the interaction between CPK28 and APX2. Similarly, the role of CDPK7 in heat stress tolerance in maize has been shown using CRISPR/Cas9. The mutants of ZmCDPK7 showed lower photosynthetic rates, ROS scavenging, and ultimately heat sensitivity. CRIPSR/Cas9 editing of another transcription factor, SlMAPK3, resulted in mutants with tolerance to heat stress reflected as less leaf wilting, membrane damage, efficient ROS scavenging, and heat shock response established the role of MAPK3 as a negative regulator of heat stress tolerance in tomato (Yu et al. 2019). Similarly in cucumber, the characterization of *CsSPT*, a basic helix-loop-helix transcription factor (bHLH), was done using CRISPR/Cas9. The Csspt mutants are sensitive to heat stress attributed to increased ROS, membrane damage, and disrupted photosynthetic machinery (Liang et al. 2024). At the gene level, though HSPs and HSFs showed similar upregulation in both mutant and wild genotypes, certain transcription factors such as bHLH, NAC, and bZIP showed downregulation.

In contrast to editing of NAC transcription factors *OsNAC006*, ONAC127, and ONAC129, CRISPR/Cas9 editing of OsNAC050 in rice resulted in mutants that displayed increased cold stress tolerance (Wang et al. 2023). Similarly, genome editing of an MYB transcription factor, OsMYB30, at two sites, resulted in enhanced tolerance to cold in mutants (Zeng et al. 2020). It was observed that OsMYB30 negatively regulates β-amylase genes by interacting with OsJAZ9 and affects the synthesis and breakdown of maltose, as an osmolyte conferring cold tolerance. Another gene, OsPRP1, encoding proline-rich proteins, was edited through CRISPR/Cas9 and the resulting mutants showed increased cold sensitivity due to poor ROS scavenging (Nawaz et al. 2019). CRISPR/Cas9-mediated knockout of OsAnn3, a rice annexin gene resulted in mutants with cold sensitivity reflecting the role of plant annexins in cold tolerance (Shen et al. 2017). CRISPR/Cas9 genome editing of SlNPR1, a master regulator of plant disease resistance, resulted in slnpr1 mutants which showed chilling injury symptoms with increased ROS accumulation and decreased expression of SlICE1 and SlCBF1 (Shu et al. 2023). Furthermore, the synthesis of ferulic acid (FA) was affected due to the decrease in the expression levels of the corresponding genes in the mutants. The role of ICE1 in drought and cold stress was also demonstrated in strawberries through CRIPSR/Cas9 editing. The role of C-repeat binding factors (CBFs) in cold response in tomatoes has been elucidated using slcbf1 mutants generated through the CRISPR/Cas9 system (Li et al. 2018). The mutants showed symptoms of chilling injury along with membrane injury and higher ROS accumulation and their poor scavenging. Cold stress induces vacuolar acid invertase (VInv) activity in potato tubers where sucrose is cleaved to hexoses, thus providing osmo-protection, known as cold-induced sweetening (CIS) response. Genome editing of VInv through CRSIPR/Cas9 showed this chilling response was still maintained through an alternative route, i.e., through the raffinose pathway, thus increasing the levels of myo-inositol and cold tolerance (Teper-Bamnolker et al. 2023). CRISPR/Cas9 genome editing of PAT1 (phytochrome A signal transduction1, VaPAT1) in grapevine resulted in a modified protein with the loss of C-terminus in addition to reduced levels of VaLOX3 and JA levels when treated with cold stress. This indicates the role of PAT1 in cold responses by regulating jasmonic acid biosynthesis (Wang et al. 2021).

6.5 Omics Technologies in Crop Improvement

The ever-growing advances in novel technologies and analytical tools for studying and analyzing, genes, transcripts, proteins, and metabolites have led to the development of omics technology. It is a multidisciplinary field that functions for the exhaustive analysis of biological data at various levels of biological organization (Talukdar and Sinjushin 2015; Yang et al. 2021). Omics technologies (genomics, transcriptomics, proteomics, metabolomics, etc.) have become an integral part of crop improvement programs around the globe due to their potential to offer a valuable understanding of the genetic and molecular mechanisms governing various traits in crops including abiotic stress tolerance (Qi et al. 2023). These technologies are capable of the analysis of the generated biological information at the genome, transcriptome, proteome, and metabolome levels, leading to a better understanding of complex traits and their interactions with the environment (Bashir et al. 2023). These technologies hold promise for identifying genes and gene products linked to specific traits and for elucidating functional relationships between genotype and phenotype. They also enable system-wide analyses from genome to phenome, introgression of superior alleles, accurate trait mapping, and the cloning of major genes associated with complex traits (Yang et al. 2021; Cortes et al. 2023). The integration of omics technologies in crop research and breeding holds great promise for accelerating the development of robust and high-throughput marker systems, leading to new opportunities and strategies for improving crop plants against various abiotic stresses.

6.5.1 Genomics

With the advent of low-cost sequencing technologies, integration of genomic data into breeding pipelines became a routine activity and genomics-assisted breeding (GAB) has aided in developing tailor-made crop varieties with desirable agromorphological features and superior stress tolerance (Wani et al. 2020; Varshney et al. 2021). A complete sequence information of the crop genome can be generated through whole-genome sequencing (WGS) technology.

Rice became the first crop to have its genome sequenced (International Rice Genome Sequencing Project 2005) followed by a diverse array of major and minor crops including cereals like maize (Schnable et al. 2009) and bread wheat (Zimin et al. 2017), pulses such as pigeon pea (Singh et al. 2022) and chickpea (Varshney et al. 2013), and oil seeds like groundnut (Bertioli et al. 2019) and soybean (Schmutz et al. 2010). This comprehensive genomic information allows researchers to study genetic variations and understand the genetic basis of several traits (Tanaka et al. 2020). This information can be further used to identify stress tolerance genes which is pivotal for selecting previously reported candidate genes for focused breeding programs and the novel genomic regions can be mapped in a diverse germplasm panel through a popular approach, known as genome-wide association studies (GWAS) (Abhijith et al. 2022). GWAS helps in mapping genomic regions associated with complex traits like abiotic stress tolerance in crops (Ravikiran et al. 2022). This approach also helps in reporting previously reported candidate genes controlling stress tolerance if the marker-trait associations (MTA) are in linkage disequilibrium (LD) with the candidate gene.

6.5.2 Transcriptomics

The transcriptome is the whole set of RNA transcripts produced by an organism's genome, which can help researchers analyze gene expression patterns, regulatory networks, and molecular mechanisms underlying stress tolerance (Mann et al. 2019; Raza et al. 2021). Gene expression profiling under various abiotic stress levels has revealed molecular responses of plants to environmental challenges. In the case of salt stress and drought, transcriptome studies have helped in identifying genes involved in water uptake, retention, and stress signaling pathways (Luo et al. 2020; Xia et al. 2020; Yang et al. 2020; Zhang et al. 2021; Mann et al. 2021). In case of salinity, transcriptomic analysis under salinity stress reveals several genes with roles in ion homeostasis and osmotic regulation, and provides keen information into salt stress tolerance mechanisms (Hu et al. 2017; Fu et al. 2019; Mann et al. 2019; Li et al. 2024). Similarly, studies have untangled heat shock proteins, cold-responsive genes, and other factors that mediate plant responses to stress-related temperature fluctuations (Shi et al. 2017; He et al. 2012; Peng et al. 2023). This approach can help in identifying key regulators of abiotic stress tolerance in crops by analyzing co-expression patterns and gene interactions, which will eventually unravel the complex molecular networks that govern plant stress responses.

6.5.3 Proteomics and Metabolomics

Proteomics and metabolomics are two other important omics disciplines that complement transcriptomics and genomics in studying the molecular mechanisms regulating stress responses and trait expression in crop plants. Proteomics is another branch of omics that studies the total protein complement in an organism and this approach generates information on the levels of protein expression, protein–protein interactions, and post-translational modifications (Alugoju and Tencomnao 2023). Profiling of protein expression under stress conditions can identify stress-responsive proteins, signaling molecules, and metabolic enzymes that play critical roles in stress tolerance and adaptation of crops to changing environments (Mann et al. 2015; Ali et al. 2023; Feng et al. 2023). Whereas the focus of metabolomics lies on studying the complete set of small molecules (metabolites) in the plant, generating valuable details of metabolic pathways, biochemical reactions, and metabolic signatures associated with specific physiological conditions (Shen et al. 2023). This approach also can identify stress-responsive metabolites, metabolic pathways, and biochemical markers that are associated with stress tolerance (Mann et al. 2018; Shunkao et al. 2023). One of such example where a combination of proteomics and metabolomics analyses of two spring-wheat cultivars under drought stress revealed that the drought-susceptible Bahar experienced significant protein and metabolic disruptions, particularly in photosynthesis and nitrogen metabolism, while the drought-tolerant Kavir showed only minor biochemical changes, providing insights into mechanisms of drought tolerance (Michaletti et al. 2018).

6.6 Integrative Approaches and Systems Biology

6.6.1 Multi-omics Integration

It is of utmost importance to analyze the multiple omics data, which are generated separately, and then reanalyze and interpret them collectively (Chao et al. 2024). This integration can shed light on the underlying biological processes and interactions that influence stress tolerance.

One such example is the case of jojoba. To investigate the molecular mechanisms underlying heat tolerance, physiological and multi-omics techniques identified key proteins, genes, and metabolites involved in the heat stress response. This research revealed the roles of protein folding, photosynthesis, signal pathways, and metabolic adjustments in enhancing heat tolerance (Zheng et al. 2022). The major benefits of this integrative approach are discussed below.

6.6.1.1 Identification of Key Regulators

Integrating multiple omics data sets allows for the identification of key regulators, biomarkers, and candidate genes associated with specific traits or stress responses in crops (RoyRoychowdhury et al. 2023). This approach can analyze correlations and interactions across omics layers, and thereby pinpoint critical genes, proteins, and metabolites that influence stress tolerance.

6.6.1.2 Prediction of Phenotypic Traits

This approach also facilitates the prediction of phenotypic traits and the identification of molecular markers linked to desirable traits in crops (Wang et al. 2023). The combined analysis of multi-omics data could help in developing predictive models for trait expression, stress tolerance, and crop performance under varying environmental conditions.

6.6.1.3 Precision Breeding and Crop Improvement

The integration of multi-omics data provides a foundation for precision breeding and targeted crop improvement strategies. Here, breeders can identify precise targets for genome editing and stacking of favorable alleles to introduce specific traits, conferring multi-stress tolerance, into elite crop varieties without any linkage drag.

6.6.1.4 Data Mining and Knowledge Discovery

Multi-omics integration involves advanced data mining and bioinformatics analyses to extract meaningful insights from complex omics data sets (Chao et al. 2024). This can aid in uncovering hidden patterns, regulatory networks, and functional relationships that drive the expression of stress tolerance with a specific response to environmental effects.

6.6.2 Systems Biology Models

Systems biology provides a powerful framework for unraveling the complexity of plant stress responses and identifying potential targets for improving stress tolerance (Chawla et al. 2011). Advances in computational modeling and network analysis have opened more avenues for uncovering key regulators and pathways that can be targeted to increase the resilience of plants to various environmental stresses.

Predictive modeling and simulation of crop responses to stress play a crucial role in understanding and improving crop resilience in challenging environmental conditions (Soltani et al. 2023). Predictive models are developed based on biological knowledge, experimental data, and mathematical algorithms to simulate crop responses to different stress factors. These models can range from simple mathematical equations to complex computational simulations that capture the dynamics of crop stress responses. These models can generate various abiotic stress scenarios to predict the way plants may respond to different environmental conditions. These models also can be used to optimize various crop management practices, such as irrigation scheduling and fertilizer application to minimize stress impact on crops.

6.7 Translating Laboratory Research

6.7.1 Field to Lab in Crop Improvement

Translating laboratory research to the field of plant breeding is a multifaceted process that involves integrating advancements in scientific research with practical applications to improve crop performance, resilience, and productivity (Langensiepen et al. 2020). This process is crucial for addressing global challenges such as food security, climate change, and sustainable agriculture. Two essential techniques in contemporary plant breeding are genomic selection and marker-assisted breeding (Nadeem et al. 2018). Genetic indicators linked to desired characteristics including disease resistance, drought tolerance, and higher production are found in lab settings (Passioura 2012). By enabling the selection of better plants at the seedling stage and eliminating the need for lengthy field trials, this speeds up the breeding process. Plant breeding has undergone a revolutionary change because of the CRISPR-Cas9 technology, which allows for precise genome editing (Dutta et al. 2024a). To impart advantageous features, scientists can introduce or eliminate particular genes in the lab. CRISPR has been utilized, for instance, to create rice cultivars with increased nutritional value or resistance to bacterial blight (Kumar et al. 2020). High-throughput phenotyping, in which many plants are automatically tested for particular features, is a common practice in laboratory research (Li et al. 2021).

Putting this into practice in the field will require creating transportable phenotyping instruments that are adaptable to various settings (Dutta et al. 2024b). Drones using multispectral sensors, for instance, can track the growth and health of crops across vast areas, giving breeding programs important information (Dutta et al. 2023). In laboratory investigations, environmental conditions are frequently simulated in order to examine plant responses. Growth chambers and greenhouses are examples of controlled environment facilities that provide researchers the ability to adjust light, humidity, and temperature (Maity et al. 2023). Extensive field experiments conducted in various locations and growing seasons are necessary to validate the reported advantages in natural and changeable settings before using these results in the field (Vanino et al. 2022). To account for the potential effects of genotype–environment interactions on plant performance, this step is crucial. The use of bioinformatics and data integration is vital to modern plant breeding (Dutta et al. 2024c). Large volumes of data are produced by environmental monitoring, phenotyping, and genetic sequencing in laboratory research. To translate these data to the field, analysis and interpretation of the data must be done using bioinformatics techniques to spot trends and correlations that might help with breeding decisions (Kusmec et al. 2021). By predicting which plant kinds are most likely to flourish under particular conditions, machine learning algorithms may help breeders choose the best candidates for field trials (Sundaramoorthi and Dong 2022). For laboratory research to be effectively applied in the field, geneticists, agronomists, ecologists, and farmers must work together (Beres et al. 2020). For example, several climate-smart varieties of rice (Nair and Shylaraj 2021), maize (Gazal et al. 2018), wheat (Gahlaut et al. 2023), and pulses (Sheoran et al. 2022) are derived using the MABB program. To ensure that breeding programs are in line with actual agricultural demands, farmers are required to provide feedback on the performance and practicality of new plant types (Sawarkar et al. 2023). Managing legal frameworks and social concerns are additional challenges in releasing laboratory-bred plants into the field. To guarantee safety and effectiveness, new breeding methods and genetically modified organisms (GMOs) are governed by strict laws (Pramanik et al. 2024). Another important component is public acceptability, which necessitates open information about the advantages and disadvantages of novel plant species (Siddiqui et al. 2022). In summary, applying lab research to the area of plant breeding is a challenging but crucial process that uses discoveries in science to create crop types that can adapt to a changing environment and an expanding population. This procedure combines phenotyping, data analysis, environmental modeling, genetic techniques, and multidisciplinary cooperation to guarantee that research findings become practical agricultural advancements.

Plant types that are more resistant to pressures brought on by climate change, such as heat, drought, and salt, can be created in laboratories (Das et al. 2023). Through the identification and manipulation of genes linked to stress tolerance, scientists can engineer plants that are more resilient to upcoming environmental challenges (Das et al. 2022). The adaptation and performance of these varieties across various climatic zones are ensured by field testing in a variety of conditions, which contributes to the security of food production in the face of climate variability. Researchers concentrate on characteristics such as nitrogen-use efficiency and insect resistance that lessen the requirement for chemical inputs such as fertilizers and pesticides (Pergner and Lippert 2023). Crops that have minimal environmental effects while maintaining high yields can be produced by breeding for these characteristics. To evaluate these novel varieties' ecological effects and long-term sustainability and make sure they support sustainable agricultural techniques, field validation is essential. Improving the nutritional quality of crops is another important aspect of modern plant breeding (Swarup et al. 2021). In the lab, biofortification strategies can enhance the levels of vitamins, minerals, and other essential nutrients in staple crops (Dutta et al. 2021). These biofortified crops undergo rigorous field trials to evaluate their agronomic performance and nutritional benefits, ensuring they can effectively address malnutrition and improve public health (Dutta et al. 2020; Ramadas et al. 2020). The population of microorganisms associated with plants, known as the plant microbiome, is the subject of growing attention in laboratory studies (Santoyo 2022). By comprehending how these microorganisms affect plant growth, health, and stress resilience, crop types that maximize advantageous microbial interactions may be developed. To promote soil health and sustainable crop production, field trials are necessary to ascertain how these interactions appear in various soil types and agricultural techniques (Tahat et al. 2020). Precision agriculture has advanced to the point that real-time data from sensors and GPS-guided equipment enable exact application of herbicides, fertilizers, and water. These technologies are developed and refined through laboratory study, and subsequently put to the test and deployed in the field. Plant breeding and precision agriculture work together to maximize agricultural production while reducing resource consumption and environmental impact. Critical input from field implementation helps to shape ongoing research. Researchers are able to generate superior varieties and enhance breeding programs by using data from real-world performance. The ongoing feedback loop that exists between the field and the lab guarantees that plant breeding is adaptable to new problems and changing demands in agriculture (Sarkar et al. 2024). In conclusion, the process of transferring laboratory research into the world of plant breeding is complex and dynamic, including many different scientific, technological, and practical factors. To create crop varieties that satisfy the demands of contemporary agriculture and support global food security, a comprehensive strategy combining cutting-edge research, field validation, sustainable practices, and stakeholder participation is needed.

6.7.2 Precision Agriculture and Digital Tools in Plant Breeding

Precision agriculture and digital tools have revolutionized plant breeding by enabling more precise, data-driven approaches to crop management and improvement. These advancements help breeders and farmers optimize resources, improve yields, and develop resilient crop varieties. Large-scale data gathering and analysis are essential to precision agriculture (Júnior et al. 2024). GPS, remote sensing, and geographic information systems (GIS) are examples of digital instruments that make accurate field mapping and crop status monitoring possible (Rajabpour and Yarahmadi 2024). Drones and satellites may produce high-resolution photos that offer comprehensive details on plant health, soil properties, and surrounding situations. Breeders may use these data to pinpoint areas for improvement and gain a better understanding of how various factors impact plant development (von Nonn et al. 2024). High-throughput phenotyping devices evaluate plant attributes quickly and precisely by utilizing sensors and imaging technology (Kumari et al. 2024). Breeding programs become more reliable when measurements are made with greater accuracy and reproducibility, thanks to digital phenotyping instruments (Araus et al. 2022). An all-encompassing picture of plant performance is made possible by the integration of genetic, phenotypic, and environmental data made easier by digital technologies (Akhtar et al. 2024). These integrated data may be analyzed by sophisticated bioinformatics tools and machine learning algorithms to find genetic markers linked to desired qualities. These data expedite the generation of better varieties by guiding the choice of parent plants. Thus, precision agriculture fills the gap that exists between genomics and real-world applications in breeding. Digital decision support systems (DSS) employ data analysis to deliver actionable insights (Qadeer and Davis 2023). These systems forecast results and suggest optimal agricultural management techniques using algorithms and models. DSS can assist breeders in plant breeding in making well-informed choices on selection, crossbreeding, and resource allocation (Tonle et al. 2024). Breeders can assess the possible effects of several tactics and select the most successful approach by modeling various situations. Breeders may create crop types that are resistant to climate change with the use of precision agricultural equipment (Xiong et al. 2022). Digital instruments facilitate the discovery of characteristics that improve climate adaptation by tracking environmental parameters and forecasting meteorological trends (Lee and Mwebaza 2021). Then, breeders may concentrate on creating cultivars resistant to harsh weather, drought, and other climate-related stressors. Furthermore, early detection of disease and pest outbreaks with precision agriculture's early warning system facilitates prompt action and risk management (Brown et al. 2022). Digital tools optimize resource utilization and reduce environmental effects, which promote sustainable agriculture. By minimizing the overuse of pesticides and fertilizers, precision farming techniques lower pollution and runoff (Gawande et al. 2023). Digital technologies and precision agriculture are revolutionizing plant breeding by facilitating focused management methods, increasing efficiency, and supplying precise data. The creation of robust, high-yielding crop types that can tackle the issues of climate change and food security is aided by these developments. Precision agriculture promotes innovation and sustainability in contemporary agriculture by fusing technology and conventional breeding techniques.

6.8 Future Directions and Perspectives

Emerging technologies and innovations are being explored to enhance breeding for abiotic stress tolerance in crops. These technologies have the potential to accelerate the development of stress-resilient crop varieties. Genome editing techniques, such as CRISPR-Cas9, enable precise modifications to the plant genome, creating crops better adapted to challenging environmental conditions. High-throughput phenotyping, genomic selection, bioinformatics, synthetic biology, nanotechnology, and climate resilience modeling are some of the technologies that can be used to develop stress-tolerant crop varieties.

Synthetic biology techniques allow researchers to design and engineer biological systems for specific purposes, introducing novel genetic elements or modifying existing metabolic pathways to improve stress response mechanisms. Nanotechnology offers novel methodologies for enhancing stress tolerance in crops, such as nanoparticles for delivery, nano-sensors, and nanostructured materials. Climatic resilience modeling integrates climatic data with crop growth models to forecast the performance of different crop varieties under future climate scenarios. This allows breeders to prioritize features and breeding goals for boosting abiotic stress tolerance.

The potential of these innovations and technologies is significant, as they can help develop crop varieties with high resilience to abiotic stresses, sustainability, and ensuring food security in the face of climate change and other environmental challenges. Artificial intelligence (AI) and machine learning (ML) algorithms can analyze large data sets to discover genetic markers linked with abiotic stress tolerance, while quantum biology studies how quantum phenomena influence biological processes. Bioinformatics advances enable the analysis of complex genomic data, facilitating the identification of key genes, QTLs, and regulatory pathways involved in stress responses. By leveraging the capabilities of synthetic biology, nanotechnology, and other emerging fields, researchers can revolutionize crop breeding for abiotic stress tolerance, contributing to global food security in the face of changing environmental conditions.

Policy frameworks and support for advanced breeding techniques are essential for their responsible development and deployment. Key aspects that policymakers should consider include regulatory oversight, risk assessment, international harmonization, public engagement and education, funding and support, ethical considerations, long-term monitoring and assessment, and adoption barriers. Regulatory oversight ensures the safety of products developed using advanced breeding techniques while facilitating innovation. Risk assessment methodologies are developed tailored to the specific characteristics of products developed using advanced breeding techniques. International harmonization works toward avoiding unnecessary barriers to trade and facilitating the global exchange of breeding materials and technologies. Public engagement and education involve engaging stakeholders in discussions about the benefits and risks of advanced breeding techniques, providing accurate and accessible information.

Funding and support involve investing in research and development to further advance and refine advanced breeding techniques, as well as providing funding and support for training programs. Ethical considerations include equity, access, and environmental sustainability. Long-term monitoring and assessment ensure ongoing safety and efficacy, and feedback from monitoring efforts is incorporated into regulatory decision-making processes to adapt to new knowledge and emerging risks over time.

Economic impacts of advanced breeding techniques include increased productivity, cost reduction, market opportunities, environmental benefits, and competitive advantage. Adoption barriers include regulatory uncertainty, lack of infrastructure, intellectual property issues, public perception and acceptance, and capacity building. Addressing these barriers requires a coordinated effort from policymakers, researchers, industry stakeholders, and civil society organizations. Investments in research, infrastructure, regulatory harmonization, public engagement, and capacity building are essential to realizing the potential economic benefits of advanced breeding techniques while ensuring their responsible and equitable adoption.

6.9 Conclusion

The vision for the future of breeding for abiotic stress tolerance in crops is one where advanced technologies and interdisciplinary collaboration are combined to address global food security and environmental sustainability. Advanced breeding techniques, such as genome editing, genomic selection, and synthetic biology, enable breeders to rapidly develop crop varieties with enhanced resilience to diverse abiotic stresses. Data-driven innovation, such as big data analytics and machine learning algorithms, unlocks valuable insights from large-scale genomic, phenotypic, and environmental data sets, enabling breeders to identify genetic markers associated with abiotic stress tolerance, optimize breeding strategies, and predict crop performance under changing climate conditions.

Global collaboration and knowledge sharing foster innovation and accelerate progress in breeding for abiotic stress tolerance, with researchers, breeders, policymakers, and industry stakeholders working across borders to develop and deploy stress-resilient crop varieties. Stress-tolerant crop varieties are integrated into sustainable agricultural systems that prioritize environmental stewardship and resilience, such as conservation agriculture, precision irrigation, and agroforestry.

Smallholder farmers in developing countries can be empowered to improve their livelihoods and adapt to climate variability by adopting advanced breeding technologies and sustainable agricultural practices. Transparent communication and consumer education efforts foster trust and acceptance of genetically modified and genome-edited crops, enabling informed choices and promoting the adoption of innovative agricultural technologies.

Long-term goals for breeding for abiotic stress tolerance in crops include developing stress-tolerant crop varieties, increasing agricultural productivity, enhancing resilience to climate change, reducing environmental impact, empowering smallholder farmers, promoting equity and social justice, and fostering sustainable food systems. By investing in research, innovation, and collaboration, we can work toward a future where all people have access to nutritious and sustainably produced food, even in the face of mounting environmental challenges.

References

Abhijith KP, Gopala Krishnan S, Ravikiran KT et al (2022) Genome-wide association study reveals novel genomic regions governing agronomic and grain quality traits and superior allelic combinations for basmati rice improvement. Front Plant Sci 13:994447

Ahmad HM, Azeem F, Tahir N et al (2018) QTL mapping for crop improvement against abiotic stresses in cereals. *J Anim* Plant Sci 28(6)

Akhtar MS, Zafar Z, Nawaz R et al (2024) Unlocking plant secrets: a systematic review of 3D imaging in plant phenotyping techniques. Comput Electron Agric 222:109033

Alam MS, Yang Z-K, Li C et al (2022) Loss-of-function mutations of OsbHLH044 transcription factor lead to salinity sensitivity and a greater chalkiness in rice (Oryza sativa L.). Plant Physiol Biochem 193:110–123

Alexandratos N, Bruinsma J (2012) World agriculture towards 2030/2050: the 2012 revision

Ali AEE, Husselmann LH et al (2023) Comparative proteomics analysis between maize and sorghum uncovers important proteins and metabolic pathways mediating drought tolerance. Life 13(1):170

Alugoju P, Tencomnao T (2023) Medicinal plants proteomics in response to abiotic stresses. In: Medicinal plants: their response to abiotic stress. Springer Nature, Singapore, pp 35–107

Amara I, Capellades M, Ludevid MD et al (2013) Enhanced water stress tolerance of transgenic maize plants over-expressing LEA Rab28 gene. J Plant Physiol 170:864–873

Amasiddha B, Ramya K et al (2016) Evaluation of marker assisted backcross breeding derived lines for morpho-physiological characters under late sown heat stress condition in bread wheat. Indian J Genet 76(03):304–311

Apse MP, Aharon GS, Snedden WA, Blumwald E (1999) Salt tolerance conferred by overexpression of a vacuolar Na+/H+ Antiport in Arabidopsis. Science 80(285):1256–1258

Araus JL, Kefauver SC et al (2022) Crop phenotyping in a context of global change: what to measure and how to do it. J Integr Plant Biol 64(2):592–618

Athar HR, Ashraf M (2009) Strategies for crop improvement against salinity and drought stress: an overview. In: Salinity and water stress: improving crop efficiency, pp 1–16

Bartholomé J, Frouin J, Brottier L et al (2023) Genomic selection for salinity tolerance in japonica rice. PLoS One 18(9):e0291833

Bashir T, Ul Haq SA, Masoom S et al (2023) Quality trait improvement in horticultural crops: OMICS and modern biotechnological approaches. Mol Biol Rep 50(10):8729–8742

Beres BL, Hatfield JL, Kirkegaard JA et al (2020) Toward a better understanding of genotype× environment× management interactions – a global wheat initiative agronomic research strategy. Front Plant Sci 11:515450

Bernardo R, Charcosset A (2006) Usefulness of gene information in marker-assisted recurrent selection: a simulation appraisal. Crop Sci 46(2):614–621

Bertioli DJ, Jenkins J, Clevenger J et al (2019) The genome sequence of segmental allotetraploid peanut Arachis hypogaea. Nat Genet 51(5):877–884

Bharadwaj C, Tripathi S, Soren KR (2021) Introgression of "QTL-hotspot" region enhances drought tolerance and grain yield in three elite chickpea cultivars. Plant Genome 14(1):e20076

Bhatnagar-Mathur P, Vadez V, Sharma KK (2008) Transgenic approaches for abiotic stress tolerance in plants: retrospect and prospects. Plant Cell Rep 27:411–424

Bo W, Zhaohui Z, Huanhuan Z et al (2019) Targeted mutagenesis of NAC transcription factor gene, OsNAC041, leading to salt sensitivity in Rice. Rice Sci 26:98–108

Brown ME, Mugo S, Petersen S et al (2022) Designing a pest and disease outbreak warning system for farmers, agronomists and agricultural input distributors in East Africa. Insects 13(3):232

Chao H, Zhang S, Hu Y et al (2024) Integrating omics databases for enhanced crop breeding. J Integr Bioinform 20(4):20230012

Chawla K, Barah P, Kuiper M et al (2011) Systems biology: a promising tool to study abiotic stress responses. In: Omics and plant abiotic stress tolerance, p 163

Cheng Z, Targolli J, Huang X, Wu R (2002) Wheat LEA genes, PMA80 and PMA1959, enhance dehydration tolerance of transgenic rice (Oryza sativa L.). Mol Breed 10:71–82

Cortés AJ, Castillejo MÁ et al (2023) 'Omics' approaches for crop improvement. Agronomy 13(5):1401

Dalal M, Tayal D, Chinnusamy V, Bansal KC (2009) Abiotic stress and ABA-inducible Group 4 LEA from Brassica napus plays a key role in salt and drought tolerance. J Biotechnol 139:137–145

Das A, Singh S, Islam Z et al (2022) Current progress in genetic and genomics-aided breeding for stress resistance in cucumber (Cucumis sativus L.). Sci Hortic 300:111059

Das A, Rout BM, Datta S et al (2023) Spinach (Spinacia oleracea L.) breeding: from classical to genomics-centric approach. In: Smart plant breeding for vegetable crops in post-genomics era. Springer Nature, Singapore, pp 117–142

Dewey DR (1962) Breeding crested wheatgrass for salt tolerance, p 403–07

Dhawan G, Kumar A, Dwivedi P et al (2021) Introgression of qDTY1.1 governing reproductive stage drought tolerance into an elite basmati Rice variety "Pusa basmati 1" through marker assisted backcross breeding. Agron 11:202

Dietz KJ, Vogelsang L (2023) A general concept of quantitative abiotic stress sensing. Trends Plant Sci

Ding F, Qiang X, Jia Z et al (2022) Knockout of a novel salt responsive gene SlABIG1 enhance salinity tolerance in tomato. Environ Exp Bot 200:104903

Dong Q-L, Wang C-R, Liu D-D et al (2013) MdVHA-A encodes an apple subunit A of vacuolar H+-ATPase and enhances drought tolerance in transgenic tobacco seedlings. J Plant Physiol

Dudley JW(1984) A method of identifying lines for use in improving parents of a single cross 1. Crop Sci 24(2):355–357

Dutta S, Muthusamy V, Hossain F et al (2020) Analysis of genetic variability for retention of kernel carotenoids in sub-tropically adapted biofortified maize under different storage conditions. J Cereal Sci 93:102987

Dutta S, Muthusamy V, Hossain F et al (2021) Effect of storage period on provitamin-A carotenoids retention in biofortified maize hybrids. Int J Food Sci Tech 56(7):3148–3156

Dutta S, Kumar Singh A, Prasad Mondal B et al (2023) Perspective chapter: digital inclusion of the farming sector using drone technology. IntechOpen. https://doi.org/10.5772/intechopen.108740

Dutta S, Paul D, Das A et al (2024a) Directed mutagenesis in plants through genome editing using guide RNA library. Nucleus:1–13

Dutta S, Mukherjee B, Sawarkar A (2024b) Enhanced agricultural productivity using hydroponics technique: a smart farming system. IntechOpen. https://doi.org/10.5772/intechopen.112780

Dutta S, Zunjare RU, Sil A et al (2024c) Prediction of matrilineal specific patatin-like protein governing in-vivo maternal haploid induction in maize using support vector machine and dipeptide composition. Amino Acids 56(1):20

Dwivedi P, Ramawat N, Dhawan G et al (2021) Drought tolerant near isogenic lines (NILs) of Pusa 44 developed through marker assisted introgression of qDTY2.1 and qDTY3.1 enhances yield under reproductive stage drought stress. Agriculture 11:64

EPA U (2023) Climate change impacts on agriculture and food supply

Feng L, Li Y, Zhou YL et al (2023) Integrative transcriptomic and proteomic analyses reveal a positive role of BES1 in salt tolerance in Arabidopsis. Front Plant Sci 14:1034393

Fisher JA, Scott BJ (1987) In: Searle PGE, Davey BG (eds) Priorities in plant soil relations research for plant production. University of Sydeny, pp 135–137

Fu L, Ding Z, Sun X et al (2019) Physiological and transcriptomic analysis reveals distorted ion homeostasis and responses in the freshwater plant Spirodelapolyrhiza L. under salt stress. Genes 10(10):743

Gahlaut V, Gautam T et al (2023) Abiotic stress tolerance in wheat (Triticum aestivum L.): molecular breeding perspectives. In: QTL mapping in crop improvement. Academic Press, pp 101–117

Gao C (2021) Genome engineering for crop improvement and future agriculture. Cell 184(6):1621–1635

Gawande V, Saikanth DRK, Sumithra BS et al (2023) Potential of precision farming technologies for eco-friendly agriculture. Int J Plant Sci 35(19):101–112

Gazal A, Dar ZA et al (2018) Molecular breeding for abiotic stresses in maize (Zea mays L.). In: Maize germplasm-characterization and genetic approaches for crop improvement

Gokidi Y, Bhanu AN (2016) Marker assisted recurrent selection: an overview. Adv Life Sci 5(17):6493–6499

Gonzalez Guzman M, Cellini F et al (2022) New approaches to improve crop tolerance to biotic and abiotic stresses. Physiol Plant 174(1):e13547

Han JH, Shin NH, Moon JH et al (2020) Genetic and phenotypic characterization of rice backcrossed inbred sister lines of Saltol in temperate saline reclaimed area. Plant Breed Biotechnol 8:58–68

Hassen BM, Bartholomé J, Valè G et al (2018) Genomic prediction accounting for genotype by environment interaction offers an effective framework for breeding simultaneously for adaptation to an abiotic stress and performance under normal cropping conditions in rice. G3: Genes Genomes Genet 8(7):2319–2332

He Y, Li W, Lv J et al (2012) Ectopic expression of a wheat MYB transcription factor gene, TaMYB73, improves salinity stress tolerance in Arabidopsis thaliana. J Exp Bot 63:1511–1522

Hoekema A, Hirsch PR, Hooykaas PJ et al (1983) A binary plant vector strategy based on separation of vir-and T-region of the agrobacterium tumefaciens Ti-plasmid. Nature 303(5913):179–180

Hu H, Dai M, Yao J et al (2006) Overexpressing a NAM, ATAF, and CUC (NAC) transcription factor enhances drought resistance and salt tolerance in rice. Proc Natl Acad Sci 103:12987–12992

Hu W, Ding Z, Tie W et al (2017) Comparative physiological and transcriptomic analyses provide integrated insight into osmotic, cold, and salt stress tolerance mechanisms in banana. Sci Rep 7(1):43007

Hu Z, Li J, Ding S et al (2021) The protein kinase CPK28 phosphorylates ascorbate peroxidase and enhances thermotolerance in tomato. Plant Physiol 186:1302–1317

Illouz-Eliaz N, Nissan I, Nir I et al (2020) Mutations in the tomato gibberellin receptors suppress xylem proliferation and reduce water loss under water-deficit conditions. J Exp Bot 71:3603–3612

Ingram J, Bartels D (1996) The molecular basis of dehydration tolerance in plants. Annu Rev Plant Biol 47(1):377–403

Ito Y, Katsura K, Maruyama K et al (2006) Functional analysis of Rice DREB1/CBF-type transcription factors involved in cold-responsive gene expression in transgenic Rice. Plant Cell Physiol 47:141–153

Joshi DR, Ghimire R, Kharel T (2021) Conservation agriculture for food security and climate resilience in Nepal. J Agron 113(6):4484–4493

Júnior MRB, de Almeida Moreira BR, dos Santos CV et al (2024) Precision agriculture in the United States: a comprehensive meta-review inspiring further research, innovation, and adoption. Comput Electron Agric 221:108993

Kalogiannidis S, Kalfas D, Chatzitheodoridis F (2022) Role of crop-protection technologies in sustainable agricultural productivity and management. Land 11(10):1680

Kamal NM, Gorafi YSA, Abdeltwab H et al (2021) A new breeding strategy towards introgression and characterization of stay-green QTL for drought tolerance in sorghum. Agriculture 11(7):598

Kamali S, Singh A (2023) Genomic and transcriptomic approaches to developing abiotic stress-resilient crops. Agron 13(12):2903

Kassahun B, Bidinger FR, Hash CT et al (2010) Stay-green expression in early generation sorghum [Sorghum bicolor (L.) Moench] QTL introgression lines. Euphytica 172(3):351–362

Kikuchi K, Ueguchi-Tanaka M, Yoshida KT et al (2000) Molecular analysis of the NAC gene family in rice. Mol Gen Genet MGG 262:1047–1051

Kishor PBK, Hong Z, Miao GH et al (1995) Overexpression of [delta]-Pyrroline-5-carboxylate Synthetase increases proline production and confers Osmotolerance in transgenic plants. Plant Physiol 108:1387–1394

Krasensky J, Jonak C (2012) Drought, salt, and temperature stress-induced metabolic rearrangements and regulatory networks. J Exp Bot 63(4):1593–1608

Kuanar SR, Ray A, Sethi SK et al (2017) Physiological basis of stagnant flooding tolerance in rice. Rice Sci 24:73–84

Kumar A, Kumar R, Sengupta D et al (2020) Deployment of genetic and genomic tools toward gaining a better understanding of rice-Xanthomonas oryzae pv. oryzae interactions for development of durable bacterial blight resistant rice. Front Plant Sci 11:1152

Kumari P, Gangwar H, Kumar V et al (2024) Crop Phenomics and high-throughput phenotyping. In: Digital agriculture: a solution for sustainable food and nutritional security, pp 391–423

Kusmec A, Zheng Z, Archontoulis S et al (2021) Interdisciplinary strategies to enable data-driven plant breeding in a changing climate. One Earth 4(3):372–383

Langensiepen M, Jansen MA, Wingler A et al (2020) Linking integrative plant physiology with agronomy to sustain future plant production. Environ Exp Bot 178:104125

Lee WJ, Mwebaza R (2021) Digitalization to achieve technology innovation in climate technology transfer. Sustain For 14(1):63

Lee Y-P, Kim S-H, Bang J-W et al (2007) Enhanced tolerance to oxidative stress in transgenic tobacco plants expressing three antioxidant enzymes in chloroplasts. Plant Cell Rep 26:591–598

Li X, Gao Q, Liang Y et al (2013) A novel salt-induced gene from sheepgrass, LcSAIN2, enhances salt tolerance in transgenic Arabidopsis. Plant Physiol Biochem 64:52–59

Li R, Zhang L, Wang L, Chen L, Zhao R, Sheng J, Shen L (2018) Reduction of tomato-plant chilling tolerance by CRISPR–Cas9-mediated SlCBF1 mutagenesis. J Agric Food Chem 66(34):9042–9051

Li D, Quan C, Song Z et al (2021) High-throughput plant phenotyping platform (HT3P) as a novel tool for estimating agronomic traits from the lab to the field. Front Bioeng Biotechnol:623705

Li N, Shao T, Xu L et al (2024) Transcriptome analysis reveals the molecular mechanisms underlying the enhancement of salt-tolerance in Melia azedarach under salinity stress. Sci Rep 14(1):10981

Liang Y, Yang C, Ming F et al (2024) A bHLH transcription factor, CsSPT, regulates high-temperature resistance in cucumber. Hortic Plant J 10:503–514

Liu Q, Kasuga M, Sakuma Y et al (1998) Two transcription factors, DREB1 and DREB2, with an EREBP/AP2 DNA binding domain separate two cellular signal transduction pathways in drought- and low-temperature-responsive gene expression, respectively, in Arabidopsis. Plant Cell 10:1391–1406

Lou D, Wang H, Liang G, Yu D (2017) OsSAPK2 confers abscisic acid sensitivity and tolerance to drought stress in Rice. Front Plant Sci 8

Lou D, Wang H, Yu D (2018) The sucrose non-fermenting-1-related protein kinases SAPK1 and SAPK2 function collaboratively as positive regulators of salt stress tolerance in rice. BMC Plant Biol 18:1–17

Luo Z, Xiong J, Xia H et al (2020) Transcriptomic divergence between upland and lowland ecotypes contributes to rice adaptation to a drought-prone agroecosystem. Evol Appl 13(9):2484–2496

Maity A, Paul D, Lamichaney A et al (2023) Climate change impacts on seed production and quality: current knowledge, implications, and mitigation strategies. Seed Sci Technol 1:7–38

Mammadov J, Buyyarapu R, Guttikonda SK et al (2018) Wild relatives of maize, rice, cotton, and soybean: treasure troves for tolerance to biotic and abiotic stresses. Front Plant Sci 9:886

Mann A, Bishi SK, Mahatma MK, Kumar A (2015) Metabolomics and salt stress tolerance in plants. In: Wani SH, Hossain MA (eds) Managing salt tolerance in plants: molecular and genomic perspectives. CRC Press, pp 251–266

Mann A, Singh S, Kumar A, Kumar S, Kumar B (2018) Plant Ionomics: an important component of functional biology. In: Metabolic adaptations in plants during abiotic stress. CRC Press, pp 147–154

Mann A, Kumar N, Lata C, Kumar A, Kumar A, Meena BL (2019) Functional annotation of differentially expressed genes under salt stress in *Dichanthium annulatum*. Plant Physiol Rep 24:104–111

Mann A, Kumar A, Sanwal SK, Sharma PC (2020) Sustainable production of pulses under saline lands in India. In: Legume crops-prospects, production and uses. IntechOpen. https://doi.org/10.5772/intechopen.90304

Mann A, Kumar N, Kumar A, Lata C, Kumar A, Meena BL, Mishra D, Grover M, Gaba S, Parameswaran C, Mantri N (2021) de novo transcriptomic profiling of differentially expressed genes in grass halophyte Urochondra setulosa under high salinity. Sci Rep 11(1):5548

Mann A, Lata C, Kumar N et al (2023a) Halophytes as new model plant species for salt tolerance strategies. Front Plant Sci 14:1137211

Mann A, Kumar N, Lata C, Kumar A, Meena BL, Kumar A (2023b) Physiological and differential gene expression reveals a trade-off between antioxidant capacity and salt tolerance in halophytes *Urochondra setulosa* and *Dichanthium annulatum*. Plant Growth Regul 102:555–570

Mann A, Ranga P, Choudhary et al (2024) Genome editing technologies for enhancing plant resilience to biotic and abiotic stresses-brief review. J Soil Sal Wat Qual 16(2):180–193

Mao H, Li S, Chen B, et al (2022) Variation in cis-regulation of a NAC transcription factor contributes to drought tolerance in wheat. Mol Plant 15:276–292

Meena HP, Bainsla NK, Yadav DK (2016) Breeding for abiotic stress tolerance in crop plants. In: Recent advances in plant stress physiology, pp 329–378

Michaletti A, Naghavi MR, Toorchi M et al (2018) Metabolomics and proteomics reveal drought-stress responses of leaf tissues from spring-wheat. Sci Rep 8(1):5710

Mishra B (1994) Breeding for salt tolerance in crops. In: Rao DLN, Singh NT, Gupta RK, Tyagi NK (eds) Salinity management for sustainable agriculture. Central Soil Salinity Research Institute, Karnal, pp 226–259

Nadeem MA, Nawaz MA, Shahid MQ (2018) DNA molecular markers in plant breeding: current status and recent advancements in genomic selection and genome editing. Biotechnol Biotechnol Equip 2:261–285

Nair MM, Shylaraj KS (2021) Introgression of dual abiotic stress tolerance QTLs (Saltol QTL and Sub1 gene) into Rice (Oryza sativa L.) variety Aiswarya through marker assisted backcross breeding. Physiol Mol Biol Plants 27(3):497–514

Nakashima K, Takasaki H, Mizoi J et al (2012) NAC transcription factors in plant abiotic stress responses. Biochim Biophys Acta Gene Regul Mech 1819:97–103

Nanjo T, Kobayashi M, Yoshiba Y et al (1999) Biological functions of proline in morphogenesis and osmotolerance revealed in antisense transgenic Arabidopsis thaliana. Plant J 18:185–193

Nawaz G, Han Y, Usman B et al (2019) Knockout of OsPRP1, a gene encoding proline-rich protein, confers enhanced cold sensitivity in rice (Oryza sativa L) at the seedling stage. 3 Biotech 9:254

Ogata T, Ishizaki T, Fujita M, Fujita Y (2020) CRISPR/Cas9-targeted mutagenesis of OsERA1 confers enhanced responses to abscisic acid and drought stress and increased primary root growth under nonstressed conditions in rice. PLoS One 15:e0243376

Park J-R, Kim E-G, Jang Y-H et al (2022) Applications of CRISPR/Cas9 as new strategies for short breeding to drought gene in Rice. Front Plant Sci 13

Passioura JB (2012) Phenotyping for drought tolerance in grain crops: when is it useful to breeders? Funct Plant Biol 39(11):851–859

Peng T, Guo C, Yang J (2023) Transcriptome analysis revealed molecular basis of cold response in Prunus mume. Mol Plant Breed 43(5):34

Pergner I, Lippert C (2023) On the effects that motivate pesticide use in perspective of designing a cropping system without pesticides but with mineral fertilizer – a review. AgronSustain Dev 43(2):24

Pham TT, Sripichitt P, Chanprame S (2006) Transfer of drought resistant character from wild rice (Oryza meridionalis and Oryza nivara) to cultivated rice (Oryza sativa L.) by backcrossing and immature embryo culture. Kasetsart University

Pramanik K, Barman M, Kundu R (2024) Scope of genetically engineered organisms in food science and pest management strategy. In: Food process engineering and technology: safety, packaging, nanotechnologies and human health. Springer Nature, Singapore, pp 147–161

Qadeer F, Davis J (2023) The role of big data analytics in decision support systems: a managerial perspective. Manag Sci Res Arch 1(02):111–120

Qi W, Chen J, Han Y et al (2023) Omics-driven crop improvement for stress tolerance. Front Plant Sci 14:1172228

Rajabpour A, Yarahmadi F (2024) Remote sensing, geographic information system (GIS), and machine learning in the pest status monitoring. In: Decision system in agricultural Pest management. Springer Nature, Singapore, pp 247–353

Ramadas S, Vellaichamy S, Ramasundaram P et al (2020) Biofortification for enhancing nutritional outcomes and policy imperatives. In: Wheat and barley grain biofortification. Woodhead Publishing, pp 309–327

Ramayya JP, Vinukonda VP, Singh UM et al (2021) Marker-assisted forward and backcross breeding for improvement of elite Indian rice variety Naveen for multiple biotic and abiotic stress tolerance. PLoS One 16(9):e0256721

Ravikiran KT, Gopala Krishnan S, Abhijith KP et al (2022) Genome-wide association mapping reveals novel putative gene candidates governing reproductive stage heat stress tolerance in rice. Front Genet 13:876522

Raza A, Tabassum J, Kudapa H et al (2021) Can omics deliver temperature resilient ready-to-grow crops? Crit Rev Biotechnol 41(8):1209–1232

Ren Y, Huang Z, Jiang H et al (2021) A heat stress responsive NAC transcription factor heterodimer plays key roles in rice grain filling. J Exp Bot 72:2947–2964

Roberts HF (1929) Plant hybridization before mendel. Princeton University Press, Princeton, NJ, p 374

RoyChoudhury A, Roy C, Sengupta DN (2007) Transgenic tobacco plants overexpressing the heterologous lea gene Rab16A from rice during high salt and water deficit display enhanced tolerance to salinity stress. Plant Cell Rep 26:1839–1859

Roychowdhury R, Das SP, Gupta A et al (2023) Multi-omics pipeline and omics-integration approach to decipher plant's abiotic stress tolerance responses. Genes 14(6):1281

Santoyo G (2022) How plants recruit their microbiome? New insights into beneficial interactions. J Adv Res 40:45–58

Sarkar S, Das S, Dutta S et al (2024) Economic analysis of the behavior of prices and market arrivals of tomatoes–a case study of West Bengal, India. Agric Rural Stud 2(2):0008–0008

Sawarkar A, Pradhan A, Raman RB et al (2023) Identification of best combiners and heterotic crosses for seedling characters in jute (Corchorus olitorius L.) under normal and drought conditions. Electron J Plant Breed 14(2):530–539

Schmutz J, Cannon SB, Schlueter J et al (2010) Genome sequence of the palaeopolyploid soybean. Nature 463(7278):178–183

Schnable PS, Ware D, Fulton RS et al (2009) The B73 maize genome: complexity, diversity, and dynamics. Science 326(5956):1112–1115

Shen C, Que Z, Xia Y et al (2017) Knock out of the annexin gene OsAnn3 via CRISPR/Cas9-mediated genome editing decreased cold tolerance in rice. J Plant Biol 60:539–547

Shen S, Zhan C, Yang C et al (2023) Metabolomics-centered mining of plant metabolic diversity and function: past decade and future perspectives. Mol Plant 16(1):43–63

Sheoran S, Thribhuvan R, Saini M et al (2022) Conventional, genomics, and post-genomics era of pulses breeding: current status and future prospects. In: Advances in legumes for sustainable intensification, pp 553–574

Shi J, Yan B, Lou X et al (2017) Comparative transcriptome analysis reveals the transcriptional alterations in heat-resistant and heat-sensitive sweet maize (Zea mays L.) varieties under heat stress. BMC Plant Biol 17:1–10

Shikha M, Kanika A, Rao AR et al (2017) Genomic selection for drought tolerance using genome-wide SNPs in maize. Front Plant Sci 8:550
Shim JS, Oh N, Chung PJ et al (2018) Overexpression of OsNAC14 improves drought tolerance in Rice. Front. Plant Sci 9
Shu P, Li Y, Xiang L et al (2023) SlNPR1 modulates chilling stress resistance in tomato plant by alleviating oxidative damage and affecting the synthesis of ferulic acid. Sci Hortic (Amsterdam) 307:111486
Shunkao S, Theerakulpisut P, Wanichthanarak K et al (2023) Integrative physiological and metabolomics study reveals adaptive strategies of wheat seedlings to salt and heat stress combination. Plant Growth Regul 100(1):181–196
Siddiqui SA, Zannou O, Karim I et al (2022) Avoiding food neophobia and increasing consumer acceptance of new food trends – a decade of research. Sustain For 14(16):10391
Singh UM, Dixit S, Alam S et al (2022) Marker-assisted forward breeding to develop a drought-, bacterial-leaf-blight-, and blast-resistant rice cultivar. Plant Genome 15(1):e20170
Soltani Z, Moghadam A, Tahmasebi A et al (2023) Integrative systems biology analysis of barley transcriptome – hormonal signaling against biotic stress. PLoS One 18(4):e0281470
Stracke R, Werber M, Weisshaar B (2001) The R2R3-MYB gene family in Arabidopsis thaliana. Curr Opin Plant Biol 4:447–456
Sun T, Ma N, Wang C et al (2021) A Golgi-localized sodium/hydrogen exchanger positively regulates salt tolerance by maintaining higher K+/Na+ ratio in soybean. Front Plant Sci 12
Sun L, Wang R, Tang W et al (2022) Robust identification of low-Cd rice varieties by boosting the genotypic effect of grain cd accumulation in combination with marker-assisted selection. J Hazard Mater 424:127703
Sundaramoorthi D, Dong L (2022) Machine learning and optimization based decision-support tool for seed variety selection. Ann Oper Res:1–35
Swarup S, Cargill EJ, Crosby K et al (2021) Genetic diversity is indispensable for plant breeding to improve crops. Crop Sci 61(2):839–852
Tahat MM, Alananbeh K et al (2020) Soil health and sustainable agriculture. Sustain For 12(12):4859
Talukdar D, Sinjushin A (2015) Cytogenomics and mutagenomics in plant functional biology and breeding. PlantOmics: The omics of plant science 113–156
Tanaka N, Shenton M, Kawahara Y et al (2020) Whole-genome sequencing of the NARO world Rice Core collection (WRC) as the basis for diversity and association studies. Plant Cell Physiol 61(5):922–932
Taunk J, Rani A, Singh R et al (2019) Genomic strategies for improving abiotic stress tolerance in crop plants. In: Genetic enhancement of crops for tolerance to abiotic stress: mechanisms and approaches, vol I, pp 205–230
Teper-Bamnolker P, Roitman M, Katar O et al (2023) An alternative pathway to plant cold tolerance in the absence of vacuolar invertase activity. Plant J 113:327–341
Thanh PT, Phan PDT, Mori N et al (2011) Development of backcross recombinant inbred lines between Oryza sativa Nipponbare and O. rufipogon and QTL detection on drought tolerance. Breed Sci 61(1):76–79
Tonle FB, Niassy S, Ndadji MM et al (2024) A road map for developing novel decision support system (DSS) for disseminating integrated pest management (IPM) technologies. Comput Electron Agric 217:108526
Tran MT, Son GH, Song YJ et al (2023) CRISPR-Cas9-based precise engineering of SlHyPRP1 protein towards multi-stress tolerance in tomato. Front Plant Sci 14
Ueno D, Yamaji N, Kono I et al (2010) Gene limiting cadmium accumulation in rice. Proc Natl Acad Sci 107(38):16500–16505
Usatov AV, Alabushev AV, Kostylev PI et al (2015) Introgression the saltol QTL into the elite rice variety of Russia by marker-assisted selection. Am J Agric Biol Sci 10(4):165–169

Usman MG, Rafii MY, Martini MY et al (2018) Introgression of heat shock protein (Hsp70 and sHsp) genes into the Malaysian elite chilli variety Kulai (Capsicum annuum L.) through the application of marker-assisted backcrossing (MAB). Cell Stress Chaperones 23(2):223–234

Vanino S, DiBene C, Piccini C et al (2022) A comprehensive assessment of diversified cropping systems on agro-environmental sustainability in three Mediterranean long-term field experiments. Eur J Agron 140:126598

Varshney RK, Song C, Saxena RK et al (2013) Draft genome sequence of chickpea (Cicer arietinum) provides a resource for trait improvement. Nat Biotechnol 31(3):240–246

Varshney RK, Bohra A, Yu J et al (2021) Designing future crops: genomics-assisted breeding comes of age. Trends Plant Sci 26(6):631–649

Verma D, Singla-Pareek SL, Rajagopal D et al (2007) Functional validation of a novel isoform of Na+/H+ antiporter from Pennisetum glaucum for enhancing salinity tolerance in rice. J Biosci 32:621–628

Vinod KK, Krishnan SG, Thribhuvan R et al (2019) Genetics of drought tolerance, mapping QTLs, candidate genes and their utilization in rice improvement. In: Genomics assisted breeding of crops for abiotic stress tolerance, vol II, pp 145–186

Vivitha P, Raveendran M, Vijayalakshmi D (2017) Introgression of QTLs controlling spikelet fertility maintains membrane integrity and grain yield in improved white Ponni derived progenies exposed to heat stress. Rice Sci 24(1):32–40

Von Nonn J, Villarreal ML, Blesius L et al (2024) An open-source workflow for scaling burn severity metrics from drone to satellite to support post-fire watershed management. Environ Model Softw 172:105903

Wang Y, Jiang J, Zhao X et al (2006) A novel LEA gene from Tamarix androssowii confers drought tolerance in transgenic tobacco. Plant Sci 171:655–662

Wang Z, Wong DCJ, Wang Y et al (2021) GRAS-domain transcription factor PAT1 regulates jasmonic acid biosynthesis in grape cold stress response. Plant Physiol 186:1660–1678

Wang K, Abid MA, Rasheed A et al (2023) DNNGP, a deep neural network-based method for genomic prediction using multi-omics data in plants. Mol Plant 16(1):279–293

Wani SH, Choudhary JR, Choudhary M et al (2020) Recent advances in genomics assisted breeding for drought stress tolerance in major cereals. J Cereal Res 12(1):1–12

Wise MJ (2003) LEAping to conclusions: a computational reanalysis of late embryogenesis abundant proteins and their possible roles. BMC Bioinformatics 4:52

Wu J, Yan G, Duan Z et al (2020) Roles of the Brassica napus DELLA protein BnaA6.RGA, in modulating drought tolerance by interacting with the ABA signaling component BnaA10. ABF2. Front Plant Sci 11

Xia H, Ma X, Xu K et al (2020) Temporal transcriptomic differences between tolerant and susceptible genotypes contribute to rice drought tolerance. BMC Genomics 21:1–18

Xiong W, Reynolds M, Xu Y (2022) Climate change challenges plant breeding. Curr Opin Plant Biol 70:102308

Yang Y, Guo Y, Zhong J et al (2020) Root physiological traits and transcriptome analyses reveal that root zone water retention confers drought tolerance to Opisthopappustaihangensis. Sci Rep 10(1):2627

Yang Y, Saand MA, Huang L et al (2021) Applications of multi-omics technologies for crop improvement. Front Plant Sci 12:563953

Yu W, Wang L, Zhao R et al (2019) Knockout of SlMAPK3 enhances tolerance to heat stress involving ROS homeostasis in tomato plants. BMC Plant Biol 19:354

Zeng Y, Wen J, Zhao W et al (2020) Rational improvement of Rice yield and cold tolerance by editing the three genes OsPIN5b, GS3, and OsMYB30 with the CRISPR–Cas9 system. Front. Plant Sci 10

Zhang M, Cao Y, Wang Z et al (2018) A retrotransposon in an HKT1 family sodium transporter causes variation of leaf Na+ exclusion and salt tolerance in maize. New Phytol 217:1161–1176

Zhang J, Huang D, Zhao X et al (2021) Evaluation of drought resistance and transcriptome analysis for the identification of drought-responsive genes in Iris germanica. Sci Rep 11(1):16308

Zheng L, Wu W, Chen Q et al (2022) Integrated transcriptomics, proteomics, and metabolomics identified biological processes and metabolic pathways involved in heat stress response in jojoba. Ind Crop Prod 183:114946

Zhou H, Zhang F, Zhai F et al (2022) Rice GLUTATHIONE PEROXIDASE1-mediated oxidation of bZIP68 positively regulates ABA-independent osmotic stress signaling. Mol Plant 15:651–670

Zimin AV, Puiu D, Hall R et al (2017) The first near-complete assembly of the hexaploid bread wheat genome, Triticum aestivum. Gigascience 6(11):gix097

Breeding Strategies for Improved Multistress-Resilient Crops

7

B. M. Lokeshkumar, Sanchika Snehi, Krishanu, Santosh Kumar, K. T. Ravikiran, Rahul Kumar, Mukesh Choudhary, Suneetha Kota, Arvind Kumar, Anita Mann, Satish Kumar Sanwal, and Nitish Ranjan Prakash

Abstract

Multiple stresses including abiotic and biotic, affecting crop plants are more common nowadays in the era of climate change. These stresses include several abiotic stresses such as drought, salinity, cold, heat, mineral deficiency, mineral toxicity, and waterlogging and biotic stresses such as diseases, pests, insects, and weeds. Breeding suitable cultivars with the capacity to endure multiple stresses are need of the hour. The approaches in the past were conventional pedigree-based selection, backcross-assisted introgression and transfer, and recombination breeding. However, in the recent past, the marker-assisted backcross breeding (MABB) approach has been used successfully to transfer multiple abiotic and biotic stress-tolerant oligogenic/major QTLs from donor genotypes into elite lines. Presently, with the advent of robust and cheaper genotyping costs, the capability to analyze big data, high computational efficiency, and the use of rapid

B. M. Lokeshkumar · Krishanu · A. Kumar · A. Mann · S. K. Sanwal · N. R. Prakash (✉)
ICAR-Central Soil Salinity Research Institute, Karnal, Haryana, India

S. Snehi
Institute of Agricultural Sciences, BHU, Varanasi, Uttar Pradesh, India

S. Kumar
ICAR-Indian Agricultural Research Institute, Hazaribagh, Jharkhand, India

K. T. Ravikiran
ICAR-Central Soil Salinity Research Institute, RRS, Lucknow, Uttar Pradesh, India

R. Kumar
ICAR Research Complex for North-Eastern Region, Lembucherra, Tripura, India

M. Choudhary
ICAR-Indian Institute of Maize Research, Ludhiana, Punjab, India

S. Kota
ICAR-Indian Institute of Rice Research, Hyderabad, Telangana, India

A. Mann et al. (eds.), *Cutting Edge Technologies for Developing Future Crop Plants*,
https://doi.org/10.1007/978-981-96-2508-6_7

generation techniques (Speed Breeding), genomic selection has emerged as a major player in breeding. Genomic selection breeding pipelines are very robust in predicting the performance of genotypes. The present chapter summarizes a thorough understanding of conventional and molecular breeding approaches to develop multistress-tolerant crop varieties.

Keywords

Multistress-tolerant crop varieties · Genomic selection · Marker-assisted back-cross breeding · Stress tolerance breeding · Biotic stress · Abiotic stress

7.1 Introduction

Plant breeding has evolved from the domestication of crops for various human needs (for food, feed, fodder, shelter, clothes, medicines, and other novel needs) to the desired precise change in crop performance using novel genomic approaches (Prakash et al. 2020). The global population is expected to approach 10 billion by 2050, necessitating a 60% increase in food crop production to ensure food security (Jaganathan et al. 2018). However, abiotic stresses such as salinity, drought, and extreme temperatures are major factors that significantly reduce crop yield (Sandhu et al. 2021; Mann et al. 2020) accounting for nearly 50% of yield losses and resulting in an estimated $14–19 million in economic losses annually (Yadav et al. 2020). The impact of these stresses is likely to worsen due to climate change, making it essential for researchers to develop strategies to manage these challenges for sustainable agriculture. Historically, researchers have employed various methods to enhance abiotic stress tolerance in plants. Traditional breeding, which involves crossing parental plants to produce offspring with desirable traits, has had limited success due to the complexity of traits involved in abiotic stress tolerance (Potts et al. 2023).

Various abiotic and biotic stresses have impacted crop plants and led to the ecological selection and development of landraces and ecotypes (Villa et al. 2005). These variabilities represented in the forms of old cultivars are great sources of tolerance to various biotic and abiotic stresses. Various abiotic stresses like drought, salinity, cold, heat, waterlogging, nutrient deficiency, and mineral toxicity impact crops by hampering their growth and development (Marone et al. 2021). Abiotic stresses in common are destabilizing cellular functions by creation of reactive oxygen species (ROS), impaired enzymatic functions, destruction of cellular membranes, distorted water potential, damage to chlorophyll, and movement of nutrients, metabolites, and photosynthetic products (Arbona et al. 2017). Biotic stresses such as the occurrence of pests and diseases in crops are impacting yield, quality, and overall growth. Plant breeding can play a substantial role in mitigating these risks through the development of climate-smart crop varieties with multistress tolerance (Prakash et al. 2020). The stress-tolerant varieties offer an economical approach to coping with stresses by reducing farmer's dependency on other management methods.

Plant breeding approaches for stress-tolerant varieties depend on the availability of variations for tolerance to stress and the identification of suitable donor genotypes (Villa et al. 2005). The breeding approaches may differ based on the availability of traits in cultivated varieties, landraces, wild forms, wild species, and crop wild relatives (Casañas et al. 2017). Traditionally, pure line selection from locally adapted cultivars (landraces) was performed to identify, uniform, stable, and stress-tolerant lines of original landraces (Snehi et al. 2022). Salt-tolerant varieties such as CSR1, CSR2, CSR3, and CSR4 were selected from traditional local landraces such as Damodar, Getu, Dasal, etc. (Prakash et al. 2024). Pedigree breeding approaches were used to combine stress tolerance with high-yield potential in various crops and lead to the development of stress-tolerant varieties (Snehi et al. 2022). This approach yielded transgressive segregants through recombination breeding. The backcross approach is more successful when the aim is to enhance the resistance/tolerance of an already popular variety by transferring monogenic/oligogenic stress tolerance from donor genotypes (in most cases, landraces and wild relatives) (Rai et al. 2018).

Climate change has severely impacted poor farmers in stress-prone regions (Karki et al. 2020). Unfortunately, these areas have rainfed cropping systems and thus are severely impacted on the livelihood and income security of farmers (Guntukula 2020). The climatic vagaries such as poor rainfall (draught), impaired rainfall patterns (untimely rainfall), heavy downpour (flood and submergence), air and water pollution (toxicity), and anthropogenic activities and their cumulative impact have led researchers to tackle it by developing multistress-tolerant cultivars (Prakash et al. 2020; Sandhu et al. 2021). There is a good interrelationship between climate change and the occurrence of change in the pattern of biotic stresses (minor pests or diseases becoming major) (van Zonneveld et al. 2020). Therefore, the occurrence of multiple stresses at a time along with their interactive effects has created avenues to develop multistress-tolerant climate-smart crop varieties (Prakash et al. 2024). The development of DNA-based molecular markers and nucleotide sequencing technology has changed the facets of breeding stress-tolerant crops (Mallick et al. 2015). Identification of molecular markers linked to genomic loci governing stress tolerance and their successful transfer into elite varieties using marker-assisted backcross breeding were the prime focus of the last two decades (Mohanavel et al. 2024). Advancements in phenomics and transcriptomics have now helped in trait dissection, characterization, and linking complex traits with several novel physiological phenomena (Zhao et al. 2019; Mann et al. 2019). These along with the ability of big data analysis have helped breeders to design novel crop varieties and assess the worth of any genotypes in the breeding population using genomic selection (Visakh et al. 2024). This chapter delves into details of advancements in breeding strategies for developing multistress-tolerant crop varieties using novel approaches, past achievements, and their future implications.

7.2 Classical Breeding Approaches

Classical breeding approaches have been the cornerstone of plant improvement for multistress tolerance, helping crops overcome drought, salinity, and heat (Prakash et al. 2020). These methods rely on exploiting genetic variation, selecting the best-performing individuals, and crossing them to produce improved varieties. Traditional breeding approaches have proven effective, albeit slower, in dealing with multi-stress resilience (Nachimuthu et al. 2017). However, combining them with modern genomic and phenomic techniques has enhanced the precision and speed of breeding pipelines (Sandhu et al. 2021).

7.2.1 Pure Line and Mass Selection

Mass selection involves selecting and propagating individuals that perform well under stress conditions and has been especially useful in crops like sorghum, where mass selection for drought and heat tolerance has led to the development of high-yielding varieties that can thrive under arid conditions (Priyadarshan 2019). Pure-line selection, on the other hand, is used to create uniform, stable varieties from a heterogeneous population (Snehi et al. 2022; Prakash et al. 2024). This approach has been crucial in developing stress-tolerant varieties of wheat and barley. Through pure-line selection, several drought-resistant wheat varieties have been developed, showing significant improvements in yield stability under water-limited environments.

7.2.2 Pedigree Breeding

Pedigree breeding and recurrent selection have been widely utilized to improve multistress resilience in crops (Yadav et al. 2024). Pedigree selection is used to track the inheritance of desired traits across generations, focusing on phenotypic traits like yield under stress and physiological adaptations like water-use efficiency and root architecture under drought, salinity, nutrient-use efficiency, mineral toxicity, and waterlogging (Numan et al. 2021). This method has been successfully employed in drought-prone environments, especially in crops like wheat, barley, and rice. Pedigree breeding in wheat has led to the development of several drought-tolerant varieties like *PBW 343*, which also exhibits heat tolerance (Singh et al. 2017). In maize, recurrent selection—a continuous process of selecting the best-performing individuals over multiple generations—has been effective in improving drought resilience in breeding programs (Mohanavel et al. 2024; Ribaut and Ragot 2007). Specifically, studies by Cooper et al. (2014) demonstrated that recurrent selection in maize significantly increased the frequency of alleles responsible for drought tolerance. Hybridization is one of the most fundamental approaches in classical breeding, combining the favorable traits of two different parents, and has been critical in enhancing crop resilience to multiple stresses (Sandhu et al. 2021). Drought-tolerant maize hybrids have been developed by hybridizing drought-resilient inbred lines with high-yielding cultivars.

7.2.3 Backcross Breeding

Hybridization, followed by backcross breeding, has also been employed to improve salinity and heat tolerance in several crops (Numan et al. 2021). In rice, backcross breeding has been utilized to introduce blast and blight-tolerant traits from wild relatives into elite cultivars (Miah et al. 2013). For instance, the introgression of drought tolerance QTLs, such as *qDTY1.1* and *qDTY3.1* into high-yielding rice varieties, has significantly improved their ability to withstand both drought and salinity stresses (Mohanavel et al. 2024). Several leaf and stem rust resistance genes from different wild relatives of wheat were introduced using a conventional approach (Ijaz et al. 2023). Similarly, in barley, genes for salt tolerance have been successfully transferred from the wild species *Hordeum spontaneum* to commercial barley varieties (Rehman et al. 2021).

7.2.4 Use of Crop Wild Relatives (CWRs)

The use of wild relatives and landraces in breeding programs offers an invaluable genetic reservoir for stress resilience (Marone et al. 2021). Wild species and landraces often exhibit traits that have been lost in modern cultivars, such as deeper root systems, osmotic adjustment, and better water-use efficiency, which are essential for coping with drought, heat, and salinity stresses (Dwivedi et al. 2016). In rice, landraces like Nagina-22 have been used to develop varieties that show both drought and heat tolerance (Mukherjee et al. 2023). Similarly, the wild relatives of wheat, such as *Aegilops tauschii*, have been exploited for their genetic traits related to salinity and drought tolerance, leading to the development of elite cultivars that can thrive under multistress conditions (Gudi et al. 2024). Despite the success of classical breeding, several challenges remain. Multistress tolerance is often governed by complex, polygenic traits, making it difficult to achieve rapid improvements (He et al. 2018). Moreover, evaluating stress tolerance in field conditions is time-consuming and costly, as environmental factors like drought and heat stress can vary significantly between seasons (Prakash et al. 2020). Additionally, breeding for multiple stresses often involves trade-offs between yield and stress tolerance. For instance, varieties that perform well under drought conditions may not necessarily thrive under heat or salinity stresses, requiring breeders to make careful selections.

7.3 Genes and Genotypes

Genebanks conserve extensive genetic diversity across a wide range of crops, providing an invaluable resource for the identification and screening of donors with desirable traits (Nachimuthu et al. 2017). To effectively leverage these resources, it is crucial first to validate the presence of specific traits and then to pinpoint the genomic regions responsible for them (Numan et al. 2021). Two prominent methods employed for this purpose are QTL mapping, which utilizes biparental mapping populations, and genome-wide association studies (GWAS), which uses germplasm collection available with the plant breeder (Sandhu et al. 2021; Numan et al. 2021; Prakash et al. 2024). These methodologies have profoundly enhanced our capacity to discern the genetic determinants underlying critical traits in crop species (Ali et al. 2021). Table 7.1 encapsulates key genes associated with major biotic and abiotic stresses, alongside their potential donor sources. Biotic stresses, including diseases and insect pests, represent a formidable threat to global crop production (Malenica et al. 2021). According to FAO estimates, plant pests and diseases are responsible for up to 40% of worldwide food crop losses annually (Junaid and Gokce 2024). Genotypes exhibiting tolerance to these biotic stresses are often located within the secondary or tertiary gene pools of crop species, rendering their transfer into cultivated varieties a complex challenge (Kapazoglou et al. 2023). Nevertheless, substantial progress has been achieved in the mapping of genes and QTLs associated with economically significant biotic stresses across various crop species (Choudhary et al. 2024). One advantageous aspect is that resistance to most pests and diseases is typically a monogenic or oligogenic trait, which simplifies their identification and introgression relative to abiotic stress traits (Li et al. 2020). However, the emergence of new pathotypes frequently renders existing resistance genes ineffective, necessitating a renewed search for tolerance sources and the identification of novel resistance genes (Singh et al. 2021a). Despite these challenges, significant strides have been made in breeding for biotic stress resistance in numerous crops.

7.3.1 Genes

For example, in rice, 49 major genes conferring resistance to bacterial leaf blight have been identified, some of which have been successfully incorporated into popular cultivars (Ellur et al. 2016). Similarly, multiple genes have been discovered that confer tolerance to the brown planthopper, a devastating pest of rice notorious for causing "hopper burn" symptoms (Yan et al. 2023). Rust diseases pose a severe threat to wheat production worldwide. It is estimated that these diseases cause an average annual yield loss of 15.04 million tons due to the effects of three rust pathogens (Huerta-Espino et al. 2020). The significance of rust diseases is highlighted by the existence of dedicated research institutions in many countries focusing on these biotic stresses. Wheat rust resistance is one of the most extensively researched areas in the field of biotic stress. To date, 69 stem rust resistance genes (Sr), 33 leaf rust

Table 7.1 List of genes/QTLs associated with various important biotic and abiotic stresses in crops

Crop	Stress	Genes/QTLs	Donor genotypes	References
Rice	Salinity	*Saltol, SKC1, SalT*	Pokkali, Nonabokra, CSR27, CSR 11, FL478	Singh et al. (2021b)
	Drought	*qDTY1.1, qDTY2.1, qDTY2.2, qDTY3.1, qDTY3.2, qDTY12.1, Dro1,*	Nagina 22, Vandana, Apo Azucena,	Hassan et al. (2023)
	Heat	*TT1, qHTSF1.1,qHTSF4.1*	Nagina 22, NL44, Dular	Visakh et al. (2024)
	Submergence	*Sub1A*	FR13A	Kuroha and Ashikari (2020)
	Cold	*qPSST3, qPSST7, qPSST9, qCTS4a,qCTS4b*	K-39, K-78, K348	Biswas et al. (2019)
	Low phosphorous	*PSTOL1*	Kasalath	Gamuyao et al. (2012)
	Deepwater	*Snorkel1, Snorkel2*	C9285	Panda and Barik (2021)
	Blast	*Pi2, Pi9, Pi21, Pi40, Pi54, Pita, Pita²*	Rasi, IR 36, IR 64, Tadukan, Tetep, zenith	Yadav et al. (2019)
	Bacterial leaf blight	*Xa4, xa5, Xa7, xa13, Xa21, Xa33, Xa38*	TKM6, IR20, IR22, IR72, DZ192, IR1545-339, DZ78, DV85, BJ1, Chinsurah Eoro II, *O. Longistaminata,* Ba7, *O. Nivara,* IRGC81825	Fiyaz et al. (2022)
	Brown plant hopper	*Bph3, Bph4, Bph14, Bph17, bph19, Bph20, Bph21, Bph26, Bph27, bph29 Bph31, Bph33, Bph34*	*Oryza officinalis, O. Minuta, O. Nivara, O. Punctata, O. Rufipogon and O. Latifolia*	Yan et al. (2023)

(continued)

Table 7.1 (continued)

Crop	Stress	Genes/QTLs	Donor genotypes	References
Wheat	Black stem rust	*Sr13, Sr21, Sr22, Sr22b, Sr26, Sr27, Sr33, Sr35, Sr45, Sr46, SrTA1662, Sr50, Sr60, Sr61, and Sr62*	*Triticum turgidum, Secale cereal, T. Comosa, Aegilops speltoides, Triticum timopheevii ssp. Araraticum, Aegilops tauschii, Thinopyrum intermedium, rye cultivar, Aegilops searsii, Dasypyrum villosum, Aegilops geniculate*	Basnet et al. (2022), Karelov et al. (2022)
	Leaf rust	*Lr1, Lr2, Lr2a-c, Lr3, LrZH84, Lr10, Lr11, Lr12, Lr13, Lr14a-b, Lr15, Lr16, Lr17a-b, Lr18, Lr9, Lr4–6, Lr19, Lr21, Lr22a-b, Lr23, Lr24, Lr25, Lr26, Lr35, Lr36, Lr37, Lr38, Lr39, Lr50, Lr63, Lr80, LrX, and LrTM16*	*Secale cereal L, Agropyron elongatum, Secale cereal, Rosen, Secale cereal Petkus, Aegilops speltoides, Aegilops ventricosa, Agropyron intermedium*	Basnet et al. (2022), Kumar et al. (2022)
	Stripe rust	*Yr1, YrA, Yr2, Yr3a-c, Yr4a-b, Yr6, Yr11, Yr12, Yr13, Yr14b, Yr18, Yr36, and Yr46, Yr7, Yr10, Yr15, YrAS2388, and YrU1*	*Secale cereale (Yr9) and Aegilops ventricosa (Yr17)*	Basnet et al. (2022)
	Salinity	*Nax1, Nax2*	*Triticum turgidum ssp. durum*	James et al. (2011)
Barley	Leaf rust	*Rph1 to 28, RphC,* Rph_{MBR102}	LV-Taganrog, DH3/6 and DH3/127, *H. Vulgare* sub sp. *spontaneum*, etc.	Singh et al. (2015a), Kavanagh et al. (2017), Fazlikhani et al. (2019), Chen et al. (2021)
Maize	Drought/ waterlogging stress	*ZmEXPA4, ZmDRO1, ZmVPP1, ZmNAC11, DRESH8*	–	Yang et al. (2023)
	Heat stress	*ZmCPK7, ZmHsfs, ZmDREB2, ZmbZIP60*	–	
	Cold stress	*ZmICE1, ZmbZIP68, and ZmRR1*		
	Salinity	*ZmSOS1, ZmSOS2, ZmSOS3, ZmHAK4, ZmNSA1, ZmPIP2s*	–	
Chickpea	Drought	*CaTIFY4b*	ICC 4958	Barmukh et al. (2022)

resistance genes (Lr), and 105 stripe rust resistance genes (Yr) have been identified (Basnet et al. 2022). However, only about one-third of these genes have been cloned and characterized. Many of these genes have been sourced from wild relatives of wheat, such as *Thinopyrum*, *Agropyron*, and *Aegilops*. Nonetheless, these genes have been successfully introgressed into elite wheat varieties through marker-assisted backcross breeding (Mallick et al. 2015, 2022).

7.3.2 Genotypes

Climate change and global warming are exerting profound threats to global agricultural production. The frequency and intensity of various abiotic stresses, such as drought, flooding, and heat waves, are escalating. Concurrently, the degradation of arable lands due to salinization and desertification is becoming increasingly prevalent. For instance, it has been documented that between 1983 and 2009, approximately three-quarters of the world's cropped areas were afflicted by drought, leading to substantial yield reductions (Kim and Jehanzaib 2020). In rice, a 1 °C increase in ambient temperature can result in a 10% decline in yield (Peng et al. 2004). Similarly, salinity levels of 8–10 dS m^{-1} can lead to yield losses of 55%, 28%, and 15% in maize, wheat, and cotton, respectively (Zörb et al. 2019). Unlike biotic stresses, where chemical management is a viable option alongside the use of resistant varieties, mitigating abiotic stresses through agronomic management is significantly more challenging. Developing and cultivating crop varieties that are resilient to abiotic stresses remains the most economically feasible solution. Although substantial progress has been made in identifying donors for tolerance to various abiotic stresses and in elucidating the underlying genetic mechanisms, advancements in this domain still lag behind those achieved in biotic stress tolerance. This is evident from Table 7.1, which illustrates that the number of significant QTLs/genes identified for abiotic stresses is comparatively lower.

7.3.3 Use of Genes/QTLs in Stress Breeding

Multistress-tolerant crop varieties were targeted either through conventional or molecular approach. For example, *Saltol*, a QTL for seedling-stage salinity tolerance, was identified from FL478, an improved line derived from the original parent Pokkali. It is a major QTL located on chromosome 1, accounting for 43–70% of phenotypic variance (Bonilla et al. 2002). Given rice's sensitivity to salt stress at the seedling and reproductive stages, this QTL has proven highly effective in safeguarding rice seedlings against salinity and has been successfully introgressed into several popular rice varieties through marker-assisted backcross breeding. Similarly, using FR13A as a donor, *Sub1*, a major QTL for submergence tolerance, was identified. This QTL has been successfully introgressed into many widely cultivated rice varieties in India, such as Swarna and BPT5204, making it a prominent example of marker-assisted selection for abiotic stress (Das et al. 2017). In wheat, two genes for salt tolerance, *Nax1* and *Nax2*, have been identified, sourced from *Triticum turgidum* ssp. *durum*. *Nax1* functions by excluding Na^+ from the root xylem, lower parts of leaves, and leaf sheaths, while *Nax2* operates exclusively in the roots (James et al. 2011). Several genes with putative roles in various abiotic stress tolerances have also been identified in maize, although most of these were discovered through comparative genomic studies. In chickpeas, a QTL hotspot for drought tolerance was identified, explaining more than 50% of the phenotypic variation for this trait (Varshney et al. 2014). This QTL has been successfully introgressed into several popular varieties, and the corresponding near-isogenic lines (NILs) have already been released as commercial varieties in India (Bharadwaj et al. 2021). While there have been several notable achievements in breeding for abiotic stress tolerance, considerable work remains to be done to keep pace with the rapidly evolving climate.

7.4 Marker-Assisted Selection

Marker-assisted selection (MAS) has emerged as a crucial tool for accelerating the development of climate-resilient crops. MAS leverages molecular markers to identify and incorporate genes linked to stress tolerance more precisely than traditional breeding methods, which typically rely on visible traits (Hassan et al. 2023). This technique enables breeders to focus on specific genes or quantitative trait loci (QTLs) related to stress tolerance before plants exhibit the desired traits (Collard et al. 2005). By allowing the simultaneous selection of multiple traits, such as drought, heat, and salinity resistance, MAS facilitates faster and more accurate breeding for multistress tolerance. This approach is vital for developing crop varieties capable of enduring the varied and unpredictable effects of climate change (Ashraf and Foolad 2013).

7.4.1 MAS for Abiotic Stresses

MAS has been used to pyramid QTLs for drought and submergence tolerance into rice varieties. The widely grown *IR64* variety was improved using MAS to combine QTLs for salinity tolerance (*Saltol*), submergence tolerance (*Sub1*), and drought tolerance (*qDTY3.1*). Similarly, in maize, marker-assisted recurrent selection (MARS) has been shown to increase the accumulation of favorable alleles for drought and heat tolerance, allowing breeders to fast-track the development of resilient hybrids (Beyene et al. 2016). The *GW7* and *DEP2* genes, located on chromosome 7, present an opportunity for simultaneous trait enhancement in rice. However, conventional breeding faces limitations like linkage drag, making it challenging to develop multistress-tolerant varieties. The discovery of major quantitative trait loci (QTLs) for various stresses, along with advances in marker genotyping, enables the stacking of multiple genes into a single background via marker-assisted breeding.

For drought tolerance, QTLs like *qDTY1.1* and *qDTY2.1* from the Philippine upland rice variety Apo have been utilized to breed drought-resistant rice varieties. In salinity tolerance, the "*Saltol*" QTL, identified in the highly salt-tolerant Pokkali rice variety, has been instrumental in improving salinity resistance. Similarly, the discovery of the "*Sub1*" QTL in the flood-tolerant rice variety FR13A has enabled submergence tolerance breeding.

7.4.2 MAS for Biotic Stresses

Biotic stresses, such as blast disease and bacterial leaf blight, severely impact rice yields. While over 500 QTLs linked to blast resistance have been identified, only a fraction have been characterized. Genes like *Pi9* and *Pi54* offer broad-spectrum resistance to diverse blast strains. In combating bacterial leaf blight, researchers have identified 43 resistance genes, and pyramiding multiple resistant genes (e.g., *xa13* + *Xa21*) has proven effective (Mohanavel et al. 2024). Simple sequence repeats (SSRs), also referred to as microsatellites, are extensively utilized as markers in plant breeding due to their reliability, polymorphism, co-dominance, and cost-effectiveness in marker-assisted selection (MAS). SSRs represent variations in the number of repeat units within specific genomic regions, with primers designed based on the nucleotide sequences flanking these repeats. Typically, SSRs provide information about a single locus per assay, though multiplexing techniques allow for the analysis of multiple loci simultaneously. These markers are readily accessible through linkage maps and public databases and can distinguish between homozygous and heterozygous genotypes. In contrast, single-nucleotide polymorphisms (SNPs) are a more recent development in high-throughput markers, valued for their abundance, stability, and ease of analysis (Yang et al. 2020). SNPs involve single base-pair variations at specific genome locations, which can be linked to particular plant traits. MAS has been effectively applied across various crops to develop varieties resistant to multiple stresses such as diseases, pests, drought, and salinity. Examples of MAS applications in different crops, including the markers used and the traits targeted, are summarized in Table 7.2.

7.4.3 Future of MAS

Marker-assisted selection (MAS) has greatly enhanced crop breeding by allowing precise selection for stress tolerance. Advances in high-throughput genotyping and genomic technologies have made it possible to identify markers associated with multistress tolerance. Although MAS has effectively addressed single-stress scenarios, the complexity of multistress environments presents ongoing challenges. Integrating quantitative trait loci (QTLs) for various stresses into breeding programs holds promise but requires further refinement to manage genetic interactions and epistatic effects (Zhao et al. 2023). Future developments should focus on combining MAS with genomic selection (GS) to improve breeding efficiency for multistress

Table 7.2 Marker-assisted selection in different crops for multiple abiotic and biotic stress tolerance

Crop	Trait	Marker	Genes/QTLs	Variety	References
Rice	Drought and submergence tolerance	Sub1C173, RM23899,	*Sub1 gene, qDTY3.1 QTL*	IR64-Sub1	Septiningsih et al. (2009)
	Drought and salinity	RM520, RM416, RM8094	*qDTY2.2 and qDTY3.1 QTLs, Saltol QTL*	BRRI dhan56	Rashid et al. (2017)
	Submergence and drought tolerance	Sub1C173, ART5, RM23899	*Sub1 gene, qDTY2.1 and qDTY3.1 QTLs*	Swarna-Sub1	Neeraja et al. (2007)
	Bacterial blight resistance, higher grain yield, and drought tolerance	pTA248, xa13-prom, RM447, RM154	*qDTY3.1 QTL, Xa21 and Xa13 genes*	DRR Dhan 44	Pradhan et al. (2015)
	Drought and submergence tolerance	RM431, RM520, RM416	*qDTY3.1 QTL, Sub1 gene*	Sahbhagi Dhan	Kumar et al. (2008)
Wheat	Resistance to drought, heat, and leaf rust	Xgwm577, Xwmc312, csLV34	*QYld.Csdh.7AL QTL, QHt. Csdh.1D QTL, Lr34 gene*	WB02	Mondal et al. (2016)
	Salinity and drought tolerance	RM10793, Xbarc68	*SalTol QTL, QYld-3B QTL*	KRL 210	Singh et al. (2015b)
Maize	Tolerance to drought, heat, northern corn leaf blight (NCLB), and gray leaf spot (GLS)	umc1066, umc1287, umc1062, umc1060	*qDTY2.1 QTL, qHT5.1 QTL, genes providing resistance to NCLB and GLS*	DKC-67-87	Babu et al. (2012)
	Tolerance to drought, heat, European corn borer (ECB), and corn rootworm (CR)	umc2025, umc1116, umc2003, umc1951	*qDTY3.1 QTL, qHT6.1QTL, genes providing resistance to ECB and CR*	Pioneer 39 N84	Cairns et al. (2013)
Soybean	Resistance to drought, heat, soybean rust, and sudden death syndrome	Satt309, Satt201, Satt160	*qDTY1.1 QTL, qHT2.1 QTL, Rpp genes, SDS genes*	S1001	Gurung et al. (2012)
Tomato	Tolerance to drought, heat, tomato leaf curl virus (TLCV), and fusarium wilt (FW)	SSR_123, SSR_456, ty-1, I-2	*QDTY6.1 QTL, QHT2.1 QTL, genes providing resistance to TLCV and FW*	Sungold	Cohen and Weingarten (2013)
	Tolerance to drought, heat, early blight (EB), and tomato powdery mildew (TPM)	SSR_789, SSR_101, Eb-1, mil-1	*QDTY4.1 QTL, QHT3.1 QTL, genes providing resistance to EB and TPM*	H9815	Adato et al. (2017)

tolerance. Insights from functional genomics and systems biology will help unravel the genetic networks involved in multistress responses, supporting the discovery of new markers (Malenica et al. 2021). Enhanced high-throughput phenotyping and comprehensive field trials are crucial for validating markers in diverse conditions. Increased collaboration and data sharing among researchers and breeders will expedite progress, ultimately leading to the creation of resilient crops capable of thriving under multiple stress conditions.

7.5 Gene Editing

In recent years, advancements in genomics and molecular biology have introduced genetic engineering as a promising approach to increasing plant stress tolerance by incorporating specific genes or regulatory elements into the genome (Jaganathan et al. 2018). However, the global adoption of genetically modified crops remains limited due to health and environmental concerns. Given the growing food demand, innovative and effective strategies are necessary to develop crops that can withstand abiotic stresses while maintaining high yields. Genome editing has emerged as a revolutionary approach in plant science and agriculture, offering precise modifications to the genome without integrating foreign DNA (Arimura and Nakazato 2024). Natural genome alterations have been part of plant evolution for thousands of years, with significant genetic variations shaping modern crops. For example, modern corn differs substantially from its wild ancestor, teosinte, due to these natural variations (Jiao et al. 2012). However, the random nature of these mutations and the rigorous selection process for desirable traits highlighted the need for more targeted approaches (Sun et al. 2023). Genome editing allows precise alterations to the DNA sequence at specific sites within the genome and uses sequence-specific endonucleases (SSEs) to create double-stranded DNA breaks (DSBs), which are then repaired by the cell's natural repair mechanisms through nonhomologous end-joining (NHEJ) or homologous recombination (HR) (Kumar et al. 2024). NHEJ is often error-prone, leading to deletions or insertions, whereas HR is an error-free process that uses a DNA template to accurately replace nucleotides (Vogt and He 2023). Over the past decade, genome editing has been used to enhance various traits in plants, with significant advancements due to the development of simpler and more effective tools like CRISPR/Cas.

7.5.1 Mechanism of Gene Editing by CRISPR/Cas

The term "CRISPR" was first introduced by Rudd Jansen and his colleagues in 2002 to describe multiple short repeat sequences ranging from 21 to 37 base pairs (Jansen et al. 2002). These sequences, known as Clustered Regularly Interspaced Short Palindromic Repeats (CRISPR), are characterized by their repetitive nature, interspersed with nonrepetitive sequences, and flanked by a leader sequence of 300–500 base pairs. Francisco Mojica later identified conserved motifs in *Streptococcus thermophilus* that were linked with protospacers and termed these motifs as protospacer adjacent motifs (PAM) (Mojica et al. 2009). PAM sequences, typically two to five nucleotides long, play a crucial role in the site-specific recognition and cleavage of DNA by Cas (CRISPR-associated) nucleases (Gleditzsch et al. 2019). The most widely used Cas protein for genome editing is Cas9 from *Streptococcus pyogenes* (*SpCas9*), which specifically recognizes the PAM sequence 5′NGG3′ (where "N" can be A, T, G, or C) to initiate cleavage at the target site. To broaden its applicability, *SpCas9* has been engineered to recognize a wider range of PAM sequences, including NGG, GAA, and GAT, enhancing its ability to target different genomic sites (Wada et al. 2020). Additionally, Cas variants from various organisms have been engineered to recognize broader PAM sequences, improving the specificity of target recognition, especially for sequences that escape recognition by *SpCas9* (Hu et al. 2018). During the CRISPR process, the transcription of the CRISPR array results in the formation of mature CRISPR RNA (crRNA), which requires the presence of a trans-activating crRNA (tracrRNA) for its maturation (Deltcheva et al. 2011). This maturation often results in a chimera of crRNA and tracrRNA, known as short guide RNA (sgRNA). This sgRNA, in complex with Cas proteins, scans the genome for PAM sequences to induce site-specific double-stranded breaks (DSBs) (Zaychikova et al. 2020). The target DNA is recognized by the crRNA-protein complex, and cleavage at defined sites is facilitated by the HNH and RuvC-like nuclease domains. The effectiveness of the CRISPR/Cas system largely depends on the recognition of PAM sequences at target sites in the genome (Adli 2018). Following the recognition of these sequences, DSBs are introduced by sequence-specific nucleases (Cas variants) upstream of the PAM site. These breaks are then repaired by either homologous direct repair (HDR), an error-free process, or nonhomologous end-joining (NHEJ), which is error-prone (Manghwar et al. 2019). HDR utilizes a template to replace the cleaved section with a user-defined sequence, while NHEJ may induce small insertions or deletions at the cleavage site. The choice between HDR and NHEJ pathways depends on the nature of the DNA break and the phase of the cell cycle (Yang et al. 2020). Although HDR is error-free, NHEJ is more commonly observed in somatic cells, helping maintain genomic integrity in complex genomes (Tang et al. 2019). In genome editing, the guide RNA's spacer sequence can be modified to direct Cas proteins to edit different targets within the genome (Jiang et al. 2013). A comprehensive understanding of DNA repair mechanisms is crucial for achieving precise modifications in the target genome, which is essential for efficiently generating modified plant genomes.

7.5.2 CRISPR-Based Editing in Crops

CRISPR/Cas has been successfully employed to edit the genomes of a wide range of crop plants, including maize, rice, tomato, soybean, and many others (Ndudzo et al. 2024). Notably, CRISPR/Cas can target complex quantitative traits linked to abiotic stress factors, making it a valuable tool for improving stress tolerance in plants. For example, in rice, CRISPR/Cas-mediated editing of genes such as *OsPDS* (*phytoene desaturase*), *OsMPK2* (*mitogen-activated protein kinase*), and *OsBADH2* (*betaine aldehyde dehydrogenase*) has been shown to confer enhanced tolerance to abiotic stresses (Shan et al. 2013; Shan et al. 2015a, b). The editing efficiency for *OsPDS* and *OsBADH2* was reported to be around 9% and 7%, respectively. Further, CRISPR/Cas has been utilized to edit other genes in rice, such as *OsMPK5* (a negative regulator of stress responses) and *SAPK2* (*stress/ABA-activated protein kinase* 2), resulting in improved stress tolerance (Lou et al. 2017). Additionally, editing genes like *SPP* (*stromal processing peptidase*), *YSA* (*young seedling albino*), and *ROC5* (*outermost cell-specific gene5*) have shown the potential to enhance abiotic stress resilience (Feng et al. 2013). The broad applicability and accuracy of CRISPR/Cas in plant genome editing make it a promising strategy for advancing agricultural practices (Mann et al. 2024). The various examples of CRISPR-based gene editing and target genes and traits are given in Table 7.3. However, one can now target gene families so that a cascade of pathways can be targeted and modified.

Table 7.3 List of crops targeted through gene editing to develop multiple stress tolerance

Crop	Genes associated with structural variation	Phenotype observed	References
Rice	*Sub1A, Sub1B,* and *Sub1C*	Submergence tolerance	Hussain et al. (2024)
	OsNR2	Yield potential and nitrogen use efficiency	Gao et al. (2019)
	HIS1 (HPPD INHIBITOR SENSITIVE 1)	Herbicide resistance	Maeda et al. (2019)
	NRT1.1B	Nitrogen use efficiency	Li et al. (2018)
Wheat	*Rca2β*	Climate resilience and efficiency of carbon assimilation	Degen et al. (2020)
	TdDof	Solid-stem	Nilsen et al. (2020)
	Photoperiod-B1 (Ppd-B1), vernalization-A1 (Vrn-A1)	Adaptation against a broad range of environments	Errum et al. (2023)
	Promoter of Ppd-A1a and Ppd-B1a	Heading time	
Cotton	*GaFZ*	Fuzz and trichome reduction	Wang et al. (2021)
Maize	*ZmFBL41*	Banded leaf and sheath blight resistance	Li et al. (2019)
	ZmTIP1	Root hair elongation and drought tolerance	Zhang et al. (2020)
	ZmCCT9	Flowering induction under a long day	Huang et al. (2018)
	ZmVPP1	Drought tolerance	Liu et al. (2022)
	ZmWAK	Quantitative resistance to head smut disease	Dai et al. (2024)
	ZmMATE1	Aluminum tolerance	Vasconcellos et al. (2021)

7.6 Genomic Selection

Genomic selection (GS) offers an advanced approach that uses genome-wide marker information for prediction without necessarily identifying specific QTLs. Introduced by Meuwissen et al., (2021), GS has revolutionized the selection process by predicting the breeding value of an individual based on genomic data, significantly reducing the breeding cycle and enhancing genetic gain. It operates by leveraging both phenotypic and genotypic data from a reference population, commonly referred to as the "training population," to build a predictive model. In this population, individuals are both phenotypically characterized and genotyped across the entire genome using molecular markers. This enables the estimation of the effects of these markers on various traits (Meuwissen et al. 2021). Once the predictive model is developed, it can be applied to a "testing population," where only genotypic data is available. The goal is to predict the genomic estimated breeding values (GEBVs) for individuals in the testing population, using the model built from the training population. The GEBV is essentially an individual's potential genetic contribution to the next generation, predicted solely based on their genome-wide marker data, without needing to perform additional phenotyping (Alemu et al. 2024). This significantly accelerates the breeding process, as phenotypic evaluation is time-consuming and expensive, especially for traits with long development times or those that require specific environmental conditions.

7.6.1 Advantage of GS Over MAS

One of the key advantages of GS over marker-assisted selection (MAS) is that it does not rely on identifying individual markers or QTLs that are associated with traits. While MAS typically focuses on major-effect QTLs or specific regions of the genome, GS incorporates information from all markers distributed across the genome, regardless of their individual effects (Hayes and Goddard 2007). This inclusion of all available markers, even those with small individual effects, increases the precision of breeding value predictions, especially for complex polygenic traits that are controlled by many small-effect loci (Heffner et al. 2009). Moreover, the use of genome-wide markers in GS allows it to capture a higher proportion of the genetic variance compared to MAS, which tends to miss out on the minor-effect loci. As a result, GS is more effective for improving traits governed by many genes with small effects, such as grain yield or drought tolerance in crops (Lorenz and Nice 2017). This comprehensive approach makes GS particularly powerful for breeding programs targeting complex traits influenced by multiple genes and environmental factors (Nakaya and Isobe 2012). GS's reliance on genome-wide markers also provides greater flexibility and scope for application. Because it does not require the identification of specific QTLs, GS can be applied to any trait, regardless of whether it has been previously studied in detail or whether major QTLs have been identified (Hayes and Goddard 2007). This has made GS a popular choice for crop improvement programs, where traits of interest often exhibit complex inheritance patterns and are influenced by both genetic and environmental factors (Crossa et al. 2017).

7.6.2 Methods of GS

Genomic selection employs a range of statistical models to estimate breeding values, such as Bayesian Ridge Regression (BRR), Bayesian LASSO (BL), and Bayesian Generalized Linear Regression (BGLR). These models assume different prior distributions for marker effects and are well-suited to handle high-dimensional genomic data (Meuwissen et al. 2021). The marker genotypic data of the training population and their phenotypic results obtained through multienvironment evaluation are used to predict models based on various statistics (Pandey et al. 2020). Using the predictive model and genotypic data of individuals from the breeding population, genomic estimated breeding values (GEBVs) will be calculated. The performance of the individuals will be characterized based on GEBVs without phenotyping (Alemu et al. 2024). Genomic BLUP (GBLUP) is a linear mixed model that assumes random marker effects and provides a straightforward way to estimate GEBVs (Robertson et al. 2019). Machine Learning (ML) approaches such as Random Forest, Support Vector Machines (SVM), and deep learning have been employed now in GS to capture complex nonlinear relationships in the genomic data (Crossa et al. 2017). The choice of model depends on factors like population size, the trait being predicted, and computational resources. The various steps in genomic selection integrated with understanding the genetics of stress tolerance and utilization of functionally validated genes in gene editing are given in Fig. 7.1.

7.6.3 GS in Breeding for Multistress Tolerance

Genomic selection (GS) has shown immense promise in crop breeding programs, particularly for traits that are difficult to select through phenotypic methods alone and can provide a robust breeding pipeline for multistress-tolerant climate-smart varieties (Crossa et al. 2017). GS has been effectively applied in cereals like wheat and maize to predict grain yield, a trait with complex inheritance and significant genotype × environment interactions (Merrick et al. 2022). In crops like rice and soybean, genomic selection has been used to improve resistance to major diseases such as bacterial leaf blight and soybean cyst nematode along with yield (Wang et al. 2024). GS has been employed to enhance tolerance to abiotic stresses like drought and heat, which are critical in the context of climate change (Shikha et al. 2017). Genomic selection (GS) has shown significant potential in improving multistress tolerance in crops, particularly in addressing complex interactions between abiotic stresses such as drought, heat, and salinity (Rashid et al. 2023). These stresses often occur simultaneously in agricultural environments, especially under climate change scenarios, making it essential to breed crops with resilience to multiple stress factors. Traditional breeding methods face challenges in simultaneously

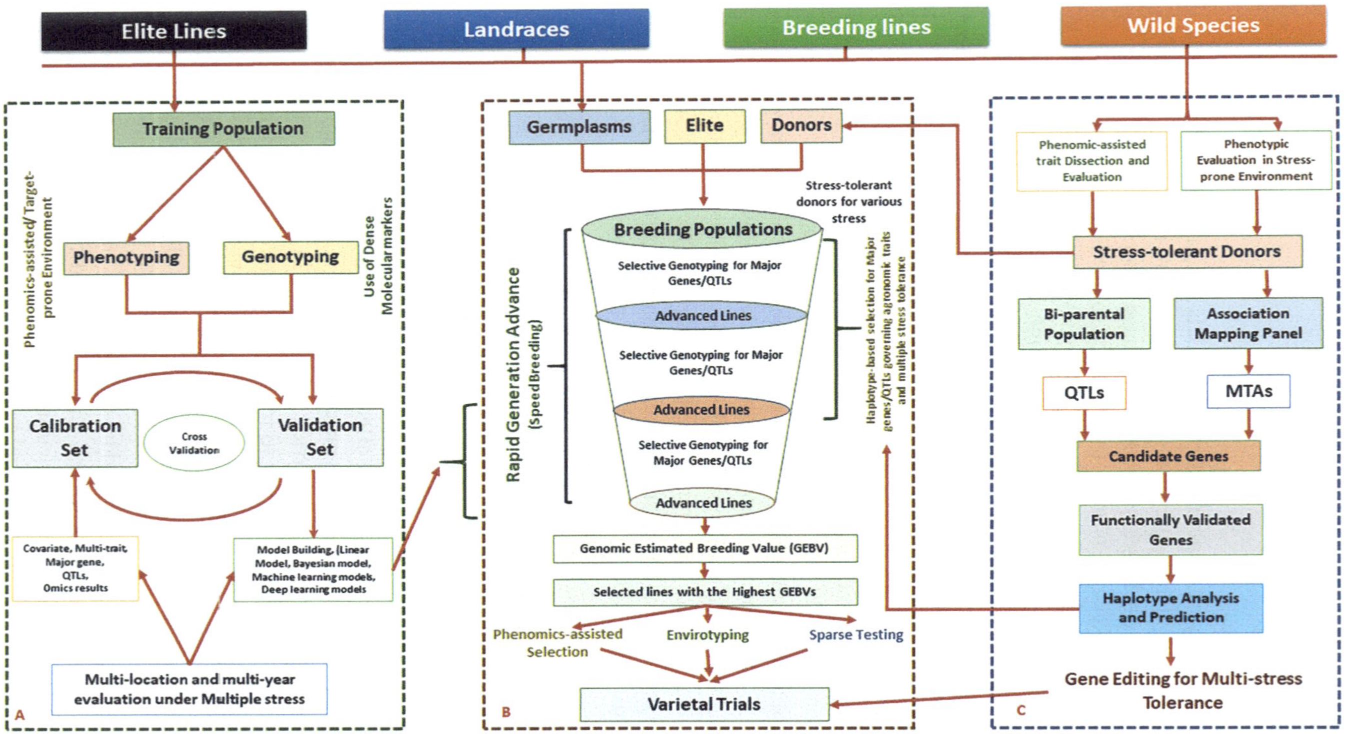

Fig. 7.1 A comprehensive breeding strategy for breeding multistress-tolerant crop varieties utilizing a genomic selection-based breeding strategy. (**a**) Development of training population from elite lines, landraces, breeding lines, and wild relatives and their extensive genotyping and phenotyping to build genomic selection model via different statistical techniques; (**b**) Use of germplasms, elite lines, and donors to develop breeding population by crossing them in definite fashion, their rapid generation advance using speed breeding and haplotype-assisted selection for multiple stress-tolerant genes/QTLs. The genotyping and predictive model used to generate genomic estimated breeding value and robust lines identified and evaluated for varietal release. (**c**) Extensive evaluation of germplasm lines using phenomics and multilocation evaluation to identify donors for stress tolerance. The donors can be used to map genes/QTLs governing stress tolerance using Bi-parental population and association studies. The QTLs can be narrowed down to functionally validated genes which can be used to perform haplotype-based selection in advanced breeding populations or for gene editing using CRISPR to develop multistress-tolerant varieties

selecting for multiple stresses due to the complex genetic architecture and genotype-by-environment interactions involved (Sandhu et al. 2021). GS, with its ability to utilize genome-wide markers, can capture the genetic variance associated with multiple stress responses more efficiently than marker-assisted selection (MAS), enabling faster genetic gains for multistress tolerance traits (Shikha et al. 2017). By using predictive models trained on data from environments that simulate various stress conditions, GS can identify genotypes with superior tolerance to multiple abiotic stresses, allowing breeders to develop resilient cultivars more rapidly (Pandey et al. 2020). This is especially relevant in staple crops such as wheat, rice, and maize, where multistress tolerance is critical for sustaining yields under increasingly erratic climatic conditions (Zhao et al. 2019). In recent years, the use of speed breeding for rapid generation advancement (RGA), sparse testing and evaluation, as well as characterization of the environment (envirotyping) is speeding up the GS approach and enhancing its efficiency (Hussain et al. 2024).

7.7 Advances in Phenotyping for Stress Breeding

The efficiency of plant breeding to develop high-yielding cultivars involves a breeding cycle of crossing, evaluation, and selection which further depends upon the genetic diversity, crossing strategy, and selection method. Genetic gain (response to selection) is an important component in the breeding cycle. Stress-related traits are largely governed by many genes each with a small effect and are highly influenced by the environment. For stress-related traits, significant genetic gain be realized only after several cycles of recurrent selection. The conventional breeding methods are time-consuming in the development of homozygous lines generated through the crossing of the selected parents.

7.7.1 Rapid Generation Advance

Rapid Generation Advance (RGA) with a range of breeding techniques emerged as a useful tool for faster development of homozygous lines from the segregating generation of the cross between divergent parents. Phenotyping is very crucial in the breeding cycle as the average phenotype collected is based on evaluated genotypes and provides the estimates of realized genetic gain. Accurate phenotypic information over the years provides the true genetic gain. A high-throughput phenotyping technology referred to as "phenomics" is considered to be an important tool for the rapid advancement of genetic gain in breeding cycles. Phenomics involves a multi-dimensional phenotyping of the growth and development process and also plant functioning as measured by physiological traits (Zhao et al. 2019). RGA coupled with phenomics empowered by bioinformatics and artificial intelligence (AI) is being extensively used to support the breeding program in terms of speed and accelerate the rate of genetic gain and breeding cycles. RGA breeding tools that involve single seed descent, speed breeding, shuttle breeding, doubled haploidy, marker-assisted breeding, and genomic selection are advantageous than conventional breeding in shortening the breeding cycle and faster development of elite lines. Speed breeding is another fast-track RGA breeding method which involves manipulating the crop growth conditions such as temperature, relative humidity, photoperiod, and light conditions to hasten the plant life cycle for flowering induction and seed set. By adopting a speed breeding strategy, at least three and up to nine generations can be advanced within a year as compared to the conventional method which allows two cycles per year for long day-to-day neutral crops (Ghosh et al. 2018). Shuttle breeding is also a kind of rapid evaluation of early-generation breeding lines (F_2) for stress tolerance in off-season nurseries at multiple locations (Prashanth Babu et al. 2022).

7.7.2 Phenomics-Assisted Evaluation

A high-throughput phenotyping platform for large-scale plant character evaluation was first initiated in Belgium (Reuzeau 2007) and in Australia. Of late, a number of national and regional Plant Phenotyping Networks have been established (Carrolla et al. 2019). For the past decade, phenomics has evolved into thriving research areas with a number of phenotyping platforms being updated at a faster pace and in gathering phenotyping information at multiple levels ranging from cellular, organ, and plant levels to the population level (Lobos et al. 2017). However, phenotyping at cellular and tissue levels, for example, root anatomical traits, is more challenging as it involves destructive samples and requires complex procedures (Hall et al. 2016). Later, most phenomics platforms focused on phenotyping individual plants in a controlled environment and automation with robotics either as sensor-to-plant or plant-to-sensor in a growth chamber or greenhouse (Gennaro et al. 2017). Several imaging techniques in high-throughput plant phenotyping platforms (HTPPs) include (1) RGB imaging for plant morphology, color, and texture; (2) Chlorophyll fluorescence imaging for photosynthesis; (3) Hyperspectral imaging for biochemical composition; (4) Thermal imaging for temperature, stomatal conductance, and transpiration rate and other advanced imaging techniques including magnetic resonance imaging for morphometric parameters/water content and laser scanning widely be employed under controlled environment conditions (Virlet et al. 2016). Field-based phenotyping platforms involving manned aircraft and unmanned aerial vehicle remote sensing platforms (UAV-RSPs) with a range of high-resolution visible, chlorophyll fluorescence and thermal infrared cameras, hyperspectral imagers, and 3D laser scanners are being employed to phenotype for abiotic and biotic stress tolerance, environmental factors, and their interaction with growth and production traits in time and space in field conditions (Furbank et al. 2019). Biotic and abiotic stresses adversely affect crop growth and productivity. Abiotic stresses such as drought, salinity and alkalinity, nutrient toxicity or deficiency, flooding and poor drainage, high and low temperatures, and heavy metals all are complex and often interacting phenomena and limit crop production worldwide (Singh et al. 2021b). Unlike biotic stresses which occur due to specific causative agents, abiotic stresses are caused by environmental factors and create a conducive environment(s) for the biotic stresses. Understanding the crop response to a given stress and their resistance mechanisms has been more challenging for the development of climate-resilient varieties. Recent innovations in molecular technology and functional genomics have enabled to dissect the complex stress traits. However, the genotype response under abiotic stress can be evaluated majorly by the manifestations of the phenotype which is more cumbersome and with limited available resources. A high-throughput phenomics facilities for the evaluation of large segregating populations and fixed lines in a faster way have been developed. Significant advances in imaging systems in combination with high-end computation tools enabled fast and nondestructive phenotyping techniques.

In breeding for stress-resistant varieties, various imaging techniques including visible light, thermal, infrared, fluorescence, spectroscopy-based imaging, and other integrated systems are in use for precise phenotyping under varied stress environments. Visible light imaging (300–700 mm) based on sensors sensitive to is commonly used for abiotic stress-associated traits. For example, LemnaTec 3D scanalyzer sensitive to visible light spectrum was employed to phenotype the variation in leaf color due to salt stress in rice (Hairmansis et al. 2014) and wheat (Meng et al. 2017); drought stress in barley (Findurová et al. 2023) and maize (Sheoran et al. 2022) to record the phenotypes of the stress-related traits. Thermal cameras with sensitivity to infrared range between 3 and 14 μm; near-infrared with 0.9–1.55 μm; and far-infrared with 7.5–13.5 μm are widely being employed to drought-associated traits including leaf water status, gas exchange, canopy temperature differences to measure drought tolerance (Li et al. 2020). Similarly, infrared red thermography is employed to detect leaf temperature differences and the transpiration rates in stress and nonstress treatment according to Planck's law for physiological traits in selecting genotypes under drought and salt stress (Al-Tamimi et al. 2016). Significant correlations between infrared image temperature patterns and physiological traits namely relative water content, stomatal conductance, and performance index were reported (Siddiqui et al. 2014). Negative correlation between image temperature and relative water content as well as stomatal conductance were observed (Siddiqui et al. 2014). Reduction in stomatal conductivity and relative leaf water content resulted in high leaf temperature detected by thermal image sensing. Hairmansis et al. (2014) studied the response of rice to varied levels of salt stress through nondestructive image-based phenotyping and quantified visible red-green-blue that discriminates shoot ion-independent stress and shoot ion-dependent stress. However, several factors like surface emittance, fluctuations in temperature and humidity, background thermal radiations, interference with the natural behavior of the leaf while handling, and not being amenable to more advanced plants due to random distribution of leaf angles can lead to errors in the measurement of absolute leaf temperature (Sirault et al. 2009). Advanced integrated imaging techniques are being employed for in situ imaging of plants. Positron emission tomography (PET), a nondestructive technique evaluates photosynthesis under stress conditions (Baker 2008). Magnetic Resonance Imaging (MRI) is another advanced technique used to study the root architecture and physiological processes in different crops (Borisjuk et al. 2012). Combination of both PET and MRI can be employed to monitor the real-time capture of the traits related to plant structure and function. Forster resonance energy transfer (FRET), is an outstanding nondestructive technology that can be employed for molecular phenotyping to study dynamic processes and pathways within the plants (Jones et al. 2014). In FRET, a high-resolution 3D laser scanner called PlantEye was used to differentiate wheat phenotype under control and salt stress (Maphosa et al. 2016). Phenomics platforms are being employed to detect biotic stress symptoms in response to disease and pest incidence. High-throughput phenotyping utilizes digital and fluorescence as well as hyperspectral cameras and sensors, laser scanners, magnetic resonance imaging to detect disease symptoms, the growth of fungal hyphae, and the quantification of disease or pest damage (Simko et al. 2017).

7.8 Conclusion and Future Perspective

Targeting multiple abiotic stress is very important in the present era of climate change and modern-day breeding utilizing genomic and phenomic selection is surely going to give a better edge to plant breeders in exercising selection and phenotyping. Potential benefits of phenomics tools for stress breeding will simplify and fasten the phenotyping and the development of climate-smart varieties. Phenomics serves as indispensable tool to assist in the genomics-assisted breeding methods capable of handling the huge information on crop and stress environments. In the era of big data, phenomics entered into mutlidimensional crop phenotyping to accurately identify the crop phenotypes that integrates genomes and phenotypes, systematically covering from cellular and tissue level to field level precise high-throughput phenotyping. Phenotyping platforms with automation, high-throughput, and high precision, will greatly improve accurate plant data collection efficiency and the efficiency of crop breeding. A number of phenotyping platforms are being updated at a faster pace. However, due to the high construction, operating, and maintenance costs of high-throughput phenotyping platforms, most academic and research institutions do not have access to these novel technologies. Increased research in affordable phenomics platforms at affordable costs will significantly increase the scope of phenomics and the advancement of genetic dissection of complex traits.

References

Adato N, Barrientos PG, Hochmuth AJ (2017) Developing multi-stress-tolerant tomato varieties using marker-assisted selection. Field Crop Res 211:103–112

Adli M (2018) The CRISPR tool kit for genome editing and beyond. Nat Commun 9(1):1911

Alemu A, Åstrand J, Montesinos-Lopez OA et al (2024) Genomic selection in plant breeding: key factors shaping two decades of progress. Mol Plant 17:552. https://doi.org/10.1016/j.molp.2024.03.007

Ali J, Anumalla M, Murugaiyan V, Li Z (2021) Green super rice (GSR) traits: breeding and genetics for multiple biotic and abiotic stress tolerance in rice. In: Rice improvement: physiological, molecular breeding and genetic perspectives. Springer International Publishing, Cham, pp 59–97

Al-Tamimi N, Brien C, Oakey H, Berger B, Saade S, Ho YS, Schmockel SM, Tester M, Negrao S (2016) Salinity tolerance loci revealed in rice using high-throughput non-invasive phenotyping. Nat Commun 7:13342

Arbona V, Manzi M, Zandalinas SI, Vives-Peris V, Pérez-Clemente RM, Gómez-Cadenas A (2017) Physiological, metabolic, and molecular responses of plants to abiotic stress. In: Stress signaling in plants: genomics and proteomics perspective, vol 2, pp 1–35

Arimura SI, Nakazato I (2024) Genome editing of plant mitochondrial and chloroplast genomes. Plant Cell Physiol 65(4):477–483

Ashraf M, Foolad MR (2013) Crop breeding for salt tolerance in the era of molecular markers and marker-assisted selection. Plant Breed 132(1):10–20

Babu R, Kumar A, Subudhi HN (2012) Marker-assisted selection for drought tolerance in maize: development of a drought-tolerant maize variety. Field Crop Res 127:115–124

Baker NR (2008) Chlorophyll fluorescence: a probe of photosynthesis in vivo. Annu Rev Plant Biol 59:89–113

Barmukh R, Roorkiwal M, Dixit GP, Bajaj P, Kholova J, Smith MR, Chitikineni A, Bharadwaj C, Sreeman SM, Rathore A, Tripathi S (2022) Characterization of 'QTL-hotspot' introgression lines reveals physiological mechanisms and candidate genes associated with drought adaptation in chickpea. J Exp Bot 73(22):7255–7272

Basnet B, Juliana P, Bhattarai K, Upreti U (2022) A review on major rust resistance gene and amino acid changes on wheat (*Triticum aestivum* L). Adv Agric 2022(1):7419326

Beyene Y, Semagn K, Crossa J, Mugo S, Atlin GN, Tarekegne A, Meisel B, Sehabiague P, Vivek BS, Oikeh S, Alvarado G, Machida L, Olsen M, Prasanna BM, Bänziger M (2016) Improving maize grain yield under drought stress and non-stress environments in sub-Saharan Africa using marker-assisted recurrent selection. Crop Sci 56(1):344–353

Bharadwaj C, Tripathi S, Soren KR, Thudi M, Singh RK, Sheoran S, Roorkiwal M, Patil BS, Chitikineni A, Palakurthi R, Vemula A (2021) Introgression of "QTL-hotspot" region enhances drought tolerance and grain yield in three elite chickpea cultivars. The Plant Genome 14(1):20076

Biswas PS, Rashid MM, Khatun H, Yasmeen R, Biswas JK (2019) Scope and progress of rice research harnessing cold tolerance. In: Advances in rice research for abiotic stress tolerance. Woodhead Publishing, pp 225–264

Bonilla P, Dvorak J, Mackill D, Deal K, Gregorio GB (2002) RFLP and SSLP mapping of salinity tolerance genes in chromosome 1 of rice (*Oryza sativa* L.) using recombinant inbred lines. Philipp Agric Sci (Philippines) 85(1)

Borisjuk L, Rolletschek H, Neuberger T (2012) Surveying the plant's world by magnetic resonance imaging. Plant J 70:129–146

Cairns JE, Prasanna BM, Orr BJ (2013) Maize research for drought and heat stress tolerance: an overview. Field Crop Res 152:81–90

Carrolla AA, Clarke J, Fahlgrenc NA, Gehanc MJ, Lawrence-Dilld C, Lorencee L (2019) NAPPN: who we ae, where we are going, and why you should join us! Plant Phenome J 2:180006

Casañas F, Simó J, Casals J, Prohens J (2017) Toward an evolved concept of landrace. Front Plant Sci 8:145

Chen C, Jost M, Clark B, Martin M, Matny O, Steffenson BJ et al (2021) Bed domain-containing nlr from wild barley confers resistance to leaf rust. Plant Biotechnol J 19:1206–1215

Choudhary M, Kumar P, Prakash NR, Gupta M, Sheoran S, Kumar P, Singh D, Jat BS, Saini DK (2024) QTLomics-based genetic dissection of key traits in major cereals: challenges and prospects. In: Omics and system biology approaches for delivering better cereals, pp 71–110

Cohen Y, Weingarten R (2013) Marker-assisted selection for multi-stress tolerance in tomatoes: a case study of Sungold. J Plant Breed Genet 12(3):112–123

Collard BC, Jahufer MZZ, Brouwer JB, Pang ECK (2005) An introduction to markers, quantitative trait loci (QTL) mapping and marker-assisted selection for crop improvement: the basic concepts. Euphytica 142:169–196

Cooper M, Gho C, Leafgren R, Tang T, Messina C (2014) Breeding drought-tolerant maize hybrids for the US corn-belt: discovery to product. J Exp Bot 65(21):6191–6204

Crossa J, Pérez-Rodríguez P, Cuevas J, Montesinos-López O, Jarquín D, De Los Campos G et al (2017) Genomic selection in plant breeding: methods, models, and perspectives. Trends Plant Sci 22(11):961–975

Dai Z, Pi Q, Liu Y, Hu L, Li B, Zhang B et al (2024) *ZmWAK02* encoding an RD-WAK protein confers maize resistance against gray leaf spot. New Phytol 241(4):1780–1793

Das G, Patra JK, Baek KH (2017) Insight into MAS: a molecular tool for development of stress resistant and quality of rice through gene stacking. Front Plant Sci 8:985

Degen GE, Worrall D, Carmo-Silva E (2020) An isoleucine residue acts as a thermal and regulatory switch in wheat rubisco activase. Plant J 103(2):742–751

Deltcheva E, Chylinski K, Sharma CM, Gonzales K, Chao Y, Pirzada ZA, Charpentier E (2011) CRISPR RNA maturation by trans-encoded small RNA and host factor RNase III. Nature 471(7340):602–607

Dwivedi SL, Ceccarelli S, Blair MW, Upadhyaya HD, Are AK, Ortiz R (2016) Landrace germplasm for improving yield and abiotic stress adaptation. Trends Plant Sci 21(1):31–42

Ellur RK, Khanna A, Yadav A, Pathania S, Rajashekara H, Singh VK, Krishnan SG, Bhowmick PK, Nagarajan M, Vinod KK, Prakash G (2016) Improvement of basmati rice varieties for resistance to blast and bacterial blight diseases using marker assisted backcross breeding. Plant Sci 242:330–341

Errum A, Rehman N, Uzair M, Inam S, Ali GM, Khan MR (2023) CRISPR/Cas9 editing of wheat *Ppd-1* gene homoeologs alters spike architecture and grain morphometric traits. Funct Integr Genomics 23(1):66

Fazlikhani L, Keilwagen J, Kopahnke D, Deising H, Ordon F, Perovic D (2019) High resolution mapping of *Rph mbr 1012* conferring resistance to *Puccinia hordei* in barley (*Hordeum vulgare* l.). Front Plant Sci 10:640. https://doi.org/10.3389/fpls.2019.00640

Feng Z, Zhang B, Ding W, Liu X, Yang DL, Wei P, Zhu JK (2013) Efficient genome editing in plants using a CRISPR/Cas system. Cell Res 23(10):1229–1232

Findurová H, Veselá B, Panzarová K, Pytela J, Trtílek M, Klem K (2023) Phenotyping drought tolerance and yield performance of barley using a combination of imaging methods. Environ Exp Bot 209:105314

Fiyaz RA, Shivani D, Chaithanya K, Mounika K, Chiranjeevi M, Laha GS, Viraktamath BC, Rao LS, Sundaram RM (2022) Genetic improvement of rice for bacterial blight resistance: present status and future prospects. Rice Sci 29(2):118–132

Furbank RT, Jimenez-Berni JA, George-Jaeggli B, Potgieter AB, Deery DM (2019) Field crop phenomics: enabling breeding for radiation use efficiency and biomass in cereal crops. New Phytol 223(4):1714–1727

Gamuyao R, Chin JH, Pariasca-Tanaka J, Pesaresi P, Catausan S, Dalid C et al (2012) The protein kinase *Pstol1* from traditional rice confers tolerance of phosphorus deficiency. Nature 488:535–539

Gao Z, Wang Y, Chen G, Zhang A, Yang S, Shang L, Wang D, Ruan B, Liu C, Jiang H, Dong G, Zhu L, Hu J, Zhang G, Zeng D, Guo L, Xu G, Teng S, Harberd NP, Qian Q (2019) The indica nitratereductase gene OsNR2 allele enhances rice yield potential and nitrogen use efficiency. Nat Commun 10(1):5207

Gennaro SFD, Rizza F, Badeck FW, Berton A, Delbono S, Gioli B et al (2017) UAV-based high-throughput phenotyping to discriminate barley vigour with visible and near-infrared vegetation indices. Int J Remote Sens 39:5330–5344

Ghosh S, Watson A, Gonzalez-Navarro OE, Ramirez-Gonzalez RH, Yanes L, Mendoza-Suárez M, Simmonds J, Wells R, Rayner T, Green P, Hafeez A (2018) Speed breeding in growth chambers and glasshouses for crop breeding and model plant research. Nat Protoc 13(12):2944–2963

Gleditzsch D, Pausch P, Müller-Esparza H, Özcan A, Guo X, Bange G, Randau L (2019) PAM identification by CRISPR-Cas effector complexes: diversified mechanisms and structures. RNA Biol 16(4):504–517

Gudi S, Jain M, Singh S, Kaur S, Srivastava P, Mavi GS et al (2024) Stress adaptive plasticity from *Aegilops tauschii* introgression lines *improves drought and heat stress tolerance in bread wheat (Triticum aestivum L.).* PeerJ 12:e17528

Guntukula R (2020) Assessing the impact of climate change on Indian agriculture: evidence from major crop yields. J Public Aff 20(1):e2040

Gurung S, Specht JE, Lee JD (2012) Marker-assisted selection for multi-stress tolerance in soybean: a case study. Field Crop Res 127:156–166

Hairmansis A, Berger B, Tester M, Roy S (2014) Image-based phenotyping for non-destructive screening of different salinity tolerance traits in rice. Rice 7:16

Hall HC, Fakhrzadeh A, Luengo Hendriks CL, Fischer U (2016) Precision automation of cell type classification and sub-cellular fluorescence quantification from laser scanning confocal images. Front Plant Sci 7:119

Hassan MA, Dahu N, Hongning T, Qian Z, Yueming Y, Yiru L, Shimei W (2023) Drought stress in rice: morpho-physiological and molecular responses and marker-assisted breeding. Front Plant Sci 14:1215371

Hayes B, Goddard M (2007) Genomic selection. J Anim Breed Genet 8:323

He M, He CQ, Ding NZ (2018) Abiotic stresses: general defenses of land plants and chances for engineering multistress tolerance. Front Plant Sci 9:1771

Heffner EL, Sorrells ME, Jannink JL (2009) Genomic selection for crop improvement. Crop Sci 49(1):1–12

Hu X, Meng X, Liu Q, Li J, Wang K (2018) Increasing the efficiency of CRISPR-Cas9-VQR precise genome editing in rice. Plant Biotechnol J 16(1):292–297

Huang C, Sun H, Xu D, Chen Q, Liang Y, Wang X et al (2018) *ZmCCT9* enhances maize adaptation to higher latitudes. Proc Natl Acad Sci 115(2):E334–E341

Huerta-Espino J, Singh R, Crespo-Herrera LA, Villaseñor-Mir HE, Rodriguez-Garcia MF, Dreisigacker S, Barcenas-Santana D, Lagudah E (2020) Adult plant slow rusting genes confer high levels of resistance to rusts in bread wheat cultivars from Mexico. Front Plant Sci 11:824

Hussain W, Anumalla M, Ismail AM, Walia H, Singh VK, Kohli A, Bhosale S, Bhardwaj H (2024) Revisiting FR13A for submergence tolerance-beyond the SUB1A gene. J Exp Bot erae299:5477

Ijaz M, Afzal A, Shabbir G, Iqbal J, Rafique M (2023) Breeding wheat for leaf rust resistance: past, present and future. Asian J Agric Biol 11(1):2021426

Jaganathan D, Ramasamy K, Sellamuthu G, Jayabalan S, Venkataraman G (2018) CRISPR for crop improvement: an update review. Front Plant Sci 9:985

James RA, Blake C, Byrt CS, Munns R (2011) Major genes for Na^+ exclusion, *Nax1* and *Nax2* (wheat HKT1; 4 and HKT1; 5), decrease Na^+ accumulation in bread wheat leaves under saline and waterlogged conditions. J Exp Bot 62(8):2939–2947

Jansen R, Greenbaum D, Gerstein M (2002) Relating whole-genome expression data with protein-protein interactions. Genome Res 12(1):37–46

Jiang W, Bikard D, Cox D, Zhang F, Marraffini LA (2013) RNA-guided editing of bacterial genomes using CRISPR-Cas systems. Nat Biotechnol 31(3):233–239

Jiao Y, Zhao H, Ren L, Song W, Zeng B, Guo J et al (2012) Genome-wide genetic changes during modern breeding of maize. Nat Genet 44(7):812–815

Jones AM, Danielson JÅ, ManojKumar SN, Lanquar V, Grossmann G, Frommer WB (2014) Abscisic acid dynamics in roots detected with genetically encoded FRET sensors. elife 3:e01741

Junaid MD, Gokce AF (2024) Global agricultural losses and their causes. Bull Biol All Sci Res 2024(1):66–66

Kapazoglou A, Gerakari M, Lazaridi E, Kleftogianni K, Sarri E, Tani E, Bebeli PJ (2023) Crop wild relatives: a valuable source of tolerance to various abiotic stresses. Plan Theory 12(2):328

Karelov A, Kozub N, Sozinova O, Pirko Y, Sozinov I, Yemets A, Blume Y (2022) Wheat genes associated with different types of resistance against stem rust (*Puccinia graminis* Pers.). Pathogens 11(10):1157

Karki S, Burton P, Mackey B (2020) The experiences and perceptions of farmers about the impacts of climate change and variability on crop production: a review. Clim Dev 12(1):80–95

Kavanagh PJ, Singh D, Bansal UK, Park RF (2017) Inheritance and characterization of the new and rare gene Rph25 conferring seedling resistance in *Hordeum vulgare* against *Puccinia hordei*. Plant Breed 136:908–912

Kim TW, Jehanzaib M (2020) Drought risk analysis, forecasting and assessment under climate change. Water 12(7):1862

Kumar A, Bernier J, Verulkar S, Lafitte HR, Atlin GN (2008) Breeding for drought tolerance: direct selection for yield, response to selection, and use of drought-tolerant donors in upland and lowland-adapted populations. Field Crop Res 107(3):221–231

Kumar K, Jan I, Saripalli G, Sharma PK, Mir RR, Balyan HS, Gupta PK (2022) An update on resistance genes and their use in the development of leaf rust resistant cultivars in wheat. Front Genet 13:816057

Kumar N, Taneja A, Ghosh M, Rothweiler U, Sundaresan NR, Singh M (2024) Harmonin homology domain-mediated interaction of RTEL1 helicase with RPA and DNA provides insights into its recruitment to DNArepair sites. Nucleic Acids Res 52(3):1450–1470

Kuroha T, Ashikari M (2020) Molecular mechanisms and future improvement of submergence tolerance in rice. Mol Breed 40(4):41

Li J, Zhang X, Sun Y, Zhang J, Du W, Guo X et al (2018) Efficient allelic replacement in rice by gene editing: a case study of the *NRT1. 1B* gene. J Integr Plant Biol 60(7):536–540

Li N, Lin B, Wang H, Li X, Yang F, Ding X et al (2019) Natural variation in Zm FBL41 confers banded leaf and sheath blight resistance in maize. Nat Genet 51(10):1540–1548

Li W, Deng Y, Ning Y, He Z, Wang GL (2020) Exploiting broad-spectrum disease resistance in crops: from molecular dissection to breeding. Annu Rev Plant Biol 71(1):575–603

Liu S, Liu X, Zhang X, Chang S, Ma C, Qin F (2022) Co-expression of *ZmVPP1* with *ZmNAC111* confers robust drought resistance in maize. Genes 14(1):8

Lobos GA, Camargo AV, Del Pozo A, Araus JL, Ortiz R, Doonan JH (2017) Editorial: plant phenotyping and phenomics for plant breeding. Front Plant Sci 8:2181

Lorenz A, Nice L (2017) Training population design and resource allocation for genomic selection in plant breeding. In: Genomic selection for crop improvement: new molecular breeding strategies for crop improvement, pp 7–22

Lou D, Wang H, Liang G, Yu D (2017) OsSAPK2 confers abscisic acid sensitivity and tolerance to drought stress in rice. Front Plant Sci 8:993

Maeda H, Murata K, Sakuma N, Takei S, Yamazaki A, Karim MR et al (2019) A rice gene that confers broad-spectrum resistance to β-triketone herbicides. Science 365(6451):393–396

Malenica N, Dunić JA, Vukadinović L, Cesar V, Šimić D (2021) Genetic approaches to enhance multiple stress tolerance in maize. Genes 12(11):1760

Mallick N, Vinod, Sharma JB, Tomar RS, Sivasamy M, Prabhu KV (2015) Marker-assisted backcross breeding to combine multiple rust resistance in wheat. Plant Breed 134(2):172–177

Mallick N, Jha SK, Agarwal P, Kumar S, Mall A, Choudhary MK, Chandra AK, Bansal S, Saharan MS, Sharma JB, Vinod (2022) Marker-assisted transfer of leaf and stripe rust resistance from *Triticum turgidum* var. *durum* cv. *Trinakria* to wheat variety HD2932. Front Genet 13:941287

Manghwar H, Lindsey K, Zhang X, Jin S (2019) CRISPR/Cas system: recent advances and future prospects for genome editing. Trends Plant Sci 24(12):1102–1125

Mann A, Kumar N, Lata C, Kumar A, Kumar A, Meena BL (2019) Functional annotation of differentially expressed genes under salt stress in *Dichanthium annulatum*. Plant Physiol Rep 24:104–111

Mann A, Kumar A, Sanwal SK, Sharma PC (2020) Sustainable production of pulses under saline lands in India. In: Legume crops-prospects, production and uses. IntechOpen. https://doi.org/10.5772/intechopen.90304
Mann A, Ranga P, Choudhary P, Yadav S, Kaul N, Dahiya A, Prakash NR, Kumar A, Kumar A, Sanwal SK (2024) Genome Editing Technologies for Enhancing Plant Resilience to Biotic and Abiotic Stresses-Brief Review. Journal of Soil Salinity and Water Quality 16(2):180–193
Maphosa L, Thoday-Kennedy E, Vkani J, Phelan A, Badenhorst P, Slater A, Spangenberg G, Kant S (2016) Phenotyping wheat under salt stress conditions using a 3D laser scanner. Israel J Plant Sci 1:1–8
Marone D, Russo MA, Mores A, Ficco DB, Laidò G, Mastrangelo AM, Borrelli GM (2021) Importance of landraces in cereal breeding for stress tolerance. Plan Theory 10(7):1267
Meng R, Saade S, Kurtek S, Berger B, Brien C, Pillen K et al (2017) Growth curve registration for evaluating salinity tolerance in barley. Plant Methods 13:1–9
Merrick LF, Herr AW, Sandhu KS, Lozada DN, Carter AH (2022) Optimizing plant breeding programs for genomic selection. Agronomy 12(3):714
Meuwissen T, van den Berg I, Goddard M (2021) On the use of whole-genome sequence data for across-breed genomic prediction and fine-scale mapping of QTL. Genet Sel Evol 53:1–15
Miah G, Rafii MY, Ismail MR, Puteh AB, Rahim HA, Asfaliza R, Latif MA (2013) Blast resistance in rice: a review of conventional breeding to molecular approaches. Mol Biol Rep 40:2369–2388
Mohanavel V, Muthu V, Kambale R, Palaniswamy R, Seeli P, Ayyenar B et al (2024) Marker-assisted breeding accelerates the development of multiple-stress-tolerant rice genotypes adapted to wider environments. Front Plant Sci 15:1402368
Mojica FJ, Díez-Villaseñor C, García-Martínez J, Almendros C (2009) Short motif sequences determine the targets of the prokaryotic CRISPR defence system. Microbiology 155(3):733–740
Mondal S, Singh RP, Mason ER, Huerta-Espino J, Autrique E, Joshi AK (2016) Grain yield, adaptation and progress in breeding for early-maturing and heat-tolerant wheat lines in South Asia. Field Crop Res 192:78–85
Mukherjee M, Behera L, Yadav MK, Mahadani P, Baksh SY, Donde R et al (2023) Improvement of grain-yield under drought stress and blast resistance in popular upland rice variety Nagina-22 using marker assisted selection. Preprint. https://doi.org/10.21203/rs.3.rs-2771415/v1
Nachimuthu VV, Sabariappan R, Muthurajan R, Kumar A (2017) Breeding rice varieties for abiotic stress tolerance: challenges and opportunities. In: Abiotic stress management for resilient agriculture, pp 339–361
Nakaya A, Isobe SN (2012) Will genomic selection be a practical method for plant breeding? Ann Bot 110(6):1303–1316
Ndudzo A, Makuvise AS, Moyo S, Bobo ED (2024) CRISPR-Cas9 genome editing in crop breeding for climate change resilience: implications for smallholder farmers in Africa. J Agric Food Res:101132
Neeraja CN, Maghirang-Rodriguez R, Pamplona A, Heuer S, Collard BC, Septiningsih EM, Mackill DJ (2007) A marker-assisted backcross approach for developing submergence-tolerant rice cultivars. Theor Appl Genet 115(6):767–776
Nilsen KT, Walkowiak S, Xiang D, Gao P, Quilichini TD, Willick IR et al (2020) Copy number variation of TdDof controls solid-stemmed architecture in wheat. Proc Natl Acad Sci 117(46):28708–28718
Numan M, Serba DD, Ligaba-Osena A (2021) Alternative strategies for multi-stress tolerance and yield improvement in millets. Genes 12(5):739
Panda D, Barik J (2021) Flooding tolerance in rice: focus on mechanisms and approaches. Rice Sci 28(1):43–57
Pandey MK, Chaudhari S, Jarquin D, Janila P, Crossa J, Patil SC, Sundravadana S, Khare D, Bhat RS, Radhakrishnan T, Hickey JM, Varshney RK (2020) Genome-based trait prediction in multi-environment breeding trials in groundnut. Theor Appl Genet 133:3101–3117

Peng S, Huang J, Sheehy JE, Laza RC, Visperas RM, Zhong X, Centeno GS, Khush GS, Cassman KG (2004) Rice yields decline with higher night temperature from global warming. Proc Natl Acad Sci U S A 6:9971–9975

Potts J, Jangra S, Michael VN, Wu X (2023) Speed breeding for crop improvement and food security. Crops 3(4):276–291

Pradhan SK, Nayak DK, Mohanty S, Behera L, Barik SR, Pandit E, Lenka S (2015) Incorporation of bacterial blight resistance and yield component QTLs through marker-assisted selection in hybrid rice parental line, DRR17A. Euphytica 205(1):183–196

Prakash NR, Sheoran S, Saini M, Punia M, Rathod NKK, Bhinda MS, Vinesh B, Choudhary MK, Sarkar B (2020) Offsetting climate change impact through genetic enhancement. In: Srinivasarao C et al (eds) Climate change and Indian agriculture: challenges and adaptation strategies. ICAR-National Academy of Agricultural Research Management, Hyderabad, pp 71–104

Prakash NR, Lokeshkumar BM, Rathor S, Warriach AS, Yadav S, Vinaykumar NM, Krishnamurthy SL, Sharma PC (2024) Mechanisms of saline and submergence tolerance in rice for coastal ecology. In: Genetic improvement of rice for salt tolerance. Singapore, Springer Nature Singapore, pp 231–256

Prashanth Babu H, Kumar M, Gaikwad KB, Kumar R, Kumar N, Palaparthi D et al (2022) Rapid generation advancement and fast-track breeding approaches in wheat improvement. In: Next-generation plant breeding approaches for stress resilience in cereal crops. Singapore, Springer Nature Singapore, pp 241–262

Priyadarshan PM (2019) Breeding for abiotic stress adaptation. In: Plant breeding: classical to modern, pp 413–455

Rai N, Bellundagi A, Kumar PK, Thimmappa KR, Rani S, Sinha N, Krishna H, Jain N, Singh GP, Singh PK, Chand S, Prabhu KV (2018) Marker-assisted backcross breeding for improvement of drought tolerance in bread wheat (*Triticum aestivum* L.). Plant Breed 137(4):514–526

Rashid MH, Bhuiyan MAR, Mahbub MAA, Hossain MT, Iftekharuddaula KM, Gregorio GB (2017) Marker-assisted backcrossing for development of high yielding salt tolerant rice varieties in Bangladesh. SABRAO J Breed Genet 49(3):264–271

Rashid MAR, Sajjad M, Gul A, Zhao Y (2023) Genomic selection and characterization in cereals. Front Genet 13:1092095

Rehman S, Amouzoune M, Hiddar H, Aberkane H, Benkirane R, Filali-Maltouf A, Al-Jaboobi M, Acqbouch L, Tsivelikas A, Verma R, Kehel Z, Birouk A, Amri A (2021) Traits discovery in *Hordeum vulgare* sbsp. *Spontaneum* accessions and in lines derived from interspecific crosses with wild *Hordeum* species for enhancing barley breeding efforts. Crop Sci 61(1):219–233

Reuzeau C (2007) TraitMill (TM): a high-throughput functional genomics platform for the phenotypic analysis of cereals. In: Vitro cellular and developmental biology – animal, vol 43, p S4

Ribaut JM, Ragot M (2007) Marker-assisted selection to improve drought adaptation in maize: the backcross approach, perspectives, limitations, and alternatives. J Exp Bot 58(2):351–360

Robertsen CD, Hjortshøj RL, Janss LL (2019) Genomic selection in cereal breeding. Agronomy 9(2):95

Sandhu N, Yadav S, Catolos M, Cruz MTS, Kumar A (2021) Developing climate-resilient, direct-seeded, adapted multiple-stress-tolerant rice applying genomics-assisted breeding. Front Plant Sci 12:637488

Septiningsih EM, Pamplona AM, Sanchez DL, Neeraja CN, Vergara GV, Heuer S, Mackill DJ (2009) Development of submergence-tolerant rice cultivars: the *Sub1* locus and beyond. Ann Bot 103(2):151–160

Shan Q, Wang Y, Li J, Zhang Y, Chen K, Liang Z, Zhang K, Liu J, Xi JJ, Qiu JL, Gao C (2013) Targeted genome modification of crop plants using a CRISPR-Cas system. Nat Biotechnol 31(8):686–688

Sheoran S, Kaur Y, Kumar S, Shukla S, Rakshit S, Kumar R (2022) Recent advances for drought stress tolerance in maize (*Zea mays* L.): present status and future prospects. Front Plant Sci 13:872566

Shikha M, Kanika A, Rao AR, Mallikarjuna MG, Gupta HS, Nepolean T (2017) Genomic selection for drought tolerance using genome-wide SNPs in maize. Front Plant Sci 8:550
Siddiqui ZS, Cho JI, Park SH, Kwon TR, Ahn BO, Lee GS, Jeong MJ, Kim KW, Lee SK, Park SC (2014) Phenotyping of rice in salt stress environment using high-throughput infrared imaging. Acta Bot Croat 73(1):149–158
Simko I, Jimenez-Berni JA, Sirault XR (2017) Phenomic approaches and tools for Phytopathologists. Phytopathology 107(1):6–17
Singh D, Dracatos P, Derevnina L, Zhou M, Park RF (2015a) *Rph23*: a new designated additive adult plant resistance gene to leaf rust in barley on chromosome 7h. Plant Breed 134:62–69
Singh K, Vinod, Sharma A, Singh RP (2015b) Salinity and drought-tolerant wheat: breeding, physiology, and molecular approaches. Indian J Plant Physiol 20(2):168–176
Singh RP, Srivastava P, Sharma A, Bains NS (2017) Bread wheat cultivar PBW 343 carries residual additive resistance against virulent stripe rust pathotype. J Crop Improv 31(2):183–191
Singh KP, Kumari P, Rai PK (2021a) Current status of the disease-resistant gene (s)/QTLs, and strategies for improvement in *Brassica juncea*. Front Plant Sci 12:617405
Singh RK, Kota S, Flowers TJ (2021b) Salt tolerance in rice: seedling and reproductive stage QTL mapping come of age. Theor Appl Genet 134:3495–3533
Sirault XRR, Jame RA, Furbank RT (2009) A new screening method for osmotic component of salinity tolerance in cereals using infrared thermography. Funct Plant Biol 36:970–977
Snehi S, Bhutia RN, Prakash NR (2022) Breeding Rice cultivars suitable for coastal regions of India. Food Sci Rep 3(4):54–58
Sun S, Wang B, Li C, Xu G, Yang J, Hufford MB et al (2023) Unraveling prevalence and effects of deleterious mutations in maize elite lines across decades of modern breeding. Mol Biol Evol 40(8):msad170
Tang XD, Gao F, Liu MJ, Fan QL, Chen DK, Ma WT (2019) Methods for enhancing clustered regularly interspaced short palindromic repeats/Cas9-mediated homology-directed repair efficiency. Front Genet 10:551
van Zonneveld M, Rakha M, Tan SY, Chou YY, Chang CH, Yen JY et al (2020) Mapping patterns of abiotic and biotic stress resilience uncovers conservation gaps and breeding potential of *Vigna* wild relatives. Sci Rep 10(1):2111
Varshney RK, Thudi M, Nayak SN, Gaur PM, Kashiwagi J, Krishnamurthy L et al (2014) Genetic dissection of drought tolerance in chickpea (*Cicer arietinum* L.). Theor Appl Genet 127:445–462
Vasconcellos RC, Mendes FF, de Oliveira AC, Guimarães LJ, Albuquerque PE, Pinto MO et al (2021) *ZmMATE1* improves grain yield and yield stability in maize cultivated on acid soil. Crop Sci 61(5):3497–3506
Villa TCC, Maxted N, Scholten M, Ford-Lloyd B (2005) Defining and identifying crop landraces. Plant Genet Resour 3(3):373–384
Virlet N, Sabermanesh K, Sadeghi-Tehran P, Hawkesford MJ (2016) Field scanalyzer: an automated robotic field phenotyping platform for detailed crop monitoring. Funct Plant Biol 44:143–153
Visakh RL, Anand S, Arya SN, Sasmita B, Jha UC, Sah RP, Beena R (2024) Rice heat tolerance breeding: a comprehensive review and forward gaze. Rice Sci 31(4):375–400
Vogt A, He Y (2023) Structure and mechanism in non-homologous end joining. DNA Repair 130:103547
Wada N, Ueta R, Osakabe Y, Osakabe K (2020) Precision genome editing in plants: state-of-the-art in CRISPR/Cas9-based genome engineering. BMC Plant Biol 20:1–12
Wang X, Miao Y, Cai Y, Sun G, Jia Y, Song S et al (2021) Large-fragment insertion activates gene *GaFZ* (*Ga08G0121*) and is associated with the fuzz and trichome reduction in cotton (*Gossypium arboreum*). Plant Biotechnol J 19(6):1110–1124
Wang J, Yang Q, Chen Y, Liu K, Zhang Z, Xiong Y et al (2024) QTL mapping and genomic selection of stem and branch diameter in soybean (*Glycine max* L.). Front Plant Sci 15:1388365

Yadav MK, Aravindan S, Ngangkham U, Raghu S, Prabhukarthikeyan SR, Keerthana U, Marndi BC, Adak T, Munda S, Deshmukh R, Pramesh D (2019) Blast resistance in Indian rice landraces: genetic dissection by gene specific markers. PLoS One 14(1):e0211061

Yadav S, Modi P, Dave A, Vijapura A, Patel D, Patel M (2020) Effect of abiotic stress on crops. Sustain Crop Prod 3(17):5–16

Yadav OP, Gupta SK, Sanjana Reddy P (2024) Pearl millet breeding for enhancing yield and stability: strategies, achievements, and perspectives. In: Pearl millet in the 21st century: food-nutrition-climate resilience-improved livelihoods, pp 331–360

Yan L, Luo T, Huang D, Wei M, Ma Z, Liu C, Qin Y, Zhou X, Lu Y, Li R, Qin G (2023) Recent advances in molecular mechanism and breeding utilization of brown planthopper resistance genes in rice: an integrated review. Int J Mol Sci 24(15):12061

Yang S, Gill RA, Zaman QU, Ulhassan Z, Zhou W (2020) Insights on SNP types, detection methods and their utilization in brassica species: recent progress and future perspectives. J Biotechnol 324:11–20

Yang Z, Cao Y, Shi Y, Qin F, Jiang C, Yang S (2023) Genetic and molecular exploration of maize environmental stress resilience: toward sustainable agriculture. Mol Plant 16(10):1496–1517

Zaychikova MV, Danilenko VN, Maslov DA (2020) CRISPR-Cas systems: prospects for use in medicine. Appl Sci 10(24):9001

Zhang X, Mi Y, Mao H, Liu S, Chen L, Qin F (2020) Genetic variation in *ZmTIP1* contributes to root hair elongation and drought tolerance in maize. Plant Biotechnol J 18(5):1271–1283

Zhao C, Zhang Y, Du J, Guo X, Wen W, Gu S, Wang J, Fan J (2019) Crop Phenomics: current status and perspectives. Front Plant Sci 10:714

Zhao X, Liu S, Zhang L (2023) Advancements in marker-assisted selection for multi-stress tolerance in crops: a review. Front Plant Sci 14:853735

Zörb C, Geilfus CM, Dietz KJ (2019) Salinity and crop yield. Plant Biol 21:31–38

Speed Breeding: A Budding Technique to Improve Crop Plants for Multistress Tolerance

8

Ramandeep Kaur, Nidhi Devi, Akanksha, Karanveer Kaur, Imran Sheikh, and Vikrant Tyagi

Abstract

The growing world population and the problems caused by climate change require the creation of crop types that are more resilient to many stresses. Several challenges that crop plants face, such as heat, salinity, drought, and mineral toxicity, may strengthen due to the current dynamic climate circumstances, ultimately affecting agricultural output. Due to resource limits and lengthy generation timeframes, traditional breeding methods sometimes encounter difficulties. A potential remedy is speed breeding, an innovative method that quickens the Plant Breeding cycle. Speed breeding allows numerous generations of crops to be produced in a shorter amount of time by adjusting environmental factors including temperature, light intensity, and photoperiod. Furthermore, speed breeding in union with high-throughput phenotyping and marker-assisted selection (MAS) techniques can enable rapid and numerous trait selections. The concepts and uses of speed breeding are examined in this chapter, with an emphasis on how it might hasten the creation of crop varieties that are more resilient to multistressors. This chapter covers the new idea of speed breeding, including its applications, methodologies, and limitations. We also talk about the difficulties and potential of this novel approach to meeting the urgent needs of contemporary agriculture.

Keywords

Speed breeding · Multistress · Marker-assisted selection (MAS)

Akanksha
Department of Genetics, Plant Breeding and Biotechnology, Dr Khem Singh Gill Akal College of Agriculture, Eternal University, Sirmour, Himachal Pradesh, India

R. Kaur · N. Devi · K. Kaur · I. Sheikh · V. Tyagi (✉)
Department of Genetics, Plant Breeding and Biotechnology, Dr. Khem Singh Gill Akal College of Agriculture, Eternal University, Sirmour, Himachal Pradesh, India

A. Mann et al. (eds.), *Cutting Edge Technologies for Developing Future Crop Plants*,
https://doi.org/10.1007/978-981-96-2508-6_8

8.1 Introduction

The world's population is growing at an unprecedented rate, which is straining agricultural supplies. The problem of guaranteeing food security is further complicated by climate change, which brings unpredictable weather patterns and rising temperatures. Over the next decades, it is expected that the world's population will increase to startling heights, creating enormous challenges in terms of feeding billions of people and addressing issues like starvation and climatic changes. To satisfy the world's food demands, some researchers have emphasized the significance of accelerating primary crop genetic progress. For plant breeders, developing resilient cultivars in a shorter amount of time using traditional methods is still a difficult issue. The primary reason for the sluggish advancement in crop improvement is the extended breeding cycles/generations. Many crop species are experiencing a rapid genetic gain through the adoption of speed breeding concepts at large and small units. This is done in an effort to protect food security and address the shortcomings of traditional approaches.

Speed breeding (SB) also known as rapid generation cycling method (Schilling et al. 2023) is a technique that involves different environmental conditions which are optimized to achieve a short generation time for a certain species. Speed breeding has recently been established for various crops, including wheat (*Triticum aestivum*), canola (*Brassica napus*), and chickpea (*Cicer arietinum*) (Ghosh et al. 2018; Watson et al. 2018). These are photoperiod-sensitive species that require long days (16 h or more) to flower, therefore they are referred to as long-day plants (Andrés and Coupland 2012). Growing long-day plants in 22–24 h of light per day is one of the main ways to achieve low generation times (rapid cycling), which is the fastest feasible transition from the vegetative to the reproductive stage of the life cycle (Ghosh et al. 2018).

"Speed breeding" helps the crop to grow and develop quickly, allowing for the cultivation of multiple generations in a single year. This approach employs controlled environmental conditions and reduces the time required for breeding cycles while increasing plant development. The concept of speed breeding is revolutionizing crop growth, particularly in terms of developing varieties that are more tolerant to multiple stresses and provide larger yields (Imam et al. 2024). Conventional breeding methods include mass selection, pure-line selection, hybridization, backcrossing, and recurrent selection depending on the mode of pollination of the crop. These approaches take more time for the creation of new crop varieties and decrease crops' genetic gain (Potts et al. 2023).

This chapter will look into the historical background, principles, methods, protocols, strategies, and challenges behind speed breeding, going over the regulated settings and environmental tricks that make it work. We will also examine the benefits and drawbacks of speed breeding, taking into account elements like expense, effectiveness, and the possibility of unforeseen outcomes.

8.2 Historical Perspective: From Traditional Breeding to Speed Breeding

The Plant Breeding methods are used to create plants with desired characteristics for crop improvement. The creation of superior crop types is complicated by the extremely long crop duration. Given the numerous stages of crossing, selection, and testing needed in the development of new plant varieties, it can take one or two decades to create a new cultivar (Cheema 2018). Traditional agricultural approaches aimed to enhance the nutritional value of various food plants. Recent scientific advancements give many chances and advances in plant breeding (Moose and Mumm 2008), develop novel techniques, including genomic selection, high-throughput phenotyping (HTP), and modern speed breeding, have accelerated plant breeding. Genetic engineering with molecular technologies helped to generate crops with desired traits through gene transformation (Majid et al. 2017; Zhang et al. 2014). In the 1980s, NASA collaborated with Utah State University to study growing wheat in space stations with constant light. This collaboration resulted in the creation of "USU-Apogee," a dwarf wheat line bred for rapid cycling (Bugbee and Koerner 1997; Bula et al. 1991). In 2003, researchers at the University of Queensland created the term "speed breeding" to characterize a collection of strategies for increasing wheat breeding productivity. Under the term "speed breeding," crop improvement programs from the past were enhanced and combined with other cutting-edge technologies to produce superior varieties more quickly. The conventional methods of single seed descent, shuttle breeding, and haploid technique were among the few manipulations used by the breeders. Over the last century, conventional breeding initiatives around the world have generated a number of significantly improved varieties. Through the use of advanced techniques including next-generation sequencing, in vitro culture, high-throughput phenotyping, genomic selection, marker-assisted selection, and gene editing in speed breeding protocols, scientists were able to develop generations more quickly (Fig. 8.1). Hickey et al. (2017) proposed "speed breeding" as a non-GMO method for selecting desirable features across multiple generations (Voss-Fels et al. 2019). The method optimizes environmental conditions and photoperiods for long-duration crops including wheat, barley, and canola which resulting in four to six generations every year (Ghosh et al. 2018; Hickey et al. 2017; Watson et al. 2018). Adopting a 22-h photoperiod and temperature-controlled regime significantly reduced generation times in durum wheat (*T. durum*), spring bread wheat (*T. aestivum*), barley (*Hordeum vulgare*), and chickpea (*Cicer arietinum*) compared to plants grown in a greenhouse without supplementary light or in the field. Rapid growing situations allowed for normal plant development, easy cross-pollination, and high seed germination rates (Domoney et al. 2013; Kumar et al. 2018). Protocols for speed breeding enable the incorporation of novel approaches and other influencing factor alterations which have been touched on in passing in this chapter.

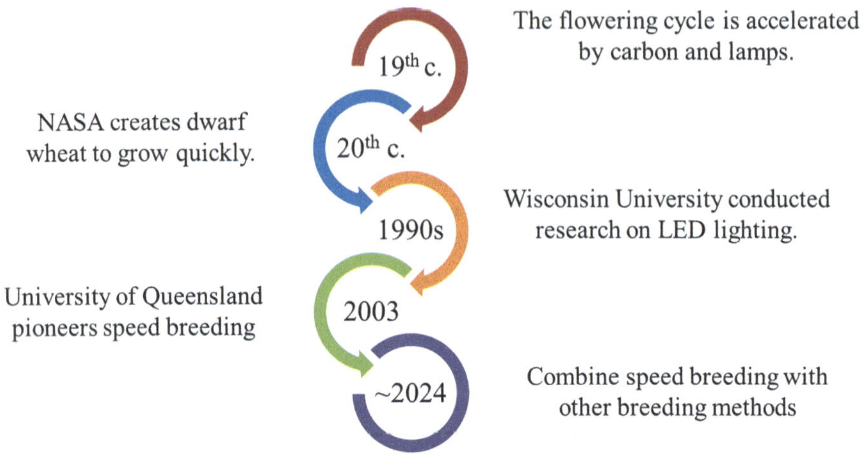

Fig. 8.1 Evolution of speed breeding technique

8.3 The Global Challenge of Food Security and Climate Change

Since the early 1900s, plant breeding has been essential to maintaining food security and safety and has substantial impact on the global food supply (Shiferaw et al. 2013). Farmers around the world are suffering significant crop losses as a result of heat and drought stress due to extreme weather patterns brought on by global climate change (Von Braun 2005). Recent developments in plant breeding and genetics are contributing to enhanced cultivars, both qualitatively and statistically. The biggest potential for speed breeding has been shown by long-day plants that flower in response to longer days; this suggests that speed breeding may be also useful for horticultural crops. Combining high-throughput genotyping and phenotyping with speed breeding can improve orphan crops and contribute to a well-nourished world population despite unpredictable environmental and socioeconomic conditions by which breeders around the world will be able to breed plants and produce improved cultivars that are better adapted to changing climates in order to feed the growing population (Ahmad et al. 2021; Mann et al. 2023). Innovative agronomic methods, in addition to traditional practices, have resulted in amazing agricultural outcomes. Sustainable crop development for global food security requires investments from commercial enterprises, extension workers, and the public sector (Ahmar et al. 2020). High-throughput technologies and speed breeding have a lot of promise for genetic research and agriculture, but they also have some drawbacks and difficulties that must be resolved for effective use and deployment.

8.4 The Need for Rapid Crop Improvement

Global food security has become a serious concern as the world's population expands and the climate changes, with the current rate of improvement of many important crops insufficient to meet future demand. Crop plants have long generation times, which contribute to the slow rate of development. However, speed breeding has transformed the entire world by shortening generation times and accelerating breeding and research operations to better agricultural types (Begna 2022). Speed breeding does not require the use of specialized laboratories it can be performed in ordinary circumstances. This allows breeding operations to save time and money by minimizing labor and field testing expenses. Speed breeding has several advantages over traditional methods, including the capacity to speed backcrossing, pyramiding, and transgenic pipelines. Researchers without access to larger areas may set up smaller speed breeding units (Shimelis and Laing 2012; Shivakumar et al. 2018). The rate of crop improvement is a major factor impeding the breeding process of long generation periods. As a solution in this direction, speed breeding is currently being used on an extensive basis to minimize generation time and enable several crop generations each year. To improve breeding efficiency, researchers are now using an integrated method that combines speed breeding with current plant breeding and genetic engineering technology (Sharma et al. 2022).

8.5 Goal and Benefits of Speed Breeding

Speed breeding is the fastest breeding technique that quickly generates homozygous lines including modifying growing variables like as temperature, photoperiod, soil moisture, population density, and carbon dioxide levels after first crossings (Potts et al. 2023). Speed breeding reduces the breeding cycle of crop research by advancing generation after generation. Genome editing can be done in any laboratory with any crop, even those with complicated genomes, and has been used with the promise of producing new crops faster with a very low chance of off-target effects (Hussain et al. 2023). A technique of "speed breeding" is used to modify the conditions in which crop genotypes are generated to boost seed development and flowering in order to reach the next breeding generation as quickly as possible (Begna 2022). Speed breeding accelerates genetic gain by using phenome and genome-assisted trait introgression, re-domestication, and plant variety registration. Also, early varietal release by speed breeding provides additional advantages over traditional breeding techniques (Gudi et al. 2022). Researchers who lack to access larger areas can still conduct speed breeding in smaller areas by setting up smaller speed breeding units (Shimelis and Laing 2012). Speed breeding has the ability to reduce the time required for cultivar development, release, and commercialization, thereby ensuring global food security. Also, speed breeding can be combined with a variety of other technologies, including marker-assisted selection, genomic selection, and CRISPR-Cas gene editing, which have the potential to reduce the long-standing bottleneck of lengthy crop breeding cycles, allowing plant breeders to keep up with an evolving climate and an ever-increasing human population (Adhikari et al. 2024). Among these approaches, double haploid (DH) was widely used in breeding efforts for a variety of plant species (Hooghvorst et al. 2020). Adapting horticulture for quicker production of new cultivars is crucial to satisfy changing consumer preferences, different climatic conditions, and rising demand. This involves extensive research across multiple disciplines (Ahmad et al. 2021).

8.6 The Fundamentals, Essential Elements, and Strategies of Speed Breeding Technology

The process of speed breeding involves creating a controlled environment that satisfies every need of the plant and guides its growth at every stage of development (Ghosh et al. 2018). The utilization of growing chambers with additional LEDs for longer photoperiods, as well as regulated humidity and temperature, are the fundamental elements of speed breeding. Environmental elements like humidity, light, and temperature have a big impact on how healthy and how quickly plants grow. Controlling these environmental variables to promote and enhance photosynthesis as well as vegetative and reproductive growth is part of optimizing plant growth. One of the core tenets of speed breeding is giving plants the photoperiod they need by adjusting the length of the light and dark cycles in accordance with the needs of the crop for quick growth. For example, crops such as wheat and chickpeas are long-day plants, flowering can be accelerated by subjecting plants to longer durations of light (Ghosh et al. 2018), usually accomplished with the aid of additional lights. It is possible to extend the day's duration in a way that goes beyond its natural duration. In switch grass, this constant exposure to light encourages quicker development, biomass output, and increased stem digestibility. For short-day crops, speed breeding has been restricted due to their need for flowering. However, Lee et al. (2023) have been working on creating protocols for short-day crops such as pigeon pea (*Cajanus cajan Millsp.*) and sorghum (*Sorghum bicolor L. Moench*). A suggested photoperiod of 13: 8: 13 h is advised at the vegetative stage at the International Crop Research Institute for the Semi-Arid Tropics (ICRISAT), under a program supported by the Bill and Melinda Gates Foundation. The pigeon pea development stage known as pod filling speeds up the breeding cycle (Ahmar et al. 2020; Gangashetty et al. 2024). Obtaining several overlapping plant generations in a short time interval is another goal of speed breeding. Speed breeding permits many generations to grow simultaneously, rather than waiting for one generation to finish its full life cycle. This is accomplished by planting and sowing successive generations at different times such that they overlap, which enables breeders to assess and choose desired features more quickly. Proper scheduling of the sowing time is necessary to guarantee consistent plant development and generational overlap. This enables plant breeders to assess and choose plants at various phases of concurrent development. Speed breeding frequently uses effective plant propagation methods in order to speed up the breeding process. This comprises the speedy

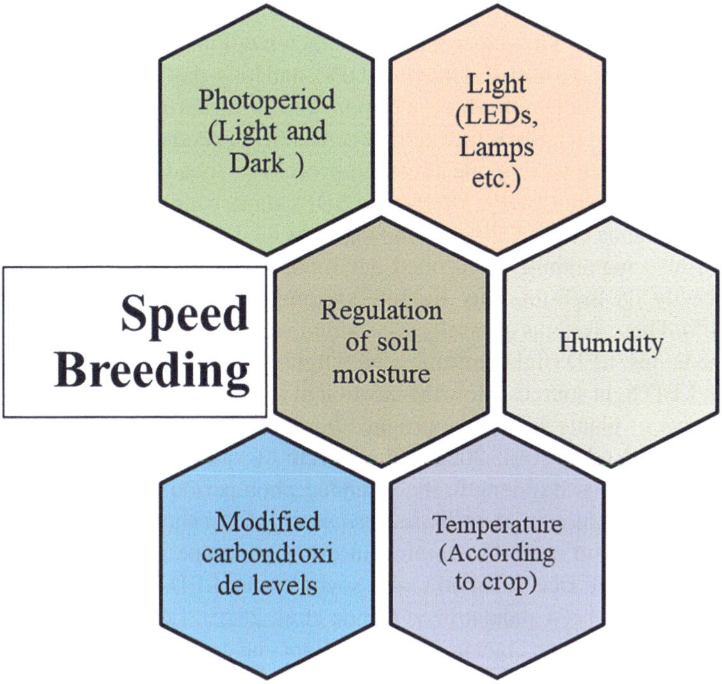

Fig. 8.2 Requirements of speed breeding

propagation of plant material and the creation of sizable populations of plants for assessment by the use of tissue culture, rapid cloning, or other techniques.

8.7 Requirements and Strategies of Speed Breeding

Speed breeding is an agricultural technique used to accelerate the growth and reproduction of crops, allowing multiple generations to be grown in a year. It is particularly valuable in plant breeding and genetic research, where researchers aim to develop new crop varieties with desirable traits (e.g., disease resistance, drought tolerance, or improved yield) in a shorter time. The key requirements and strategies of speed breeding are presented in Fig. 8.2 and explained in below sections.

8.7.1 Light Requirements and Photoperiod

For plants to grow and flourish, the quality of the light is essential. Plants sense and respond to red and far-red light through phytochromes and to blue light through phytotropins and cryptochromes. Phytochromes and cryptochromes have a major impact on plant growth and developmental processes, including shoot and root growth, plant height, flowering, maturity, circadian rhythm, photomorphogenesis, and leaf senescence. The experimental characterization of plants provides day-neutral plants (DNPs), short-day plants (SDPs), and long-day plants (LDPs) depending upon critical night length which is responsible for flowering initiation. DNPs are not impacted by the length of the night, whereas SDPs need a longer night than the critical night. The flowering and breeding process in speed breeding programs is accelerated by granting LDPs, DNPs, and SDPs shorter night periods (Garner and Allard 1920; Jähne et al. 2020), optimizing light quality and light exposure period type, and also maintaining controlled environmental conditions. Speed breeding relies heavily on high-intensity lighting systems to control flowering (Giliberto et al. 2005). These systems generally use sodium vapor lamps (SVLs), high-intensity discharge lamps, LED (light emitting diode) lights, or other specialist lighting technologies. LED light sources allow the creation of light arrays in accordance with the requirements of plants since they produce linear photon output with an electrical current input (Pandey et al. 2022). The effects of various photoperiod regimes, including the normal day length, the extended photoperiod (24 h), and the eight-hour (short day) photoperiod on *Hydrangea macrophylla* show that a longer photoperiod contributes to early blossoming in every genotype (Eid et al. 2016). For short-day crops like rice, amaranth, and soybean, the LED-based speed breeding procedure has just been standardized (Jähne et al. 2020). Legendre and van Iersel (2021) studied the lettuce plant under far-red light and concluded that increased leaf length and width eventually led to greater photosynthesis. Li et al. (2023) studied the effectiveness of far-red light on Chinese kale which ensued an increase in the budding rate of flower, internode length, stem diameter, plant stature, plant display, and also rate of photosynthesis. These lighting systems provide the necessary light spectrum and intensity to support photosynthesis, maximize plant growth rates, and improve plant yield, as in basil (Rihan et al. 2020). Croser et al. (2016) developed early- and late-flowering genotypes for peas (*Pisum sativa* L.), chickpeas (*Circer arietinum* L.), faba beans (*Vicia faba* L.), and lupins (*Lupinus albus* L.).

8.7.2 Temperature and Humidity

For adequate vegetative and reproductive growth, seedling establishment, and satisfactory germination, ideal soil and air temperatures are essential. Plant species and their phenological stages influence how different plants react to temperature (Sandip et al. 2015). Plants develop more vegetatively when the temperature rises, as long as it stays within the necessary ideal range. The vegetative phase of most plant species often demands a higher ideal temperature than the reproductive period (Hatfield and Prueger 2015). According to Mcclung et al. (2016), flowering requires an ambient temperature that is neither too high nor too low for induction in both long-day and short-day plants. Due to the potential to impact meiosis and in particular, pollen viability, many crops are susceptible to high temperatures during their reproductive cycles (Draeger and Moore 2017). Lower temperatures may be maintained during reproductive growth in order to maximize speed breeding processes, whereas higher temperatures may be used to accelerate vegetative growth (Atkin and Tjoelker 2003; Kabade et al. 2024). Temperatures of 20–22 °C were necessary for the germination of immature seed produced from wheat and barley embryo culture. After germination, seedlings were shifted to a temperature regime of 25/22 °C, synchronized with a 16/8-h light/dark photoperiod, for faster plant growth and early flowering (Zhang et al. 2014).

Relative humidity directly impacts the water relations of plants and indirectly affects leaf growth, photosynthesis, and pollination. In controlled-environment chambers, humidity is the most challenging environmental component to regulate; yet, a suitable range of 60–70% is ideal, and lower humidity levels might be preferable for crops that are more suited to dry circumstances (Ghosh et al. 2018). The number of plants available for assessment and selection has increased thanks to the application of effective plant multiplication techniques, such as tissue culture and fast cloning (Wanga et al. 2021).

8.7.3 Speed Breeding Protocols and Techniques

Most crop breeding initiatives are not increasing yield quickly enough to fulfill the increased demand for food for fast expanding global population. The development of superior crop types through plant breeding is hindered by the extremely long crop duration. One way to lessen the problems associated with food scarcity and improve food security is the quick development of improved plant varieties. Speed breeding shortens the breeding cycle and expedites crop research through quick generation progress. One technique for speed breeding is to harvest seeds early and extend the daily light exposure of plants. This will help the plants cycle quickly from seed to seed and reduce generation times for certain long-day or day-neutral crops. Protocols for speed breeding are now being developed for a number of crops, in contrast to twofold haploid technology, which creates fully homozygous lines from haploid embryos, speed breeding works with a variety of germplasm types, and does not require specialized in vitro culture tools (Ghosh et al. 2018). Until recently, speed breeding was thought to cut generation times in most crops by lengthening photoperiods yet, some crop species like pepper, radish and green vegetables like amaranth, reacted well to longer days. In crops such as cereals, legumes, and oilseeds, speed breeding has become common practice (Ghosh et al. 2018; Watson et al. 2018). Several variants have been released to date by the use of speed breeding (Table 8.1). Gaur et al. (2007) showed that three seed-to-seed generations might occur in a year in chickpeas. They have raised three generations: one during the regular growing season, one during the spring, and one under artificial lighting with incandescent bulbs in a rain shelter. With a 22-h photoperiod, it was feasible to grow seven generations of chickpeas in a year (Samineni et al. 2022). According to Cazzola et al. (2021), peas grown in a hydroponic system in a growth chamber with

Table 8.1 The standardized protocols for different crops

Sr. no.	Crop	Generation number per year	Conditions for speed breeding			References
			Photoperiod	Temperature	Relative humidity and other	
1	Wheat	5.6	22 h. Light	22/17 °C	70%, immature seed harvest	Watson et al. (2018)
2	Rice and amaranth	4–5	10 h. With blue light enriched (LEDs)	30 °C/25 °C day/night	70%	Jähne et al. (2020)
3	Barley	5.3	22 h. Light	22/17 °C temperature	70%, immature seed harvest	Watson et al. (2018)
4	Chickpea	4–6	22 h. Light	(25 ± 1) °C temperature	70%, immature seed harvest	Samineni et al. (2022)
5	Soybean	5	10 h. Photoperiod with blue light enriched (LEDs)		–	Jähne et al. (2020)
6	Oat & triticale	6–7.6	20 h. Photoperiod	25/22 °C day/night temperature	65/85% day/night RH, in vitro culture of immature embryos	Liu et al. (2016)
7	Pea	5.3	20 h. Photoperiod, 500 μM m − 2 s − 1 light intensity	21 °C/16 °C light/dark temperature	70%, hydroponic system	Mobini and Warkentin (2016)
8	Lentil	5	20 h. Photoperiod	–	100 μM gibberellin application and immature seed harvest	Lulsdorf and Banniza (2018)
9	Canola	3.7	22 h. Light	22/17 °C temperature	70%, immature seed harvest	Watson et al. (2018)
		5.1–5.9	20 h. Photoperiod	25/22 (±1) °C		Yao et al. (2016)

(continued)

Table 8.1 (continued)

Sr. no.	Crop	Generation number per year	Conditions for speed breeding			References
			Photoperiod	Temperature	Relative humidity and other	
10	Faba bean		20 h. Photoperiod	21 °C light/16 °C dark temperature	10–5 M BAP application	Mobini et al. (2020)
11	Peanut	4	24 h. Light	28 ± 3 °C max. 17 ± 3 °C min. Temperature	65%	O'Connor et al. (2013)

regulated temperature, day length, flurpimidol and antigibberellin use, and improbable seed harvesting can yield five generations annually. Similarly, Mobini et al. (2015) employed immature seed germination to produce seven generations of faba beans annually and eight generations of lentils using plant growth hormones to induce in vitro flowering.

O'Connor et al. (2016) showed that a single seed decent approach (SSD) combined with enhanced photoperiod and optimal temperature reduced the length of the crop growth period in peanuts from 145 to 89 days. When 35-day-old pigeon pea seeds were harvested, a single seed decent method combined with an immature seed germination approach produced three to four generations annually, as demonstrated by Mobini et al. (2020) and Saxena et al. (2019) extended the crop growing period for faba beans (*Vicia faba*) by applying cytokines in addition to cold treatments. The pigeon pea plant has short days and rigorous photoperiodic requirements. It can only produce one generation annually. Using a speed breeding strategy, the F_2 population of *Lens culinaris* × *Lens ervoides* is advanced with a screening protocol for aphanomyces root rot. Using an improved single seed decent method, optimum growing conditions, and gibberellin treatment for immature seed germination, Lulsdorf and Banniza (2018) effectively obtained five generations annually.

Cereals were also subjected to speed breeding techniques aimed at reducing the crop growing period. Watson et al. (2018) established speed breeding procedures for model grass *Brachypodium distachyon*, durum wheat (*T. durum*), barley (*Hordeum vulgare*), and spring bread wheat (*Triticum aestivum*). In a controlled setting, they provided a 22-h photoperiod and a 2-h dark period, measured the time of anthesis, and then concluded that the wheat plant took 35–39 days and Barley required 37–38 days to complete its initial flower opening. In contrast, the heading in *B. distachyon* takes up to 26 days. In other observations, such as the viability and number of seeds resulting in both normal and extended photoperiod conditions, there is no discernible impact on seed count or seed germination. Alahmad et al. (2018) used a speed breeding technique in durum wheat and a phenotyped, biparental population of *outrob 4* × *Caoaroi* for several characteristics such as leaf rust, crown rot, plant stature, root angle, and root number. They obtained one complete generation in 77 days. By González-Barrios et al. (2021), the oat breeding cycle was accelerated through the use of speed breeding techniques. Instead of 114 days under typical circumstances, they were able to produce one whole generation in 98 days by using a 22-h photoperiod and an early seed harvest. In barley, employed multi-trait phenotyping in conjunction with speed breeding technology introduce disease resistance to leaf rust, net and spot variants of net blotch, and spot blotch from four donor lines into the well-known scarlet variety (Hickey et al. 2017). Heuschele et al. (2019) used photoperiod extension and foliar mineral supplementation to expedite the oat breeding program. They showed that the initial flower opening happened 15 ± 3 days earlier, but under controlled conditions, the number of seeds decreased by three times and the weight of the inflorescence decreased by two times.

8.7.4 Speed Breeding Combined with Genomic Selection and Marker-Assisted Breeding

Utilizing speed breeding along with high-throughput technology is a state-of-the-art method for crop breeding and research. It shortens the time needed to generate improved crop types, which speeds up the process. Plant breeding has undergone a revolution due to sophisticated high-throughput technologies, which have made it possible for researchers to follow genes and even create predictive breeding strategies like genomic selection (Wang et al. 2018). Genomic selection shortens the breeding timescale and predicts plant performance using sequence data, SNP markers, and algorithms all without the need for field testing. To forecast breeding, speed breeding and genomic selection have been coupled value of quantitative characteristics in wheat, such as yield. Moreover, prior to field trials, recombinant inbred lines were created with the use of integrated genomic selection and a speed-breeding technique, which allowed for indirect phenotypic selection to increase key traits including height and flowering time (Watson et al. 2019).

Genome editing is the budding technology of making precise, targeted sequence changes in the deoxyribonucleic acid of living cells and organisms. Recent advances have made genome editing widely applicable, offering the opportunity to rapidly advance basic and applied biology. Until 2013, the dominant genome editing tools were zinc finger nucleases and transcription activator-like effector nucleases (Wada et al. 2020).

8.7.5 Challenges and Considerations in Speed Breeding Implementation

Implementing speed breeding methods to improve crops necessitates a sophisticated infrastructure with controls for humidity, photoperiod, temperature, and other variables. Maintaining experimental crops under controlled conditions demands the skills of competent specialists. The absence of contemporary instruments and methods in developing nations, along with intermittent financial support and unaccommodating policies, contribute to the apprehension surrounding the practical implementation of the speed breeding protocol. Access to a steady supply of electricity has been limited in many experimental fields. Energy-efficient LED lighting and solar-powered air conditioners with battery backup could be beneficial for tiny infrastructures. Due to the high expenses of infrastructure input and maintenance, rapid breeding maintains a limited number of crossings and population. In long-day crops, speed breeding is simple since it only needs a maximum photoperiod greater than the essential day length. But it is challenging with short-day plants as the public plant breeding programs have infrastructure available, and private corporations have largely supplanted them in recent years (Delannay et al. 2012; Huynh et al. 2013; Lindner 2004). Due to limited resources present in countries, it is mandatory to develop needed infrastructure before starting breeding activities and also required skilled labor to assist fast-track research. A lot of electricity is consumed in the temperature, humidity, and photoperiod system of a breeding program which costs higher (O'Connor et al. 2013). The solar grids should connect to the greenhouses for a continuous supply of electricity. Environmental issues also occur in speed breeding. The continuous exposure of light affects photoperiod which leads to severe injury in some plant species. Ikkonen and Kaznina (2022) reported that photodamage and respiration activity happen due to an increase in photoperiod. The experimental costs may be higher due to environmental factors and increasing risk of pest and disease attacks. These are some of the challenges that should be tackled in short time intervals to get more benefits from this technique.

8.8 Speed Breeding for Multistress Tolerance

Stress events can impact the ability of crops to reach their yield potential, i.e., their maximum possible yield if all inputs are nonlimiting. The susceptibility of a crop to have negative deviations away from the yield potential is quantified through a measure of yield stability. The difference between the yield potential and the actual yield of a site is often referred to as the yield gap (Mueller et al. 2012). Crops rarely reach their yield potential in most broad-acre agricultural systems due to stress events, that can be biotic (living, i.e., herbivorous insects, pathogenic virus, nematodes, bacteria or fungi), abiotic (e.g., lack of water, or extreme temperatures) or through suboptimal management of nutrition or water.

Biotic factors can be devastating to crops: in contrast to breeding for improved abiotic stress tolerance, there are multiple examples of specific breeding for improved tolerance to biotic pests, and attempts are being made to translate fundamental research relevant to this area into breeding (Boyd et al. 2013; Dangl et al. 2013). It should be noted that biotic factors and a bit of abiotic stress do not necessarily decrease yield. In some circumstances, they will improve crop productivity and quality. Plant water use solutions through agronomy, for example where stress is implemented to improve plant drought tolerance for improvements in temperature or salt tolerance gained by interactions with endophytic or rhizospheric microorganisms (Ait Barka et al. 2006; Dodd and Pérez-Alfocea 2012).

8.8.1 Abiotic Stress

Abiotic stress comprises about 50% of yield loss worldwide (Shinwari et al. 2020). Several abiotic stresses such as high and low temperatures, flood, salinity, heavy metal toxicity, mineral deficiency, and drought frequently affect crop plants (Mickelbart et al. 2015). About 70% of global yield reduction of food occurs due to drought, salinity, cold, and heat stress (Kaur et al. 2009; Thakur et al. 2010). Among these, yield losses of 40% occur due to high temperature, 20% due to salinity, 17% due to drought, and 15% due to low temperature (Ashraf et al. 2008), and several other stresses also lead to yield losses. When developing crop-improvement techniques for stress tolerance, it is crucial to comprehend the mechanisms of tolerance to abiotic stress, from the perception of environmental signals to the cellular reaction and the presentation of the adaptive response (Suprasanna et al. 2016). Speed breeding requires controlled conditions in the greenhouse for better incorporation with other techniques to serve in case of multistress experimentations as reduction in errors and other complexities. Controlled environment consent observing all the data, stresses, and other threats at early stages can cure at time and also reduce the cost of experiments because of greenhouses and pot experiments. Advance breeding techniques such as genomic selection, marker-assisted selection (MAS), and new-generation sequencing can be combined with speed breeding which facilitates the development of stress-tolerance cultivars. Target stress and environments can be defined with the help of speed breeding. It also helps to evaluate breeding lines in target stress environments by identifying superior parents, standardizing stress tolerance screens, clarifying the mechanisms underlying stress tolerance, identifying and characterizing genes imparting tolerance, pyramiding different levels of tolerance in elite lines through marker-assisted breeding, etc. Speed breeding is still in its infancy as a tool for abiotic-stress tolerance.

8.8.1.1 Drought

Drought is one of the main obstacles to sustainable crop production (Berger et al. 2010; Pennisi 2008). Plants experience water-deficit stress when there is not enough water in the soil. The reduction in plant height, stem diameter, plant biomass, and leaf area in different plants is caused by drought stress (Zheng et al. 2016). The speed breeding technique used by breeders helps them to overcome obstacles which they face to develop drought stress-tolerant cultivars (Mitra 2001). Speed breeding provides homogenous conditions to identify and manage drought stress. Reproducible and precise screening techniques can be used for the identification of superior genotypes in speed breeding, thereby eliminating the selection of lines having a negative association of stress tolerance traits with yield. It is usually observed that there is an effect of multiple stresses on plants; however, speed breeding helps to study each stress independently and elucidates the mechanism of stress tolerance.

8.8.1.2 Salinity

About 20% of agricultural land affected by abiotic stress namely salinity, and arid and semiarid regions have low-quality water which enhances soil salinity (Flowers 2004). Salinity causes water-deficit stress, which has a variety of effects on plant growth similar to drought. There is enough water in the soil but the difference in water potential caused by the salinity prevents the roots of plants from absorbing it and also suppresses growth, output, and metabolism due to nutrient deficiency. Efforts to develop salt-tolerant cultivars have been made through many breeding approaches to meet the increasing food demands. Speed breeding with association with the marker-assisted breeding approach is used to develop salt-tolerant lines in rice in a shorter time (Rana et al. 2019). Speed breeding offers scope to improve salinity tolerance and develop better cultivars in other crops like cereals, pulses, oilseeds, and vegetables at a faster rate.

Temperature

Global agriculture productivity was significantly impacted negatively by extreme temperatures, both high and low. Along with other stresses, high temperature generates oxidative stress in plants which causes significant yield losses (Mittler 2006). Low temperature comprises two types: chilling stress and freezing stress which are also very injurious abiotic stress (Mboup et al. 2012). Crop yield and other plant growth aspects are negatively affected by low temperatures (Hasanuzzaman et al. 2013). Improvement of crops to these major abiotic stresses (drought salinity and temperature) can be fast-tracked with the integration of speed breeding.

8.8.2 Biotic Stress

The environmental circumstances play a vital role in the susceptibility of plants to biotic stresses. Global warming is the major cause of changing conditions of the environment which helps to evolve several new races of pathogens. Therefore, there is the necessity of developing new resistant varieties to overcome major losses and for better crop improvement. Recently, several technologies have been developed in regard to crop improvement and speed breeding is one of them. Speed breeding protocol has been developed to shorten the breeding time in peas (Ghosh et al. 2018; Watson et al. 2018; Cazzola et al. 2020) and with new biotechnological tools, it assists in developing new resistant cultivars against new emerging pathogen races.

8.9 Case Studies of Speed Breeding in Action

Recent advancements in crops such as wheat, rice, and maize including breeding techniques genetic engineering, genomic selection, and doubled haploid technologies have resulted in shorter breeding cycles and higher genetic gain rates. Combining these techniques with speed breeding technology which involves regulated photoperiod and temperature conditions to accelerate plant growth and development, can have a significant impact on crop improvement (Ristaino 2006; Temesgen 2024).

8.9.1 Wheat

Speed breeding was successfully implemented in bread wheat (*Triticum aestivum L.*) to achieve six generations per year while applying phenotypic selection for foliar disease resistance and grain dormancy (Alahmad et al. 2018). An artificial environment with increased light duration allows for longer daylight, which aids in the management of photo-insensitive crop life cycles. A new variety takes 8–10 years to generate using traditional methods, but speed breeding can decrease the generation cycle (2–3 times). Speed breeding methods are available for a variety of plant crops that depend on chromosomes and are exposed to optimal light quality, light intensity, and temperature to improve photosynthesis and promote growth and reproduction. This procedure produces at least six generations of individual species (Shivakumar et al. 2018). Insufficient water availability is the primary barrier to production, as changing in-season rainfall can substantially alter productivity and grain quality (Able and Sissons 2014). The "speed breeding" requires continuous light and a controlled temperature of 22/17 °C (day/night) to accelerate plant development. Under such conditions, plants reach the adult growth stage (stem elongation) in roughly 3 weeks (Riaz et al. 2016). Cha et al. (2022) reported that the speed vernalization (SV) technique, with a higher vernalization temperature of 8–10 °C and 22-h photoperiod, has allowed up to five generations of spring and winter wheat. While Speed breeding systems help to reduce breeding times, molecular breeding, such as marker-assisted selection (MAS) and marker-assisted backcrossing (MABC), has emerged as a powerful tool for introducing targeted genes or major effect Quantitative Trait Loci (QTLs) into elite varieties which are widely used in global wheat breeding efforts (Cha et al. 2024; Varshney et al. 2020). Molecular breeding reduces population size and breeding efforts by allowing for the selection of lines with target genes or loci before harvesting (Varshney et al. 2021). MAS and MABC have been successfully applied in a number of situations, most notably in introducing disease resistance to diseases including powdery mildew, stripe, and stem rust and enhancing quality attributes like grain protein content and high-molecular-weight glutenin subunits (HMW-GS) (Gupta et al. 2022; Randhawa et al. 2019; Zheng et al. 2020).

8.9.2 Rice

A study published in 2019 combined speed breeding with marker-assisted selection to rapidly develop rice lines with improved salinity tolerance (Watson et al. 2018). The researchers used a speed breeding protocol with continuous light at 28 °C to advance generations quickly. The SNP markers associated with salinity tolerance QTLs were used to select desirable plants in the BC_3F_3 generation. The integrated approach allowed for the introgression of salinity tolerance QTLs into an elite rice variety in just 1–2 backcross generations.

8.9.3 Maize

The global population has been expanding for several years and is predicted to grow by at least 25% geometrically, but food supplies remain restricted (Ray et al. 2013). Improving plant phenotype for a particular desirable characteristic involves artificial selection and breeding of that trait by a plant breeder. In general, breeders choose to focus on features of diploid or diploid-like crops, i.e., maize rather than polyploid crops, which have more complex genetics. Breeders thus prefer to use crops with shorter reproductive cycles, which allow the production of several generations in a single year, resulting in faster production of the desired phenotypes by artificial breeding than crops that only reproduce annually or perennial plants that only reproduce every few years (Abreu et al. 2010; Ahmar et al. 2020; Kandemir and Saygili 2015). In a field trial, ten surveyors visually chose a maize population, five in complete areas, and five in stratified regions. The weight of the ear after harvest was then compared to the plants visibly selected. The environmental effects were simulated with related and unrelated errors. Stratification enhances efficiency for features with low heritability. When there is a correlation of mistakes in the field, using a moving average does not improve the efficiency of mass selection (Bernardo 2001).

8.9.4 Pulses

Pulses are an important part of the human diet, supply animal feed, and improve soil fertility through biological nitrogen fixation. However, pulse breeding is a time-consuming operation. Most traditional breeding efforts require 10–15 years to produce an improved cultivar (Bermejo et al. 2016). The most effective parameter for boosting genetic gain is cycle time, which is found in the breeder's equation, a model of the expected change in a trait in response to selection. Speed breeding appears as a technology that increases program efficiency while lowering costs and labor requirements. The technique uses ideal light quality, intensity, day length, and temperature control to promote photosynthesis and flowering, as well as early seed harvest. Different improvements in pulse breeding programs are raised and a speed breeding method for pea (*Pisum sativum L.*) is proposed, which involves hybridization and generation advancement in a growth chamber. Days to flowering decreased significantly when compared to field-sown cultivars under a 20-hour photoperiod with florescent tubes (500 IM m − 2 s − 1) and temperature of 20 ± 2 °C ($F = 34.9$; $p < 0.001$) (Cazzola et al. 2021; Cazzola et al. 2020). According to recent studies, speed breeding protocols are required for long-duration crops (6–9 months), such as pigeon pea, which require 12–13 years for conventional methods, 7 years for the single seed decent approach, and 4 years for speed breeding using photoperiodism. It is advised to use a photoperiod of 13: 8: 13 h at the vegetative stage. Broad spectrum light (5700 K LED) promotes early vegetative growth and the generation of pods. On the other hand, far-red light (735 nm) helped in early flowering (Ahmar et al. 2020; Gangashetty et al. 2024).

8.10 Future Perspectives and Research Needs

Speed breeding techniques are combined with other techniques to improve genetic gain. Crop enhancement, genetic research, and agriculture could all be further revolutionized by the opportunities and developments in speed breeding. Advancement in functional genomics allows direct and indirect functional gene manipulation and several other improvements such as stress tolerance and quality improvement (Cushman and Bohnert 2000; Gaikwad et al. 2020). The use of advanced imaging technologies, including drones, hyperspectral imaging, and 3D scanning, will improve high-resolution phenotyping (Paulus 2014). These advanced techniques and tools with speed breeding provide detailed insights into crop growth, genome editing, and genome modification with desired trait introduction. Advanced technologies, international cooperation, and the integration of multiple scientific fields are the hallmarks of the future of speed breeding and high-throughput technologies. These developments have the potential to significantly improve our capacity to grow crops that are more robust, productive, and sustainable, so addressing the issues of climate change, food and nutritional security, and both (Pasala et al. 2024). Protocols for short-day crops need to be optimized in various crops and require the development of crop-specific, genotype-specific, and species-specific protocols in orphan crops. In horticultural crops, long generation time is a major problem in breeding new varieties. Therefore, standardized protocols for horticultural crops should be developed in speed breeding.

8.11 Limitations of Speed Breeding

As discussed above, speed breeding is a powerful technique to get genetic gain in plant species. The key limitation is that the speed breeding settings become very expensive if sophisticated controlled environment facilities are not properly available and when combined with other techniques. It also includes the continuous supply of light, water, and temperature maintenance. The continuous use of speed breeding for research and breeding remains a challenge in resource-poor countries due to limited infrastructures, poor expertise, and limited collaborations with international organizations. Agronomic trait phenotyping can be done in combination with speed breeding, although caution is needed because phenotypic expression might be skewed in controlled conditions circumstances. To confirm that trait expression is related to the field environment, field validation of under-speed breeding is necessary. An observation of plant height and time of flowering in oats reveals that due to cross-over interaction between the genotype of the plant and the growth system of the plant, the several characters may not be consistently phenotyped (Barrios 2020). Energy-efficient LED lighting and solar-powered air conditioners with battery backup could be beneficial for tiny infrastructures. Due to the high expenses of infrastructure input and maintenance, rapid breeding maintains a limited number of crossings and population. To save resource expenditures, prevent job duplication, and facilitate the sharing and support of specialized equipment, integrated research involving scientists from several organizations is required.

8.12 Conclusion

Speed breeding has ability to increase agricultural output, by improving photosynthetic efficiency which helps to accelerate crop development. This approach allows reducing many generations of the same crop which is faster than using traditional breeding. Reducing the generation time and speeding up breeding is an innovative approach to rapidly creating new long-day plant varieties. This technique allows plant breeders to quickly develop superior crop varieties. Incorporating speed breeding into a crop development program requires streamlined operations to reduce labor and facility costs. By overcoming the issues caused by population growth, combining speed breeding with high-throughput phenotyping, genotyping, and gene editing technologies creates new, vital tools for attaining global food security. These technologies also contribute to a more sustainable and safe food supply in the future. Ranging from 30 to 35 days, most of the crop varieties are in reproductive and maturation stages. Faster genetic gain and effective trait improvement are among the unique advantages of integrating speed breeding with current breeding technologies, including genomic selection (GS), high-throughput phenotyping (HTP), marker-assisted selection (MAS), genome editing, and precision phenotyping. The combination of speed breeding with conventional, MAS and GE breeding procedures can improve the successful selection of elite genotypes and lines with innovative features such as improved yield and nutritional quality, as well as biotic and abiotic stress resistance.

References

Able J, Sissons M (2014) Durum wheat for the future: challenges, research and prospects in the 21st century. Crop Pasture Sci 65:i–i

Abreu GB, Ramalho MP, Toledo FRB, Souza J (2010) Strategies to improve mass selection in maize. Maydica 55:219–225

Adhikari S, Kumar A, Joshi A, Bharati A, Prasad P, Gangwar O, Lata C, Kumar S, Bhardwaj S (2024) Speed breeding: rapid generation tool for breeding climate smart crops. In: Singh A, Pandey S (eds) Climate-resilient agriculture. Apple Academic Press, pp 303–342. https://doi.org/10.1201/9781003455271

Ahmad S, Makhmale S, Bosamia TC, Sangh C, Nawade B (2021) Speed breeding: a potential tool for mitigating abiotic stresses. In: Rai AV, Rai A, Rai KK, Rai VP, Kumar A (eds) Stress tolerance in horticultural crops. Woodhead Press, Elsevier, pp 51–61. https://doi.org/10.1016/B978-0-12-822849-4.00012-7

Ahmar S, Gill RA, Jung K-H, Faheem A, Qasim MU, Mubeen M, Zhou W (2020) Conventional and molecular techniques from simple breeding to speed breeding in crop plants: recent advances and future outlook. Int J Mol Sci 21:2590

Ait Barka E, Nowak J, Clément C (2006) Enhancement of chilling resistance of inoculated grapevine plantlets with a plant growth-promoting rhizobacterium, Burkholderia phytofirmans strain PsJN. Appl Environ Microbiol 72:7246–7252

Alahmad S, Dinglasan E, Leung KM, Riaz A, Derbal N, Voss-Fels KP, Able JA, Bassi FM, Christopher J, Hickey LT (2018) Speed breeding for multiple quantitative traits in durum wheat. Plant Methods 14:1–15

Andrés F, Coupland G (2012) The genetic basis of flowering responses to seasonal cues. Nat Rev Genet 13:627–639

Ashraf M, Athar HR, Harris PJC, Kwon TR (2008) Some prospective strategies for improving crop salt tolerance. Adv Agron 97:45–110

Atkin OK, Tjoelker MG (2003) Thermal acclimation and the dynamic response of plant respiration to temperature. Trends Plant Sci 8:343–351

Barrios PG (2020) Resource allocation optimization for multi-environment trials in cereals breeding. The University of Wisconsin-Madison

Begna T (2022) Speed breeding to accelerate crop improvement. Int J Agric Sci Food Technol 8:178–186

Berger B, Parent B, Tester M (2010) High-throughput shoot imaging to study drought responses. J Exp Bot 61:3519–3528

Bermejo C, Gatti I, Cointry E (2016) In vitro embryo culture to shorten the breeding cycle in lentil (Lens culinaris Medik). Plant cell Tissue Organ Cult (PCTOC) 127:585–590

Bernardo R (2001) What if we knew all the genes for a quantitative trait in hybrid crops? Crop Sci 41:1–4

Boyd IL, Freer-Smith PH, Gilligan CA, Godfray HCJ (2013) The consequence of tree pests and diseases for ecosystem services. Science 342:1235773

Bugbee B, Koerner G (1997) Yield comparisons and unique characteristics of the dwarf wheat cultivar 'USU-apogee'. Adv Space Res 20:1891–1894

Bula RJ, Morrow RC, Tibbitts T, Barta D, Ignatius R, Martin T (1991) Light-emitting diodes as a radiation source for plants. Hortscience: Publ Am Soc Hortcultural Sci 26:203–205

Cazzola F, Bermejo CJ, Guindon MF, Cointry E (2020) Speed breeding in pea (*Pisum sativum* L.), an efficient and simple system to accelerate breeding programs. Euphytica 216:178

Cazzola F, Bermejo CJ, Gatti I, Cointry E (2021) Speed breeding in pulses: an opportunity to improve the efficiency of breeding programs. Crop Pasture Sci 72:165–172

Cha J-K, O'Connor K, Alahmad S, Lee J-H, Dinglasan E, Park H, Lee S-M, Hirsz D, Kwon S-W, Kwon Y (2022) Speed vernalization to accelerate generation advance in winter cereal crops. Mol Plant 15:1300–1309

Cha J-K, Park H, Kwon Y, Lee S-M, Jang S-G, Kwon S-W, Lee J-H (2024) Synergizing breeding strategies via combining speed breeding, phenotypic selection, and marker-assisted backcrossing for the introgression of Glu-B1i in wheat. Front Plant Sci 15:1402709

Cheema KS (2018) Plant Breedits applications and future prospects. Int J Eng Sci Technol 5:88–94

Croser JS, Pazos-Navarro M, Bennett RG, Tschirren S, Edwards K, Erskine W, Creasy R, Ribalta FM (2016) Time to flowering of temperate pulses in vivo and generation turnover in vivo–in

vitro of narrow-leaf lupin accelerated by low red to far-red ratio and high intensity in the far-red region. Plant cell Tissue Organ Cult (PCTOC) 127:591–599
Cushman JC, Bohnert HJ (2000) Genomic approaches to plant stress tolerance. Curr Opin Plant Biol 3:117–124
Dangl JL, Horvath DM, Staskawicz BJ (2013) Pivoting the plant immune system from dissection to deployment. Science 341:746–751
Delannay X, McLaren G, Ribaut J-M (2012) Fostering Mol Breedin developing countries. Mol Breed 29:857–873
Dodd IC, Pérez-Alfocea F (2012) Microbial amelioration of crop salinity stress. J Exp Bot 63:3415–3428
Domoney C, Knox M, Moreau C, Ambrose M, Palmer S, Smith P, Christodoulou V, Isaac PG, Hegarty M, Blackmore T (2013) Exploiting a fast neutron mutant genetic resource in *Pisum sativum* (pea) for functional genomics. Funct Plant Biol 40:1261–1270
Draeger T, Moore G (2017) Short periods of high temperature during meiosis prevent normal meiotic progression and reduce grain number in hexaploid wheat (*Triticum aestivum* L.). Theor Appl Genet 130:1785–1800
Eid GM, Albatal N, Haddad S (2016) Effect of photoperiod on the flowering of some cultivars of hydrangea (*Hydrangea macrophylla*). Int J Hortic 6:1–7
Flowers TJ (2004) Improving crop salt tolerance. J Exp Bot 55:307–319
Gaikwad KB, Rani S, Kumar M, Gupta V, Babu PH, Bainsla NK, Yadav R (2020) Enhancing the nutritional quality of major food crops through conventional and genomics-assisted breeding. Front Nutr 7:533453
Gangashetty PI, Belliappa SH, Bomma N, Kanuganahalli V, Sajja SB, Choudhary S, Gaviyappanavar R, Bomireddy D, Anil Kumar V, Pranati J (2024) Optimizing speed breeding and seed/pod chip based genotyping techniques in pigeonpea: a way forward for high throughput line development. Plant Methods 20:27
Garner WW, Allard HA (1920) Effect of the relative length of day and night and other factors of the environment on growth and reproduction in plants. Mon Weather Rev 48:415–415
Gaur P, Pande S, Sharma H, Gowda C, Sharma K, Crouch J, Vadez V, Kashiwagi J (2007) Genetic enhancement of stress tolerance in chickpea: present status and future prospects. In: Crop production in stress environments: genetic and management options: proceedings of the international conference on sustainable crop production in stress environment: management and genetic options, pp 85–94
Ghosh S, Watson A, Gonzalez-Navarro OE, Ramirez-Gonzalez RH, Yanes L, Mendoza-Suárez M, Simmonds J, Wells R, Rayner T, Green P (2018) Speed breeding in growth chambers and glasshouses for crop breeding and model plant research. Nat Protoc 13:2944–2963
Giliberto L, Perrotta G, Pallara P, Weller JL, Fraser PD, Bramley PM, Fiore A, Tavazza M, Giuliano G (2005) Manipulation of the blue light photoreceptor cryptochrome 2 in tomato affects vegetative development, flowering time, and fruit antioxidant content. Plant Physiol 137:199–208
González-Barrios FJ, Cabral-Tena RA, Alvarez-Filip L (2021) Recovery disparity between coral cover and the physical functionality of reefs with impaired coral assemblages. Glob Chang Bio l27:640–651
Gudi S, Kumar P, Singh S, Tanin MJ, Sharma A (2022) Strategies for accelerating genetic gains in crop plants: special focus on speed breeding. Physiol Mol Biol Plants 28:1921–1938
Gupta PK, Balyan HS, Chhuneja P, Jaiswal JP, Tamhankar S, Mishra VK, Bains NS, Chand R, Joshi AK, Kaur S (2022) Pyramiding of genes for grain protein content, grain quality, and rust resistance in eleven Indian bread wheat cultivars: a multi-institutional effort. Mol Breed 42:21
Hasanuzzaman M, Nahar K, Alam MM, Roychowdhury R, Fujita M (2013) Physiological, biochemical, and molecular mechanisms of heat stress tolerance in plants. Int J Mol Sci 14:9643–9684
Hatfield JL, Prueger JH (2015) Temperature extremes: effect on plant growth and development. Weather Clim Extrem 10:4–10

Heuschele J, Lode T, Andersen T, Borgå K, Titelman J (2019) An affordable and automated imaging approach to acquire highly resolved individual data – an example of copepod growth in response to multiple stressors. PeerJ 7:e6776

Hickey LT, Germán SE, Pereyra SA, Diaz JE, Ziems LA, Fowler RA, Platz GJ, Franckowiak JD, Dieters MJ (2017) Speed breeding for multiple disease resistance in barley. Euphytica 213:1–14

Hooghvorst I, Torrico O, Hooghvorst S, Nogués S (2020) In situ parthenogenetic doubled haploid production in melon "Piel de Sapo" for breeding purposes. Front Plant Sci 11:378

Hussain K, Mahrukh NRT, Zaid A, Mushtaq M (2023) The utilization of speed breeding and genome editing to achieve zero hunger. In: Prakash CS, Fiza S, Nadeem MA, Baloch FS, Qayyum A (eds) Sustainable agriculture in the era of the OMICs revolution. Springer, Cham, Switzerland, pp 1–15. https://doi.org/10.1007/978-3-031-15568-0_1

Huynh BL, Ehlers JD, Close TJ, Cissé N, Drabo I, Boukar O, Lucas MR, Wanamaker S, Pottorff M, Roberts PA (2013) Enabling tools for modern breeding of cowpea for biotic stress resistance. In: Translational genomics for crop breeding: biotic stress, vol 1, pp 183–199

Ikkonen E, Kaznina N (2022) Physiological responses of lettuce (Lactuca sativa L.) to soil contamination with Pb. Horticulture 8:951

Imam Z, Sultana R, Parveen R, Swapnil SD, Sinha S, Sahoo JP (2024) Understanding the concept of speed breeding in crop improvement: opportunities and challenges towards global food security. Trop Plant Biol 17:1–23

Jähne F, Hahn V, Würschum T, Leiser WL (2020) Speed breeding short-day crops by LED-controlled light schemes. Theor Appl Genet 133:2335–2342

Kabade PG, Dixit S, Singh UM, Alam S, Bhosale S, Kumar S, Singh SK, Badri J, Varma NRG, Chetia S (2024) SpeedFlower: a comprehensive speed breeding protocol for indica and japonica rice. Plant Biotechnol J 22:1051–1066

Kandemir N, Saygili I (2015) Apomixis: new horizons in plant breeding. Turk J Agric For 39:549–556

Kaur S, Gupta A, Kaur N, Sandhu J, Gupta S (2009) Antioxidative enzymes and sucrose synthase contribute to cold stress tolerance in chickpea. J Agron Crop Sci 195:393–397

Kumar M, Yusuf MA, Nigam M, Kumar M (2018) An update on genetic modification of chickpea for increased yield and stress tolerance. Mol Biotechnol 60:651–663

Lee D, Han K, Kim JH, Jun T-H, Lee JS (2023) Development of speedbreeding system for Korean soybean varieties [Glycine max (L.) Merr] using LED light source. Plant Breed Biotech 11:49–55

Legendre R, van Iersel MW (2021) Supplemental far-red light stimulates lettuce growth: disentangling morphological and physiological effects. Plan Theory 10:166

Li Y, Jiang H, Gao M, He R, Liu X, Su W, Liu H (2023) Far-red-light-induced morphology changes, phytohormone, and transcriptome reprogramming of Chinese kale (*Brassica alboglabra Bailey*). Int J Mol Sci 24:5563

Lindner B (2004) Privatised provision of essential plant Breedinfrastructure. Aust J Agric Resour Econ 8:301–321

Liu H, Zwer P, Wang H, Liu C, Lu Z, Wang Y, Yan G (2016) A fast generation cycling system for oat and triticale breeding. Plant Breed 135:574–579

Lulsdorf MM, Banniza S (2018) Rapid generation cycling of an F2 population derived from a cross between *Lens culinaris* Medik. And *lens ervoides* (Brign.) Grande after aphanomyces root rot selection. Plant Breed 137:486–491

Majid A, Parray G, Wani SH, Kordostami M, Sofi N, Waza SA, Shikari A, Gulzar S (2017) Genome editing and its necessity in agriculture. Int J Curr Microbiol Appl Sci 6:5435–5443

Mann A, Mirza S, Chandra P, Kumar A, Kumar A, Sujata K, Pooja N, Sanwal SK, Kulshrestha N, Banerjee BP (2023) Root system architecture and phenotyping for improved resource use efficiency in crops. In: Translating physiological tools to augment crop breeding. Springer Nature Singapore, Singapore, pp 229–255. https://doi.org/10.1007/978-981-19-7498-4_11

Mboup M, Bahri B, Leconte M, De Vallavieille-Pope C, Kaltz O, Enjalbert J (2012) Genetic structure and local adaptation of European wheat yellow rust populations: the role of temperature-specific adaptation. Evol Appl 5:341–352

Mcclung CR, Lou P, Hermand V, Kim JA (2016) The importance of ambient temperature to growth and the induction of flowering. Front Plant Sci 7:1266

Mickelbart MV, Hasegawa PM, Bailey-Serres J (2015) Genetic mechanisms of abiotic stress tolerance that translate to crop yield stability. Nat Rev Genet 16:237–251

Mitra J (2001) Genetics and genetic improvement of drought resistance in crop plants. Curr Sci:758–763

Mittler R (2006) Abiotic stress, the field environment and stress combination. Trends Plant Sci 11:15–19

Mobini SH, Warkentin TD (2016) A simple and efficient method of in vivo rapid generation technology in pea (*Pisum sativum* L.). In Vitro Cell Dev Biol Plant 52:530–536

Mobini SH, Lulsdorf M, Warkentin TD, Vandenberg A (2015) Plant growth regulators improve in vitro flowering and rapid generation advancement in lentil and faba bean. In Vitro Cell Dev Biol Plant 51:71–79

Mobini S, Khazaei H, Warkentin TD, Vandenberg A (2020) Shortening the generation cycle in faba bean *(Vicia faba)* by application of cytokinin and cold stress to assist speed breeding. Plant Breed 139:1181–1189

Moose SP, Mumm RH (2008) Mol Plantbreeding as the foundation for 21st century crop improvement. Plant Physiol 147:969–977

Mueller ND, Gerber JS, Johnston M, Ray DK, Ramankutty N, Foley JA (2012) Closing yield gaps through nutrient and water management. Nature 490:254–257

O'Connor C, Kelleher M, Hourihane JOB (2016) Calculating the effect of population-level implementation of the learning early about Peanut allergy (LEAP) protocol to prevent peanut allergy. J Allergy Clin Immunol 137:1263–1264

O'Connor D, Wright G, Dieters M, George D, Hunter M, Tatnell J, Fleischfresser D (2013) Development and application of speed breeding technologies in a commercial peanut breeding program. Peanut Sci 40:107–114

Pandey S, Singh A, Parida SK, Prasad M (2022) Combining speed breeding with traditional and genomics-assisted breeding for crop improvement. Plant Breed 141:301–313

Pasala R, Chennamsetti M, Patil B, Kadirvel P, Geethanjali S, Nagaram S, Sajja S, Vennapusa AR, Prasad PV, Mathur RK (2024) Revolutionizing crop production: the imperative of speed breeding Technology in Modern Crop Improvement. Crop Breed Genet Genom 6:1–23

Paulus M (2014) The early origins of human charity: developmental changes in preschoolers' sharing with poor and wealthy individuals. Front Psychol 5:344

Pennisi E (2008) The blue revolution, drop by drop, gene by gene. Science 320:171–173

Potts J, Jangra S, Michael VN, Wu X (2023) Speed breeding for crop improvement and food security. Crops 3:276–291

Rana S, Lemoine E, Granger JP, Karumanchi SA (2019) Preeclampsia: pathophysiology, challenges, and perspectives. Circ Res 124:1094–1112

Randhawa MS, Bains NS, Sohu VS, Chhuneja P, Trethowan RM, Bariana HS, Bansal U (2019) Marker assisted transfer of stripe rust and stem rust resistance genes into four wheat cultivars. Agronomy 9:497

Ray DK, Mueller ND, West PC, Foley JA (2013) Yield trends are insufficient to double global crop production by 2050. PLoS One 8:e66428

Riaz A, Periyannan S, Aitken E, Hickey L (2016) A rapid phenotyping method for adult plant resistance to leaf rust in wheat. Plant Methods 12:1–10

Rihan HZ, Aldarkazali M, Mohamed SJ, McMulkin NB, Jbara MH, Fuller MP (2020) A novel new light recipe significantly increases the growth and yield of sweet basil (Ocimum basilicum) grown in a plant factory system. Agronomy 10:934

Ristaino JB (2006) Tracking the evolutionary history of the potato late blight pathogen with historical collections. Outlooks Pest Management 17:228

Samineni S, Mahendrakar MD, Hotti A, Chand U, Rathore A, Gaur PM (2022) Impact of heat and drought stresses on grain nutrient content in chickpea: genome-wide marker-trait associations for protein, Fe and Zn. Environ Exp Bot 194:104688

Sandip M, Makwana A, Barad A, Nawade B (2015) Physiology of flowering-the case of mango. Int J Appl Res 1:1008–1012

Saxena K, Saxena R, Hickey L, Varshney R (2019) Can a speed breeding approach accelerate genetic gain in pigeonpea? Euphytica 215:1–7

Schilling S, Melzer R, Dowling CA, Shi J, Muldoon S, McCabe PF (2023) A protocol for rapid generation cycling (speed breeding) of hemp (Cannabis sativa) for research and agriculture. Plant J113:437–445

Sharma S, Kumar A, Dhakte P, Raturi G, Vishwakarma G, Barbadikar KM, Das BK, Shivaraj S, Sonah H, Deshmukh R (2022) Speed breeding opportunities and challenges for crop improvement. J Plant Growth Regul:1–14

Shiferaw B, Smale M, Braun H-J, Duveiller E, Reynolds M, Muricho G (2013) Crops that feed the world 10. Past successes and future challenges to the role played by wheat in global food security. Food Secur 5:291–317

Shimelis H, Laing M (2012) Timelines in conventional crop improvement: pre-breeding and breeding procedures. Aust J Crop Sci 6:1542–1549

Shinwari ZK, Jan SA, Nakashima K, Yamaguchi-Shinozaki K (2020) Genetic engineering approaches to understanding drought tolerance in plants. Plant Biotechnol Rep 14:151–162

Shivakumar M, Nataraj V, Kumawat G, Rajesh V, Chandra S, Gupta S, Bhatia V (2018) Speed breeding for Indian agriculture: a rapid method for development of new crop varieties. Curr Sci 115:1241

Suprasanna P, Nikalje G, Rai A (2016) Osmolyte accumulation and implications in plant abiotic stress tolerance. In: Iqbal N, Nazar R, Khan NA (eds) Osmolytes and plants acclimation to changing environment: emerging omics technologies. Springer, New Delhi, pp 1–12. https://doi.org/10.1007/978-81-322-2616-1_1

Temesgen B (2024) Speed breeding. Adv Crop Sci Tech 12:680

Thakur P, Kumar S, Malik JA, Berger JD, Nayyar H (2010) Cold stress effects on reproductive development in grain crops: an overview. Environ Exp Bot 67:429–443

Varshney RK, Sinha P, Singh VK, Kumar A, Zhang Q, Bennetzen JL (2020) 5Gs for crop genetic improvement. Curr Opin Plant Biol 56:190–196

Varshney RK, Barmukh R, Roorkiwal M, Qi Y, Kholova J, Tuberosa R, Reynolds MP, Tardieu F, Siddique KH (2021) Breeding custom-designed crops for improved drought adaptation. Adv Genet 2:e202100017

Von Braun J (2005) The world food situation: an overview. international Food Policy Research Institute.U.S.A, Washington, DC

Voss-Fels KP, Stahl A, Hickey LT (2019) Q&a: modern crop breeding for future food security. BMC Biol 17:18

Wada N, Ueta R, Osakabe Y, Osakabe K (2020) Precision genome editing in plants: state-of-the-art in CRISPR/Cas9-based genome engineering. BMC Plant Biol 20:1–12

Wang L, Zhu G, Johnson W, Kher M (2018) Three new approaches to genomic selection. Plant Breed 137:673–681

Wanga MA, Shimelis H, Mashilo J, Laing MD (2021) Opportunities and challenges of speed breeding: a review. Plant Breed 140:185–194

Watson A, Ghosh S, Williams MJ, Cuddy WS, Simmonds J, Rey MD, Asyraf Md Hatta M, Hinchliffe A, Steed A, Reynolds D, Adamski NM (2018) Speed breeding is a powerful tool to accelerate crop research and breeding. Nat plants 4:23–29

Watson A, Hickey LT, Christopher J, Rutkoski J, Poland J, Hayes BJ (2019) Multivariate genomic selection and potential of rapid indirect selection with speed breeding in spring wheat. Crop Sci 59:1945–1959

Yao Y, Zhang P, Wang H, Lu Z, Liu CJ, Liu H, Yan GJ (2016) How to advance up to seven generations of canola (*Brassica napus* L.) per annum for the production of pure line populations? Euphytica 209:113–119

Zhang F, Wen Y, Guo X (2014) CRISPR/Cas9 for genome editing: progress, implications and challenges. Hum Mol Genet 23:R40–R46

Zheng M, Tao Y, Hussain S, Jiang Q, Peng S, Huang J, Cui K, Nie L (2016) Seed priming in dry direct-seeded rice: consequences for emergence, seedling growth and associated metabolic events under drought stress. Plant Growth Regul 78:167–178

Zheng W, Li S, Liu Z, Zhou Q, Feng Y, Chai S (2020) Molecular marker assisted gene stacking for disease resistance and quality genes in the dwarf mutant of an elite common wheat cultivar Xiaoyan22. BMC Genet 21:1–8

Epigenetics Regulation of Abiotic Stress in Crop Plants

9

Hemant Dasila, Damini Maithani, Pragati Srivastava, Divjot Kour, Viabhav Kumar Upadhayay, Seema Bisht, Shubham Pandey, Manisha Kabdwal, and Naresh Kumar

Abstract

Plants are sessile in nature and develop numerous ways to counter biotic and abiotic stresses. One of the salient mechanisms developed by plants is epigenetics. Epigenetics involves structural changes at the molecular level that can be passed from one generation to another and thus have the ability to form transgenerational memories. Most common methods employed by plant epigenetics include DNA methylation, histone modification, RNA silencing, and some minor modification at posttranscriptional level. Modification at the molecular level not only counters abiotic stress but also promotes immunological memory against biotic stress like pathogens. Epigenetics mechanisms induce modification which allows plants to survive against temperature stress (extreme hot and cold tem-

H. Dasila (✉) · D. Kour · N. Kumar
Department of Microbiology, Akal College of Basic Sciences, Eternal University, Baru Sahib, Himachal Pradesh, India
e-mail: drhemantmicro@eternaluniversity.edu.in

D. Maithani
Department of Microbiology, School of Allied Sciences, Dev Bhoomi Uttarakhand University,
Dehradun, Uttarakhand, India

P. Srivastava
ICAR National Institute of Plant Biotechnology, IARI PUSA, New Delhi, Delhi, India

V. K. Upadhayay · S. Bisht
Departmrnt of Microbiology, G.B. Pant University of Agriculture and Technology, Pantnagar, Uttarakhand, India

S. Pandey
Department of Microbiology, University of Delhi, New Delhi, Delhi, India

M. Kabdwal
Pantanjali Research Foundation Trust, Haridwar, Uttarakhand, India

A. Mann et al. (eds.), *Cutting Edge Technologies for Developing Future Crop Plants*,
https://doi.org/10.1007/978-981-96-2508-6_9

peratures). Regulation of cold-responsive genes is also very critical; their upregulation and downregulation depend upon the temperature stress induced on plants. The role of abscisic acid is also very critical against abiotic stress.

Keywords

Epigenetics · DNA methylation · Abiotic stress · Abscisic acid

9.1 Introduction

Plants are sessile in nature and are constantly challenged with various abiotic and biotic stresses. In order to encounter, plants have to adapt at both molecular and physiological levels. Epigenetic regulation at the molecular level provides efficient and effective controls which improve the plants' survival rate during abiotic and biotic stresses (Hiramaya and Shinozaki 2010; Richards 2006). According to Conrad Waddington, epigenetics is the study of mitotically and meiotically heritable changes in gene function that cannot be explained by changes in DNA sequence. However, today's epigenetics mainly refers to chemical modification that can be inherited from one generation to the next irrespective of changes in DNA sequence (Fujimoto et al. 2012; Feng et al. 2010). Three prominent epigenetics changes that allow plants to survive in adverse conditions are DNA methylation, RNA interference, and histone modification. In eukaryotes, DNA is organized in the form of nucleosomes which are wrapped around by histone proteins. Histone proteins compromise of conserved globular proteins whose N-terminal region is exposed on the surface nucleosome octamer hotspot site for chemical modification. Histones are active sites where posttranscriptional modification takes place which strictly regulates the accessibility of specific translation machinery to genomic regions making loci less or more permissive for transcription (Kouzarides 2007).

9.2 Plant Epigenetic Mechanisms

The current state of soil where the overuse of fertilizers and pesticides have already drenched away all the nutrients in the soil, adverse climate conditions, and abiotic stress can significantly limit the yield of crop plants. In response to such challenges, plants undergo genetic modifications at the epigenetic level to enhance their chances of survival. To cope with these challenges, the plants inherit various modifications without affecting the gene expression. These modifications include DNA methylation, RNA-associated silencing, and histone modification which play a crucial role in helping plants adapt to their environment (Czajka et al. 2022; Dahiya et al. 2023; Kumar et al. 2023a, b). Developing better crop varieties requires a prior understanding of these various regulatory changes. Understanding these intricate genetic pathways is essential for plant breeders, as it can lead to the development of plant varieties better suited to thriving in stressful conditions. By incorporating these epigenetic regulations into breeding programs, breeders may be able to improve the productivity and quality of crop yields in challenging environments. In the discovery of miRNA and their ability to regulate various genes involved in the stress response of the plant, the involvement of 75 miRNA was discovered to target a total of 861 miRNA for the salinity stress (Zeeshan et al. 2021). The evolution of the various epigenomic profiling assays such as DNase-seq, ChIP-seq, FAIRE-seq, ATAC-seq, CUT and RUN, MeRIP-Seq, miCLIP, and Hi-C have helped in the construction of epigenomic maps and developed a better understanding of the epigenome of various crop plants (Dinkar et al. 2024). The term "epigenetics" refers to the study of heritable changes in gene expression that occur without changes to the underlying DNA sequence (Holliday 2006). These changes can be influenced by various factors, including environmental stressors. In plants, epigenetic changes play a crucial role in enabling them to adapt to different environmental conditions. While plants may exhibit temporary responses to stress, such as alterations in their internal balance (homeostasis), the long-term adaptation involves changes in gene expression patterns. In the presence of long-lasting environmental cues, known as "Epigenators," specific regions of the plant's chromosomes are marked, initiating a cascade of epigenetic modifications (Adel and Carels 2023). These markings can be passed on to subsequent generations, ensuring that the offspring are equipped to deal with similar environmental challenges. This inheritance of epigenetic modifications contributes to the plant's ability to adapt and thrive in its changing environments (Grativol et al. 2012). Different methods of epigenetics have been employed to protect plants against different biotic and abiotic stresses.

9.2.1 Plant DNA Methylation

DNA methylation is one of the specific chemical reactions that include the addition of methyl groups to DNA. CpG sites are prime examples of DNA methylation which forms when cytosine is placed near guanine and which are linked together by phosphodiester bond between them. These sites are hotspots for methylation and methyl groups are added by DNA methyltransferase. Cytosine methylation is considered to be a stable one which can be inherited (Law and Jacobsen 2010). Methylation reaction provide epigenetic control in the genome and seems to regulate noncoding and coding elements. This can be verified by the fact that several important processes like response to pathogen attack, genome stability, transposable element regulation, genome imprinting, expression of genes, and heterosis all are affected by DNA methylation (Vidalis et al. 2016). These all reports suggest that DNA methylation plays a key role in plant evolution and adaptation. DNA methylation has been extensively studied and categorized into two major types that include de-novo methylation and maintenance DNA methylation respectively.

5-Methylcytosine (5mC) is essential for heterochromatin and epigenetic gene silencing in both plants and animals. In differentiated cells of mammals, 5mC is mostly found at CG sites (CpG sites), but cytosines in plants can be methylated within CG, CHG, or CHH patterns (Kumar et al. 2018). Plants have cytosine methylation in a variety of genomic locations, such as silent rRNA gene repeats, centromeric repeats, and repetitive elements. It also happens in the protein-coding sections of highly expressed genes and at certain differentially regulated promoters (Ramakrishnan et al. 2021). A methyl group is added to the DNA molecule's CpG site during DNA methylation, which helps control chromatin and DNA stability. This methylation is distributed along the entire chromosome and serves to repress gene expression and maintain the integrity of the genome, particularly in transposable elements (Acharjee et al. 2023). DNA methyltransferases (DNMTs) are enzymes that add methyl groups to cytosine residues, thereby regulating gene expression. In plants, there are four types of DNMTs: MET1 (CG methylation), Dnmt2, DRM (de novo methylation), and CMT (CHG methylation). Rice has ten DNMTs, while *Arabidopsis* exhibits 24% CG, 6.7% CHG, and 1.7% CHH methylation. Active demethylation is carried out by ROS1, DML2, DML3, and DME through glycosylase activity, which directly removes methylated cytosine (Singh and Prasad 2022).

9.2.2 Histone Modification

Histone modification involves covalent modification like acetylation, methylation, sumoylation, phosphorylation, and ubiquitinoylation. These all mechanisms regulate gene expression. DNA methylation is highly influenced by histone modification in plants. For instance, methylation is common at repeat sequences in H3K9 (Saze et al. 2012). In Arabidopsis, protein homologs of H3K9 methyltransferase have been reported (Baumbusch et al. 2001). Histone helps in the packaging of the DNA inside the cell by an octamer made up of H2A, H2B, H3, and H4. Various posttranslational modifications take place in the histone tail, at serine, threonine, lysine, and arginine residues. These PTMs lead to various alterations in the expression of genes. The major histone modifications that may take place are acetylation and methylation (Peterson and Laniel 2004). Histone acetylation marks active chromatin and transcription, whereas nonactive sequences usually lack this modification. Histone acetylation occurs at the N-terminal tail of histones, this addition leads to neutralization of positive charges making the chromatin structure accessible to the transcription machinery thus leading to transcriptional activation. Conversely, removal of acetyl groups leads to gene repression due to a more condensed chromatin structure (Annunziato and Hansen 2000). Histone acetyltransferases (HATs) mediate the addition of acetyl groups to histones, while histone deacetylases (HDACs) reverse these modifications, making acetylation reversible. Specific lysine residues targeted by HATs and HDACs include H3K9, H3K14, H3K23, H3K27, H4K5, H4K8, H4K12, and H4K16 (Shen et al. 2015).

Methylation of lysine and arginine has been added by histone lysine methyltransferases (HKMTs) and protein arginine methyltransferases (PRMTs), respectively. Methylation of H3K9, H3K27, and H4K20 typically represses transcription, while H3K4 and H3K36 methylation activates it (Wang et al. 2016; Yu et al. 2009). *Arabidopsis* has 14 domain-containing HKMTs, and rice has 43. These enzymes are categorized into four classes: SU(VAR) 3–9, TRX, E(Z), and ASH1. PRMTs methylate residues like H3R2 and H4R3, with nine PRMTs identified in *Arabidopsis* (Singh and Prasad 2022). Histone demethylation, which reverses methylation, is performed by histone lysine demethylases (HDMs). There are two main types of HDMs in plants: Jumonji C (JmjC) domain-containing proteins, which use hydroxylation, and lysine-specific demethylase1 (LSD1), which uses amine oxidation. Under high-temperature conditions, a higher expression of these genes was found in the leaves of *Solanum tuberosum* L. (Mali et al. 2023). *Arabidopsis* has four KDM1/LSD1-like and twenty-one JmjC domain-containing HDMs, classified into five subclasses (Singh and Prasad 2022). These modifications enable plants to dynamically regulate gene expression, which is critical for development and environmental responses. Understanding these mechanisms offers the potential for improving crop resilience and productivity through targeted epigenetic interventions.

9.2.3 RNA Silencing

RNA or gene expression can be achieved by antisense transcript or noncoding RNAs by a process known as RNA interference. In simple words, RNA expression can be downregulated by RNA interference and this phenomenon is also called posttranscriptional gene silencing. Small interfering RNAs (siRNAs) mediate gene expression by histone modification, heterochromatin modification, or DNA methylation. siRNAs are highly susceptible and responsive to stress and thus considered to be the most stress-rearranged molecules (Alonso et al. 2019; Dowen et al. 2012; Pieterse 2012).

9.3 Epigenetics of Plant-Microbe Interaction

Plants do not have adaptive immune systems like animals to encounter microbial pathogens. However, there are physical barriers like cuticle which provides first-line defense for plant protection. Plant pathogen interaction led to the activation of plant communication signaling and the active components of this signaling include jasmonic acid, reactive oxygen species, and salicylic acid. Microbial attacks led to systemic activation of the plant defense system. The systemic response involves induced systemic response (ISR) and systemic acquired resistance (SAR). SAR forms earlier immune response traps against pathogens and involves two major components: SA and nonexpressor of pathogenesis-related genes (Loake and Grant 2007). SAR-mediated pathway led to transcriptional activation of SA-related genes. SA-related genes promote genes encoding pathogenesis-related proteins (PR). SA is also responsible for targeting gene chromatin modification (Butterbroadt et al. 2006). For instance, at the PR1 promoter level of H3, H4 acetylation and methylation at H3K4 methylation are enhanced. There is some other gene like WRKY which encodes transcription factors that can be "turned on" upon pathogen encounter or during SA-induced treatment. Similar evidence was recorded when tobacco mosaic virus infected *Nicotina tabacum* and this led to resistance against TMV for two subsequent generations (Kathiria et al. 2010).

9.4 Epigenetics of Plant Insect Interaction

Plants use intelligent strategies to send signals that can attract pollinators but at the same time deter herbivores. During that process, the plant requires a constant check on the insect community for identification which leads to quick responses to any potential threat (Kellenberger et al. 2016). DNA methylation-mediated response is heritable in nature whereas CG methylation is around 80% in plants which is organized in tissue-specific manner (Widman et al. 2014). Plants have evolved both physical and chemical barriers that are designed to act against herbivory. Classically it is an evolutionary race between plants and herbivores which led to the progression of plant in-built mechanisms to fight against insects and in turn insects also develop new mechanisms to invade plants (Crava et al. 2016). This tussle between plants and insects led to the activation of certain responsive genes. Epigenetic changes like DNA methylation act as a floral signal after activation of certain genes which has been observed in *Brassica rapa* L. (Kellenberger et al. 2016). Transgenerational memory is also one of the mechanisms which led to resistant seedlings of plants against insects mediated by DNA methylation and has been observed in radish (*Raphanus raphanistrum*) and the key point of the study was resistant seedlings were produced by those which has been attacked by insects. It was observed that when the caterpillar chewed the leaf first, the density of trichomes increased from third to seventh leaf which prevented the attack of the caterpillar in younger or new leaves. An increase in trichomes has also been reported in yellow monkey flowers (*Mimulus guttatus*) (Holeski 2007). Similar reports have been reported in *Arabidopsis* and *Solanum lycopersicum* L. where subsequent generation of plants are resistant to attack against herbivory after their first encounter (Rasmann et al. 2012). In *Arabidopsis*, the phenomenon of transgenerational resistance has been well reported. It includes priming of jasmonic acid-related genes and also induces siRNA-mediated pathways.

9.5 Immune System and Memory Epigenetics of Plants

Plants are sessile in nature but can endure challenging environmental conditions and retain memory through subsequent generations. Plants have ability to sense, store, and recall the earlier experiences of stressful conditions. Environmental-induced chromatin changes in various responsive genes contribute to plant memory (He and Li 2018). Like humans, plants too have neural mechanisms which can be studied under plant neurobiology which provides great insights into plant behavior. Gene expression regulation under stressful conditions is very critical and a strategy to promote effective response against abiotic stress without delay is known as priming. In plants, priming system and somatic memory against pathogens have been

Table 9.1 Biotic factors affecting different plant parameters

Biotic factor	Host plant	Physiological parameters affected	References
Herbivory	*Impatiens capensis* *Raphanus raphanistrum* *Brassica rapa*	Plant height and biomass Trichome density and seed mass Variation in floral scent	Steets and Ashman (2010), Agrawal (2002), Agrawal et al. (1999)
Herbivory viral infection (tobacco mosaic virus)	*Nicotiana tabacum*	Variation in homologous frequency	Katharia et al. (2010)
Bacterial infection *Pseudomonas syringae* *Xanthomonas oryzae*	*Arabidopsis thaliana* *Oryza sativa*	Variation in DNA methylation Resistance to blight	Dowen et al. (2012), Alonso et al. (2019)
Fungal infection *Alternaria brassicicola* *Botrytis cinerea*	*Arabidopsis thaliana*	Resistance to fungal infection	Zhou et al. (2012)

reported (Crisp et al. 2016; Lamke and Baurle 2017). Thus, these reports clearly suggest that plants are first primed during their first encounter with stress and have the ability to memorize past environmental experiences to prepare themselves for any recurring event (Cazzonelli et al. 2009). Plants depend upon two types of immunity, e.g., vertical disease resistance and horizontal disease resistance, and in modern days, these terminologies are referred to as effector-triggered immunity and PAMP (pathogen-associated molecular pattern)-triggered immunity. Some of the gene regulations mediated by plants against biotic factors are given in Table 9.1.

9.6 Epigenetic Regulation in Crop Plants Under Drought Stress

Plants respond to drought stress by making epigenetic changes, such as DNA methylation, histone modifications like acetylation and methylation, and nucleosome remodeling. These modifications help regulate gene expression in response to drought, allowing plants to adapt to the stress. Epigenetic modifications play a crucial role in plant stress memory, aiding in rapid adaptive responses during subsequent drought stress.

9.6.1 DNA Methylation

Different crop plants showed varying levels of methylation when subjected to drought stress. In *Triticum aestivum* L., drought priming using PEG 6000 (10, 15, 20%) for 72 h at the six-leaf stage reduced CG and CHG methylation in TaP5CS and TaBADH promoters significantly decreased plant dry weight and leaf area but maintained photosynthetic efficiency (ΦPSII) and increased nonphotochemical quenching (ΦNPQ). Additionally, photosynthetic rate and stomatal conductance were elevated, suggesting improved water-use efficiency. These changes highlight wheat's adaptive epigenetic and physiological responses to drought, enhancing tolerance by regulating key stress-related genes (Li et al. 2023). Treatment of *Malus pumila Mill* with PEG8000 increased DNA methylation levels in the MdRFNR1–1 promoter. Consequently, the fresh weights of all callus decreased, and peroxidase (POD) and catalase (CAT) activities were lower in MdRFNR1 RNAi lines compared to GL-3 plants suggesting that increased DNA methylation of MdRFNR1–1 under drought stress impacts stress response mechanisms, affecting both growth and antioxidant enzyme activities (Niu et al. 2022).

Linseed, one of the oil crop plants, was found to have increased methylation when given drought stress with an increase in intron retention and alternative splice sites. This leads to alternative splicing regulation in specific genes (Wang et al. 2023). Wang et al. (2011) investigated DNA methylation patterns in *Arabidopsis thaliana*, revealing that genes near repetitive elements undergo epigenetic silencing, countered by mechanisms like ROS1-mediated demethylation. ASI1 and IBM1 were found to prevent CHG methylation and help plants adapt to drought stress (Wang et al. 2011). Drought-induced DNA methylation changes in *Oryza sativa* are reversible at 70% of sites, with 29% persisting after recovery. Sensitive rice varieties exhibit hypomethylation under drought, while tolerant ones show hypermethylation, influencing gene expression. Genes like RD29A and RAP2.4 show increased H3K4me3 and H3K9Ac during drought, aiding drought tolerance and recovery. JMJ17 plays a role in demethylating H3K4me1/2/3, affecting stress-responsive genes (Matsui et al. 2008).

The epigenetic changes are not only responsible for modulating the molecular changes but also help in creating a memory of it to be passed on to future generations. Drought-tolerant genotypes of rice plants were found to exhibit a notable rise in global 5-methylcytosine (5-mC) content under stress. This epigenetic modification helps regulate gene expression critical for stress tolerance. The elevated 5-mC levels persist even after the stress is withdrawn and in subsequent generations, indicating the role of DNA methylation in maintaining stress memory and contributing to transgenerational stress tolerance (Kumar et al. 2023a, b).

9.6.2 Histone Modification

Barley adapts to drought stress through complex mechanisms, including histone modifications like H3K9 acetylation, which upregulates ABA-related genes involved in ABA action regulation. H3K9ac is also responsible for short-term stress memory by responding more flexibly to drought than H3K4 trimethylation, while also serving as a template for H3K4 trimethylation (Ost et al. 2023). The function of WHIRLY1, a ssDNA-binding protein was studied in Barley. The protein acts as a retrograde signal transmitting a stress signal from the chloroplast to the nucleus and triggering a stress-related gene expression. Overexpression of HvWHIRLY1 caused a delay in drought stress-induced senescence, and also suppression in ABA-related genes (*HvNCED1*, *CCD5*, and *CCD8*). Variable expression of HAT genes is responsible for the acetylation of the histone. In rice, the expression of *OsHAG702/703, OsHAM701, OsHDA704/705/706/711/712/713, OsHDT701, and OsSRT702* was significantly increased after drought treatment which is responsible for histone acetylation (Hou et al. 2021). These findings highlight the importance of epigenetic mechanisms in plant adaptation to drought stress, informing future breeding strategies for resilience.

9.7 Epigenetic Changes in Plants Under Cold Stress

Most important markers in plants under cold stress are 5-methyl cytosine (5mC) and N6-methyl adenine (6 mA) and these two markers are reported to regulate the transcriptional activity of plants under severe cold conditions (Verma et al. 2022; Kidokoro et al. 2020). Out of these two markers, modification in 5mC is most studied so far in plants under cold stress. However, 6 mA is widely distributed across the rice genome and *Arabidopsis*has been associated with gene expression under stress response (Liu and He 2020). Plants do have the property to transfer cold stress-induced DNA methylation patterns within the generation in the form of transgenerational epigenetic memories so that subsequent generations have the ability to respond to cold stress during the second time (Sharma et al. 2022). Variations in DNA methylation patterns in response to cold acclimation have been reported in *Populus simonii* (Song et al. 2016), *Chorispora bungeana* (Song et al. 2015), and *Phyllostachys edulis* (Ding et al. 2022). Cold stress induces both DNA methylation and demethylation which regulates freezing tolerance in *Arabidopsis*. In cucumbers, reports suggest that cold stress impacts RNA-directed DNA methylation process which results in DNA demethylation (RdDM). However, the lowered activity of MSH1 and RDM4 which plays a key role in the RdDM pathway reduced cold response in Arabidopsis (Xie et al. 2021). Cold stress reported to reduce 5mC expression in vernalized plants and stimulate the expression of casein kinase II A subunit (CKA2) and B-subunit (CKB4) which is known to elevate CK2 activity resulting in a reduction of clock genes which is an essential player in photoperiod perception (Duan et al. 2017).

9.8 Gene Involved in Cold Stress in Plants

During cold stress, plants experience membrane stiffening which leads to calcium influx into cells. This calcium influx triggers signaling cascade mechanisms which are regulated by multiple genes (Ding et al. 2019). Cold stress induces the expression of numerous transcription factors such as CBF. CBF signaling pathway is reported to induce transcriptional cold stress memory in plants (Hwarari et al. 2022). In *Arabidopsis*, CBF proteins bind to cis-acting element CRT/DRE which is present at the promoter region of cold-responsive genes such as COR genes

(Adhikari et al. 2022). Cold stress rapidly induces the expression of CBF1, CBF2, and CBF3 along with the COR gene. The expression of these genes increased plant tolerance against cold stress (Guo et al. 2018). Depending upon epigenetic changes, CBF genes are either positively or negatively regulated by different transcription factors or genes. For example, the cbf2 mutant in Arabidopsis was reported to increase freezing tolerance and cold acclimation but negatively regulates CBF1 and CBF3 (Novillo et al. 2004). COLD-RESPONSIVE PROTEIN KINASE 1 (CRPK1) is a cold-activated plasma membrane protein but negatively regulates the CBF pathway. CRPK1 phosphorylates proteins which interact and destabilize CBF proteins in the nucleus and thus plays a negative role in cold stress (Liu et al. 2017). On the other hand, expression of CBFs is positively regulated by ICE1 and ICE2, thus maintaining the activation of CBF/COR genes (Guo et al. 2018). Nonexpressor of pathogenesis-related genes 1 (NPR1) gene reported to be master of plant pathogen response. NPR1 is reported to be epigenetically regulated where SET DOMAIN GROUP 8 (SDG8) induces H3K36 methylation upon pathogen infection (Zhang et al. 2020). However, recent genetic and molecular evidence has shown that NPR1 plays an essential role in cold acclimation by independently regulating cold-induced gene expression of salicylic acid and TGA TFs (Olate et al. 2018).

9.9 Abscisic Acid Role in Epigenetic Regulation of Crop Plants

Abscisic acid (ABA) regulates growth, development, and stress responses in plants. ABA was first isolated from young cotton fruits as a substance accelerating abscission (Schroeder et al. 2001). ABA along with gibberellic acid (GA) maintains the equilibrium between seed dormancy and germination. Water deficit impacts the root system which alters the levels of ABA and subsequently, the changes in the root environment have both local and systemic effects on ABA-mediated responses (Puértolas et al. 2015). Further, under the conditions of drought stress, ABA induces the expression of genes for responses and tolerance to stress (Fujita et al. 2011). It has been revealed through transcriptomic approaches that approximately 50% of the genes regulated by ABA are governed by water stress or salinity (Sah et al. 2016). ABA controls many essential physiological roles in plants including closure of the stomata, germination of the seeds, and accumulation of the cuticular wax, bud dormancy, and osmotic regulation. ABA is present in algae, cyanobacteria, fungi, and higher plants. Four species of cyanobacteria showed enhanced levels of ABA under saline conditions. Further, ABA has been revealed to play a defensive role in the lower plants as it protects cells of *Chlamydomonas reinhardtii* from oxidative damage and photoinhibition under different abiotic stress (Saradhi et al. 2000). In mosses and sporophytes of hornworts, ABA regulates stomatal opening and closing defining another important role of ABA.

ABA additionally controls downstream responses to biotic and abiotic stresses through transcriptional and posttranscriptional mechanisms. Over the span of 40 years, the core components involved in the biosynthesis of ABA and signaling have been identified through diverse approaches. Genetic screens for *viviparous* mutants in *Arabidopsis* and maize as well as the sugar, salt, and ABA insensitive mutants during germination helped in the identification of a number of the components in the biosynthesis and signaling of ABA (Chen et al. 2020). The harsh environmental conditions lead to an increase in the ABA biosynthesis which then binds to its receptor for the initiation of the signal transduction and cellular responses to stressful environmental conditions and thus is described as a stress hormone (Mehrotra et al. 2014; Ng et al. 2014).

Epigenetic mechanisms regulate the expression of stress-responsive genes in the fight against abiotic stresses (Chung et al. 2022). During seedling development, histone modifications play a key role in ABA responses (Ryu et al. 2014). ABA signaling consists of three major components including pyrabactin resistance (PYR)/PYR1-like (PYL)/regulatory component of ABA receptor (RCAR), protein phosphatase 2C (PP2C), and SNF1-related protein kinase 2 (SnRK2). The components are assembled to form a signaling complex called an ABA signalosome (Mehrotra et al. 2014). In the presence of ABA, the structural changes occur in PYR/PYL/RCAR which exposes an active surface followed by binding and inactivation of PP2Cs that prevents the dephosphorylation of SnRK2s and activates SnRK2 kinases (Soon et al. 2012). The activated SnRK2 leads to the phosphorylation of downstream effectors and causes alterations in several biological processes including epigenetic modifications, transcription, and response to abiotic stress (Tan et al. 2018).

The effects of ABA and salinity stress on the histone acetylation and methylation of abiotic stress response genes have been studied (Chen et al. 2010). Both ABA and salinity induce histone H3K9K14 acetylation and H3K4 trimethylation on the contrary decreasing the H3K9 di-methylation of some ABA and abiotic stress-responsive genes, signifying coordinated regulation of functionally related gene groups through histone modifications in plant cells under abiotic stress (Luo et al. 2012). ABA triggers the cytosine methylation consequently increasing the total DNA methylation in *Spirodela polyrhiza* (Zhao et al. 2015). Under drought stress, the decreased levels of ABA in ABA-deficient maize mutant *viviparous 10* showed a reduction in CG and CHG methylation (Sallam and Moussa 2021). *Arachis hypogea* treatment with ABA increased the acetylation of H3K14 (Su et al. 2015). The exogenous application of ABA in rice elevated the levels of H3K9me2 in meristematic cells (Polosoro et al. 2019). Tea showed increased expression of the biosynthesis genes due to dehydration stress during postharvest processing (Gu et al. 2021).

Stomatal regulation by ABA is another important phenomenon to avoid drought stress. In the presence of unfavorable environmental conditions, the regulation of ion channels in guard cells by ABA plays a critical role in the closure of the stomata. It has been evidenced that the ABA might mediate stomatal response by chromatin remodeling at certain loci. Histone deacetylation is known to be involved in the regulation of guard cells (Chinnusamy et al. 2008). Furthermore, ABA has been established as a gene expression regulator as it induces changes in methylation and histone acetylation. The extent of methylation was observed to differ considerably across the rice cultivars exposed to different abiotic stresses (Akhter et al. 2021).

The methylated regions were correlated with differential expression of genes necessary for abiotic stress response with a possible link among hypermethylated and small RNA (Deng et al. 2018). ABA thus acts as the central regulator of different plant stresses. The advancement of molecular technologies has deepened the understanding of the fundamental mechanisms responsible for the contribution of ABA in stress tolerance. Further, an understanding of ABA signal transduction pathways and epigenetic modifications may help in producing transgenic plants with improved stress tolerance.

References

Acharjee S, Chauhan S, Pal R et al (2023) Mechanisms of DNA methylation and histone modifications. Prog Mol Biol Transl Sci 197:51–92. https://doi.org/10.1016/bs.pmbts.2023.01.001

Adel S, Carels N (2023) Plant tolerance to drought stress with emphasis on wheat. Plan Theory 12(11):2170. https://doi.org/10.3390/plants12112170

Adhikari L, Baral R, Paudel D et al (2022) Cold stress in plants: strategies to improve cold tolerance in forage species. Plant Stress 4:100081. https://doi.org/10.1016/j.stress.2022.100081

Agrawal AA (2002) Herbivory and maternal effects: mechanisms and consequences of transgenerational induced plant resistance. Ecology 83(12):3408–3415

Agrawal AA, Laforsch C, Tollrian R (1999) Transgenerational induction of defences in animals and plants. Nature 401(6748):60–63. https://doi.org/10.1038/43425

Akhter Z, Bi Z, Ali K et al (2021) In response to abiotic stress, DNA methylation confers EpiGenetic changes in plants. Plan Theory 10:1096

Alonso C, Ramos-Cruz D, Becker C (2019) The role of plant epigenetics in biotic interactions. New Phytol 221(2):731–737. https://doi.org/10.1111/nph.15408

Annunziato AT, Hansen JC (2000) Role of histone acetylation in the assembly and modulation of chromatin structures. Gene Expr 9(1–2):37–61. https://doi.org/10.3727/000000001783992687

Baumbusch LO, Thorstensen T, Krauss V et al (2001) The Arabidopsis thaliana genome contains at least 29 active genes encoding SET domain proteins that can be assigned to four evolutionarily conserved classes. Nucleic Acids Res 29(21):4319–4333. https://doi.org/10.1093/nar/29.21.4319

Butterbrodt T, Thurow C, Gatz C (2006) Chromatin immunoprecipitation analysis of the tobacco PR-1a-and the truncated CaMV 35S promoter reveals differences in salicylic acid-dependent TGA factor binding and histone acetylation. Plant Mol Biol 61:665–674. https://doi.org/10.1007/s11103006-0039-2

Cazzonelli CI, Millar T, Finnegan EJ et al (2009) Promoting gene expression in plants by permissive histone lysine methylation. Plant Signal Behav 4(6):484–488. https://doi.org/10.4161/psb.4.6.8316

Chen LT, Luo M, Wang YY et al (2010) Involvement of Arabidopsis histone deacetylase HDA6 in ABA and salt stress response. J Exp Bot 61:3345–3353

Chen K, Li G-J, Bressan RA et al (2020) Abscisic acid dynamics, signaling, and functions in plants. J Integr Plant Biol 62:25–54. https://doi.org/10.1111/jipb.12899

Chinnusamy V, Gong Z, Zhu JK (2008) Abscisic acid-mediated epigenetic processes in plant development and stress responses. J Integr Plant Biol 50:1187–1195

Chung S, Kwon C, Lee J-H (2022) Epigenetic control of abiotic stress signaling in plants. Genes Genomics 44:267–278. https://doi.org/10.1007/s13258-021-01163-3

Crava CM, Brütting C, Baldwin IT (2016) Transcriptome profiling reveals differential gene expression of detoxification enzymes in a hemimetabolous tobacco pest after feeding on jasmonate-silenced *Nicotiana attenuata* plants. BMC Genomics 17:1–15. https://doi.org/10.1186/s12864-016-3348-0

Crisp PA, Ganguly D, Eichten SR et al (2016) Reconsidering plant memory: intersections between stress recovery, RNA turnover, and epigenetics. Sci Adv 2(2):e1501340
Czajka K, Mehes-Smith M, Nkongolo K (2022) DNA methylation and histone modifications induced by abiotic stressors in plants. Genes Genomics 44(3):279–297. https://doi.org/10.1007/s13258-021-01191-z
Dahiya A, Kumar A, Kumar A, Sanwal SK, Mann A, Kulshreshtha N (2023) Current status and future perspectives of epigenetic gene regulation for salt tolerance in wheat. In: Kumar A, Dhansu P, Mann A (eds) Salinity and drought tolerance in plants. Springer, Singapore, pp 333–345. https://doi.org/10.1007/978-981-99-4669-3_17
Deng J, Kou S, Zhang C, Zou Q, Li P, Yuan P (2018) DNA methylation and plant stress responses. J Plant Physiol Pathol 6:4. https://doi.org/10.4172/2329-955X.1000182
Ding Y, Shi Y, Yang S (2019) Advances and challenges in uncovering cold tolerance regulatory mechanisms in plants. New Phytol 222:1690–1704. https://doi.org/10.1111/nph.15696
Ding Y, Zou LH, Wu J et al (2022) The pattern of DNA methylation alteration, and its association with the expression changes of non-coding RNAs and mRNAs in moso bamboo under abiotic stress. Plant Sci 325:111451. https://doi.org/10.1016/j.plantsci.2022.111451
Dinkar V, Pandey S, Kumar A et al (2024) Epigenetic regulations under plant stress: a cereals perspective. Env Exp Bot 220:105688. https://doi.org/10.1016/j.envexpbot.2024.105688
Dowen RH, Pelizzola M, Schmitz RJ et al (2012) Widespread dynamic DNA methylation in response to biotic stress. Proc Natl Acad Sci 109(32):E2183–E2191. https://doi.org/10.1073/pnas.1209329109
Duan W, Zhang H, Zhang B et al (2017) Role of vernalization-mediated demethylation in the floral transition of Brassica rapa. Planta 245:227–233. https://doi.org/10.1007/s00425-016-2622-3
Feng S, Jacobsen SE, Reik W (2010) Epigenetic reprogramming in plant and animal development. Science 330(6004):622–627
Fujimoto R, Sasaki T, Ishikawa R et al (2012) Molecular mechanisms of epigenetic variation in plants. Int J Mol Sci 13(8):9900–9922
Fujita Y, Fujita M, Shinozaki K et al (2011) ABA-mediated transcriptional regulation in response to osmotic stress in plants. J Plant Res 124:509–525. https://doi.org/10.1007/s10265-011-0412-3
Grativol C, Hemerly AS, Ferreira PCG (2012) Genetic and epigenetic regulation of stress responses in natural plant populations. Biochim et Biophys Acta (BBA) Gene Regul Mech 1819(2):176–185. https://doi.org/10.1016/j.bbagrm.2011.08.010
Gu D, Yang J, Wu S et al (2021) Epigenetic regulation of the phytohormone abscisic acid accumulation under dehydration stress during postharvest processing of tea (Camellia sinensis). J Agric Food Chem 69:1039–1048
Guo X, Liu D, Chong K (2018) Cold signaling in plants: insights into mechanisms and regulation. J Integr Plant Biol 60:745–756. https://doi.org/10.1111/jipb.12706
He Y, Li Z (2018) Epigenetic environmental memories in plants: establishment, maintenance, and reprogramming. Trends Genet 34(11):856–866. https://doi.org/10.1016/j.tig.2018.07.006
Hirayama T, Shinozaki K (2010) Research on plant abiotic stress responses in the post-genome era: past, present and future. Plant J 61(6):1041–1052
Holeski LM (2007) Within and between generation phenotypic plasticity in trichome density of Mimulus guttatus. J Evol Biol 20(6):2092–2100
Holliday R (2006) Epigenetics: a historical overview. Epigenetics 1(2):76–80. https://doi.org/10.4161/epi.1.2.2762
Hou J, Ren R, Xiao H et al (2021) Characteristic and evolution of HAT and HDAC genes in Gramineae genomes and their expression analysis under diverse stress in Oryza sativa. Planta 253(3):2. https://doi.org/10.1007/s00425-021-03589-1
Hwarari D, Guan Y, Ahmad B et al (2022) ICE-CBF-COR signaling cascade and its regulation in plants responding to cold stress. Int J Mol Sci 23:1549. https://doi.org/10.3390/ijms23031549
Kathiria P, Sidler C, Golubov A et al (2010) Tobacco mosaic virus infection results in an increase in recombination frequency and resistance to viral, bacterial, and fungal pathogens in the progeny of infected tobacco plants. Plant Physiol 153(4):1859–1870

Kellenberger RT, Schlüter PM, Schiestl FP (2016) Herbivore-induced DNA demethylation changes floral signalling and attractiveness to pollinators in *Brassica rapa*. PLoS One 11(11):e0166646. https://doi.org/10.1371/journal.pone.0166646

Kidokoro S, Kim JS, Ishikawa T, Suzuki T, Shinozaki K, Yamaguchi-Shinozaki K (2020) DREB1A/CBF3 is repressed by transgene-induced DNA methylation in the Arabidopsis ice1-1 mutant. The Plant Cell. 32(4):1035–1048.

Kouzarides T (2007) Chromatin modifications and their function. Cell 128(4):693–705

Kumar S, Chinnusamy V, Mohapatra T (2018) Epigenetics of modified DNA bases: 5-Methylcytosine and beyond. Front Genet 9. https://doi.org/10.3389/fgene.2018.00640

Kumar S, Seem K, Mohapatra T (2023a) Biochemical and epigenetic modulations under drought: remembering the stress tolerance mechanism in Rice. Life 13(5):1156. https://doi.org/10.3390/life13051156

Kumar P, Kumar P, Suniti K, U., Avni, Mann, A. (2023b) Transcriptional regulatory network involved in drought and salt stress response in Rice. In: Kumar A, Dhansu P, Mann A (eds) Salinity and drought tolerance in plants. Springer, Singapore. https://doi.org/10.1007/978-981-99-4669-3_13

Lämke J, Bäurle I (2017) Epigenetic and chromatin-based mechanisms in environmental stress adaptation and stress memory in plants. Genome Biol 18:1–11

Law JA, Jacobsen SE (2010) Establishing, maintaining and modifying DNA methylation patterns in plants and animals. Nat Rev Genet 11(3):204–220. https://doi.org/10.1038/nrg2719

Li Q, Wang X, Sun Z et al (2023) DNA methylation levels of TaP5CS and TaBADH are associated with enhanced tolerance to PEG-induced drought stress triggered by drought priming in wheat. Plant Physiol Biochem 200:107769. https://doi.org/10.1016/j.plaphy.2023.107769

Liu J, He Z (2020) Small DNA methylation, big player in plant abiotic stress responses and memory. Front Plant Sci 11. https://doi.org/10.3389/fpls.2020.595603

Liu Z, Jia Y, Ding Y et al (2017) Plasma membrane CRPK1-mediated phosphorylation of 14-3-3 proteins induces their nuclear import to fine-tune CBF signaling during cold response. Mol Cell 66:117–128.e115. https://doi.org/10.1016/j.molcel.2017.02.016

Loake G, Grant M (2007) Salicylic acid in plant defence – the players and protagonists. Curr Opin Plant Biol 10(5):466–472. https://doi.org/10.1016/j.pbi.2007.08.008

Luo M, Liu X, Singh P et al (2012) Chromatin modifications and remodeling in plant abiotic stress responses. Biochim et Biophys Acta (BBA) Gene Regul Mech 1819:129–136. https://doi.org/10.1016/j.bbagrm.2011.06.008

Mali S, Dutta M, Goel K et al (2023) High temperature triggers differential expression of JUMONJI C (JmjC) domain-containing histone demethylase genes in leaf and stolon tissues of potato (Solanum tuberosum L.) genotypes. J Plant Growth Regul. https://doi.org/10.1007/s00344-023-11094-w

Matsui A, Ishida J, Morosawa T et al (2008) Arabidopsis transcriptome analysis under drought, cold, high-salinity and ABA treatment conditions using a tiling Array. Plant Cell Physiol 49(8):1135–1149. https://doi.org/10.1093/pcp/pcn101

Mehrotra R, Bhalothia P, Bansal P et al (2014) Abscisic acid and abiotic stress tolerance – different tiers of regulation. J Plant Physiol 171:486–496. https://doi.org/10.1016/j.jplph.2013.12.007

Ng LM, Melcher K, Teh BT et al (2014) Abscisic acid perception and signaling: structural mechanisms and applications. Acta Pharmacol Sin 35:567–584

Niu C, Jiang L, Cao F et al (2022) Methylation of a MITE insertion in the *MdRFNR1-1* promoter is positively associated with its allelic expression in apple in response to drought stress. Plant Cell 34(10):3983–4006. https://doi.org/10.1093/plcell/koac220

Novillo F, Alonso JM, Ecker JR et al (2004) CBF2/DREB1C is a negative regulator of CBF1/DREB1B and CBF3/DREB1A expression and plays a central role in stress tolerance in Arabidopsis. Proc Natl Acad Sci USA 101:3985–3990. https://doi.org/10.1073/pnas.0303029101

Olate E, Jimenez-Gomez JM, Holuigue L, Salinas J (2018) NPR1 mediates a novel regulatory pathway in cold acclimation by interacting with HSFA1 factors. Nat Plants 4:811–823. https://doi.org/10.1038/s41477-018-0254-2

Ost C, Cao HX, Nguyen TL et al (2023) Drought-stress-related reprogramming of gene expression in barley involves differential histone modifications at ABA-related genes. Int J Mol Sci 24(15):12065. https://doi.org/10.3390/ijms241512065

Peterson CL, Laniel MA (2004) Histones and histone modifications. Curr Biol 14(14):546–551. https://doi.org/10.1016/j.cub.2004.07.007

Pieterse CM (2012) Prime time for transgenerational defense. Plant Physiol 158(2):545–545. https://doi.org/10.1104/pp.112.900430

Polosoro A, Enggarini W, Ohmido N (2019) Global epigenetic changes of histone modification under environmental stresses in rice root. Chromosom Res 27:287–298. https://doi.org/10.1007/s10577-019-09611-3

Puértolas J, Conesa MR, Ballester C et al (2015) Local root abscisic acid (ABA) accumulation depends on the spatial distribution of soil moisture in potato: implications for ABA signalling under heterogeneous soil drying. J Exp Botany 66:2325–2334

Ramakrishnan M, Satish L, Kalendar R et al (2021) The dynamism of transposon methylation for plant development and stress adaptation. Int J Mol Sci 22(21):11387. https://doi.org/10.3390/ijms222111387

Rasmann S, De Vos M, Casteel CL et al (2012) Herbivory in the previous generation primes plants for enhanced insect resistance. Plant Physiol 158(2):854–863. https://doi.org/10.1104/pp.111.187831

Richards EJ (2006) Inherited epigenetic variation – revisiting soft inheritance. Nat Rev Genet 7(5):395–401. https://doi.org/10.1038/nrg1834

Ryu H, Cho H, Bae W et al (2014) Control of early seedling development by BES1/TPL/HDA19-mediated epigenetic regulation of ABI3. Nat Commun 5:4138. https://doi.org/10.1038/ncomms5138

Sah SK, Reddy KR, Li J (2016) Abscisic acid and abiotic stress tolerance in crop plants. Frontiers Plant Sci 7:571. https://doi.org/10.3389/fpls.2016.00571

Sallam N, Moussa M (2021) DNA methylation changes stimulated by drought stress in ABA-deficient maize mutant vp10. Plant Physiol Biochem PPB 160:218–224. https://doi.org/10.1016/j.plaphy.2021.01.024

Saradhi PP, Suzuki I, Katoh A et al (2000) Protection against the photo-induced inactivation of the photosystem II complex by abscisic acid. Plant Cell Environ 23(7):711–718. https://doi.org/10.1046/j.1365-3040.2000.00579.x

Saze H, Tsugane K, Kanno T et al (2012) DNA methylation in plants: relationship to small RNAs and histone modifications, and functions in transposon inactivation. Plant Cell Physiol 53(5):766–784. https://doi.org/10.1093/pcp/pcs008

Schroeder JI, Kwak JM, Allen GJ (2001) Guard cell abscisic acid signalling and engineering drought hardiness in plants. Nature 410(6826):327–330. https://doi.org/10.1038/35066500

Sharma M, Kumar P, Verma V et al (2022) Understanding plant stress memory response for abiotic stress resilience: molecular insights and prospects. Plant Physiol Biochem 179:10–24. https://doi.org/10.1016/j.plaphy.2022.03.004

Shen Y, Wei W, Zhou DX (2015) Histone acetylation enzymes coordinate metabolism and gene expression. Trends Plant Sci 20(10):614–621. https://doi.org/10.1016/j.tplants.2015.07.005

Singh RK, Prasad M (2022) Delineating the epigenetic regulation of heat and drought response in plants. Crit Rev Biotechnol 42(4):548–561. https://doi.org/10.1080/07388551.2021.1946004

Song Y, Liu L, Feng YW et al (2015) Chilling- and freezing-induced alterations in cytosine methylation and its association with the cold tolerance of an alpine Subnival plant, *Chorispora bungeana*. PLoS One 10(8):e0135485. https://doi.org/10.1371/journal.pone.0135485

Song Y, Ci D, Tian M et al (2016) Stable methylation of a non-coding RNA gene regulates gene expression in response to abiotic stress in Populus simonii. J Exp Bot 67(5):1477–1492. https://doi.org/10.1093/jxb/erv543

Soon FF, Ng LM, Zhou XE et al (2012) Molecular mimicry regulates ABA signaling by SnRK2 kinases and PP2C phosphatases. Science 335(6064):85–88. https://doi.org/10.1126/science.1215106

Steets JA, Ashman TL (2010) Maternal effects of herbivory in Impatiens capensis. Int J Plant Sci. 171(5):509–518

Su LC, Deng B, Liu S et al (2015) Isolation and characterization of an osmotic stress and ABA induced histone deacetylase in *Arachis hygogaea*. Frontiers Plant Sci. 6:512. https://doi.org/10.3389/fpls.2015.00512

Tan W, Zhang D, Zhou H et al (2018) Transcription factor HAT1 is a substrate of SnRK2.3 kinase and negatively regulates ABA synthesis and signaling in Arabidopsis responding to drought. PLoS Genet 14(4):e1007336. https://doi.org/10.1371/journal.pgen.1007336

Verma N, Giri SK, Singh G, Gill R, Kumar A (2022) Epigenetic regulation of heat and cold stress responses in crop plants. Plant Gene. 29:100351.

Vidalis A, Živković D, Wardenaar R et al (2016) Methylome evolution in plants. Genome Biol 17(1):264. https://doi.org/10.1186/s13059-016-1127-5

Wang WS, Pan YJ, Zhao XQ et al (2011) Drought-induced site-specific DNA methylation and its association with drought tolerance in rice (*Oryza sativa* L.). J Exp Bot 62(6):1951–1960. https://doi.org/10.1093/jxb/erq391

Wang J, Meng X, Yuan C et al (2016) The roles of cross-talk epigenetic patterns in Arabidopsis thaliana. Brief Funct Genomics 15(4):278–287. https://doi.org/10.1093/bfgp/elv025

Wang L, Wang L, Tan M et al (2023) The pattern of alternative splicing and DNA methylation alteration and their interaction in linseed (Linum usitatissimum L.) response to repeated drought stresses. Biol Res 56(1):12. https://doi.org/10.1186/s40659-023-00424-7

Widman N, Feng S, Jacobsen SE et al (2014) Epigenetic differences between shoots and roots in Arabidopsis reveals tissue-specific regulation. Epigenetics 9(2):236–242. https://doi.org/10.4161/epi.26869

Xie W, Tang Q, Yan F et al (2021) Transcriptional memory and response to adverse temperatures in plants. J Zhejiang Univ Sci 22(10):791–804. https://doi.org/10.1631/jzus.B2100287

Yu YU, Zhongyuan B, Wen-Hui S et al (2009) Aiwu dong, an update on histone lysine methylation in plants. Prog Nat Sci 19(4):407–413. https://doi.org/10.1016/j.pnsc.2008.07.015

Zeeshan M, Qiu CW, Naz S et al (2021) Genome-wide discovery of miRNAs with differential expression patterns in responses to salinity in the two contrasting wheat cultivars. Int J Mol Sci 22(22):12556. https://doi.org/10.3390/ijms222212556

Zhang X, Ménard R, Li Y et al (2020) Arabidopsis SDG8 potentiates the sustainable transcriptional induction of the pathogenesis-related genes PR1 and PR2 during plant defense response. Front Plant Sci 11:277. https://doi.org/10.3389/fpls.2020.00277

Zhou LL, Shi MZ, Xie DY (2012) Regulation of anthocyanin biosynthesis by nitrogen in TTG1–GL3/TT8–PAP1-programmed red cells of Arabidopsis thaliana. Planta. 236:825–837.

Zhao Z, Shi HJ, Wang ML et al (2015) Analysis of DNA methylation of *Spirodela polyrhiza* (grater duckweed) in response to abscisic acid using methylation-sensitive amplied polymorphism. Russ J Plant Physiol 62:127–135. https://doi.org/10.1134/S1021443715010197

Gene Editing Prospective for Engineering Climate-Smart Plants

10

Poonam Ranga, Priti Choudhary, Noyonika Kaul, Sujata Yadav, Avni Dahiya, Ashwani Kumar, Arvind Kumar, Satish Kumar Sanwal, Anita Mann, and Rajender Kumar Yadav

Abstract

Global warming and its consequential climate change pose significant challenges for plant agriculture, limiting crop productivity. This issue is further compounded by the urgent necessity of feeding the growing global population amidst an increasingly unpredictable climate. Concept of climate-smart agriculture focuses on cultivating robust crops that can thrive in adverse conditions while preserving essential micronutrient levels. These challenges necessitate the development of resilient crop plants to withstand these stresses while maintaining sustainable yields. Although traditional methods of plant transformation have provided valuable insights into plant biology and upscaled the agricultural practices, still the regeneration and genetic modification across different crop species varies in different environmental conditions. The recent advancement of genome editing technologies offers promising solutions by enabling precise alterations in plant genomes to get the desired traits. The advantage of these alterations includes suppression or elimination of specific genes without permanently introducing foreign DNA. In this chapter, we have briefly discussed about the evolution of genome editing technologies from meganucleases (MNs), transcription activator-like effector nucleases (TALENs), zinc finger nucleases (ZFNs), to clustered regularly interspaced short palindromic repeat—Cas (CRISPR/Cas) systems and their utilization in developing climate-resilient plants. These advance technologies may be utilized as smart agricultural practices to develop climate-smart crops and ensure a sustainable global food production.

P. Ranga · P. Choudhary · N. Kaul · S. Yadav · A. Dahiya · A. Kumar · A. Kumar
S. K. Sanwal · A. Mann (✉) · R. K. Yadav
ICAR-Central Soil Salinity Research Institute, Karnal, Haryana, India

A. Mann et al. (eds.), *Cutting Edge Technologies for Developing Future Crop Plants*,
https://doi.org/10.1007/978-981-96-2508-6_10

Keywords

Climate change · Gene editing · MNs · TALENs · ZFNs · CRISPR-Cas9

10.1 Introduction

According to the Intergovernmental Panel on Climate Change (IPCC) Sixth Assessment Report in 2023, the adverse impacts of anthropogenic climate change are expected to intensify, leading further to water scarcity and challenges in food production (IPCC Report 2023) due to rapidly growing global population (Tilman et al. 2011). Cumulatively, these factors will significantly impact agricultural crops as a consequence imposing selection pressure on plant metabolic processes. All the abiotic factors affect plant physiological, biochemical, molecular processes and thereby, affecting the plant growth and development resulting in reduced yields (Fig. 10.1). Therefore, developing climate-smart crops is crucial for addressing both United Nations Sustainable Development Goals (SDGs): SDG 2, which aims to

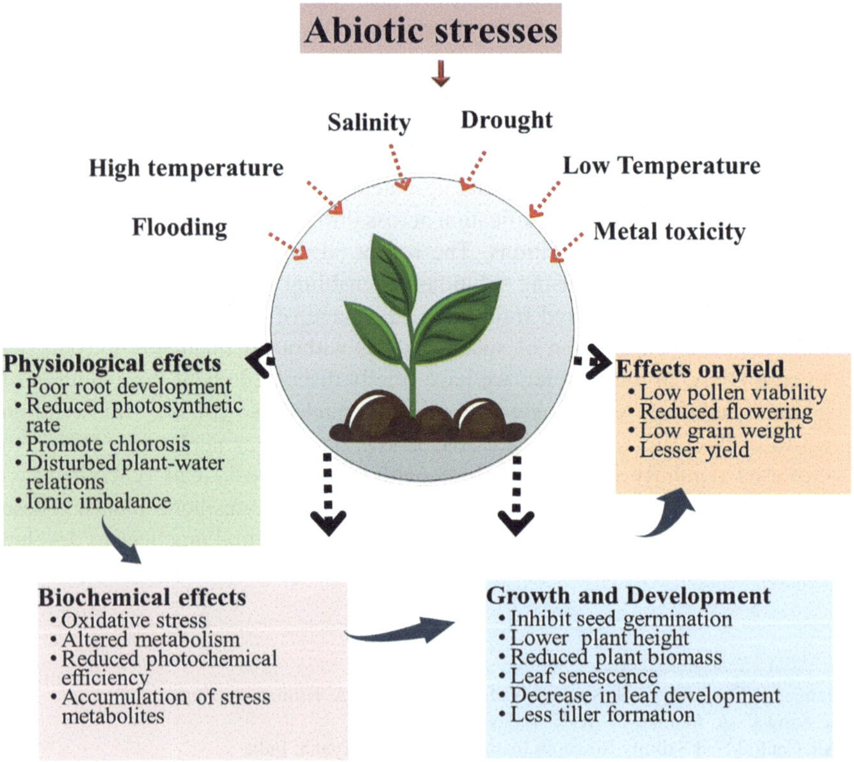

Fig. 10.1 Impact of abiotic stresses on plant functional traits

achieve Zero Hunger (Cappelli et al. 2021); and SDG 13, focused on Climate Action (Campbell et al. 2018; United Nations 2020a, b).

Availability of climate stress-resilient or multistress-tolerant crops can ensure stable food production despite changing environmental conditions. Agricultural practices and technologies have improved from conventional breeding to selective breeding, mutagenesis, speed breeding, transgenic technology, and genome editing (Kharkwal 2012; de Vries 1901; Prado et al. 2014). Conventional breeding is known for being labor-intensive and time-consuming, often taking years to develop improved/new varieties. To expedite this process, genetically modified (GM) crops have been created using transgenic technologies to introduce desirable traits (Qui et al. 2021). Transgenic Bt cotton serves as a notable example, being the only transgenic crop approved for widespread commercial cultivation. Overexpression of numerous responsive genes for abiotic stress tolerance has also been reported in many crops. For example, in cotton, salt-tolerant transgenic plants were developed with coexpression of Arabidopsis AtNHX1 and the H^+-pyrophosphatase gene AVP1. This dual-gene approach proved more effective in improving stress tolerance compared to single-gene expression (Zhang et al. 2011; Shen et al. 2015). Further improvement in salt tolerance was achieved in cotton by coexpressing AtNHX1 with the TsVP gene from *Thellungiella halophila* (Cheng et al. 2018). Additionally, overexpression of OsSIZ1 and AVP1 significantly boosted tolerance to combined salt/drought and heat/drought stresses along with enhancing fiber yield (Esmaeili et al. 2021). In tomato, transgenic plants with overexpression of both LeNHX2 and SlSOS2 exhibited improved NaCl tolerance and better growth under saline conditions compared to those overexpressing only one of these genes (Baghour et al. 2019). In wheat, overexpression of the abscisic acid-stress-ripening (ASR) protein TaASR1-D increased antioxidant capacity and ABA sensitivity, thereby enhancing abiotic stress tolerance (Qiu et al. 2021).

Notably, the extensive adoption of GM crops has been hindered by stringent government regulations and environmental biosafety laws (Prado et al. 2014). Japan was the first country to exempt use of gene editing, and now, Indian Department of Biotechnology has also permitted the use of SDN technology for edits pertaining to SDN-1 and SDN-2 category and is deemed as non-transgenic. Concerns over unverified health and environmental impacts observed during field trials have contributed to these regulatory challenges. The classic method for phenotyping and understanding gene functions involves disrupting normal gene activity. Experiments involving loss and gain of function (known as forward and reverse genetics) have been conducted since Thomas Morgan's discovery that genes residing on chromosomes and gene mutations are responsible for distinct traits. This pivotal insight spurred scientists worldwide to systematically induce gene mutations using chemicals, radiation, and random virus integration (DeMarini et al. 1989). This approach facilitated the linking of specific mutant genes to particular phenotype. However, when applied extensively across genomes, these techniques can raise intricate challenges, particularly in the development of mutant or transgenic plants.

Genome editing is defined as group of genetic engineering techniques which involves the modification of single or many nucleotide base pairs of gene by insertion, deletion, transitions, and transversions at specific sites in plant genome (Dunn and Pinkert 2014; Mann et al. 2024). Genome editing technologies have evidenced advantageous over GM crops as no foreign gene is transferred to the host genome. The emergence of gene editing now enables the creation of climate-smart crops with exceptional precision and specificity (Ahmar et al. 2020), and the plants, thus, developed are termed as gene-edited plants (GEP). The genetic material of every living organism is made of DNA which contains all information about the genes for each character/trait. Gene functions and sequences have been known for particular traits through various scientific findings which has made easy for editing these sequences. Gene editing resembles like editing a book with all information available and need to precisely cut or change or add the required part of gene sequences on particular chromosome or genome of plant with specific enzymes. GEP differ from GM plants that in later new gene sequences were inserted from same or other organism, while in the GEP, gene sequences of the same organism are edited.

10.2 Types of Genome Editing Technologies

Basically, genome editing is of three types depending on the type of insertion or deletion of a particular gene sequences and insertion of foreign DNA or not. These are SDN1, SDN2, and SDN3.

SDN1, cuts the double-stranded DNA without adding any foreign DNA. This break leads to any change like gene knock-out or silencing or deletion.

SDN2, the cut in double strand of DNA is further ligated using a small nucleotide template. This template contains few gene sequences which are incorporated in the target genome creating modifications.

SDN3, the template inserted in the double-stranded break of DNA contains gene sequences of other organism, hence, may contain foreign DNA.

The basic principle of genome editing is through site-specific recombinase or site-specific nuclease systems. The developmental progress of various technologies used for crop improvement over a period of time is depicted in Fig. 10.2. Starting from creating mutations, during 1901, plant improvement tools have reached to using CRISPR-Cas till date and still far way to go. Various path-throughs used till date are briefly described here.

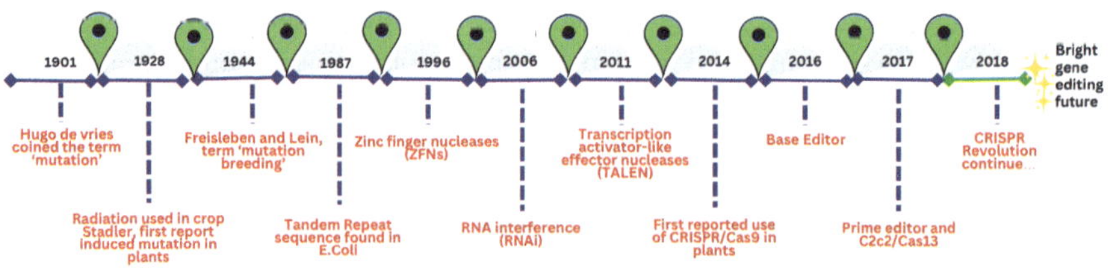

Fig. 10.2 Progressive timescale of scientifically used techniques for crop improvement

10.2.1 Site-Specific Recombinases (SSRs)

Site-specific recombinases (SSRs) are the enzymes which identify brief target DNA sequences spanning approximately 35–40 nucleotides and facilitate specific DNA editing via recombination. Site-specific recombinase systems were initially identified in bacteria and yeast, where they play crucial roles in various biological processes. It is a common genetic engineering strategy effecting a permanent modification of the target genome (Wang et al. 2011). The application of site-specific recombination technology in manipulating plant genomes has proven effective in several key aspects: resolving complex transgene insertions to single copies, eliminating undesired DNA segments, and accurately integrating DNA into specific genomic loci. The mechanism of site-specific recombinases is illustrated pictorally in Fig. 10.3. Recombination between two DNA sites begins when the recombinase enzyme recognizes and binds to these sites, either on separate DNA molecules or distant segments of the same molecule. The sites are then brought together in a synaptic complex, where controlled transesterification reactions allow DNA strands to be cut and rejoined. The recombinase enzyme forms a temporary covalent bond with the DNA, conserving the energy from the cleavage and using it to rejoin the strands without needing external energy like ATP (Wang et al. 2011).

Moreover, recombinases have shown versatility by enabling targeted recombination not only within nuclear genomes but also within non-nuclear targets like the plastid genome of tobacco. Wang et al. (2011) explore strategies for using multiple site-specific recombinase systems to precisely insert transgenes into predetermined genomic loci and remove unwanted selectable marker genes after genetic modification.

These strategies aim to precisely insert transgenes into specific genomic locations and remove unwanted genes (Wang et al. 2011). Site-specific recombination

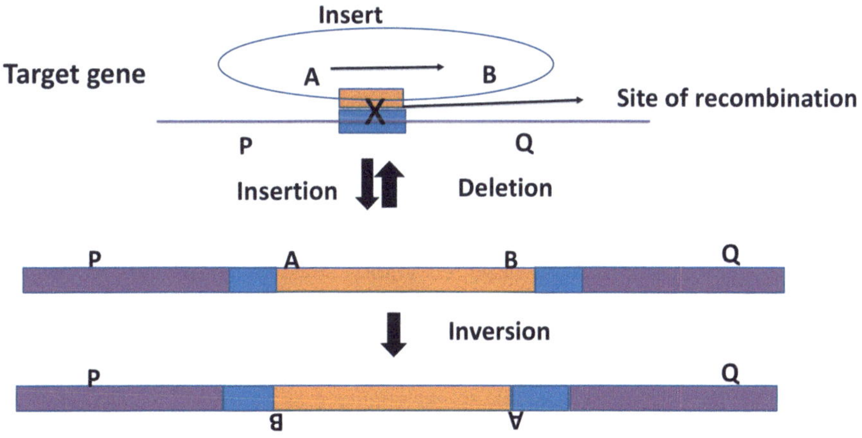

Fig. 10.3 Schematic representation of the mechanism of action of site-specific recombinases (SSRs)

(SSR) has been successfully applied to various plants, including Arabidopsis (Zuo et al. 2001), tomato (Zhang et al. 2006), maize (Zhang et al. 2003), rice, aspen (Ebinuma and Komamine 2001; Sreekala et al. 2005), and tobacco (Liu et al. 2005). Recombinases are widely used in genome manipulation across different organisms such as plants, bacteria, yeast, and mammals, enabling both knockouts and knock-ins in eukaryotic genomes. Their main advantage lies in their ability to function independently of intracellular repair mechanisms and without causing double-stranded breaks. Additionally, recombinases simplify the detection of gene insertion events by facilitating the integration of the desired gene segments. However, a notable limitation of these chimeric enzymes is their tendency to cause off-target effects.

10.2.2 Site-Specific Nucleases (SSNs)

Site-specific nucleases (SSNs) are constructed endonucleases designed to cleave DNA at specific sequences within the genome. These nucleases possess a DNA-binding domain that precisely binds to the target sequences (Gaj et al. 2013). Site-specific nuclease cleaves the target sequence and then modifies the gene at targeted site followed by cellular DNA repair at the site of gene modification. Currently, there are four main families of engineered nucleases used in genome editing namely, Engineered meganucleases (MegaN); Zinc finger nucleases (ZFNs); Transcription activator-like effector nucleases (TALENs) and Clustered regularly interspaced short palindromic repeat (CRISPR)/CRISPR-associated protein9 (cas9) nuclease system. These are defined briefly in next sections.

10.2.2.1 Engineered Meganucleases (MegaN)

Engineered Meganucleases (MegaN)—Meganucleases are natural endonucleases discovered in the late 1980s (Stoddard 2005). They are part of an endonuclease family capable of recognizing and cleaving large sequences of DNA ranging from 15 to 40 base pairs in length across most genomes (Gallagher et al. 2014). Meganucleases have been adapted for genome editing by companies like Cellectis modifying the regions responsible for DNA cleavage and DNA binding to target different sequences while maintaining high specificity. Precision BioSciences developed a method called directed nuclease editor (DNE) for designing engineered meganucleases that precisely target and modify specific genomic locations (Ashworth et al. 2010). After cleavage, natural DNA repair mechanisms, including homologous recombination, facilitate the integration of modified sequences or new genes.

Due to their ability to target large recognition sites, meganucleases are highly advantageous for genome engineering. However, their natural occurrence is limited, which restricts their applicability to all potentially interesting genetic loci. One drawback of meganucleases is the high cost and time required to construct sequence-specific enzymes for every possible target sequence as compared to another site-specific nucleases. Creating a target-specific meganuclease involves an initial protein engineering step, making them technically challenging to work with.

10.2.2.2 Zinc Finger Nucleases (ZFNs)

The zinc finger protein (ZFP) was first demonstrated as a site-specific nuclease in 1980. ZFN is an artificial endonuclease which involves pairing of two engineered proteins, each equipped with carefully selected zinc finger DNA-binding domains, with the DNA-cleavage domain (catalytic domain) of the FokI endonuclease. These proteins recognize two DNA sequences that are separated by only a few nucleotides. This arrangement enables the zinc finger nucleases to dimerize and make double-stranded cleavage on the target site of the gene. The DNA-binding domain of ZFNs consists of three to six zinc finger repeats, with each repeat able to recognize 9–18 base pairs of DNA. A three-finger array can recognize a 9 base pair target sequence if each zinc finger domain precisely binds to a 3 base pair DNA sequence (Scott 2005).

Numerous studies have highlighted the utility of ZFNs in generating genome-edited crops, including tobacco (Cai et al. 2009); Arabidopsis (Zhang et al. 2010; Osakabe et al. 2010), and corn (Shukla et al. 2009). A study by Osakabe et al. (2010) has evident the role of engineered ZFNs to target the regulator of ABA-INSENSITIVE4 (ABI4) stress-response gene by cleaving at specific recognition sites. These ZFN genes, under the control of a heat-shock promoter, were then introduced into the Arabidopsis genome. In addition, ZFN-mediated cleavage resulted in gene mutations, including deletions and substitutions in the ABI4 gene after heat-shock to somatic cells, leading to increased salt tolerance (Osakabe et al. 2010).

However, a significant challenge with this approach is that the specificity overlapping of individual zinc fingers is affected by the adjacent zinc fingers. This "context dependence" poses a problem, as the standard modular assembly method may fail to reliably predict and achieve the desired DNA recognition and cleavage (Gaj et al. 2013). ZFN-based editing technology has certain drawbacks, such as the intricate and costly construction of protein domains tailored for each specific genome locus. Additionally, there is a risk of inaccuracy in cleavage of target DNA (off-target effects) due to single nucleotide substitutions or improper interactions between domains (Puchta and Hohn 2010). Improving ZFN-mediated targeting in plant genomes requires addressing several challenges: reducing nuclease toxicity, selecting appropriate plant tissues for targeting, ensuring enzyme activity introduction, minimizing off-target mutations, and reliably detecting mutations (Puchta and Hohn 2010).

10.2.2.3 Transcription Activator-Like Effector Nucleases (TALENs)

In 2011, a new method of genome editing was developed as a substitute for ZFNs known as the TALEN (transcription activator-like effector nucleases) system to enhance the efficiency and accessibility (Clarke et al. 2012). This system evolved from transcription activator-like effectors (TALEs) originally found in phytopathogenic bacteria of the *Xanthomonas* genus (Boch et al. 2009). Phytopathogenic bacteria use the type III secretion pathway to deliver proteins into plant cells during infection (Bogdanove et al. 2010). After entry into the plant cell, TALEs (transcription activator-like effectors) move into the nucleus and bind to specific DNA sequences in the nucleus which activate the target gene expression. These effector proteins belong to a family of DNA-binding proteins that can activate the expression of specific target genes, functioning similarly to eukaryotic transcription factors.

TALENs are synthetic nucleases created by fusing the catalytic domain of the restriction endonuclease FokI with customized monomers from DNA-binding domains (Moscou and Bogdanove 2009). Transcriptional activator-like effectors (TALEs) comprise 13–28 nucleotide repeats. Each TALE repeat targets only one nucleotide, allowing flexible target design. The DNA-binding ability of TALEs was first defined in 2007, and later, the recognition code of TALE target DNA sequences was decoded (Romer et al. 2007). This engineering enables TALENs to target virtually any specific nucleotide sequence within the genome.

TALENs were first applied in crop improvement on *Oryza sativa*. The bacterial blight susceptibility gene OsSWEET14 was interrupted using TALENs, resulting in a mutant *Oryza sativa* that was resistant to blight disease. TALENs initiate selective and efficient manipulation of target DNA by identifying and activating specific DNA sequences upstream of the transcription start site (TSS) through a set of tandem repeats (Jankele and Svoboda 2014). TALENs have been implicated in crop improvement for plants such as *Nicotiana*, *Arabidopsis*, rapeseed, potato, flax, soybean, sugarcane, barley, rice, maize, tomato, and wheat (Gaj et al. 2013; Martinez-Fortun et al. 2017; Ran et al. 2017).

Gene mutations induced in several plant species, such as Arabidopsis (Cermak et al. 2011), Brachypodium (Shan et al. 2014), tobacco (Zhang et al. 2013), and many other organisms (Streubel et al. 2012; Tong et al. 2015), are among the most significant types of targeted genetic modifications (TGMs) mediated by TALENs reported to date. Despite their many advantages, there are some disadvantages associated with using TALENs in crop improvement: (i). **Size of cDNA**: The cDNA encoding TALENs is approximately 3 kb, making it difficult to deliver and express a pair of TALENs in a cell; (ii). **TALE Repeat Composition**: The composition of TALE repeats remains a bottleneck, affecting the efficiency of TALEN targeting; (iii). **Variable Efficacy**: The efficacy of TALENs targeting a specific gene is inconsistent. (iv). **High Repetitiveness**: The highly repetitive nature of TALENs limits their compatibility with certain viral vectors. Therefore, researchers were searching for more efficient genome editing tools to generate transgenics.

10.2.2.4 CRISPR/Cas9

CRISPR/Cas9 technology, a Nobel Prize-winning innovation by Emmanuelle Charpentier and Jennifer Doudna during 2020, is the sharpest gene technology tools. This genetic scissor, derived from *Streptococcus pyogenes*, utilizes RNA-guided nucleases which consists of Cas9 endonuclease and a single guide RNA (sgRNA). This system achieves precise site-specific recognition through the complementary interaction between a non-coding RNA and the target DNA. This interaction forms a complex of non-coding RNA and Cas proteins with nuclease activity. The sgRNA is formed by combining two small RNA molecules: CRISPR RNA (crRNA) and trans-activating CRISPR RNA (tracrRNA). These components together direct the Cas9 nuclease to a specific site on the DNA. Once the target site is located, the Cas9 protein introduces a double-stranded break in the DNA. This break activates the cell's repair machinery, which can either introduce desired changes through non-homologous end joining (NHEJ) or incorporate specific genetic material via homology-directed repair (HDR). Initially discovered as part of the bacterial immune system, CRISPR-Cas systems have been harnessed for precise genome editing in various organisms, including crops. In agriculture, the CRISPR-Cas system enables scientists to modify crops with unprecedented precision. Unlike traditional breeding methods that rely on chance genetic recombination, CRISPR-Cas allows for the targeted modification of specific genes associated with desirable traits.

Working of CRISPR-Cas 9 Technology

To understand the working of CRIPS system, we have designed a simple workflow (Fig. 10.4). Firstly, the desired gene sequence for a specific trait is downloaded from NCBI (https://www.ncbi.nlm.nih.gov/) or another relevant database. Next, gene specific primers are designed and amplified through polymerase chain reaction (PCR). This PCR product is confirmed through sequencing. Guide RNA (sgRNA) is designed from this validated sequence using online available tools. Extract the Cas9 and gRNA sequences from their respective sources and transform it into *E. coli*. The colonies are further screened for successful incorporation of gene cassette into the vector using colony PCR or sanger sequencing or restriction digestion. The positive colonies are then transformed to plant or explant using suitable crop-specific methods. Finally, PCR-based confirmation of putative transformants is done by PCR-based CASP assay, sanger sequencing, or Indel PCR assay.

Construction of Guide RNA (sgRNA) for CRISPR-Cas9 Technology

The sgRNAs direct Cas9 nuclease to recognize the target DNA sequence to interrupt transcriptional regulation. The gRNA-Cas9 complex identifies the target sequence by gRNA-DNA pairing between the 5′-end sequence of gRNA spacer and one DNA strand (complementary stand of protospacer). Cas9 requires the PAM sequence at the target site. The approximately 20 nucleotide-long gRNA spacer sequence could be readily programmed to target DNA sites with PAM using various tools available online (Table 10.1). Computational tools play a crucial role in designing CRISPR/Cas experiments for targeted genome modification. These tools assist in creating guide RNAs (gRNAs) tailored for specific nucleases, ensuring optimal binding to the desired genomic site while minimizing off-target effects by avoiding similar sequences in the organism's genome. Ensure high on-target and low off-target effects using BLAST. Structural characteristics of sgRNAs can be analyzed by ensuring the integrity of the four-stem loop structures, which are crucial for efficient processing and binding to the Cas9 protein (http://rna.tbi.univie.ac.at/cgi-bin/RNAWebSuite/RNAfold.cgi). "WeReview: CRISPR Tools" is a live, user-updatable online repository of computational tools for designing CRISPR/Cas experiments. Users can find and suggest updates or new tools (Torres-Perez et al. 2019).

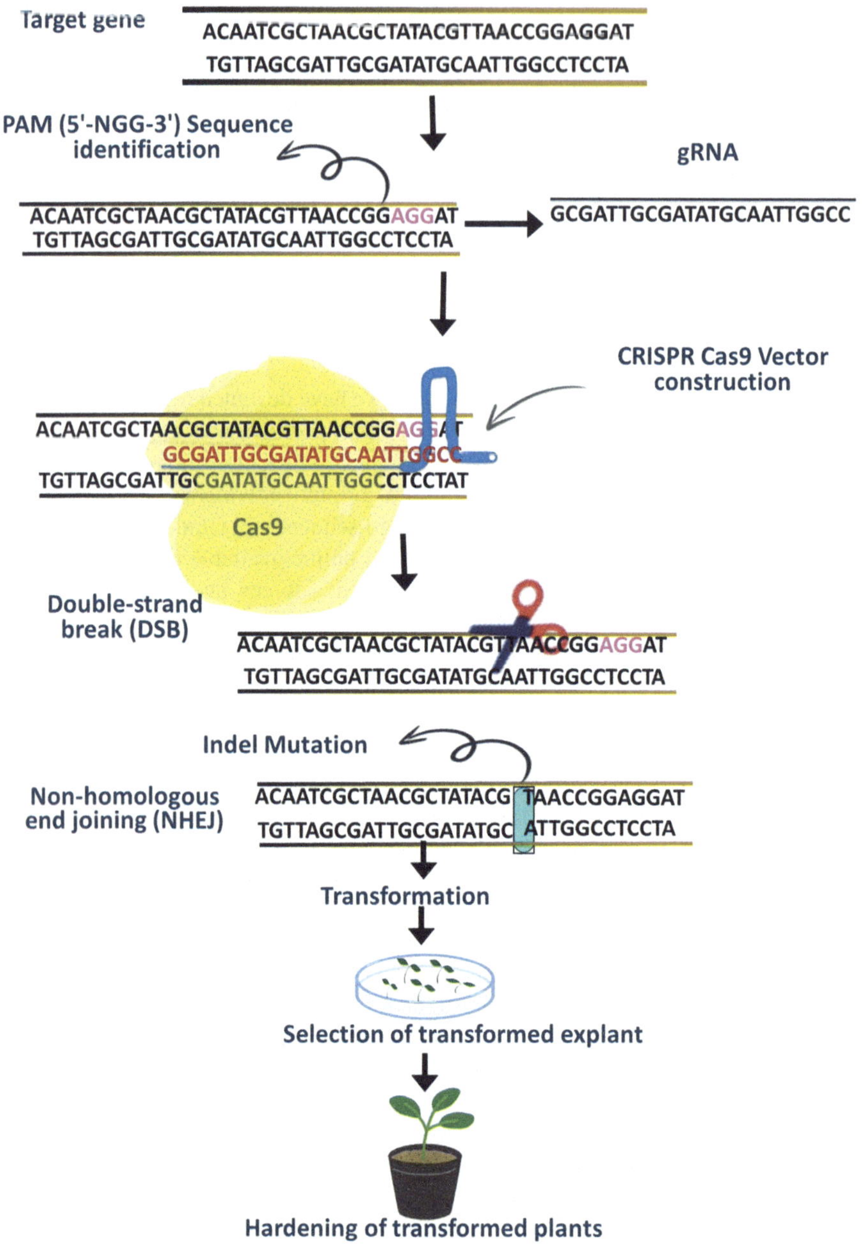

Fig. 10.4 Brief workflow illustrating the CRISPR/Cas9 editing procedure

Classification of CRISPR/Cas9 System

Based on signature genes and the typical organization of loci, CRISPR-Cas system is classified into four major types (I, II, III, and IV).

Table 10.1 Bioinformatics softwares used for designing of guide RNA

S. no	Online tools	Website link	References
1.	CRISPR direct	https://crispr.dbcls.jp	Naito et al. (2015)
2.	CRISPR-P 2.0	http://cbi.hzau.edu.cn/cgi-bin/CRISPR	Liu et al. (2017)
3.	CRISPR-PLANT	http://omap.org/crispr/index.html	Xie et al. (2014)
4.	CRISPR-GE	http://skl.scau.edu.cn/	Xie et al. (2017)
5.	Breaking-Cas	http://bioinfogp.cnb.csic.es/tools/breakingcas/	Oliveros et al. (2016)
6.	CRISPOR.Org	http://crispor.org	Concordet and Haeussler (2018)
7.	CRISPR-BETs	https://bioinfor.yzu.edu.cn/crisprbets/	Wu et al. (2022)
8.	CHOPCHOP	https://chopchop.cbu.uib.no/	Montague et al. (2014)
9.	CRISPR Multitargeter	https://multicrispr.net/	Prykhozhij et al. (2015)
10.	E-crisp	http://www.e-crisp.org/E-CRISP/	Heighwer et al. (2014)

Type I CRISPR-Cas systems contain the signature gene cas3, encoding a helicase with ssDNA-stimulated ATPase activity and often fused to an HD domain with endonuclease activity. In *Methanocaldococcus jannaschii*, the HD domain also shows exonuclease activity. Type I systems typically have a single operon with cas1, cas2, and genes for the cascade complex subunits. Type I systems are divided into six subtypes (I-A to I-F), each with unique signature genes and operon organization. Subtypes I-E and I-F lack the cas4 gene. Some variants of these subtypes and standalone effector complexes associated with transposon-related genes have also been identified.

Type II CRISPR-Cas systems are characterized by the signature gene cas9, which encodes a multidomain protein essential for crRNA maturation and target DNA cleavage. Known as "HNH" systems or Streptococcus-like, they contain cas9 along with the ubiquitous cas1 and cas2 genes and typically one or two tracrRNA genes. Cas9, a large protein (~800–1400 amino acids), includes two nuclease domains (RuvC-like and HNH) required for DNA cleavage. Recent crystal structures of Cas9 reveal a two-lobed structure critical for PAM recognition and sgRNA interaction. Type II systems are classified into three subtypes (II-A, II-B, and II-C). Type II-A includes the csn2 gene, forming homotetrameric rings that bind dsDNA, while II-B lacks csn2 but has a Cas4 family gene. The recently proposed II-C subtype only contains cas1, cas2, and cas9 genes. These systems have been developed into powerful genome editing tools with significant biotechnological potential.

Type III CRISPR-Cas systems are characterized by the signature gene cas10, which encodes a multidomain protein with a palm domain similar to those in cyclases and polymerases. Cas10 includes four distinct domains: a cyclase-like N-terminal domain, a helical domain with a Zn-binding motif, an active palm domain, and a C-terminal alpha helical domain. Cas10 is the large subunit of effector complexes in type III systems. Type III systems also encode other effector complex subunits, including a small subunit, a Cas5 group RAMP protein, and several Cas7 group RAMP proteins. Unlike type I systems, type III systems often lack their own cas1 and cas2 genes, instead using crRNAs from type I or II systems. Two subtypes exist within type III: III-A, which often targets DNA, and III-B, which targets RNA. Both subtypes are typically found alongside other CRISPR-Cas systems in a genome.

Type IV CRISPR-Cas systems, commonly found on plasmids and in some bacterial genomes, are typified by the CRISPR-Cas locus in *Acidithiobacillus ferrooxidans* ATCC 23270. These systems lack cas1 and cas2 genes and are often not associated with CRISPR arrays. The effector complex includes a reduced large subunit (csf1), RAMP proteins from the Cas5 (csf3) and Cas7 (csf2) groups, and sometimes a small subunit. There are two subtypes: one with a DinG family helicase (csf4) and another without DinG but with a small alpha helical protein. Type IV systems may use crRNA from other CRISPR arrays or generate crRNA directly from foreign RNA.

Among all types, Cas9, an integral part of class 2 (type II) system, is the major effector protein for CRISPR-mediated genome editing revolution. Still, the classification of CRISPR-Cas systems is evolving as new variants are discovered like Cas12 and 13.

10.3 Advantages of CRISPR/Cas9 in Agriculture

CRISPR/Cas9 surpasses traditional breeding and genetic engineering in precision by allowing the modification of a few nucleotides, entire alleles, or the insertion of new genes. It holds the potential to create nutritionally enhanced, pest-resistant, and climate-resilient crops while potentially avoiding stringent GMO regulations. Scientists believe that gene-edited plants will gain greater public acceptance and can be developed rapidly with minimal off-target effects, even in complex genomes. Experts have ranked genome editing as the most prominent new breeding technique with CRISPR as having the greatest potential to enhance crop agronomic performance and contribute to closing the food security gap by 2050 (Lassoued et al. 2018). Whole-genome sequencing has demonstrated that CRISPR/Cas9 induces a low frequency of off-target mutations in plants (Li et al. 2019).

Numerous recent proof-of-concept studies in plant breeding highlight the technological promise of genome editing, particularly CRISPR-Cas (CRISPR-associated proteins) technology, in producing a wide range of targeted and diverse traits due to its efficiency in terms of cost, time, and precision in editing genomes, compared to earlier, more resource-intensive tools like zinc finger nucleases (ZFN) and transcription activator-like effector nucleases (TALEN) (Alok et al. 2020; Wada et al. 2020). The CRISPR/Cas9 genome editing system was used in wheat protoplasts to target the editing of stress-responsive transcription factor genes, specifically the wheat dehydration responsive element binding protein 2 (TaDREB2) and the wheat ethylene responsive factor 3 (TaERF3). Targeted editing of TaDREB2 and TaERF3 was achieved through the transient expression of small guide RNA and Cas9 protein in wheat protoplasts, increasing drought resistance without affecting growth parameters (Kim et al. 2018). High oleic soybean and high GABA tomato were the first gene-edited plants followed by high oil canola, non-browning mushroom, bacterial blight resistant rice are typical examples of gene editing beneficial for farmers. Further, mutant alleles of drought and salt tolerance (DST) gene has been created for *indica* rice cv. MTU1010 by using CRISPR-Cas9 gene editing (Kumar et al. 2020) and now these lines are in field trail.

CRISPR has been successfully applied to several crops and plants, including wheat (Ogatta et al. 2020), corn (Liu et al. 2024), soybeans, tomatoes, potatoes, oranges, bananas, cassava, and flowers, with rice being the most extensively studied (Huang et al. 2019; Alam et al. 2022; Li et al. 2022; Kumar et al. 2020), maize (Ahn et al. 2020; Ansari et al. 2020; Bouzroed et al. 2020; Kim et al. 2018; Haque et al. 2018; Jaganathan et al. 2018; Ricroch et al. 2017). A brief summary of work carried out using various gene editing technologies has been tabulated in Table 10.2.

The crops currently being genome-edited are predominantly cereals (61%), followed by oilseed crops (52%), vegetables (47%), fruits and nuts (34%), and pulses (27%) (Lassoued et al. 2021). With the increasing food demand and the inevitable challenges posed by climate change, there is a growing need for greater flexibility and innovation in crop resilience and production systems. For decades, traditional plant breeding methods were the primary choice for crop improvement programs. However, genome editing now offers a superior tool for precise and targeted modifications in any crop plant, allowing for both a deeper understanding of fundamental biological systems and significant improvements in agricultural traits. CRISPR/Cas9 has outpaced its predecessors, and ongoing advancements continue to enhance its functional capabilities. The successful utilization of these systems requires collaboration among scientists, breeders, ethicists, governments, and stakeholders. The technology's efficiency, specificity, simplicity, and flexibility for multiplexing have made it an essential tool in crop improvement.

10.4 Conclusions and Way Forward

In the quest for climate-resilient crops, it is essential to delve into the genetic foundations of traits influenced by climate. This involves discovering genes associated with climate resilience, pinpointing CRISPR-Cas targets, and presenting case studies illustrating effective gene editing for climate adaptation. The advent of genome editing has sparked significant interest among agricultural scientists due to its straightforwardness, precision, and potential to develop improved crop varieties by adding valuable traits or removing undesirable ones. Researchers are actively working to enhance crop productivity, stress tolerance, disease and pest resistance, reduce input costs, and boost nutritional content. Gene editing offers diverse options: from altering a few nucleotides in the vast genomes of living cells to modifying entire alleles or inserting new genes into targeted genomic regions. Beyond enhancing nutritional value, gene editing presents an effective strategy for developing crops resistant to pests and adaptable to harsh environments.

Table 10.2 Applications of gene editing technologies to enhance climate resilience in plants

S. no.	Crop	Targeted protein	Candidate gene (s)*	Abiotic stress tolerance	Genome editing type	References
1.	Arabidopsis	ABA-INSENSITIVE4	(ABI4) gene	Salt tolerance	Site-directed mutagenesis via ZFN	Osakabe et al. (2010)
2.	Cotton	Na^+ /H^+ antiporter H^+ pyrophosphatase	AVP1/AtNHX1	Salt and drought tolerance	Transgenic	Shen et al. (2015)
3.	Cotton	Na^+ /H^+ antiporter (V-NHX) and tonoplast H^+ translocating inorganic pyrophosphatase (VP)	TsVP/AtNHX1	Salt tolerance	Transgenic	Cheng et al. (2018)
4.	Cotton	Arabidopsis vacuolar pyrophosphatase/ ligase gene	AVP1/OsSIZ1	Combined heat and drought and combined drought and salt tolerance	Transgenic	Esmaeili et al. (2021)
5.	Maize	Choline dehydrogenase from Escherichia coli)/ V-H^+ -PPase from *Thellungiella halophila*	betaA/TsVP	Drought tolerance	–	Wei et al. (2011)
6.	Maize	Nuclear factor-Y subunit A	*ZmNF-YA*	Drought and salt stress	miRNA	Luan et al. (2015)
7.	Maize	GA biosynthesis enzyme	*ZmGA20ox3* *Loss of function*	Drought stress	CRISPR-Cas9	Liu et al. (2024)
8.	Potato	Superoxide dismutase/ ascorbate peroxidase	APX/SOD	Heat and oxidative stress tolerance	–	Tang et al. (2006)
9.	Potato	Choline oxidase/ superoxide dismutase/ ascorbate peroxidase	codA/SOD/APX	Salt and drought tolerance	Transgenic	Ahmad et al. (2010)
10.	Potato	Superoxide dismutase/ ascorbate peroxidase	PaSOD/RaAPX	Salt tolerance	Transgenic	Shafi et al. (2017)
11.	Rice	Dehydration responsive element B1	OsDREB1B	Salt and drought tolerance	SDN3	Datta et al. (2012)
12.	Rice	Late embryogenesis abundant (LEA) proteins	OsLEA gene	Salt and drought tolerance	Transgenic	Duan and Cai (2012)

S. no.	Crop	Targeted protein	Candidate gene (s)*	Abiotic stress tolerance	Genome editing type	References
13.	Rice	abZIP transcription factor/ protein kinase	OsbZIP46CA1/ SAPK6	Drought, heat, and cold tolerance	Double transgenic	Chang et al. (2017)
14.	Rice	Glutamine synthetase	OsGS1;1/OsGS2	Osmotic, salt, and drought tolerance	Transgenic	James et al. (2018)
15.	Rice	MYB transcription factor	OsMYB6	Salt and drought tolerance	Transgenic	Tang et al. (2019)
16.	Rice	9-cis-epoxycarotenoid dioxygenase 5	OsNCED5	Regulates salt and water stress	CRISPR-Cas9	Huang et al. (2019)
17.	Rice	MYB transcription factor	OsMYB6	Salt and drought tolerance	Transgenic	Tang et al. (2019)
18.	Rice	Vacuolar H^+-pyrophosphatase 1	OVP1	Salt tolerance	Transgenic	Kim et al. (2020)
19.	Rice	Enhanced response to ABA1	OsERA1	Drought tolerance	CRISPR/Cas9-induced frameshift mutations	Ogata et al. (2020)
20.	Rice	Basic helix-loop-helix (bHLH) proteins	OsbHLH044 (loss of function)	Salinity sensitivity	CRISPR/Cas9 system with two sgRNAs	Alam et al. (2022)
21.	Rice	Reduction of grain chalkiness and improvement of grain quality	OsRR30	–	CRISPR/Cas9	Li et al. (2022)
22.	Tobacco	Late embryogenesis abundant 7(LEA) g8ene/ basic leucine zipper (bZIP) gene	LEA/bZIP	Salt and osmotic stress tolerance	Single-gene transgenic	Qu et al. (2012)
23.	Tobacco	Betaine synthesis gene/ vacuolar Na+ / H+ antiporter gene	AhBADH/ SeNHX1	Salt tolerance	Transgenic	Zhou et al. (2008)
24.	Tomato	K^+, Na^+ /H^+ 3 antiporter gene/regulatory kinase gene	LeNHX2/SlSOS2	Salt tolerance	Double transgenic	Baghour et al. (2019)

(continued)

Table 10.2 (continued)

S. no.	Crop	Targeted protein	Candidate gene (s)*	Abiotic stress tolerance	Genome editing type	References
25.	Tomato	Mitogen-activated protein kinase 3	SlMAPK3	High temperature	–	Yu et al. (2019)
26.	Tomato	Auxin response factor 4 down regulation	SlARF4	Salinity and osmotic stress	CRISPR/Cas9 mutant	Bouzroud et al. (2020)
27.	Tomato	Dehydration responsive element binding1A/zinc finger proteins transcription factor genes	AtDREB1A/ BcZAT12	Drought tolerance	Double transgenic	Krishna et al. (2021)
28.	Wheat	Dehydration responsive element binding protein 2 ethylene responsive factor 3	*TaDREB2, TaERF3*	Drought tolerance	CRISPR/Cas9	Kim et al. (2018)
29.	Wheat	Abscisic acid-stress-ripening (ASR) protein	*TaASR1-D*	Osmotic and drought tolerance	Transgenic	Qiu et al. (2021)

Unlike traditional genetic modification, genome editing often results in minimal DNA alterations, potentially circumventing rigorous regulatory procedures. Many experts believe crops improved through precise gene editing techniques may be more readily accepted by the public compared to genetically modified organisms. Genome editing enables rapid development of new crop varieties with minimal risk of unintended genetic changes. Its versatility allows it to be applied in any laboratory setting to virtually any crop, including those with complex genomes that are challenging to breed using conventional methods.

Thus, genome editing holds promise for revolutionizing agricultural practices and addressing global food challenges efficiently and sustainably. The way forward involves continuous advancements in genome editing technologies, comprehensive understanding of genome editing consequences, and collaboration among scientists, breeders, ethicists, governments, and stakeholders to ensure these technologies are fully and responsibly utilized for the benefit of global agriculture.

References

Ahmad R, Kim YH, Kim MD, Kwon SY, Cho K, Lee HS, Kwak SS (2010) Simultaneous expression of choline oxidase, superoxide dismutase and ascorbate peroxidase in potato plant chloroplasts provides synergistically enhanced protection against various abiotic stresses. Physiol Plant 138(4):520–533

Ahmar S, Gill RA, Jung K-H, Faheem A, Qasim MU, Mubeen M, Zhou W (2020) Conventional and molecular techniques from simple breeding to speed breeding in crop plants: recent advances and future outlook. Int J Mol Sci 21:2590

Ahn CH, Ramya M, An HR, Park PM, Kim YJ, Lee SY, Jang S (2020) Progress and challenges in the improvement of ornamental plants by genome editing. Plan Theory 9:687. https://doi.org/10.3390/plants9060687

Alam MS, Yang ZK, Li C, Yan Y, Liu Z, Nazir MM, Xu JH (2022) Loss-of-function mutations of OsbHLH044 transcription factor lead to salinity sensitivity and a greater chalkiness in rice (Oryza sativa L.). Plant Physiol Biochem 193:110–123

Alok A, Sandhya D, Jogam P, Rodrigues V, Bhati KK, Sharma H, Kumar J (2020) The rise of the CRISPR/Cpf1 system for efficient genome editing in plants. Front Plant Sci 11:264. https://doi.org/10.3389/fpls.2020.00264

Ansari WA, Chandanshive SU, Bhatt V, Nadaf AB, Vats S, Katara JL, Sonah H et al (2020) Genome editing in cereals: approaches, applications and challenges. Int J Mol Sci 21:4040. https://doi.org/10.3390/ijms21114040

Ashworth J, Taylor GK, Havranek JJ, Quadri SA, Stoddard BL, Baker D (2010) Computational reprogramming of homing endonuclease specificity at multiple adjacent base pairs. Nucleic Acids Res 38:5601–5608

Boch J, Scholze H, Schornack S, et al (2009) Breaking the code of DNA binding specificity of TAL-type III effectors. Science 326(5959):1509–1512. https://doi.org/10.1126/science.1178811

Bogdanove AJ, Schornack S, Lahaye T (2010) TAL effectors: finding plant genes for disease and defense. Curr Opin Plant Biol 3(4):394–401. https://doi.org/10.1016/j.pbi.2010.04.010

Baghour M, Gálvez FJ, Sánchez ME, Aranda MN, Venema K, Rodríguez-Rosales MP (2019) Overexpression of LeNHX2 and SlSOS2 increases salt tolerance and fruit production in double transgenic tomato plants. Plant Physiol Biochem 135:77–86

Bouzroud S, Gasparini K, Hu G, Barbosa MAM, Rosa BL, Fahr M et al (2020) Down regulation and loss of auxin response factor 4 function using CRISPR/Cas9 alters plant growth, stomatal function and improves tomato tolerance to salinity and osmotic stress. Genes 11(3):272

Cai CQ, Doyon Y, Ainley WM et al (2009) Targeted transgene integration in plant cells using designed zinc finger nucleases. Plant Mol Biol 69:699–709

Campbell BM et al (2018) Urgent action to combat climate change and its impacts (SDG 13): transforming agriculture and food systems. Curr Opin Environ Sustain 34:13–20. https://doi.org/10.1016/j.cosust.2018.06.005

Cappelli A, Lupori L, Cini E (2021) Baking technology: a systematic review of machines and plants and their effect on final products, including improvement strategies. Trends Food Sci Technol 115:275–284

Cermak T, Doyle EL, Christian M, et al (2011) Efficient design and assembly of custom TALEN and other TAL effector-based constructs for DNA targeting [published correction appears in Nucleic Acids Res. 2011 Sep 1;39(17):7879]. Nucleic Acids Res 39(12):e82. https://doi.org/10.1093/nar/gkr218

Chang Y, Nguyen BH, Xie Y, Xiao B, Tang N, Zhu W et al (2017) Co-overexpression of the constitutively active form of OsbZIP46 and ABA-activated protein kinase SAPK6 improves drought and temperature stress resistance in rice. Front Plant Sci 8:1102

Cheng C, Zhang Y, Chen X, Song J, Guo Z, Li K, Zhang K (2018) Co-expression of AtNHX1 and TsVP improves the salt tolerance of transgenic cotton and increases seed cotton yield in a saline field. Mol Breed 38:1–15

Clarke L, Zheng-Bradley X, Smith R et al (2012) The 1000 genomes project: data management and community access. Nat Methods 9:459–462

Concordet JP, Haeussler M (2018) CRISPOR: intuitive guide selection for CRISPR/Cas9 genome editing experiments and screens. Nucleic Acids Res 46(W1):W242–W245

Datta K, Baisakh N, Ganguly M, Krishnan S, Yamaguchi Shinozaki K, Datta SK (2012) Overexpression of Arabidopsis and rice stress genes' inducible transcription factor confers drought and salinity tolerance to rice. Plant Biotechnol J 10(5):579–586

de Vries H (1901) Die mutations theorie I. Versuche und beobachtungen über die entstehung von arten im pfanzenreich. Verlag Von & Veit Comp, Leipzig

DeMarini DM, Brockman HE, de Serres FJ, Evans HH, Stankowski LF, Hsie AW (1989) Specific-locus mutations induced in eukaryotes (especially mammalian cells) by radiation and chemicals: a perspective. Mutat Res Rev Genet Toxicol 220(1):11–29

Duan J, Cai W (2012) OsLEA3-2, an abiotic stress induced gene of rice plays a key role in salt and drought tolerance

Dunn DA, Pinkert CA (2014) Gene editing. Elsevier, Transgenic Animal Technology, pp 229–248

Ebinuma H, Komamine A (2001) MAT (multi-auto-transformation) vector system. The oncogenes of agrobacterium as positive markers for regeneration and selection of marker-free transgenic plants. In Vitro Cell Dev Biol 37:103–113

Esmaeili N, Cai Y, Tang F, Zhu X, Smith J, Mishra N et al (2021) Towards doubling fibre yield for cotton in the semiarid agricultural area by increasing tolerance to drought, heat and salinity simultaneously. Plant Biotechnol J 19(3):462–476

Gaj T, Gersbach CA, Barbas CF (2013) ZFN, TALEN, and CRISPR/Cas-based methods for genome engineering. Trends Biotechnol 31(7):397–405

Gallagher RR, Li Z, Lewis AO, Isaacs FJ (2014) Rapid editing and evolution of bacterial genomes using libraries of synthetic DNA. Nat Protoc 9(10):2301–2316. https://doi.org/10.1038/nprot.2014.082

Haque E, Taniguchi H, Hassan MM, Bhowmik P, Karim MR, Śmiech M, Islam T (2018) Application of CRISPR/Cas9 genome editing technology for the improvement of crops cultivated in tropical climates: recent progress, prospects, and challenges. Front Plant Sci 9. https://doi.org/10.3389/fpls.2018.00617

Heigwer F, Kerr G, Boutros M (2014) E-CRISP: fast CRISPR target site identification. Nat Methods 11:122–123

Huang Y, Jiao Y, Xie N, Guo Y, Zhang F, Xiang Z et al (2019) OsNCED5, a 9-cis-epoxycarotenoid dioxygenase gene, regulates salt and water stress tolerance and leaf senescence in rice. Plant Sci 287:110188

IPCC (2023) Summary for policymakers. In: Core Writing Team, Lee H, Romero J (eds) Climate change 2023: synthesis report. Contribution of working groups I, II and III to the sixth assessment report of the intergovernmental panel on climate change. IPCC, Geneva, pp 1–34

Jaganathan D, Ramasamy K, Sellamuthu G, Jayabalan S, Venkataraman G (2018) CRISPR for crop improvement: an update review. Front Plant Sci 9:985. https://doi.org/10.3389/fpls.2018.00985

James D, Borphukan B, Fartyal D, Ram B, Singh J, Manna M et al (2018) Concurrent overexpression of OsGS1; 1 and OsGS2 genes in transgenic rice (*Oryza sativa L.*): impact on tolerance to abiotic stresses. Front Plant Sci 9:786

Jankele R, Svoboda P (2014) TAL effectors: tools for DNA targeting. Brief Funct Genomics 13(5):409–419

Kharkwal MC (2012) A brief history of plant mutagenesis. In: Shu QJ, Forster BP, Nakagawa H (eds) Plant mutation breeding and biotechnology. CABI, Wallingford, pp 21–30

Kim D, Alptekin B, Budak H (2018) CRISPR/Cas9 genome editing in wheat. Funct Integr Genomics 18:31–41

Kim JJ, Park SI, Kim YH, Park HM, Kim YS, Yoon HS (2020) Overexpression of a proton pumping gene OVP1 enhances salt stress tolerance, root growth and biomass yield by regulating ion balance in rice (*Oryza sativa L.*). Environ Exp Bot 175:104033

Krishna R, Ansari WA, Jaiswal DK, Singh AK, Verma JP, Singh M (2021) Co-overexpression of AtDREB1A and BcZAT12 increases drought tolerance and fruit production in double transgenic tomato (*Solanum lycopersicum*) plants. Environ Exp Bot 184:104396

Kumar SVV, Verma RK, Yadav SK, Yadav P, Watts A, Rao MV, Chinnusamy V (2020) CRISPR-Cas9 mediated genome editing of drought and salt tolerance (OsDST) gene in indica mega rice cultivar MTU1010. Physiol Mol Biol Plants 26:1099–1110

Lassoued R, Hesseln H, Phillips PWB, Smyth SJ (2018) Top plant breeding techniques for improving food security: an expert Delphi survey of the opportunities and challenges, vol 14. Int J Agric Resour, Governance Ecol, p 14

Lassoued R, Phillips PWB, Macall DM, Hesseln H, Smyth SJ (2021) Expert opinions on the regulation of plant genome editing. Plant Biotechnol J 19(6):1104–1109

Li J, Manghwar H, Sun L, Wang P, Wang G, Sheng H, Zhang J, Liu H, Qin L, Rui H et al (2019) Whole genome sequencing reveals rare off-target mutations and considerable inherent genetic or/and somaclonal variations in CRISPR/Cas9-edited cotton plants. Plant Biotechnol J 17:858–868

Li C, Gong C, Wu J, Yang L, Zhou L, Wu B et al (2022) Improvement of rice agronomic traits by editing type-B response regulators. Int J Mol Sci 23(22):14165

Liu HK, Yang C, Wei ZW (2005) Heat shock-regulated site-specific excision of extraneous DNA in transgenic plants. Plant Sci 168:997–1003

Liu H, Ding Y, Zhou Y, Jin W, Xie K, Chen LL (2017) CRISPR-P 2.0: An improved CRISPR-Cas9 tool for genome editing in plants. Mol Plant 10(3):530–532

Liu Y, Chen Z, Zhang C, Guo J, Liu Q, Yin Y et al (2024) Gene editing of ZmGA20ox3 improves plant architecture and drought tolerance in maize. Plant Cell Rep 43(1):18

Luan M, Xu M, Lu Y, Zhang L, Fan Y, Wang L (2015) Expression of zma-miR169 miRNAs and their target ZmNF-YA genes in response to abiotic stress in maize leaves. Gene 555(2):178–185

Mann A, Ranga P, Choudhary P, Yadav S, Kaul N, Dahiya A, Prakash NR, Kumar A, Kumar A, Sanwal SK (2024) Genome Editing Technologies for Enhancing Plant Resilience to Biotic and Abiotic Stresses-Brief Review. Journal of Soil Salinity and Water Quality 16(2):180–193

Martínez-Fortún J, Phillips DW, Jones HD (2017) Potential impact of genome editing in world agriculture. Emerg Topics Life Sci 1(2):117–133

Montague TG, Cruz JM, Gagnon JA, Church GM, Valen E (2014) CHOPCHOP: a CRISPR/Cas9 and TALEN web tool for genome editing. Nucleic Acids Res 42:W401–W407

Moscou MJ, Bogdanove AJ (2009) A simple cipher governs DNA recognition by TAL effectors. Science 326(5959):1501. https://doi.org/10.1126/science.1178817

Naito Y, Hino K, Bono H, Ui-Tei K (2015) CRISPR direct: software for designing CRISPR/Cas guide RNA with reduced off-target sites. Bioinformatics 31(7):1120–1123

Ogata T, Ishizaki T, Fujita M, Fujita Y (2020) CRISPR/Cas9-targeted mutagenesis of OsERA1 confers enhanced responses to abscisic acid and drought stress and increased primary root growth under nonstressed conditions in rice. PLoS One 15(12):e0243376

Oliveros JC, Franch M, Tabas-Madrid D, San-León D, Montoliu L, Cubas P, Pazos F (2016) Breaking-Cas-interactive design of guide RNAs for CRISPR-Cas experiments for ENSEMBL genomes. Nucleic Acids Res 44(W1):W267–W271

Osakabe K, Osakabe Y, Toki S (2010) Site-directed mutagenesis in Arabidopsis using custom-designed zinc finger nucleases [published correction appears in Proc Natl Acad Sci U S A. 2011 Jan 4;108(1):433]. Proc Natl Acad Sci USA 107(26):12034–12039. https://doi.org/10.1073/pnas.1000234107

Prado JR, Segers G, Voelker T, Carson D, Dobert R, Phillips J, Cook K, Cornejo C, Monken J, Grapes L (2014) Genetically engineered crops: from idea to product. Annu Rev Plant Biol 65(2014):769–790

Prykhozhij SV, Rajan V, Gaston D, Berman JN (2015) Correction: CRISPR multitargeter: a web tool to find common and unique CRISPR single guide RNA targets in a set of similar sequences. PLoS One 10(9):e0138634

Puchta H, Hohn B (2010) Breaking news: plants mutate right on target. Proc Natl Acad Sci U S A 107(26):11657–11658. https://doi.org/10.1073/pnas.1006364107. PMID: 20554917; PMCID: PMC2900667.

Qiu D, Hu W, Zhou Y, Xiao J, Hu R, Wei Q et al (2021) TaASR1-D confers abiotic stress resistance by affecting ROS accumulation and ABA signalling in transgenic wheat. Plant Biotechnol J 19(8):1588–1601

Qu GZ, Zang L, Xilin H, Gao C, Zheng T, Li KL (2012) Co-transfer of LEA and bZip genes from Tamarix confers additive salt and osmotic stress tolerance in transgenic tobacco. Plant Mol Biol Report 30:512–518

Ran Y, Liang Z, Gao C (2017) Current and future editing reagent delivery systems for plant genome editing. Sci China Life Sci 60:490–505

Ricroch A, Clairand P, Harwood W (2017) Use of CRISPR systems in plant genome editing: toward new opportunities in agriculture. Emerging Top Life Sci 1:169–182. https://doi.org/10.1042/etls20170085

Römer P, Hahn S, Jordan T, Strauss T, Bonas U, Lahaye T (2007) Plant pathogen recognition mediated by promoter activation of the pepper Bs3 resistance gene. Science 318(5850):645–648. https://doi.org/10.1126/science.1144958

Scott C (2005) The zinc finger nuclease monopoly. Nat Biotechnol 23:915–918

Shafi A, Pal AK, Sharma V, Kalia S, Kumar S, Ahuja PS, Singh AK (2017) Transgenic potato plants overexpressing SOD and APX exhibit enhanced lignification and starch biosynthesis with improved salt stress tolerance. Plant Mol Biol Report 35:504–518

Shan Q, Wang Y, Li J, Gao C (2014) Genome editing in rice and wheat using the CRISPR/Cas system. Nat Protoc 9(10):2395–2410

Shen G, Wei J, Qiu X, Hu R, Kuppu S, Auld D et al (2015) Co-overexpression of AVP1 and AtNHX1 in cotton further improves drought and salt tolerance in transgenic cotton plants. Plant Mol Biol Report 33:167–177

Shukla V, Doyon Y, Miller J et al (2009) Precise genome modification in the crop species *Zea mays* using zinc-finger nucleases. Nature 459:437–441

Sreekala C, Wu L, Gu K, Wang D, Tian D, Yin Z (2005) Excision of a selectable marker in transgenic rice (Oryza sativa L.) using a chemically regulated Cre/loxP system. Plant Cell Rep 24:86–94

Stoddard BL (2005) Homing endonuclease structure and function. Q Rev Biophys 38(1):49–95

Streubel J, Blücher C, Landgraf A, Boch J (2012) TAL effector RVD specificities and efficiencies. Nat Biotechnol 30(7):593–595

Tang L, Kwon SY, Kim SH et al (2006) Enhanced tolerance of transgenic potato plants expressing both superoxide dismutase and ascorbate peroxidase in chloroplasts against oxidative stress and high temperature. Plant Cell Rep 25:1380–1386

Tang Y, Bao X, Zhi Y, Wu Q, Guo Y, Yin X et al (2019) Overexpression of a MYB family gene, OsMYB6, increases drought and salinity stress tolerance in transgenic rice. Front Plant Sci 10:168
Tilman D, Balzer C, Hill J, Befort BL (2011) Global food demand and the sustainable intensification of agriculture. Proc Natl Acad Sci USA 108(50):20260–20264
Tong Y, Charusanti P, Zhang L, Weber T, Lee SY (2015) CRISPR-Cas9 based engineering of actinomycetal genomes. ACS Synth Biol 4(9):1020–1029
Torres-Perez R, Garcia-Martin JA, Montoliu L, Oliveros JC, Pazos F (2019) WeReview: CRISPR tools-live repository of computational tools for assisting CRISPR/Cas experiments. Bioengineering (Basel) 6(3):63
United Nations (2020a) The sustainable development goals report 2020
United Nations (2020b) Goal 13: take urgent action to combat climate change and its impacts
Wada N, Ueta R, Osakabe Y, Osakabe K (2020) Precision genome editing in plants: state-of-the-art in CRISPR/Cas9-based genome engineering. BMC Plant Biol 20:234
Wang Y, Yau Y, Perkins-Balding D, Thomson JG (2011) Recombinase technology: applications and possibilities. Plant Cell Rep 30(3):267–285
Wei A, He C, Li B, Li N, Zhang J (2011) The pyramid of transgenes TsVP and BetA effectively enhances the drought tolerance of maize plants. Plant Biotechnol J 9(2):216–229
Wu Y, He Y, Sretenovic S, Liu S, Cheng Y, Han Y, Liu G, Bao Y, Fang Q, Zheng X, Zhou J, Qi Y, Zhang Y, Zhang T (2022) CRISPR-BETS: a base-editing design tool for generating stop codons. Plant Biotechnol J 20(3):499–510
Xie K, Zhang J, Yang Y (2014) Genome-wide prediction of highly specific guide RNA spacers for the CRISPR-Cas9 mediated genome editing in model plants and major crops. Mol Plant 7:923
Xie X, Ma X, Zhu Q, Zeng D, Li G, Liu YG (2017) CRISPR-GE: a convenient software toolkit for CRISPR-based genome editing. Mol Plant 10(9):1246–1249
Yu W, Wang L, Zhao R, Sheng J, Zhang S, Li R, Shen L (2019) Knockout of SlMAPK3 enhances tolerance to heat stress involving ROS homeostasis in tomato plants. BMC Plant Biol 19:1–13
Zhang W, Subbarao S, Addae P, Shen A, Armstrong C, Peschke V, Gilbertson L (2003) Cre/lox-mediated marker gene excision in transgenic maize (*Zea mays L.*) plants. Theor Appl Genet 107:1157–1168
Zhang Y, Li H, Ouyang B, Lu Y, Ye Z (2006) Chemical-induced autoexcision of selectable markers in elite tomato plants transformed with a gene conferring resistance to lepidopteran insects. Biotechnol Lett 28:1247–1253
Zhang F, Maeder ML, Unger-Wallace E, Hoshaw JP, Reyon D, Christian M et al (2010) High frequency targeted mutagenesis in Arabidopsis thaliana using zinc finger nucleases. Proc Natl Acad Sci 107(26):12028–12033
Zhang H, Shen G, Kuppu S, Gaxiola R, Payton P (2011) Creating drought-and salt-tolerant cotton by overexpressing a vacuolar pyrophosphatase gene. Plant Signal Behav 6(6):861–863
Zhang Y, Zhang F, Li X, Baller JA, Qi Y, Starker CG et al (2013) Transcription activator-like effector nucleases enable efficient plant genome engineering. Plant Physiol 161(1):20–27
Zhou S, Chen X, Zhang X, Li Y (2008) Improved salt tolerance in tobacco plants by co-transformation of a betaine synthesis gene BADH and a vacuolar Na+/H+ antiporter gene SeNHX1. Biotechnol Lett 30:369–376
Zuo J, Niu QW, Møller SG, Chua NH (2001) Chemical-regulated, site-specific DNA excision in transgenic plants. Nat Biotechnol 19(2):157–161

Next-Gen Strategies in Host Plant Resistance to Insects: Breakthroughs and Future Horizons

11

Poonam Jasrotia, Pritam Kumari, and Prem Lal Kashyap

Abstract

Insect pest resistance is a critical aspect of modern agriculture, as pest infestations can lead to significant crop losses and impact food security. This chapter reviews the key approaches developed to enhance insect pest resistance in crops, focusing on both traditional and advanced methods. Conventional breeding techniques, such as selective breeding and hybridization, have long been used to develop pest-resistant varieties. However, these methods are time-consuming and limited by the genetic diversity available within a species. In recent years, biotechnological advancements have enabled more precise and efficient approaches. Genetic modification (GM) has introduced traits such as insect resistance in crops like Bt cotton and Bt corn, which produce proteins toxic to specific pests. Gene editing tools, such as CRISPR/Cas9, offer further refinement by allowing targeted modifications to plant genomes, enhancing pest resistance without introducing foreign DNA. RNA interference (RNAi) is another promising approach, where gene silencing is used to disrupt vital processes in pest species, thereby reducing their impact on crops. In addition to genetic approaches, integrated pest management (IPM) strategies combine biological control methods with pest-resistant varieties to reduce reliance on chemical pesticides, promoting environmentally sustainable agriculture.

P. Jasrotia (✉) · P. L. Kashyap
ICAR-Indian Institute of Wheat and Barley Research, Karnal, Haryana, India
e-mail: poonam.jasrotia@icar.gov.in

P. Kumari
ICAR-Indian Institute of Wheat and Barley Research, Karnal, Haryana, India

CCS Haryana Agricultural University, Hisar, Haryana, India

A. Mann et al. (eds.), *Cutting Edge Technologies for Developing Future Crop Plants*,
https://doi.org/10.1007/978-981-96-2508-6_11

The chapter discusses the progress and challenges of these approaches, highlighting the need for ongoing research to address pest resistance evolution and optimize these technologies for global food production.

Keywords
Genetic modification · RNA interference · CRISPR/Cas9 · high-throughput screening · Bt crops · precision agriculture · sustainable pest management

11.1 Introduction

Pest resistance in crops is a crucial aspect of modern agriculture, directly influencing food security, economic stability, and environmental sustainability. Insect pests cause significant damage to crops, leading to reduced yields and increased costs for pest management (Ali et al. 2023). Inadequate management of agricultural pests and plant diseases can lead to approximately 20–40% of global crop losses, resulting in an economic decline of about $220 billion annually (Sarkozi 2019). The development of insect pest-resistant crop varieties plays a vital role in mitigating these losses. Traditional pest control methods, such as chemical pesticides, often face limitations, including environmental contamination and the development of resistance by pests themselves (Deguine et al. 2021). In contrast, insect pest-resistant crops offer a sustainable alternative by incorporating genetic traits that render them less susceptible to specific pests. This approach reduces the reliance on chemical pesticides, which is beneficial for both environmental health and human safety (Sharma and Bakshi 2022). Moreover, pest-resistant crops contribute to economic stability for farmers. By reducing the need for costly pesticides and decreasing crop losses, these varieties can enhance profitability and reduce the financial risks associated with pest infestations. For instance, the adoption of Bt cotton, which expresses a protein toxic to certain pests, has been associated with increased yields and reduced pesticide use in various countries (Verma et al. 2022). Furthermore, pest-resistant crops support biodiversity and ecological balance by minimizing the impact of pest control methods on non-target species and ecosystems. By mitigating the need for broad-spectrum insecticides, these crops help preserve beneficial insects and maintain ecological stability (Angon et al. 2023). Therefore, pest resistance is crucial for improving agricultural productivity, promoting environmental sustainability, and ensuring economic viability.

11.2 Current Challenges in Developing Insect-Resistant Varieties

Breeding crops for insect resistance present several contemporary challenges that complicate efforts to enhance agricultural sustainability and productivity (Smith 2021). One major issue is the rapid evolution of pest resistance, which can undermine the effectiveness of resistant crop varieties (Wani et al. 2022). Insect pests can quickly adapt to new resistance traits through genetic mutations, necessitating continuous development and deployment of novel resistance mechanisms (Tabashnik et al. 2023). Additionally, the complexity of plant-pest interactions makes it difficult to predict and manage resistance traits. For instance, while genetic modifications can confer resistance to specific pests, they may inadvertently affect other ecological interactions, such as the relationships between plants and beneficial insects or soil microorganisms (Abdul et al. 2022). Another significant challenge is the trade-off between developing resistance and maintaining other desirable crop traits, such as yield, quality, and adaptability to varying environmental conditions (Dwivedi et al. 2021). This balance requires sophisticated breeding techniques and an understanding of the genetic architecture of resistance traits. Furthermore, regulatory and public acceptance issues can impede the adoption of genetically modified resistant crops. In many regions, stringent regulations and consumer concerns about genetically modified organisms (GMOs) create hurdles for the commercialization and widespread use of pest-resistant varieties (Vega Rodríguez et al. 2022). Addressing these challenges demands a multidisciplinary approach that integrates advances in genomics, pest management strategies, and stakeholder engagement to ensure the development of effective, sustainable, and socially acceptable pest-resistant crops.

11.3 Approaches for Insect Pest Resistance

Genetic approaches to insect pest resistance have revolutionized crop protection by leveraging advanced techniques to develop varieties that can withstand insect infestations. At the forefront of these approaches is genetic modification, where genes from various sources are introduced into crops to confer resistance to specific pests.

11.3.1 Genetic Modification and Transgenic Crops

At the forefront of various approaches for insect pest resistance is the genetic modification, where genes from various sources are introduced into crops to confer resistance to specific insect pests. Genetic modification and the development of transgenic crops have been pivotal in advancing insect pest resistance, offering targeted and effective solutions to pest-related challenges in agriculture (Kamatham et al. 2021). Genetic modification involves introducing foreign genes into a plant's genome to confer resistance to specific pests. One of the most prominent examples of this technology is Bt (*Bacillus thuringiensis*) crops, which are engineered to produce proteins toxic to certain insects but safe for humans and other organisms. For example, Bt cotton and Bt corn have been engineered to produce the Bt toxin, which targets pests like the cotton bollworm and European corn borer. These crops have significantly reduced the need for chemical insecticides, enhancing both environmental and economic sustainability (Tabashnik et al. 2023). Studies have shown that Bt cotton significantly reduces the need for chemical insecticides, leading to increased yields and environmental benefits (Peshin et al. 2021). A list of genetically modified crops containing Bt genes (Kumari et al. 2022) for resistance to insects is given in Table 11.1.

Another notable example is Bt maize, which incorporates genes from *B. thuringiensis* to target pests such as the European corn borer. The widespread adoption of Bt maize has not only decreased pest-related damage but also reduced the frequency of chemical pesticide applications, contributing to more sustainable farming practices (García et al. 2023).

Recent advancements have expanded the scope of genetic modification beyond traditional Bt crops. For example, the development of RNA interference (RNAi) technology has enabled the creation of crops that produce double-stranded RNA to silence specific genes in pests (Li et al. 2022). This approach has been applied in developing pest-resistant potatoes that target the genes of the Colorado potato beetle, significantly reducing pest infestation and crop damage (Pardo-López et al. 2013).

Genetic modification and transgenic crops serve as effective tools in combating insect pests, providing notable advantages in pest control and reducing the need for chemical inputs. A list of genetically modified crops containing Bt genes for resistance to insects is given in Table 11.1. Despite these advancements, challenges persist, including the potential for pests to develop resistance to Bt proteins and regulatory hurdles associated with GM crops. The success of such modified crops requires careful management, including strategies to delay pest resistance, such as planting non-Bt refuge crops (Kumar et al. 2021). Ongoing research focuses on enhancing the durability of resistance traits and improving public acceptance of genetically modified organisms (GMOs) (Bradshaw and Bradshaw 2021).

11.3.2 CRISPR and Gene Editing Technologies

CRISPR and gene editing technologies represent cutting-edge approaches for developing insect pest-resistant crops, providing opportunities to enhance agricultural productivity and sustainability through precise genetic modifications (Lallawmkimi et al. 2024). CRISPR and other gene editing technologies have emerged as transformative tools in developing insect pest-resistant crops, offering unprecedented precision and efficiency in genetic modification. CRISPR-Cas9, a powerful genome editing technique, allows for specific alterations to an organism's DNA, enabling researchers to introduce, delete, or modify genes with high accuracy. This technology has significant potential for enhancing pest resistance in crops by targeting

Table 11.1 Genetically modified crops containing *Bt* genes for resistance to insects

S. no.	Crop	Target insects	Transgene	References
1.	Rice	*Scirpophaga incertulas*, *Cnaphalocrocis medinalis*	*cry2AX1*	Rajadurai et al. (2018)
		Cnaphalocrocis medinalis	*cry2A*	Gunasekara et al. (2017)
		Cnaphalocrocis medinalis, *Scirpophaga incertulas*	*cry 2a/cry 1a(c)*	Maqbool et al. (2001)
		Scirpophaga incertulas	*cry 1a(b)/ cry1a(c)*	Tu et al. (2000)
		Scirpophaga incertulas, *Cnaphalocrocis medinalis*	*cry 1 a(b)*	Ghareyazie et al. (1997)
		Scirpophaga incertulas	*cry 1a(c)*	Nayak et al. (1997)
		Scirpophaga incertulas and *Chilo suppressalis*	*cry 1 a(b)*	Wünn et al. (1996)
		Chilo suppressalis, *Cnaphalocrocis medinalis*	*cry 1a(b)*	Fujimoto et al. (1993)
2.	Cotton	*Helicoverpa armigera*	*cry2AX1*	Jadhav et al. (2020)
		Spodoptera litura	*cry1AC + cry2Ab*	Siddiqui et al. (2019)
		Anthonomus grandis	*cry1Aa*	Ribeiro et al. (2006)
		Helicoverpa armigera	*cry2AX*	Sakthi et al. (2015)
		Helicoverpa armigera	*cry1Ab*	Ali Khan et al. (2013)
		Helicoverpa armigera	*cry1Ab + NptII*	Khan et al. (2011)
3.	Pigeon pea	*Helicoverpa armigera*	*cry2Aa*	Baburao and Sumangala (2018)
		Helicoverpa armigera	*cry2Aa*	Singh et al. (2018)
		Helicoverpa armigera	*cry1AC+ cry2Aa*	Ghosh et al. (2017)
4.	Chickpea	*Helicoverpa armigera*	*cryIIAa*	Sawardekar et al. (2017)
5.	Tomato	*Tuta absoluta*	*cry1Ab*	Soliman et al. (2021)
		Tuta absoluta	*cry1Ac*	Selale et al. (2017)
6.	Castor	*Achaea janata*, *Spodoptera litura*	*cry1AC*	Muddanuru et al. (2019)
7.	Soyabean	*Holotrichia parallela*	*cry 8 like*	Qin et al. (2019)
8.	Sweet potato	*Spodoptera litura*	*cry1Aa*	Zhong et al. (2019)

Source: Kumari et al. (2022)

genes involved in pest interactions or defense mechanisms (Kumari et al. 2022; Mathur and Arora 2024) (Table 11.2).

One notable application of CRISPR in insect pest resistance is the development of rice varieties with enhanced resistance to the brown planthopper, a major pest in rice cultivation. By using CRISPR-Cas9 to edit the rice genome and introduce resistance genes, researchers have created rice plants with reduced susceptibility to this pest, demonstrating the technology's potential for improving crop resilience (Mehta

Table 11.2 Insects genetically modified for pest control using the CRISPR/Cas9 technology

S. no.	Target insects	Target gene	References
1.	*Ostrinia furnacalis*	*ABCC2*	Wang et al. (2020a, b)
2.	*Helicoverpa armigera*	α-6-nicotinic acetylcholine receptor (*nAchR*)	Zuo et al. (2020)
3.	*Rhopalosiphum padi*	*ß-1-3glucanase* in maize	Kim et al. (2020)
4.	*Plutella xylostella*	*PxABCC2, PxABCC3*	Guo et al. (2019)
5.	*Helicoverpa armigera*	*Tetraspanin*	Jin et al. (2018)
6.	*Spodoptera exigua*	Ryanodine receptor	Zuo et al. (2020, 2017)
7.	*Ceratitis capitata*	Eye Pigmentation Gene White Eye (*We*)	Meccariello et al. (2017)
8.	*Plutella xylostella*	Abdominal-A homeotic gene (*Pxabd-A*)	Huang et al. (2016)
9.	*Drosophila melanogaster*	Chitin synthase 1	Douris et al. (2016)
10.	*Agrotis ipsilon*	*Yellow-Y* gene	Chen et al. (2016)
11.	*Locusta migratoria*	Odorant receptor co-receptor (*Orco*) gene	Li et al. (2016)
12.	*Spodoptera litura*	Abdominal-A (*Slabd-A*) gene	Bi et al. (2016)
13.	*Spodoptera littoralis*	Olfactory receptor co-receptor (*Orco*) gene	Koutroumpa et al. (2016)
14.	*Helicoverpa armigera*	*HaCad*	Wang et al. (2016)
15.	*Tribolium castaneum*	E-cadherin gene, *EGFP*	Gilles et al. (2015)

Source: Kumari et al. (2022)

et al. 2020). Another example is the modification of the maize genome to enhance resistance to the fall armyworm. Scientists have utilized CRISPR to target and disrupt genes in maize that are critical for the insect's feeding and survival, leading to reduced damage and improved yield (Chen and Palli 2023; Kumari et al. 2024).

Additionally, CRISPR has been employed to develop crops with enhanced resistance to multiple pests simultaneously. For instance, researchers have used CRISPR to modify tomato plants, making them resistant to both the tomato leafminer and the whitefly, two common pests that affect tomato yields. This dual resistance not only improves the plants' defense mechanisms but also reduces the need for chemical pesticides (D'Esposito et al. 2023; Kumar et al. 2023).

While CRISPR and other gene editing technologies offer substantial benefits, challenges remain, including potential off-target effects and regulatory hurdles (Han et al. 2020). Ensuring the accuracy and safety of gene edits and navigating complex regulatory landscapes are critical for the successful deployment of these technologies in agriculture (Molinari et al. 2024).

11.3.3 Gene Silencing and RNA Interference

Gene silencing and RNA interference (RNAi) are advanced techniques used to develop insect pest-resistant crops by targeting and suppressing specific genes in pests, thereby reducing their ability to damage crops (Saakre et al. 2024). RNAi involves introducing double-stranded RNA (dsRNA) into a plant that matches the sequence of target pest genes. This dsRNA is then taken up by the pest when it feeds on the plant, leading to the degradation of the target gene's mRNA and subsequent silencing of the gene's expression (Nitnavare et al. 2021).

One prominent example of RNAi technology applied in pest resistance is the development of genetically modified crops expressing dsRNA targeting key genes in the Colorado potato beetle. By incorporating dsRNA that targets essential genes for beetle development and survival, researchers have created potato plants that are highly resistant to this notorious pest. Studies have demonstrated that RNAi-expressing potatoes can significantly reduce beetle feeding and damage, thereby improving crop yield and reducing the need for chemical pesticides (Mehlhorn et al. 2020).

Another successful application is in cotton, where RNAi technology has been used to develop varieties resistant to the cotton bollworm. Researchers introduced dsRNA targeting genes crucial for the bollworm's larval development. This strategy resulted in cotton plants that are less susceptible to bollworm infestations, leading to reduced reliance on insecticides and better crop protection (Jaiwal et al. 2020).

Additionally, RNAi has been used in corn to target the western corn rootworm, a major pest that damages maize roots and reduces crop yield. Corn plants engineered to express dsRNA targeting rootworm genes have shown significant reductions in pest populations and damage, highlighting RNAi's potential for managing this destructive pest (Darlington et al. 2022). The list of genetically modified crops utilizing RNA interference for insect resistance (Kumari et al. 2022) is given in Table 11.3.

Gene silencing and RNA interference represent powerful tools in developing insect pest-resistant crops, offering precise and effective solutions to manage pest infestations and reduce environmental impacts. Despite its promise, these approaches face challenges such as ensuring effective delivery of dsRNA and potential off-target effects. Though, ongoing research aims to optimize RNAi methods and address these issues to enhance the effectiveness and safety of RNAi-based pest control strategies (Saakre et al. 2024).

11.3.4 Pest-Resistant Proteins

The use of pest-resistant proteins represents a significant advancement in crop protection, utilizing proteins with insecticidal properties to provide inherent resistance to pests. These proteins, often derived from natural sources, are incorporated into crops through genetic engineering to confer resistance against specific insects,

Table 11.3 Genetically modified crops utilizing RNA interference for insect resistance

S. no.	Crop	Target insects	Silenced gene	References
1.	Rice	*Scirpophaga incertulas*	*AchE*-Acetylcholine esterase	Kola et al. (2019)
		C. suppressalis	Aminopeptidase N genes *APN1* + *APN2*	Qiu et al. (2017)
2.	Maize	Lepidopteran	dsRNA-Spray	Li et al. (2015)
		Diabrotica virgifera LeConte	*Dvvgr, dvbol*	Niu et al. (2017)
		Diabrotica v. virgifera	hunchback (hb) and brahma (brm) gene	Khajuria et al. (2015)
		Diabrotica virgifera virgifera LeConte	Suppression of target mRNA	Baum et al. (2007)
3.	Potato	*Leptinotarsa decemlineata*	*ECR* gene	Hussain et al. (2019)
		Leptinotarsa decemlineata	*β-actin* gene	Zhang et al. (2015)
4.	Tomato, tobacco	*H. armigera*	Chitinase gene-*HaCHI*	Mamta et al. (2016)
5.	Soyabean	*Aphis glycines*	*TREH, ATPD, ATPE, CHSI*	Yan et al. (2020)
		Leguminivora glycinivorella	SpbP0-dsRNA	Meng et al. (2017)
6.	Cotton	*Helicoverpa armigera*	Juvenile hormone methyl transferase (JHMT)	Ni et al. (2017)
7.	Tobacco	*Spodoptera littoralis*	*Sl 102 immune* gene	Di Lelio et al. (2022)
		Bemisia tabaci	Phenolic glucoside malonyltransferase	Xia et al. (2021)
		Bemisia tabaci	*BtACTB gene*	Dong et al. (2020)
		Manduca sexta,	v-*ATPaseA* gene	Burke et al. (2019)

Source: Kumari et al. (2022)

reducing the need for chemical pesticides and enhancing crop yields (Kumari et al. 2022).

One of the most well-known examples of pest-resistant proteins is the Cry proteins produced by the bacterium *Bacillus thuringiensis* (Bt). These proteins are toxic to various insect pests when ingested. Bt crops, such as Bt cotton and Bt maize, are engineered to express Cry proteins, which target pests like the cotton bollworm and the European corn borer. The Cry proteins bind to receptors in the insect gut, causing cell damage and ultimately leading to the insect's death (Koul 2020). The successful implementation of Bt cotton has led to substantial reductions in pesticide use, increased yields, and environmental benefits (Peshin et al. 2021).

Another example involves the use of proteinase inhibitors to confer pest resistance. For instance, research has demonstrated that introducing plant-derived proteinase inhibitors, such as those from potatoes or soybeans, into crops can deter feeding by pests like the Colorado potato beetle. These inhibitors disrupt the digestion of proteins in the insect gut, leading to reduced growth and survival rates (Divekar et al. 2023).

Additionally, the development of plants expressing lectins—proteins that bind to specific carbohydrates on the surface of insect cells—has shown promise in pest resistance (Roy and Chakraborty 2021). The use of lectins from plants like snowdrop (*Galanthus nivalis*) has been tested in transgenic crops to target pests such as aphids and beetles. These lectins interfere with the insect's ability to feed and digest, thereby providing a protective effect against pest damage (He et al. 2022; Kumari et al. 2022) (Table 11.4). Besides, vegetative insecticidal protein (Vip) genes have also been utilized for resistance against cotton bollworm (*H. armigera*) and tobacco budworm (*Heliothis virescens*) and other insect pests listed in Table 11.5. In addition, to overcome the resistance problem with Bt proteins (Kumari et al. 2022), fusion proteins can provide enhanced specificity and efficacy, targeting pests while minimizing harm to non-target organisms (Table 11.6).

Recent advancements also include the use of antimicrobial peptides (AMPs) as pest-resistant proteins. AMPs, which have been identified in various organisms including plants and insects, exhibit broad-spectrum insecticidal activity (Bakare et al. 2022; George et al. 2023). For example, the expression of AMPs derived from the venom of certain predatory insects has been tested in crops to control a range of pests, offering a new avenue for pest management (Azmiera et al. 2022). The use of pest-resistant proteins in crops can provide effective solutions for managing pest populations, reducing reliance on chemical pesticides, and promoting sustainable agriculture. As it is a new concept, research is being carried out to optimize these proteins and explore new sources to enhance their effectiveness and durability.

11.3.5 Marker-Assisted Selection (MAS)

Marker-assisted selection (MAS) is a powerful tool in plant breeding that enhances the efficiency of developing insect pest-resistant crop varieties (Joshi et al. 2024). By using molecular markers linked to pest resistance traits, breeders can more accurately and rapidly select plants with the desired genetic traits, significantly speeding up the breeding process compared to traditional methods. MAS accelerates the breeding process and enhances the development of crop varieties that can better withstand pest pressures, ultimately supporting more sustainable agricultural practices (Razzaq et al. 2021).

One prominent example of MAS in action is the development of rice varieties resistant to the brown planthopper (BPH), a major pest in rice cultivation. Researchers have identified quantitative trait loci (QTLs) associated with BPH resistance and developed molecular markers linked to these QTLs. Using MAS, breeders have been able to incorporate these resistance traits into high-yielding rice varieties, improving both pest resistance and overall crop performance (Kumar et al. 2020a, b; Chen et al. 2020; Wang et al. 2021).

Table 11.4 Crops with lectin genes for insect resistance

S. no.	Crop	Target insects	Transgene	References
1.	Rice	*Nilaparvata lugens, Sogatella furcifera and Nephotettix nigropictus*	*Allium sativum* leaf agglutinin (ASAL) and *Galanthus nivalis* lectin (GNA)	Bharathi et al. (2011)
		Sap sucking pests	Snowdrop lectin (*Galanthus nivalis* agglutinin; GNA)	Sudhakar et al. (1998)
2.	Wheat	*Metopolophium dirhodum, Schizaphis graminum, Rhopalosiphum padi*, and *Sitobion avenae*	*Pinellia pedatisecta* agglutinin (PPA)	Duan et al. (2018)
		Sitobion avenae	Snowdrop lectin (*Galanthus nivalis* agglutinin; GNA)	Stoger et al. (1999)
3.	Rice, sugarcane	*Eoreuma loftini* (Dyar) and *Diatraea saccharalis*	Snowdrop lectin (*Galanthus nivalis* agglutinin; GNA)	Sétamou et al. (2002)
4.	Tomato	*Lacanobia oleracea*	GNA-neuropeptide-allatostatin	Fitches et al. (2004)
		Lacanobia oleracea	GNA-lepidopteran-specific toxin (ButaIT)	Trung et al. (2006)
5.	Rice and potato	*Nilaparvata lugens and Myzus persicae)*	GNA-spider-venom toxin I (SFI1)	Down et al. (2006)
6.	Potato	*Myzus persicae*	*Galanthus nivalis* agglutinin (GNA)	Mi et al. (2017)
		Myzus persicae and *Sitobion avenae*	Hv1a/GNA	Nakasu et al. (2014)
7.	Transgenic *Brassica juncea*	*Lipaphis erysimi*	Lentil lectin (LL) and chickpea protease inhibitor (CPPI) genes	Rani et al. (2017a)
8.	Mustard	*Lipaphis erysimi*	*Colocasia esculenta* tuber agglutinin (CEA) + *Galanthus nivalis* agglutinin (GNA)	Das et al. (2018)
9.	Cotton	*Aphis gossypii* and *Spodoptera litura*	Insect gut binding lectin from *Sclerotium rolfsii*	Vanti et al. (2018)
10.	Cowpea	*Callosobruchus chinensis*	Arcl on APA locus from *Phaseolus vulgaris*	Grazziotin et al. (2020)
11.	Common bean	*Callosobruchus chinensis*	Arcelin	Hilda et al. (2022)
		Callosobruchus chinensis	Arcelin-5, Leucoagglutinin, Erythroagglutinin	Caroline et al. (2022)

(continued)

Table 11.4 (continued)

S. no.	Crop	Target insects	Transgene	References
12.	Stored grains	*Sitophilus oryzae*	*Polygonum persicaria* L. (PPA) Lectin	Khoobdel et al (2022)
		Sitophilus. zeamais	*Microgramma vacciniifolia* rhizome lectin (MvRL)	de Albuquerque et al. (2020)
		Sitophilus zeamais	Water-soluble *Moringa oleifera* lectin (WSMoL)	de Oliviera et al. (2020)
		Sitophilus zeamais	*Schinus terebinthifolius* leaf lectin (SteLL)	de Santana Souza et al. (2018)
		Sitophilus zeamais	*M. urundeuva* leaf lectin (MuLL)	Napoleão et al. (2013)

Source: Kumari et al. (2022)

Table 11.5 Crops with *VIP* genes for insect resistance

S. no.	Crop	Target insects	Transgene	References
1.	Sugarcane	*Chilo infuscatellus*	*Vip3A*	Riaz et al. (2020)
2.	Cotton	*Helicoverpa armigera*	*Vip3AcAaa* (*Vip3Aa1* + *Vip3Ac1*)	Chen et al. (2018)
3.	Cowpea	*Maruca vitrata*	*Vip3Ba1*	Bett et al. (2017)
4.	Cotton	*Heliothis. zea* and *H. virescens*	*Vip3A* + *cry1Ab*	Bommireddy et al. (2011)

Source: Kumari et al. (2022)

Another example is the development of maize varieties with resistance to the European corn borer (ECB). In this case, MAS was used to integrate resistance genes such as the *Htp* (high temperature requirement A) gene, which provides resistance to ECB larvae. By utilizing markers linked to these resistance genes, breeders have been able to accelerate the development of ECB-resistant maize varieties, thereby reducing reliance on chemical pesticides and improving maize yield and quality (Abel et al. 2000).

In the case of cotton, MAS has been employed to enhance resistance to the cotton bollworm. Researchers identified several QTLs associated with bollworm resistance and used molecular markers to select for these traits in new cotton varieties. This approach has helped in developing cotton plants with improved resistance to bollworm infestations, leading to reduced pest damage and increased profitability for cotton growers (Kushanov et al. 2021).

Additionally, MAS has been used in the development of potato varieties resistant to the Colorado potato beetle. By identifying and using markers linked to resistance traits, breeders have been able to develop potato varieties with enhanced resistance to this damaging pest, which has contributed to more sustainable potato production (Cooper et al. 2004).

Table 11.6 Crops having fusion proteins for insect resistance

S. no.	Crop	Target insects	Transgene	References
1.	Rice	*Chilo suppressalis*	*cry2Aa + cry1Ca*	Qiu et al. (2019)
		Ostrinia furnacalis, Cnaphalocrocis medinalis	*cry1Ab + vip3A*	Xu et al. (2018)
		Scirpophaga incertulas, Cnaphalocrocis medinalis, Nilaparvata lugens	*cry1AC + ASAL*	Boddupally et al. (2018)
		Scirpophaga incertulas, Cnaphalocrocis medinalis	*cry2AX1 (cry2Aa + cry2Ac)*	Chakraborty et al. (2016)
2.	Cotton	*Helicoverpa armigera, Spodoptera litura*	*cry2Ab + cry1F + cry1AC*	Katta et al. (2020)
		Spodoptera litura, Ostrinia nubilalis	*cry1Be + cry1Fa*	Meade et al. (2017)
		Lygus spp.	*cry51Aa2*	Gowda et al. (2016)
3.	Maize	*Spodoptera exigua, Harmonia axyridis*	*cry1Ab/cry2Aj*	Chang et al. (2017)
4.	Sugarcane	*Scirpophaga excerptalis*	*cry2Aa + cry1Ca, cry1Ab + cry1Ac*	Koerniati et al. (2020)
5	*Brassica juncea*	*Lipaphis erysimi*	Lentil lectin (LL) and chickpea protease inhibitor (CPPI) genes	Rani et al. (2017b)

Source: Kumari et al. (2022)

11.3.6 Genomic Selection and High-Throughput Screening

Genomic selection and high-throughput screening are advanced techniques that have significantly enhanced the ability to develop insect pest-resistant crop varieties. These methods leverage genomic data and technological innovations to improve the efficiency and accuracy of breeding programs aimed at combating pest infestations (Hafeez et al. 2023).

Genomic selection involves predicting the breeding value of plants based on their genomic information. By using a set of genetic markers that are associated with traits of interest, such as insect resistance, breeders can select plants with the desired traits more effectively. This approach utilizes genome-wide markers to estimate the genetic potential of plants, allowing for more precise selection even before the plants are fully grown (Tyagi et al. 2020). For example, genomic selection has been applied in wheat to enhance resistance to the wheat stem sawfly, a significant pest. Researchers used high-density SNP (single nucleotide polymorphism) markers to identify and select wheat lines with improved pest resistance, leading to the development of more resilient varieties (Bathini et al. 2023).

High-throughput screening enables the rapid and efficient assessment of large numbers of plants for specific traits, such as pest resistance, through the use of automated technologies. This approach combines both phenotyping and genotyping to quickly pinpoint plants with favorable characteristics (Song et al. 2021). For example, high-throughput screening has been applied in maize to evaluate resistance to the fall armyworm. By merging automated phenotyping tools with genomic data, researchers were able to swiftly identify and select maize lines with greater resistance, greatly accelerating the breeding process (Kamweru et al. 2022).

An example of both genomic selection and high-throughput screening in action is the development of insect-resistant soybean varieties. Researchers have employed genomic selection to identify genetic markers associated with resistance to the soybean cyst nematode and other pests (Shaibu et al. 2020). High-throughput screening techniques have been used to rapidly evaluate large soybean populations for these markers, facilitating the development of varieties with enhanced pest resistance and better agronomic performance (Kwon et al. 2024).

Furthermore, in cotton, genomic selection has been used to improve resistance to multiple pests, including the cotton bollworm and the pink bollworm. High-throughput phenotyping platforms have allowed researchers to rapidly assess pest resistance traits across large cotton breeding populations, accelerating the development of new resistant varieties (Tarazi and Vaslin 2022).

Zhang et al. (2024) utilized GWAS to uncover QTLs for aphid resistance in wheat, offering valuable targets for breeding programs. Synthetic biology is another promising area, where researchers are incorporating novel insecticidal proteins into crops. Bowen et al. (2021) engineered transgenic corn with synthetic peptides that exhibit insecticidal properties, providing new avenues for pest control.

11.4 Integrating Pest Resistance into Crop Improvement Programs

11.4.1 Designing Durable Resistance Strategies

Designing durable insect pest resistance strategies involves integrating multiple approaches to ensure long-lasting protection against pests while minimizing resistance development (Ali et al. 2023). One of the key methods is the use of genetically modified crops through genome editing technologies like CRISPR/Cas9. These tools allow for precise modifications in pest resistance genes, enabling crops to better defend themselves by expressing insecticidal proteins or enhancing their natural defense mechanisms. Such modifications often involve resistance to multiple pest species, making it harder for pests to adapt and evolve resistance quickly (Gupta et al. 2021). In addition, breeding for traits like thickened cell walls, higher concentrations of defensive compounds (e.g., flavonoids and phenolics), and increased production of herbivore-induced volatiles has proven to be effective in various crops. These traits help plants either directly repel pests or attract natural predators (Al-Khayri et al. 2023). An example is the increased resistance of rice and cowpea to specific pests, driven by such biochemical changes (Togola et al. 2020; Sharma et al. 2024). However, ensuring durability also means rotating crops and using biopesticides to delay the onset of pest resistance (Singh et al. 2023). Recent research suggests that a diversified use of biopesticides, integrated with advanced genetic solutions, is crucial in maintaining the efficacy of pest control strategies (Archana et al. 2022). Combining these techniques, alongside regular monitoring of pest populations and resistance patterns, can help maintain effective insect pest resistance over time.

11.4.2 Balancing Pest Resistance with Other Agronomic Traits

Balancing insect pest resistance with other agronomic traits is a crucial challenge in modern crop breeding, as enhancing resistance can sometimes inadvertently compromise important factors like yield, nutritional quality, or tolerance to environmental stress (Bijlwan et al. 2024). Traditional breeding often results in trade-offs between pest resistance and productivity. For instance, some resistant varieties developed through conventional breeding have shown lower yields compared to susceptible varieties due to the allocation of plant resources to defense mechanisms rather than growth (Chawla et al. 2023). Recent advances in genetic technologies, including genome editing, marker-assisted selection, and genetic mapping, have made it easier to balance pest resistance with agronomic traits (Rato et al. 2021) For instance, CRISPR/Cas9 has been used to precisely target genes responsible for pest resistance while minimizing the negative effects on yield and quality (Rao and Wang 2021). By focusing on quantitative trait loci (QTLs) that influence both resistance and agronomic traits, scientists are better able to introduce favorable alleles that offer pest protection without reducing plant vigor or crop output (Wani et al. 2022). Moreover, breeding programs are increasingly looking at multi-trait selection strategies, where pest resistance is combined with traits like drought tolerance and nutrient efficiency. This holistic approach ensures that the plants are not only pest-resistant but also capable of thriving in various environmental conditions (Chaudhary et al. 2024). Notably, in maize and rice, high-resistance varieties have been developed that also maintain strong yields and resilience to abiotic stresses like drought. Through these integrative approaches, modern breeding can create crops that are robust across multiple dimensions, ensuring long-term agricultural sustainability while maintaining high productivity (Tian et al. 2021).

11.5 Ecological and Environmental Considerations

11.5.1 Impact of Resistant Varieties on Ecosystems

The introduction of insect-resistant crop varieties has significantly impacted ecosystems, both positively and negatively. On the positive side, insect-resistant varieties, such as genetically modified (GM) crops like Bt cotton and Bt maize, reduce the need for chemical pesticides (Gadaleta 2020). This leads to lower environmental contamination, promotes biodiversity, and protects non-target organisms like beneficial insects and pollinators. For example, studies have shown that Bt crops decrease pesticide use by up to 37%, reducing harm to important arthropods such as ladybugs and predatory beetles, which play crucial roles in pest control (Krishnamoorthi et al. 2024). However, the impact of insect-resistant crops on ecosystems is not entirely positive. One of the main concerns is the potential development of pest resistance over time. As insect populations are exposed to resistant crops, they may evolve to overcome the engineered defenses, leading to a resurgence of pest populations or the rise of secondary pests.This can disrupt the ecological balance and increase the need for alternative pest control strategies, including chemical inputs, which undermine the initial environmental benefits (Ali et al. 2023).

Moreover, the widespread use of a few resistant varieties can lead to a reduction in crop genetic diversity (Salgotra and Chauhan 2023). A decline in diversity can make ecosystems more vulnerable to pests, diseases, and environmental stresses, which could ultimately threaten agricultural sustainability (Skendžić et al. 2021). Additionally, the potential for cross-breeding between GM crops and wild relatives raises concerns about unintended ecological consequences, such as the transfer of resistance genes into non-crop species, which could affect biodiversity (Verma and Saxena 2021).

11.5.2 Managing Resistance to Prevent Resistance Evolution

Managing insect resistance is essential to preventing the evolution of resistance, which can undermine pest control strategies. One of the most effective methods is the high-dose/refuge strategy, commonly used in Bt crops. This involves planting non-Bt refuge areas alongside Bt crops to maintain a population of non-resistant pests. These susceptible pests mate with any resistant individuals, diluting the resistance genes in the population (Brewer and Bonsall 2021). The success of this approach has been observed in controlling pests like the European corn borer in Bt maize fields (García et al. 2023). Another key strategy is crop rotation, which disrupts the life cycles of pests by altering their food sources and habitats. By rotating different crops, the continuous exposure of insect pests to the same selective pressures is avoided, reducing the likelihood of resistance development (Han et al. 2022). For example, alternating Bt crops with non-Bt crops can help manage pest populations and slow resistance evolution (Gassmann and Reisig 2023).

Additionally, integrating biological controls, such as introducing natural predators or parasitoids, can help manage pest populations without over-reliance on chemical or genetic control measures. By maintaining natural pest control agents in the ecosystem, insecticide resistance can be mitigated, as these predators help regulate pest populations. Finally, monitoring pest populations for early signs of resistance is critical (Parkash et al. 2023). Field surveillance and molecular diagnostics can detect resistance before it becomes widespread, allowing for timely adjustments in pest management strategies. A proactive resistance management plan that combines genetic, biological, and cultural practices ensures the long-term efficacy of pest-resistant crops (Jambagi et al. 2024).

11.6 Economic and Regulatory Aspects

11.6.1 Cost-Benefit Analysis of Pest-Resistant Crops

A cost-benefit analysis of insect-pest-resistant crops reveals both economic and environmental advantages, although certain costs and risks must also be considered (Singh et al. 2023). On the benefit side, these crops, including genetically modified varieties like Bt cotton and Bt maize, significantly reduce the need for chemical pesticides, cutting down on production costs for farmers. Studies show that pest-resistant crops can decrease pesticide use by up to 37%, resulting in lower expenditures on chemicals and less environmental pollution. In addition, higher yields and improved crop quality lead to greater profitability, with farmers reporting yield increases ranging from 10% to 30% depending on the crop and region (Shekhawat and Hasmi 2023).

Environmentally, pest-resistant crops help conserve biodiversity by reducing harm to non-target species, including beneficial insects like pollinators and natural predators of pests. This also lowers the risk of pesticide runoff into waterways, which can damage aquatic ecosystems. Furthermore, by integrating these crops into sustainable farming systems, farmers can contribute to long-term soil health and resilience against pest outbreaks. However, there are costs and risks associated with pest-resistant crops (Mukhtar and Shankar 2023). The development and adoption of these varieties require significant investment in research, regulatory approval, and intellectual property rights. Farmers often face higher seed costs for genetically engineered crops compared to conventional seeds (Brookes 2022). Additionally, over-reliance on resistant crops can lead to the evolution of pest resistance, potentially necessitating additional pest management strategies, including chemical interventions. There are also concerns about the long-term effects on biodiversity if resistance genes transfer to wild plant species. Overall, the benefits of insect-pest-resistant crops, especially in reducing chemical use and enhancing yields, outweigh the costs when integrated into broader sustainable farming practices, but careful management is essential to mitigate risks (Parkash et al. 2023).

11.6.2 Intellectual Property and Licensing Issues

Intellectual property (IP) and licensing issues are critical aspects in the development and commercialization of insect-resistant crop varieties. These crops, often genetically engineered, involve the use of patented technologies like Bt (*Bacillus thuringiensis*) genes, genome editing tools such as CRISPR/Cas9, and other proprietary breeding methods (Kedisso et al. 2022). Companies that develop these technologies hold patents that protect their inventions and provide them exclusive rights to commercialize the seeds for a specified period. This has led to a concentration of market power in a few multinational companies, such as Monsanto (now Bayer), which control a large share of the seed market. The licensing of these patented technologies often involves agreements where farmers purchase seeds under specific terms (Kiran and Pandey 2020).

Farmers may be required to sign contracts that prohibit the saving or replanting of seeds from one growing season to the next, which ensures continuous revenue streams for the patent holders but can increase costs for farmers, especially in developing countries. This creates ethical concerns regarding seed sovereignty and access to technology, particularly for smallholder farmers who may struggle to afford these technologies (Ajoykumar et al. 2021). Additionally, the regulatory frameworks governing the licensing and commercialization of genetically modified organisms (GMOs) vary by country, adding complexity to the IP landscape. In some regions, governments and non-governmental organizations have raised concerns about the dominance of corporate patents in agricultural biotechnology, advocating for open-access technologies that promote innovation and food security. While IP protection incentivizes innovation by rewarding the developers of pest-resistant crops, it also raises challenges in terms of equitable access and market competition. Ensuring that licensing agreements are fair and that public research institutions play a role in developing non-patented varieties can help balance these concerns (Rozas et al. 2022).

11.6.3 Regulatory Frameworks and Approval Processes

The regulatory frameworks and approval processes for insect-resistant crop varieties, particularly genetically modified organisms (GMOs), are comprehensive and vary by country. These processes are designed to ensure that such crops are safe for human consumption, animal feed, and the environment. In many countries, regulatory oversight is shared by multiple government agencies. For example, in the United States, the USDA, FDA, and EPA each have roles in evaluating GM crops: the USDA oversees plant health and environmental safety, the FDA assesses food and feed safety, and the EPA evaluates pesticide-producing crops like Bt varieties for their environmental impacts. In the European Union, the approval process is more stringent. GM crops must pass through a rigorous assessment by the European Food Safety Authority (EFSA), which evaluates risks to human health, the environment, and biodiversity. Public consultations are also a key part of the EU's regulatory system, reflecting its precautionary stance on GMOs. The approval process can take years and may face political and public opposition, as evidenced by the limited adoption of GM insect-resistant crops in many EU countries. In developing countries, regulatory systems are often influenced by international agreements like the Cartagena Protocol on Biosafety, which provides guidelines for the safe transfer and handling of GMOs. National biosafety authorities typically evaluate insect-resistant varieties based on their potential risks to local ecosystems and agriculture. Despite differences across regions, the common goal of these frameworks is to ensure that insect-resistant crops are safe and beneficial, while addressing concerns related to biodiversity and sustainable agriculture.

11.7 Future Directions and Emerging Technologies

The future of insect pest resistance in agriculture is poised for transformative breakthroughs, emerging technologies and scientific advancements. Systems biology and functional genomics are playing a pivotal role in understanding complex plant-pest interactions at a molecular level (Enders and Begcy 2021). By mapping gene networks and their functions, scientists can identify key resistance genes and pathways that regulate pest defense mechanisms. Advances in functional genomics, including RNA sequencing and CRISPR-based genome editing, allow precise manipulation of these pathways, leading to the development of crop varieties that are not only pest-resistant but also optimized for other agronomic traits like drought tolerance and nutrient efficiency (Bhat and Kumar 2022).

Artificial intelligence (AI) and machine learning (ML) are becoming integral tools in the field of crop protection. These technologies can analyze vast datasets generated from field trials, genomic studies, and environmental monitoring to predict pest outbreaks, assess resistance patterns, and optimize breeding strategies. AI-driven models help breeders and researchers identify the most effective genetic combinations for resistance, accelerating the development of new varieties. ML algorithms are also being used to refine integrated pest management (IPM) systems, making them more responsive and adaptive by predicting pest pressures and suggesting timely interventions (Mesías-Ruiz et al. 2023).

Looking further ahead, innovations such as nanotechnology are emerging as game-changers in pest resistance. Nanoparticles can be engineered to deliver pesticides in a more targeted and controlled manner, reducing environmental impact and enhancing pest control efficiency (Jiang et al. 2024). These nano-scale solutions may also be used to strengthen plant immunity or deliver RNA interference (RNAi) molecules that disrupt pest gene function, offering a novel approach to pest management. Beyond nanotechnology, future research may explore synthetic biology, enabling the design of entirely new plant defense systems, and microbial solutions that leverage the soil microbiome to bolster plant resistance (Wang et al. 2022).

Together, these advances in systems biology, AI, and emerging technologies like nanotechnology are set to redefine how we approach insect pest resistance in sustainable agriculture. They promise to make pest control more precise, efficient, and environmentally friendly, aligning with the broader goals of agricultural sustainability and food security (Danai-Varsou et al. 2023).

11.8 Conclusion and Way Forward

Breeding for insect-resistant crops has become an essential pillar in modern agriculture, offering sustainable solutions to pest management challenges. The integration of traditional breeding methods, genetic engineering, and emerging biotechnologies has led to significant progress in reducing crop losses, decreasing reliance on chemical pesticides, and improving agricultural productivity. Key insights from the development of insect-resistant crops, particularly through genetic modifications like Bt crops, highlight their capacity to enhance crop resilience while simultaneously promoting environmental health by reducing pesticide use. The future of breeding for insect resistance will likely involve a multidisciplinary approach combining advances in genomics, molecular biology, and data-driven technologies. Functional genomics and systems biology will be crucial in unraveling the complex genetic pathways responsible for plant-pest interactions, enabling more precise manipulation of resistance traits. Additionally, gene editing tools such as CRISPR/Cas9 will continue to revolutionize crop breeding, allowing breeders to target specific genes related to pest resistance without the trade-offs often associated with traditional breeding. Another emerging trend is the incorporation of artificial intelligence (AI) and machine learning (ML) to optimize pest-resistant breeding programs by analyzing large datasets and predicting resistance outcomes. The next decade is expected to see more innovations in nanotechnology, which could offer novel pest control strategies by delivering targeted RNA interference (RNAi) molecules or enhancing plant immunity through nano-enabled delivery systems. Beyond that, synthetic biology might enable the design of entirely new pest-resistant traits, while research into harnessing the microbiome could offer natural solutions to bolstering plant resistance.

As we look forward, the vision for pest-resistant crop development involves creating crops that are not only resilient to a wide range of pests but also adaptive to changing environmental conditions, thus contributing to global food security. Achieving this vision will require continuous collaboration among public and private sectors to ensure that the benefits of these technologies are accessible globally, especially to smallholder farmers in developing regions. With sustained research investment and careful management of resistance, pest-resistant crops have the potential to shape the future of sustainable agriculture.

References

Abdul Aziz M, Brini F, Rouached H, Masmoudi K (2022) Genetically engineered crops for sustainably enhanced food production systems. Front Plant Sci 13:1027828

Abel CA, Berhow MA, Wilson RL, Binder BF, Hibbard BE (2000) Evaluation of conventional resistance to European corn borer (Lepidoptera: Crambidae) and western corn rootworm (Coleoptera: Chrysomelidae) in experimental maize lines developed from a backcross breeding program. J Econ Entomol 93(6):1814–1821

Ajoykumar KN, Singh G, Shackira AM (2021) Socio-ethical aspect of genetically modified organisms: a critical analysis. In: Policy issues in genetically modified crops, pp 421–450

Ali Khan G, Bakhsh A, Ghazanfar M, Riazuddin S, Husnain T (2013) Development of transgenic cotton lines harboring a pesticidal gene (cry1Ab). Emir J Food Agric 25(6):434. https://doi.org/10.9755/ejfa.v25i6.13133

Ali MA, Abdellah IM, Eletmany MR (2023) Towards sustainable management of insect pests: protecting food security through ecological intensification. Int J Chem Biochem Sci 24(4):386–394

Al-Khayri JM, Rashmi R, Toppo V, Chole PB, Banadka A, Sudheer WN et al (2023) Plant secondary metabolites: the weapons for biotic stress management. Meta 13(6):716

Angon PB, Mondal S, Jahan I, Datto M, Antu UB, Ayshi FJ, Islam MS (2023) Integrated pest management (IPM) in agriculture and its role in maintaining ecological balance and biodiversity. Adv Agric 2023(1):5546373

Archana HR, Darshan K, Lakshmi MA, Ghoshal T, Bashayal BM, Aggarwal R (2022) Biopesticides: a key player in agro-environmental sustainability. In: Trends of applied microbiology for sustainable economy. Academic Press, pp 613–653

Azmiera N, Krasilnikova A, Sahudin S, Al-Talib H, Heo CC (2022) Antimicrobial peptides isolated from insects and their potential applications. J Asia Pac Entomol 25(2):101892

Baburao TM, Sumangala B (2018) Development and molecular characterization of transgenic pigeon pea carrying cry2Aa for pod borer resistance. J Pharm Phytochem 75:1581–1585

Bakare OO, Gokul A, Fadaka AO, Wu R, Niekerk LA, Barker AM et al (2022) Plant antimicrobial peptides (PAMPs): features, applications, production, expression, and challenges. Molecules 27(12):3703

Bathini A, Mendu L, Singh NP, Cook J, Weaver D, Sherman J, Hager M, Mondal S, Mendu V (2023) A "solid" solution for wheat stem sawfly (hymenoptera: Cephidae) resistance: genetics, breeding and development of solid stem wheat. Crop Sci 63(4):1934–1948

Baum JA, Bogaert T, Clinton W, Heck GR, Feldmann P, Ilagan O et al (2007) Control of coleopteran insect pests through RNA interference. Nat Biotechnol 25(11):1322–1326. https://doi.org/10.1038/nbt1359

Bett B, Gollasch S, Moore A, James W, Armstrong J, Walsh T et al (2017) Transgenic cowpeas (*Vigna unguiculata* L. Walp) expressing *bacillus thuringiensis* Vip3Ba protein are protected against the Maruca pod borer (*Maruca vitrata*). Plant Cell Tissue Organ Cult 131(2):335–345. https://doi.org/10.1007/s11240-017-1287-3

Bharathi Y, Kumar SV, Pasalu IC, Balachandran SM, Reddy VD, Rao KV (2011) Pyramided rice lines harbouring *Allium sativum* (asal) and *Galanthus nivalis* (gna) lectin genes impart enhanced resistance against major sap-sucking pests. J Biotechnol 152(3):63–71. https://doi.org/10.1016/j.jbiotec.2011.01.021

Bhat S, Kumar S (2022) Improvement of resistance in plants against insect-pests using genome editing tools. In: Genome editing: current technology advances and applications for crop improvement. Springer International Publishing, Cham, pp 237–249

Bi HL, Xu J, Tan AJ, Huang YP (2016) CRISPR/Cas9-mediated targeted gene mutagenesis in *Spodoptera litura*. Insect Sci 23(3):469–477. https://doi.org/10.1111/1744-7917.12341

Bijlwan A, Ranjan R, Kumar M, Gupta S, Jha A (2024) Managing crop adaptation to changing environment. In: Plant functional traits for improving productivity. Singapore, Springer Nature Singapore, pp 257–285

Boddupally D, Tamirisa S, Gundra SR, Vudem DR, Khareedu VR (2018) Expression of hybrid fusion protein (Cry1Ac:: ASAL) in transgenic rice plants imparts resistance against multiple insect pests. Sci Rep 8(1):1–10

Bommireddy, P. L., Leonard, B. R., Temple, J., Price, P., Emfinger, K., Cook, D., & Hardke, J. T. (2011). Field performance and seasonal efficacy profiles of transgenic cotton lines expressing Vip3A and VipCot against *Helicoverpa zea* (Boddie) and *Heliothis virescens* (F.). https://doi.org/10.18474/0749-8004-43.4.349

Bowen D, Yin Y, Flasinski S, Chay C, Bean G, Milligan J et al (2021) Cry75Aa (Mpp75Aa) insecticidal proteins for controlling the Western corn rootworm, Diabrotica virgifera virgifera LeConte (Coleoptera: Chrysomelidae), isolated from the insect-pathogenic bacterium Brevibacillus laterosporus. Appl Environ Microbiol 87(5):e02507–e02520

Bradshaw JE, Bradshaw JE (2021) Gene editing and genetic transformation of potatoes. In: Potato breeding: theory and practice, pp 505–551

Brewer TR, Bonsall MB (2021) Combining refuges with transgenic insect releases for the management of an insect pest with non-recessive resistance to Bt crops in agricultural landscapes. J Theor Biol 509:110514

Brookes G (2022) Farm income and production impacts from the use of genetically modified (GM) crop technology 1996-2020. GM Crops Food 13(1):171–195

Burke WG, Kaplanoglu E, Kolotilin I, Menassa R, Donly C (2019) RNA interference in the tobacco hornworm, *Manduca sexta*, using plastid-encoded long double-stranded RNA. Front Plant Sci 10:313. https://doi.org/10.3389/fpls.2019.00313

Caroline NM, Deogracious PM, George MT, James RM, Joel WD, Paul MK (2022) Identification of potential seed storage protein responsible for bruchid resistance in common bean landraces from Tanzania and Malawi. Afr J Biotechnol 21(1):35–45

Chakraborty M, Reddy PS, Mustafa G, Rajesh G, Narasu VM, Udayasuriyan V, Rana D (2016) Transgenic rice expressing the cry2AX1 gene confers resistance to multiple lepidopteran pests. Transgenic Res 25(5):665–678. https://doi.org/10.1007/s11248-016-9954-4

Chang X, Lu Z, Shen Z, Peng Y, Ye G (2017) Bitrophic and tritrophic effects of transgenic cry1Ab/cry2Aj maize on the beneficial, non target *Harmonia axyridis* (Coleoptera: Coccinellidae). Environ Entomol 46(5):1171–1176. https://doi.org/10.1093/ee/nvx113

Chaudhary D, Pal N, Arora A, Prashant BD, Venadan S (2024) Plant functional traits in crop breeding: advancement and challenges. In: Plant functional traits for improving productivity. Singapore, Springer Nature Singapore, pp 169–202

Chawla R, Poonia A, Samantara K, Mohapatra SR, Naik SB, Ashwath MN et al (2023) Green revolution to genome revolution: driving better resilient crops against environmental instability. Front Genet 14:1204585

Chen X, Palli SR (2023) Development of multiple transgenic CRISPR/Cas9 methods for genome editing in the fall armyworm, Spodoptera frugiperda. J Pest Sci 96(4):1637–1650

Chen B, Hu J, Almeida R, Liu H, Balakrishnan S, Covill-Cooke C et al (2016) Expanding the CRISPR imaging toolset with *Staphylococcus aureus* Cas9 for simultaneous imaging of multiple genomic loci. Nucleic Acids Res 44(8):e75–e75. https://doi.org/10.1093/nar/gkv1533

Chen W, Liu C, Lu G, Cheng H, Shen Z, Wu K (2018) Effects of Vip3AcAa+ Cry1Ac cotton on midgut tissue in *Helicoverpa armigera* (Lepidoptera: Noctuidae). J Insect Sci 18(4):13. https://doi.org/10.1093/jisesa/iey075

Chen Q, Zeng G, Hao M, Jiang H, Xiao Y (2020) Improvement of rice blast and brown planthopper resistance of PTGMS line C815S in two-line hybrid rice through marker-assisted selection. Mol Breed 40(2):21

Cooper SG, Douches DS, Grafius EJ (2004) Combining genetic engineering and traditional breeding to provide elevated resistance in potatoes to Colorado potato beetle. Entomol Exp Appl 112(1):37–46

D'Esposito D, Guadagno A, Amoroso CG, Cascone P, Cencetti G, Michelozzi M, Guerrieri E, Ercolano MR (2023) Genomic and metabolic profiling of two tomato contrasting cultivars for tolerance to Tuta absoluta. Planta 257(3):47

Danai-Varsou D, Zhang P, Afantitis A, Guo Z, Lynch I, Melagraki G (2023) Nanoinformatics and artificial intelligence for nano-enabled sustainable agriculture. In: Nano-enabled sustainable and precision agriculture. Academic Press, pp 503–531

Darlington M, Reinders JD, Sethi A, Lu AL, Ramaseshadri P, Fischer JR, Boeckman CJ, Petrick JS, Roper JM, Narva KE, Vélez AM (2022) RNAi for western corn rootworm management: lessons learned, challenges, and future directions. Insects 13(1):57

Das A, Ghosh P, Das S (2018) Expression of *Colocasia esculenta* tuber agglutinin in Indian mustard provides resistance against *Lipaphis erysimi* and the expressed protein is non-allergenic. Plant Cell Rep 37(6):849–863

de Albuquerque LP, Procópio TF, da Silva Guedes CC, Pontual EV, Paiva PMG, Napoleão TH (2020) Antinutritional effects of the chitin-binding lectin from *Microgramma vacciniifolia* rhizome (MvRL) on *Sitophilus zeamais*. J Stored Prod Res 88:101652. https://doi.org/10.1016/j.jspr.2020.101652

de Oliveira APS, Agra-Neto AC, Pontual EV, de Albuquerque Lima T, Cruz KCV, de Melo KR et al (2020) Evaluation of the insecticidal activity of *Moringa oleifera* seed extract and lectin (WSMoL) against *Sitophilus zeamais*. J Stored Prod Res 87:101615. https://doi.org/10.1016/j.jspr.2020.101615

de Santana Souza C, Procópio TF, do Rego Belmonte B, Paiva PMG, de Albuquerque LP, Pontual EV, Napoleão TH (2018) Effects of Opuntia ficus-indica lectin on feeding, survival, and gut enzymes of maize weevil, Sitophilus zeamais. Appl Biol Chem 61:337–343

Deguine JP, Aubertot JN, Flor RJ, Lescourret F, Wyckhuys KA, Ratnadass A (2021) Integrated pest management: good intentions, hard realities. A review. Agron Sustain Dev 41(3):38

Di Lelio I, Barra E, Coppola M, Corrado G, Rao R, Caccia S (2022) Transgenic plants expressing immunosuppressive dsRNA improve entomopathogen efficacy against *Spodoptera littoralis* larvae. J Pest Sci 1-16:1413. https://doi.org/10.1007/s10340-021-01467-z

Divekar PA, Rani V, Majumder S, Karkute SG, Molla KA, Pandey KK, Behera TK, Govindharaj GPP (2023) Protease inhibitors: an induced plant defense mechanism against herbivores. J Plant Growth Regul 42(10):6057–6073

Dong Y, Yang Y, Wang Z, Wu M, Fu J, Guo J et al (2020) Inaccessibility to double-stranded RNAs in plastids restricts RNA interference in *Bemisia tabaci* (whitefly). Pest Manag Sci 76(9):3168–3176. https://doi.org/10.1002/ps.5871

Douris V, Steinbach D, Panteleri R, Livadaras I, Pickett JA, Van Leeuwen T et al (2016) Resistance mutation conserved between insects and mites unravels the benzoylurea insecticide mode of action on chitin biosynthesis. Proc Natl Acad Sci 113(51):14692–14697. https://doi.org/10.1073/pnas.1618258113

Down RE, Fitches EC, Wiles DP, Corti P, Bell HA, Gatehouse JA, Edwards JP (2006) Insecticidal spider venom toxin fused to snowdrop lectin is toxic to the peach-potato aphid, *Myzus persicae* (Hemiptera: Aphididae) and the rice brown planthopper, *Nilaparvata lugens* (Hemiptera: Delphacidae). Pest Manag Sci 62(1):77–85. https://doi.org/10.1002/ps.1119

Duan X, Hou Q, Liu G, Pang X, Niu Z, Wang X et al (2018) Expression of *Pinellia pedatisecta* lectin gene in transgenic wheat enhances resistance to wheat aphids. Molecules 23(4):748. https://doi.org/10.3390/molecules23040748

Dwivedi SL, Reynolds MP, Ortiz R (2021) Mitigating tradeoffs in plant breeding. Iscience 24(9):102965

Enders L, Begcy K (2021) Unconventional routes to developing insect-resistant crops. Mol Plant 14(9):1439–1453

Fitches E, Wilkinson H, Bell H, Bown DP, Gatehouse JA, Edwards JP (2004) Cloning, expression and functional characterisation of chitinase from larvae of tomato moth (*Lacanobia oleracea*): a demonstration of the insecticidal activity of insect chitinase. Insect Biochem Mol Biol 34(10):1037–1050. https://doi.org/10.1016/j.ibmb.2004.06.012

Fujimoto H, Itoh K, Yamamoto M, Kyozuka J, Shimamoto KO (1993) Insect resistant rice generated by introduction of a modified δ-endotoxin gene of *Bacillus thuringiensis*. Bio/Technology 11(10):1151–1155. https://doi.org/10.1038/nbt1093-1151

Gadaleta PG (2020) Risk assessment of insect-resistant genetically modified crops on non-target arthropods and benefits to associated biodiversity of agro-ecosystems. In: GMOs: implications for biodiversity conservation and ecological processes, pp 343–357

García M, García-Benítez C, Ortego F, Farinós GP (2023) Monitoring insect resistance to Bt maize in the European Union: update, challenges, and future prospects. J Econ Entomol 116(2):275–288

Gassmann AJ, Reisig DD (2023) Management of insect pests with Bt crops in the United States. Annu Rev Entomol 68(1):31–49

George B, Pooja S, Suchithra TV, Sebastian D (2023) Historical developments of antimicrobial peptide research. In: Antimicrobial peptides. Academic Press, pp 1–16

Ghareyazie B, Alinia F, Menguito CA, Rubia LG, de Palma JM, Liwanag EA et al (1997) Enhanced resistance to two stem borers in an aromatic rice containing a synthetic cryIA (b) gene. Mol Breed 3(5):401–414. https://doi.org/10.1023/a:1009695324100

Ghosh G, Ganguly S, Purohit A, Chaudhuri RK, Das S, Chakraborti D (2017) Transgenic pigeonpea events expressing Cry1Ac and Cry2Aa exhibit resistance to *Helicoverpa armigera*. Plant Cell Rep 36(7):1037–1051. https://doi.org/10.1007/s00299-017-2133-0

Gilles AF, Schinko JB, Averof M (2015) Efficient CRISPR-mediated gene targeting and transgene replacement in the beetle *Tribolium castaneum*. Development 142(16):2832–2839. https://doi.org/10.1242/dev.125054

Gowda A, Rydel TJ, Wollacott AM, Brown RS, Akbar W, Clark TL et al (2016) A transgenic approach for controlling Lygus in cotton. Nat Commun 7(1):12213. https://doi.org/10.1038/s41467-020-14789-w

Grazziotin MA, Cabral GB, Ibrahim AB, Machado RB, Aragao FJ (2020) Expression of the Arcelin 1 gene from *Phaseolus vulgaris* L. in cowpea seeds (*Vigna unguiculata* L.) confers bruchid resistance. Ann Appl Biol 176(3):268–274. https://doi.org/10.1111/aab.12568

Gunasekara JMA, Jayasekera GAU, Perera KLNS, Wickramasuriya AM (2017) Development of a Sri Lankan rice variety Bg 94-1 harbouring Cry2A gene of *bacillus thuringiensis* resistant to rice leaf folder [*Cnaphalocrocis medinalis* (Guenée)]. J Natl Sci Found 45(2):143. https://doi.org/10.4038/jnsfsr.v45i2.8180

Guo Z, Li Y, Ding SW (2019) Small RNA-based antimicrobial immunity. Nat Rev Immunol 19(1):31–44. https://doi.org/10.1038/s41577-018-0071-x

Gupta S, Kumar A, Patel R, Kumar V (2021) Genetically modified crop regulations: scope and opportunity using the CRISPR-Cas9 genome editing approach. Mol Biol Rep 48(5):4851–4863

Hafeez U, Ali M, Hassan SM, Akram MA, Zafar A (2023) Advances in breeding and engineering climate-resilient crops: a comprehensive review. Int J Res Adv Agric Sci 2(2):85–99

Han HA, Pang JKS, Soh BS (2020) Mitigating off-target effects in CRISPR/Cas9-mediated in vivo gene editing. J Mol Med 98(5):615–632

Han P, Lavoir AV, Rodriguez-Saona C, Desneux N (2022) Bottom-up forces in agroecosystems and their potential impact on arthropod pest management. Annu Rev Entomol 67(1):239–259

He P, Jia H, Xue H, Zeng Y, Tian L, Hu X et al (2022) Expression of modified snowdrop lectin (Galanthus nivalis agglutinin) protein confers aphids and Plutella xylostella resistance in Arabidopsis and cotton. Genes 13(7):1169

Hilda K, Bhuvaragavan S, Kamatchi R, Meenakumari M, Janarthanan S (2022) Cloning, expression and characterization of arcelin and its impact on digestive enzymes of the stored product insect pest, *Callosobruchus maculatus* (F.). Pestic Biochem Physiol 180:104982. https://doi.org/10.1016/j.pestbp.2021.104982

Huang Y, Chen Y, Zeng B, Wang Y, James AA, Gurr GM et al (2016) CRISPR/Cas9 mediated knockout of the abdominal-A homeotic gene in the global pest, diamondback moth (*Plutella xylostella*). Insect Biochem Mol Biol 75:98–106. https://doi.org/10.1016/j.ibmb.2016.06.004

Hussain T, Aksoy E, Çalışkan ME, Bakhsh A (2019) Transgenic potato lines expressing hairpin RNAi construct of molting-associated EcR gene exhibit enhanced resistance against Colorado potato beetle (*Leptinotarsa decemlineata*, say). Transgenic Res 28(1):151–164. https://doi.org/10.1007/s11248-018-0109-7

Jadhav MS, Rathnasamy SA, Natarajan B, Duraialagaraja S, Varatharajalu U (2020) Study of expression of indigenous Bt cry2ax1 gene in t3 progeny of cotton and its efficacy against Helicoverpa armigera (Hubner). Braz Arch Biol Technol 63

Jaiwal A, Natarajaswamy K, Rajam MV (2020) RNA silencing of hormonal biosynthetic genes impairs larval growth and development in cotton bollworm, Helicoverpa armigera. J Biosci 45:1–14

Jambagi SR, Gaddanakeri S, Dileep KN, Chaitra DS (2024) Insecticide resistance management strategies for sustainable insect control. In: Entomology redefined. CRC Press, pp 263–285

Jiang X, Yang F, Jia W, Jiang Y, Wu X, Song S et al (2024) Nanomaterials and nanotechnology in agricultural pesticide delivery: a review. Langmuir 40:18806–18820

Jin L, Wang J, Guan F, Zhang J, Yu S, Liu S, Wu Y (2018) Dominant point mutation in a tetraspanin gene associated with field-evolved resistance of cotton bollworm to transgenic Bt cotton. Proc Natl Acad Sci 115(46):11760–11765

Joshi A, Kumar A, Adhikari S, Saroj R (2024) Marker-assisted selection: strategy for biotic and abiotic stress tolerance in plants. In: Climate-resilient agriculture. Apple Academic Press, pp 137–191

Kamatham S, Munagapati S, Manikanta KN, Vulchi R, Chadipiralla K, Indla SH, Allam US (2021) Recent advances in engineering crop plants for resistance to insect pests. Egypt J Biol Pest Control 31:1–14

Kamweru I, Anani BY, Beyene Y, Makumbi D, Adetimirin VO, Prasanna BM, Gowda M (2022) Genomic analysis of resistance to fall armyworm (Spodoptera frugiperda) in CIMMYT maize lines. Genes 13(2):251

Katta S, Talakayala A, Reddy MK, Addepally U, Garladinne M (2020) Development of transgenic cotton (Narasimha) using triple gene Cry2Ab-Cry1F-Cry1Ac construct conferring resistance to lepidopteran pest. J Biosci 45(1):1–11

Kedisso EG, Maredia K, Guenthner J, Koch M (2022) Commercialization of genetically modified crops in Africa: opportunities and challenges. Afr J Biotechnol 21(5):188–197

Kiran U, Pandey NK (2020) Transgenic food crops: public acceptance and IPR. In: Transgenic technology based value addition in plant biotechnology. Academic Press, pp 273–307

Koul B (2020) Genetically modified (GM) crops harbouring bacillus thuringiensis (BT) gene (s) to combat biotic stress caused by insect pests. In: Phyto-microbiome in stress regulation, pp 21–61

Krishnamoorthi A, Patel PK, Kumar S, Khambhu CV, Kumar V, Selvam H (2024) The future of farming with advances in biological control techniques for crop health. Microbiol Res J Int 34(7):93–112

Kumar K, Gambhir G, Dass A, Tripathi AK, Singh A, Jha AK, Yadava P, Choudhary M, Rakshit S (2020a) Genetically modified crops: current status and future prospects. Planta 251(4):91

Kumar K, Kaur P, Kishore A, Vikal Y, Singh K, Neelam K (2020b) Recent advances in genomics-assisted breeding of brown planthopper (Nilaparvata lugens) resistance in rice (Oryza sativa). Plant Breed 139(6):1052–1066

Kumar P, Adarsh A, Kumari R, Jakhar DS, Kumar A, Kumari A (2021) Molecular approaches for control of sucking Pest in tomato. In: Biotic stress management in tomato. Apple Academic Press, pp 205–226

Kumari P, Jasrotia P, Kumar D, Kashyap PL, Kumar S, Mishra CN, Kumar S, Singh GP (2022) Biotechnological approaches for host plant resistance to insect pests. Front Genet 13:914029

Kumari R, Saha T, Kumar P, Singh AK (2024) CRISPR/Cas9-mediated genome editing technique to control fall armyworm (Spodoptera frugiperda) in crop plants with special reference to maize. In: Physiology and molecular biology of plants, vol 30, pp 1–13

Kushanov FN, Turaev OS, Ernazarova DK, Gapparov BM, Oripova BB, Kudratova MK, Rafieva FU, Khalikov KK, Erjigitov DS, Khidirov MT, Kholova MD (2021) Genetic diversity, QTL mapping, and marker-assisted selection technology in cotton (Gossypium spp.). Front Plant Sci 12:779386

Kwon KM, Viana JP, Walden KK, Usovsky M, Scaboo AM, Hudson ME, Mitchum MG (2024) Genome scans for selection signatures identify candidate virulence genes for adaptation of the soybean cyst nematode to host resistance. Mol Ecol e17490

Kumar S, Korra T, Thakur R, Arutselvan R, Kashyap AS, Nehela Y, Keswani C (2023) Role of plant secondary metabolites in defence and transcriptional regulation in response to biotic stress. Plant Stress 8:100154

Khan GA, Bakhsha A, Riazuddin S, Husnain T (2011) Introduction of cry1Ab gene into cotton (Gossypium hirsutum) enhances resistance against Lepidopteran pest (Helicoverpa armigera). Span J Agric Res 9:296–302. https://doi.org/10.5424/sjar/20110901-136-10

Khajuria C, Vélez AM, Rangasamy M, Wang H, Fishilevich E, Frey ML, Siegfried BD (2015) Parental RNA interference of genes involved in embryonic development of the western corn rootworm, Diabrotica virgifera virgifera LeConte. Insect Biochem Mol Biol 63:54–62. https://doi.org/10.1016/j.ibmb.2015.05.011

Kim SY, Bengtsson T, Olsson N, Hot V, Zhu LH, Åhman I (2020) Mutations in two aphid-regulated β-1, 3-glucanase genes by CRISPR/Cas9 do not increase barley resistance to Rhopalosiphum padi L. Front Plant Sci 1043. https://doi.org/10.3389/fpls.2020.01043

Koerniati S, Sukmadjaja D, Samudra IM (2020) C synthetic gene of CryIAb-CryIAc fusion to generate resistant sugarcane to shoot or stem borer. In IOP Conference Series: Earth and Environmental Science. IOP Publishing, Vol. 418(1), p. 012069. https://doi.org/10.1088/1755-1315/418/1/012069

Kola VSR, Pichili R, Padmakumari AP, Mangrauthia SK, Balachandran SM, Madhav MS (2019) Knockdown of acetylcholinesterase (AChE) gene in rice yellow stem borer, Scirpophaga incertulas (Walker) through RNA interference. Agri Gene 11:100081

Koutroumpa FA, Monsempes C, François MC, de Cian A, Royer C, Concordet JP, Jacquin-Joly E (2016) Heritable genome editing with CRISPR/Cas9 induces anosmia in a crop pest moth. Sci Rep 6(1):1–9. https://doi.org/10.1038/srep29620

Lallawmkimi MC, Veda DS, Yadav A, Kumar M, Rout A (2024) Innovative approaches in crop genetic engineering for sustainable agriculture: a review. J Adv Biol Biotechnol 27(8):615–631

Li H, Guan R, Guo H, Miao X (2015) New insights into an RNAi approach for plant defence against piercing-sucking and stem-borer insect pests. Plant, Cell & Environ 38(11):2277–2285. https://doi.org/10.1111/pce.12546.

Li X, Liu X, Lu W, Yin X, An S (2022) Application progress of plant-mediated RNAi in pest control. Front Bioeng Biotechnol 10:963026

Li Y, Zhang J, Chen D, Yang P, Jiang F, Wang X, Kang L (2016) CRISPR/Cas9 in locusts: successful establishment of an olfactory deficiency line by targeting the mutagenesis of an odorant receptor co-receptor (Orco). Insect Biochem Mol Biol 79:27–35

Maqbool SB, Riazuddin S, Loc NT, Gatehouse AM, Gatehouse JA, Christou P (2001) Expression of multiple insecticidal genes confers broad resistance against a range of different rice pests. Mol Breed 7(1):85–93

Mamta RKRK, Rajam MV (2016) Targeting chitinase gene of Helicoverpa armigera by host-induced RNA interference confers insect resistance in tobacco and tomato. Plant Mol Biol 90(3):281–292. https://doi.org/10.1007/s11103-015-0414-y. Epub 2015 Dec 10

Mathur F, Arora A (2024) CRISPR-based genetic control strategies for insect pests to mitigate classical insecticidal approaches. In: Gene editing in plants: CRISPR-Cas and its applications. Singapore, Springer Nature Singapore, pp 667–707

Meade T, Narva K, Storer, NP, Sheets JJ, Burton SL, Woosley AT (2017) U.S. Patent No. 9,556,453. Washington, DC: U.S. Patent and Trademark Office

Meccariello A, Monti SM, Romanelli A, Colonna R, Primo P, Inghilterra MG, Saccone G (2017) Highly efficient DNA-free gene disruption in the agricultural pest Ceratitis capitata by CRISPR-Cas9 ribonucleoprotein complexes. Sci Rep 7:1–11

Mehlhorn SG, Geibel S, Bucher G, Nauen R (2020) Profiling of RNAi sensitivity after foliar dsRNA exposure in different European populations of Colorado potato beetle reveals a robust response with minor variability. Pestic Biochem Physiol 166:104569

Mehta S, Lal SK, Sahu KP, Venkatapuram AK, Kumar M, Sheri V, Varakumar P, Vishwakarma C, Yadav R, Jameel MR, Ali M (2020) CRISPR/Cas9-edited rice: a new frontier for sustainable agriculture. In: New frontiers in stress management for durable agriculture, pp 427–458

Meng F, Li Y, Zang Z, Li N, Ran R, Cao Y, Li W (2017) Expression of the double-stranded RNA of the soybean pod borer Leguminivora glycinivorella (Lepidoptera: Tortricidae) ribosomal protein Po gene enhances the resistance of transgenic soybean plants. Pest Manag Sci 73(12):2447–2455

Mesías-Ruiz GA, Pérez-Ortiz M, Dorado J, De Castro AI, Peña JM (2023) Boosting precision crop protection towards agriculture 5.0 via machine learning and emerging technologies: a contextual review. Front Plant Sci 14:1143326

Mi X, Liu X, Yan H, Liang L, Zhou X, Yang J, Zhang N (2017) Expression of the Galanthus nivalis agglutinin (GNA) gene in transgenic potato plants confers resistance to aphids. Comptes Rendus Biol 340(1):7–12. https://doi.org/10.1016/j.crvi.2016

Molinari MDC, Pagliarini RF, Florentino LH, Lima RN, Arraes FBM, Abbad SV, de Farias MP, Mertz-Henning LM, Rech E, Nepomuceno AL, Molinari HBC (2024) Navigating the path from lab to market: regulatory challenges and opportunities for genome editing Technologies for Agriculture. In: Plant genome editing technologies: speed breeding, crop improvement and sustainable agriculture, pp 25–63

Mukhtar Y, Shankar U (2023) Integrated pest and pollinator management in India: a way forward to sustainable agriculture. Indian J Agric Sci 93(9):939–947

Muddanuru T, Polumetla AK, Maddukuri L, Mulpuri S (2019) Development and evaluation of transgenic castor (Ricinus communis L.) expressing the insecticidal protein Cry1Aa of Bacillus thuringiensis against lepidopteran insect pests. Crop Prot 119:113–125. https://doi.org/10.1016/j.cropro.2019.01.016

Nakasu EY, Edwards MG, Fitches E, Gatehouse JA, Gatehouse AM (2014) Transgenic plants expressing ω-ACTX-Hv1a and snowdrop lectin (GNA) fusion protein show enhanced resistance to aphids. Front Plant Sci 5:673. https://doi.org/10.3389/fpls.2014.00673

Napoleão TH, do Rego Belmonte B, Pontual EV, de Albuquerque LP, Sá RA, Paiva LM, Paiva PMG (2013) Deleterious effects of Myracrodruon urundeuva leaf extract and lectin on the maize weevil, Sitophilus zeamais (Coleoptera, Curculionidae). J Stored Prod Res 54:26–33. https://doi.org/10.1016/j.jspr.2013.04.002

Nayak P, Basu D, Das S, Basu A, Ghosh D, Ramakrishnan NA, Sen SK (1997) Transgenic elite indica rice plants expressing CryIAc∂-endotoxin of Bacillus thuringiensis are resistant against yellow stem borer (Scirpophaga incertulas). Proc Natl Acad Sci 94(6):2111–2116. https://doi.org/10.1073/pnas.94.6.2111

Ni M, Ma W, Wang X, Gao M, Dai Y, Wei X, Zhu Z (2017) Next-generation transgenic cotton: pyramiding RNAi and Bt counters insect resistance. Plant Biotechnol J 15(9):1204–1213. https://doi.org/10.1111/pbi.12709

Nitnavare RB, Bhattacharya J, Singh S, Kour A, Hawkesford MJ, Arora N (2021) Next generation dsRNA-based insect control: success so far and challenges. Front Plant Sci 12:673576

Niu X, Kassa A, Hu X, Robeson J, McMahon M, Richtman NM, Wu G (2017) Control of western corn rootworm (Diabrotica virgifera virgifera) reproduction through plant-mediated RNA interference. Sci Rep 7(1):1–13. https://doi.org/10.1038/s41598-017-12638-3

Pardo-López L, Soberón M, Bravo A (2013) Bacillus thuringiensis insecticidal three-domain Cry toxins: mode of action, insect resistance and consequences for crop protection. FEMS Microbiol Rev 37(1):3–22

Parkash J, Rajat AK, Jeevan BG (2023) Integrated pest management: a comprehensive overview. Adv Entomol 3

Peshin R, Hansra BS, Singh K, Nanda R, Sharma R, Yangsdon S, Kumar R (2021) Long-term impact of Bt cotton: an empirical evidence from North India. J Clean Prod 312:127575

Qin D, Liu XY, Miceli C, Zhang Q, Wang PW (2019) Soybean plants expressing the Bacillus thuringiensis cry8-like gene show resistance to Holotrichia parallela. BMC Biotech 19(1):1–12. https://doi.org/10.1186/s12896-019-0563-1

Qiu L, Fan J, Zhang B, Liu L, Wang X, Lei C et al (2017) RNA interference knockdown of aminopeptidase N genes decrease the susceptibility of Chilo suppressalis larvae to Cry1Ab/Cry1Ac and Cry1Caexpressing transgenic rice. J Invertebr Pathol 145:9–12

Qiu L, Sun Y, Jiang Z, Yang P, Liu H, Zhou H, Ma W (2019) The midgut V-ATPase subunit A gene is associated with toxicity to crystal 2Aa and crystal 1Ca-expressing transgenic rice in Chilo suppressalis. Insect Mol Biol 28(4):520–527. https://doi.org/10.1111/imb.12570

Rajadurai G, Kalaivani A, Varanavasiyappan S, Balakrishnan N, Udayasuriyan V, Sudhakar D, Natarajan N (2018) Generation of insect resistant marker-free transgenic rice with a novel cry2AX1 gene. Electron J Plant Breed 9(2):723–732

Rani S, Sharma V, Hada A, Koundal KR (2017a) Efficient genetic transformation of Brassica juncea with lectin using cotyledons explants. Int J Food Sci Technol Res 7:1–12

Rani S, Sharma V, Hada A, Bhattacharya RC, Koundal KR (2017b) Fusion gene construct preparation with lectin and protease inhibitor genes against aphids and efficient genetic transformation of Brassica juncea using cotyledons explants. Acta Physiol Plant 39(5):1–13. https://doi.org/10.1007/s11738-017-2415-8

Rao MJ, Wang L (2021) CRISPR/Cas9 technology for improving agronomic traits and future prospective in agriculture. Planta 254:1–16

Rato C, Carvalho MF, Azevedo C, Oblessuc PR (2021) Genome editing for resistance against plant pests and pathogens. Transgenic Res 30(4):427–459

Razzaq A, Kaur P, Akhter N, Wani SH, Saleem F (2021) Next-generation breeding strategies for climate-ready crops. Front Plant Sci 12:620420
Riaz S, Nasir IA, Bhatti MU, Adeyinka OS, Toufiq N, Yousaf I, Tabassum B (2020) Resistance to Chilo infuscatellus (Lepidoptera: Pyraloidea) in transgenic lines of sugarcane expressing Bacillus thuringiensis derived Vip3A protein. Mole Biol Rep 47(4):2649–2658. https://doi.org/10.1007/s11033-020-05355-0
Ribeiro ADO, Pereira EJG, Galvan TL, Picanco MC, Picoli EDT, Da Silva DJH, Otoni WC (2006) Effect of eggplant transformed with oryzacystatin gene on Myzus persicae and Macrosiphum euphorbiae. J Appl Entomol 130(2):84–90
Roy A, Chakraborty A (2021) Natural insecticidal proteins and their potential in future IPM. In: Plant-pest interactions: from molecular mechanisms to chemical ecology: chemical ecology, pp 265–303
Rozas P, Kessi-Pérez EI, Martínez C (2022) Genetically modified organisms: adapting regulatory frameworks for evolving genome editing technologies. Biol Res 55:31
Saakre M, Jaiswal S, Rathinam M, Raman KV, Tilgam J, Paul K, Sreevathsa R, Pattanayak D (2024) Host-delivered RNA interference for durable pest resistance in plants: advanced methods, challenges, and applications. Mol Biotechnol 66(8):1786–1805
Salgotra RK, Chauhan BS (2023) Genetic diversity, conservation, and utilization of plant genetic resources. Genes 14(1):174
Sakthi AR, Naveenkumar A, Deepikha PS, Balakrishnan N, Kumar KK, Devi EK, Balasubramanian P (2015) Expression and inheritance of chimeric cry2AX1 gene in transgenic cotton plants generated through somatic embryogenesis. In Vitro Cellular and Developmental Biology. Plant 51(4):379–389
Sarkozi A (2019) New standards to curb the global spread of plant pests and diseases. Food and Agriculture Organization of the United Nations, Roma
Sawardekar SV, Katageri IS, Salimath PM, Kumar PA, Kelkar VG (2017) Standardization of in-vitro genetic transformation technique in chickpea (Cicer arietinum L.) for pod-borer resistance. Adv Agric Res Tech 1(2)
Selale H, Dağlı F, Mutlu N, Doğanlar S, Frary A (2017) Cry1Ac-mediated resistance to tomato leaf miner (Tuta absoluta) in tomato. Plant Cell, Tissue and Organ Culture (PCTOC) 131(1):65–73
Sétamou M, Bernal JS, Legaspi JC, Mirkov TE, Legaspi BC Jr (2002) Evaluation of lectin-expressing transgenic sugarcane against stalk borers (Lepidoptera: Pyralidae): effects on life history parameters. J Econ Entomol 95(2):469–477. https://doi.org/10.1603/0022-0493-95.2.469
Shaibu AS, Li B, Zhang S, Sun J (2020) Soybean cyst nematode-resistance: gene identification and breeding strategies. Crop J 8(6):892–904
Sharma N, Bakshi A (2022) Integrated pest management through genetically modified crops. In: Genetically modified crops and food security. Routledge, pp 229–246
Sharma KR, Raju SVS, Singh SK, Singh R, Meena RS (2024) Rice genotypes and the biochemical basis of resistance against brown planthopper, Nilaparvata lugens (Stål). Entomon 49(2):215–220
Shekhawat SS, Hasmi SK (2023) Safety and benefits of Bt and Bt cotton: factures, refute, and allegations. In: Genetic engineering. Apple Academic Press, pp 23–52
Singh M, Pandey N, Sharma OP (2023) IPM concept and strategies for sustainable agriculture. In: Integrated pest management in diverse cropping systems. Apple Academic Press, pp 31–59
Skendžić S, Zovko M, Živković IP, Lešić V, Lemić D (2021) The impact of climate change on agricultural insect pests. Insects 12(5):440
Smith CM (2021) Conventional breeding of insect-resistant crop plants: still the best way to feed the world population. Curr Opin Insect Sci 45:7–13
Song P, Wang J, Guo X, Yang W, Zhao C (2021) High-throughput phenotyping: breaking through the bottleneck in future crop breeding. Crop J 9(3):633–645
Siddiqui HA, Asif M, Asad S, Naqvi RZ, Ajaz S, Umer N, Mansoor S (2019) Development and evaluation of double gene transgenic cotton lines expressing Cry toxins for protection against chewing insect pests. Sci Rep 9(1):1–7. https://doi.org/10.1038/s41598-019-48188-z

Singh S, Kumar NR, Maniraj R, Lakshmikanth R, Rao KYS, Muralimohan NS, R. (2018) Expression of Cry2Aa, a Bacillus thuringiensis insecticidal protein in transgenic pigeon pea confers resistance to gram pod borer. Helicoverpa armigera. Sci Rep 8(1):1–12. https://doi.org/10.1038/s41598-018-26358-9

Soliman, H. I., Abo-El-Hasan, F. M., El-Seedy, A. S., & Mabrouk, Y. M. (2021). Agrobacterium-mediated transformation of tomato (Lycopersicon esculentum mill.) using a synthetic cry1ab gene for enhanced resistance against Tuta absoluta (Meyrick). J Microbiol Biotechnol Food Sci 67–74. https://doi.org/10.15414/jmbfs.2017.7.1.67-74

Stoger E, Williams S, Christou P, Down RE, Gatehouse JA (1999) Expression of the insecticidal lectin from snowdrop (Galanthus nivalis agglutinin; GNA) in transgenic wheat plants: effects on predation by the grain aphid Sitobion avenae. Mol Breed 5(1):65–73. https://doi.org/10.1023/a:1009616413886

Sudhakar D, Fu X, Stoger E, Williams S, Spence J, Brown DP, Christou P (1998) Expression and immunolocalisation of the snowdrop lectin, GNA in transgenic rice plants. Transgenic Res 7(5):371–378. https://doi.org/10.1023/a:1008856703464

Tabashnik BE, Fabrick JA, Carrière Y (2023) Global patterns of insect resistance to transgenic Bt crops: the first 25 years. J Econ Entomol 116(2):297–309

Tarazi R, Vaslin MF (2022) The viral threat in cotton: how new and emerging technologies accelerate virus identification and virus resistance breeding. Front Plant Sci 13:851939

Tian Z, Wang JW, Li J, Han B (2021) Designing future crops: challenges and strategies for sustainable agriculture. Plant J 105(5):1165–1178

Togola A, Boukar O, Servent A, Chamarthi S, Tamo M, Fatokun C (2020) Identification of sources of resistance in cowpea mini core accessions to Aphis craccivora Koch (Homoptera: Aphididae) and their biochemical characterization. Euphytica 216(6):88

Trung NP, Fitches E, Gatehouse JA (2006) A fusion protein containing a lepidopteran-specific toxin from the South Indian red scorpion (Mesobuthus tamulus) and snowdrop lectin shows oral toxicity to target insects. BMC Biotech 6(1):1–12. https://doi.org/10.1186/1472-6750-6-18

Tu J, Zhang G, Datta K, Xu C, He Y, Zhang Q, Datta SK (2000) Field performance of transgenic elite commercial hybrid rice expressing Bacillus thuringiensis δ-endotoxin. Nat Biotechnol 18(10):1101–1104. https://doi.org/10.1038/80310

Tyagi S, Kesiraju K, Saakre M, Rathinam M, Raman V, Pattanayak D, Sreevathsa R (2020) Genome editing for resistance to insect pests: an emerging tool for crop improvement. ACS Omega 5(33):20674–20683

Vanti GL, Katageri IS, Inamdar SR, Hiremathada V, Swamy BM (2018) Potent insect gut binding lectin from Sclerotium rolfsii impart resistance to sucking and chewing type insects in cotton. J Biotech 278:20–27. https://doi.org/10.1016/j.jbiotec.2018.04.018

Vega Rodríguez A, Rodríguez-Oramas C, Sanjuán Velázquez E, Hardisson de la Torre A, Rubio Armendáriz C, Carrascosa Iruzubieta C (2022) Myths and realities about genetically modified food: a risk-benefit analysis. Appl Sci 12(6):2861

Verma S, Saxena S (2021) Genetically modified crops changing the food insecurity landscape of the undernourished regions of the world. In: Policy issues in genetically modified crops. Academic Press, pp 143–160

Verma V, Negi S, Kumar P, Srivastava DK (2022) Global status of genetically modified crops. In: Agricultural biotechnology: latest research and trends. Singapore, Springer Nature Singapore, pp 305–322

Wang J, Zhang H, Wang H, Zhao S, Zuo Y, Yang Y, Wu Y (2016) Functional validation of cadherin as a receptor of Bt toxin Cry1Ac in Helicoverpa armigera utilizing the CRISPR/Cas9 system. Insect Biochem Mol Biol 76:11–17. https://doi.org/10.1016/j.ibmb.2016.06.008

Wang J, Ma, H, Zuo Y, Yang Y, Wu Y (2020a) CRISPR-mediated gene knockout reveals nicotinic acetylcholine receptor (nAChR) subunit α6 as a target of spinosyns in Helicoverpa armigera. Pest Manag Sci 76(9):2925–2931. https://doi.org/10.1002/ps.5889

Wang X, Xu Y, Huang J, Jin W, Yang Y, Wu Y (2020b) CRISPR-mediated knockout of the ABCC2 gene in Ostrinia furnacalis confers high-level resistance to the Bacillus thuringiensis Cry1Fa toxin. Toxins 12(4):246. https://dx.doi.org/10.3390%2Ftoxins12040246

Wang X, Guo X, Ma X, Luo L, Fang Y, Zhao N, Han Y, Wei Z, Liu F, Qin B, Li R (2021) Development of new rice (*Oryza sativa* L.) breeding lines through marker-assisted introgression and pyramiding of brown planthopper, blast, bacterial leaf blight resistance, and aroma genes. Agronomy 11(12):2525

Wang D, Saleh NB, Byro A, Zepp R, Sahle-Demessie E, Luxton TP et al (2022) Nano-enabled pesticides for sustainable agriculture and global food security. Nat Nanotechnol 17(4):347–360

Wani SH, Choudhary M, Barmukh R, Bagaria PK, Samantara K, Razzaq A, Jaba J, Ba MN, Varshney RK (2022) Molecular mechanisms, genetic mapping, and genome editing for insect pest resistance in field crops. Theor Appl Genet 135(11):3875–3895

Wünn J, Klöti A, Burkhardt PK, Biswas GCG, Launis K, Iglesias VA, Potrykus I (1996) Transgenic indica rice breeding line IR58 expressing a synthetic cryIA (b) gene from Bacillus thuringiensis provides effective insect pest control. Biotechnology 14(2):171–176

Xia J, Guo Z, Yang Z, Han H, Wang S, Xu H, Zhang Y (2021) Whitefly hijacks a plant detoxification gene that neutralizes plant toxins. Cell 184(7):1693–1705. https://doi.org/10.1016/j.cell.2021.02.014

Xu C, Cheng J, Lin H, Lin C, Gao J, Shen Z (2018) Characterization of transgenic rice expressing fusion protein Cry1Ab/Vip3A for insect resistance. Sci Rep 8:1–8. https://doi.org/10.1038/s41598-018-34104-4

Yan S, Qian J, Cai C, Ma Z, Li J, Yin M, Shen J (2020) Spray method application of transdermal dsRNA delivery system for efficient gene silencing and pest control on soybean aphid Aphis glycines. J Pest Sci 93(1):449–459

Zhang J, Khan SA, Hasse C, Ruf S, Heckel DG, Bock R (2015) Full crop protection from an insect pest by expression of long double-stranded RNAs in plastids. Science 347(6225):991–994. https://doi.org/10.1126/science.1261680

Zhang M, Chen Z, Wu H, Kong F, Guo B, Wang Y et al (2024) Identification of candidate genes for English grain aphid resistance from QTLs using a RIL population in wheat. Agronomy 14(3):637

Zhong Y, Ahmed S, Deng G, Fan W, Zhang P, Wang H (2019) Improved insect resistance against Spodoptera litura in transgenic sweet potato by over expressing Cry1Aa toxin. Plant Cell Rep 38(11):1439–1448. https://doi.org/10.1007/s00299-019-02460-8

Zuo Y, Wang H, Xu Y, Huang J, Wu S, Wu Y, Yang Y (2017) CRISPR/Cas9 mediated G4946E substitution in the ryanodine receptor of Spodoptera exigua confers high levels of resistance to diamide insecticides. Insect Biochem Mol Biol 89:79–85. https://doi.org/10.1016/j.ibmb.2017.09.005

Zuo Y, Xue Y, Lu W, Ma H, Chen M, Wu Y, Hu Z (2020) Functional validation of nicotinic acetylcholine receptor (nAChR) α6 as a target of spinosyns in Spodoptera exigua utilizing the CRISPR/Cas9 system. Pest Manag Sci 76(7):2415–2422

Part III

Physiological and Microbial Signatures for Combating Plant Stresses

Underlying Survival Mechanisms in Model Trees for Enhanced Abiotic Stress Tolerance

12

Manish Kumar, Raj Kumar, S. Sarvade, Kamini Gautam, Gajender Yadav, R. Banyal, Anita Mann, and Sneha Dobhal

Abstract

Abiotic stresses like drought, salinity, and extreme temperatures present formidable obstacles to plant vitality, critically affecting their growth and yield. Grasping the physiological and molecular adaptations plants employ to endure stress is crucial for understanding their survival and thriving in challenging conditions. This chapter delves into the critical mechanisms and physiological strategies plants use to adapt to abiotic stress, examining how they manage ionic imbalances, osmotic adjustments, oxidative damage, and the intricate signaling pathways that orchestrate these responses. Ionic stress arises from the buildup of detrimental ions, especially sodium (Na^+), which interferes with cellular processes and causes ion imbalances within the plant. Plants counteract this through ion homeostasis mechanisms involving ion transporters and selective ion uptake. Osmotic stress, or osmoregulation, entails managing cellular water balance by producing osmoprotectants such as proline and glycine betaine. These compounds play a vital role in preserving cell turgor and stabilizing proteins and cellular structures when water is scarce. Oxidative stress results from the buildup of reactive oxygen species (ROS), which can cause significant damage to cellular

M. Kumar (✉) · R. Kumar · G. Yadav · R. Banyal · A. Mann
ICAR- Central Soil Salinity Research Institute, Karnal, Haryana, India
e-mail: manish.kumar2@icar.gov.in

S. Sarvade
Department of Forestry, Jawaharlal Nehru Krishi VishwaVidhyalaya, Jabalpur, Madhya Pradesh, India

K. Gautam
ICAR-Indian Grassland and Fodder Research Institute, Jhansi, India

S. Dobhal
VCSGUUHF, College of Forestry, Ranichauri, Tehri Garhwal, Uttarakhand, India

A. Mann et al. (eds.), *Cutting Edge Technologies for Developing Future Crop Plants*, https://doi.org/10.1007/978-981-96-2508-6_12

components. To combat oxidative stress, plants produce antioxidants like ascorbate peroxidase and glutathione, which neutralize reactive oxygen species (ROS) and shield cellular components from oxidative harm. The perception of stress and the subsequent signaling pathways rely on intricate networks of receptors and secondary messengers that trigger stress responses and activate genes responsible for managing stress. Plants detect stress signals using specialized receptors and transmit these signals through intricate intracellular cascades, incorporating crucial elements such as mitogen-activated protein kinases (MAPKs) and calcium ions. Regulated by diverse transcription factors, these genes coordinate the plant's adaptive mechanisms to navigate environmental challenges effectively. Water use efficiency (WUE) is a vital trait that indicates how effectively a plant utilizes water to produce biomass, especially when water resources are scarce. Improving water use efficiency (WUE) involves fine-tuning stomatal control and boosting photosynthetic performance to enhance both water uptake and retention. Comprehending the mechanisms of survival and physiological adaptations offers valuable insights for enhancing crop resilience and productivity amid rising abiotic stressors.

Keywords
Abiotic stress · Oxidative stress · Osmoregulation · Osmotic stress · Stress signaling · Water use efficiency

12.1 Introduction

Abiotic stressors like drought, salinity, extreme temperatures, heavy metals, and pollution significantly impact the growth, development, and survival of tree species (IPCC 2007). These environmental challenges can result in diminished photosynthesis, impaired growth, damaged foliage, and heightened vulnerability to pests and diseases (Sinclair and Muchow 2001; Grift et al. 2011). These nonliving stress factors frequently serve as the main constraints on tree growth and development, adversely affecting plant health by creating osmotic stress. This disrupts the equilibrium of inorganic and organic compounds within the plant, causing problems like ion toxicity, the buildup of reactive oxygen species (ROS), and damage to cell walls and plasma membranes (Zhang et al. 2022; Praveen et al. 2023). These extreme environmental factors not only limit plant survival but also affect their distribution in nature. Abiotic stress encompasses a range of challenges, including extreme temperatures, flooding, salinity, metal toxicity, and nutrient deficiencies, each of which elicits different responses in plants (Zhang et al. 2020). These stresses severely impact crop production by disrupting both vegetative and reproductive development, thereby reducing productivity. In response to these environmental challenges, plants have developed intricate mechanisms that allow them to sense stress and adapt their growth strategies to ensure survival and reproduction. Recent research highlights the multifaceted nature of plant responses to abiotic stress, revealing that these responses involve multiple processes such as sensing, signaling, transcription, and protein modification. Trees, for instance, may develop adaptive traits like deeper root systems, altered membrane compositions, and the production of protective compounds to withstand stress. While adaptive changes enhance resistance, non-adaptive responses, such as protein misfolding and cell wall disruption, can weaken plants (Radha et al. 2023). The molecular mechanisms underlying these stress reactions are intricate, involving several genes, signaling pathways, and metabolic processes (Dong et al. 2022).

Plant functional characters have become a research hotspot in physiological and community ecology, normally used to predict species distributions, dynamics, and reactions to environmental change (Adler et al. 2014). The hydraulic stratagem of a plant, for instance, is indicative of its ability to survive drought stress (Brodribb 2017). The water status of plants is meticulously linked to xylem physiognomies, with xylem physiology often regarded as the basis of terrestrial plant productivity because of its role in balancing carbon uptake and transpiration through the stomata (Brodribb 2009). While high growth and development rates are usually associated with amplified hydraulic efficacy (Tyree 2003; Kumar et al. 2022; Kumar 2024), studies recommend the link between xylem safety and efficiency characters are not continually correlated and varies by species to species (Kumar et al. 2022; Kumar 2024). Species with more rigorous stomatal control tend to close their stomata at higher water potentials, whereas those with lower stomatal stringency exhibit greater hydraulic vulnerability, highlighting a complex interplay between stomatal behavior and xylem functionality under drought conditions (waite et al. 2024). Xylem embolism, which diminishes photosynthetic activity (Sperry and Tyree 1988), impacts forest productivity by affecting hydraulic efficiency in trees and slowing the transport of nutrients and water essential for growth (Brodribb and Cochard 2009). Ultimately, the failure to restore water in xylem conduits leads to hydraulic failure and, consequently, death of plant (Brodribb and Cochard 2009; Choat et al. 2012; Anderegg et al. 2015).

To mitigate the impacts of abiotic stress, effective management strategies are essential. These include selecting stress-resistant species, improving soil and water management, and utilizing phytoremediation techniques (Salt et al. 1995; Sah et al. 2016; Sheoran et al. 2021). Enhancing our indulgent of these adaptive reactions is vital for maintaining resilient forests and urban green spaces in the expression of climate change and environmental degradation. This chapter explores plant responses to abiotic stress and the mechanisms and physiological adaptations that allow tolerance in plant species.

12.1.1 Importance of Studying Model Tree Plants

Revising abiotic stress in model tree plants is essential for numerous reasons. First, it improves our understanding of the vital biological processes that trees perform to cope with adverse environmental conditions. By examining these stress responses, researchers can identify significant genes, proteins, and metabolic pathways involved in resilience. This knowledge is priceless for developing strategies to develop the tolerance of economically and ecologically important tree species (Sharma and Dietz 2006; Mann et al. 2023; Banyal et al. 2018). Second, model tree plants serve as proxies for more complex and less-studied species, providing insights that can be applied across various forest ecosystems. This approach fosters the development of more robust forests, better equipped to endure the escalating impacts of climate change, including extended droughts, extreme heat, and rising soil salinity. Third, investigating how model tree species respond to abiotic stress can significantly enhance urban planning and management strategies. Urban trees encounter distinct challenges such as pollution, restricted water resources, and compacted soils. By studying how model tree species cope with these conditions, we can make more informed choices about which species to plant and how to manage them effectively, promoting healthier and more sustainable urban forests. Additionally, the knowledge derived from these studies can inform practical solutions like advanced breeding techniques, genetic modifications, and biotechnological advancements.

Research into how plants handle abiotic stress has drawn on a diverse array of both model and nonmodel species. Pioneering studies with model organisms such as Arabidopsis, maize, and bryophytes like *Physcomitrium patens* have been instrumental in uncovering the biochemical, genetic, molecular, and physiological mechanisms of stress resilience (Rensing et al. 2020; Yu et al. 2020). Halophytes such as *Mesembryanthemum cristallinum* and *Thellungiella halophila*, resurrection plants like *Craterostigma plantagineum* and *Selaginella lepidophylla* (Pampurova and van Dijck 2014; Giarola and Bartels 2015; Oh et al. 2015; Himabindu et al. 2016; Ríos-Meléndez et al. 2021), and major crops including barley, potato, soybean, common bean, and tomato (Bitocchi et al. 2017; Rothan et al. 2019) have all played crucial roles in revealing mechanisms of stress tolerance. Advanced techniques such as genome sequencing, gene annotation, functional omics, and gene validation in these species have been pivotal in deepening our understanding of plant stress responses. Advancements in enhancing abiotic stress tolerance in plants have been achieved through a variety of methods. Traditional techniques such as induced mutations and inter-generic or inter-specific crosses have proven effective. Molecular and biotechnological innovations, including quantitative trait loci (QTL) mapping, marker-assisted selection, haplotype analysis, genome-wide association studies, genetic engineering, and genome/epigenome editing, have further expanded our capabilities. The incorporation of beneficial microbes has also contributed to these efforts (Villalobos-López et al. 2022). Nonetheless, to attain lasting stress tolerance, it is essential to adopt a holistic approach that considers the unique physiology, ecology, and breeding traits of each plant species within their natural environments.

12.2 Mechanisms and Physiological Adaptations for Abiotic Stress Tolerance

12.2.1 Ionic Stress

Soil salinity represents a major global concern, impacting over one billion hectares of land (Montanarella et al. 2015). This environmental issue drastically hampers crop growth and yields. Research by Munns and Tester (2008) highlights that the ratio of sodium to potassium is crucial in assessing a plant's salt tolerance. Plants have developed strategies to minimize the uptake of detrimental sodium ions, thus alleviating some of the negative impacts of salinity. The productivity of agriculture is profoundly influenced by the osmotic and ion-specific effects present in saline soils (Yadav et al. 2007; Munns and Tester 2008).

Plants primarily address salt stress through ion transporters embedded in their cell membranes. For instance, Na^+/H^+ antiporters are essential for either expelling excess sodium (Na^+) from cells or sequestering it in vacuoles to reduce its harmful effects. This mechanism is crucial for maintaining a proper sodium-to-potassium (K^+) balance within the cell. In trees, however, salt stress can diminish osmotic potential, often leading to what is known as physiological drought, which can ultimately be fatal to the plant. Furthermore, high concentrations of Na^+ and Cl^- ions can cause ionic toxicity, triggering secondary oxidative stress and damaging cellular organelles, which disrupts tree growth and development (Qi et al. 2018; Zhang et al. 2023). Plants employ three basic strategies to enhance their salt tolerance (Fig. 12.1):

1. Salt avoidance or ion selectivity: This strategy involves restricting the absorption of detrimental ions to prevent their accumulation in the plant.
2. Salt accumulation or ion sequestration: Storing excess ions in vacuoles to prevent cellular damage.
3. Maintaining ionic and osmotic balance: Stabilizing Na^+/K^+ levels and managing cellular water content.

Elevated ion levels from excessive uptake can disrupt the nutrient balance within plants. Salt tolerance mechanisms are generally divided into two main categories: avoidance and tolerance. Avoidance involves limiting the absorption of salt ions, while tolerance encompasses the sequestration of these ions within the plant's cells (Greenway and Munns 1980). Effective management of salt stress also relies on the plant's ability to perceive, integrate, and process stress signals to regulate cellular functions and development. Beyond these fundamental mechanisms, plants employ a range of adaptive strategies to thrive in saline environments. These strategies include ion compartmentalization, osmotic adjustment, osmolyte production, succulence development, and selective ion uptake and transport (Koyro et al. 2016; Banyal et al. 2023; Mann et al. 2023).

Some plants have evolved specialized structures, such as salt glands, to excrete excess salts and protect their cells from damage. Examples include species like

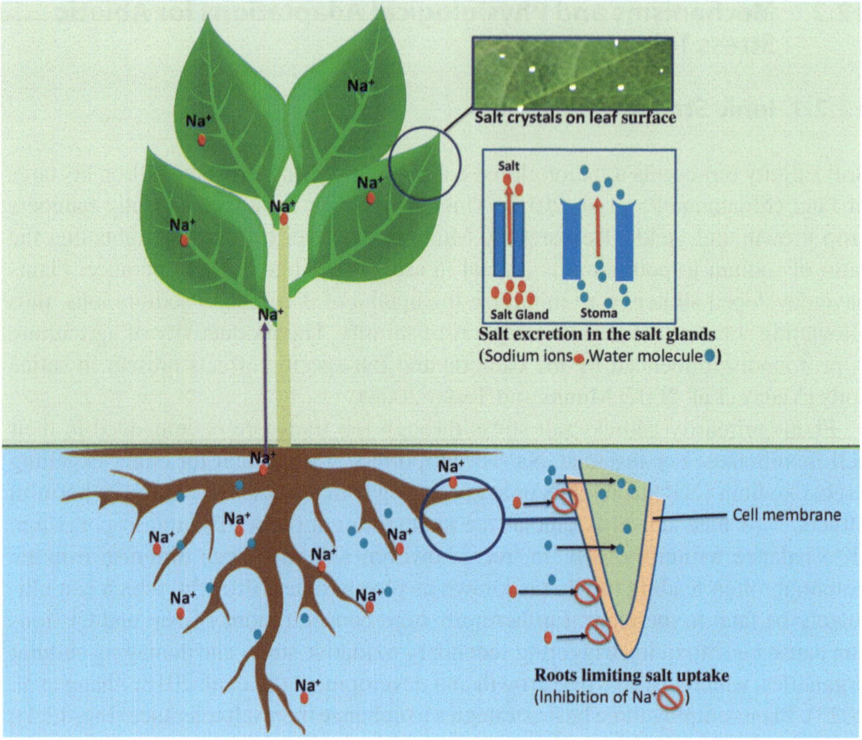

Fig. 12.1 Ion selectivity, avoidance, and sequestration to make ionic and osmotic balance as tree adaptive mechanism under stress condition

Limonium spp., *Tamarix* spp., and *Acacia ampliceps* (salt wattle). Other species, such as *Salvadora persica* and *Suaeda nudiflora*, have developed the ability to accumulate high salt concentrations in their vacuoles without compromising their physiological functions and growth, thus allowing them to survive in extremely saline conditions (Banyal et al. 2017, 2018).

12.2.2 Osmotic Stress (Osmoregulation)

Abiotic stresses such as high salinity and drought impose significant hyperosmotic challenges on plants, threatening their growth and survival (Munns et al. 2020; Mann et al. 2020). To counter these extreme conditions, plants engage in osmoregulation, a process where they elevate the concentration of solutes within their cells. This increase in solutes helps to lower the osmotic potential, thereby enhancing their ability to retain water more effectively (Gong et al. 2020). The solutes involved in osmoregulation are categorized into two main groups: organic and inorganic osmolytes. Organic solutes include sugars like sucrose, sugar alcohols such as mannitol and sorbitol, amino acids like proline and glycine betaine, and organic acids such as malate and citrate. Inorganic solutes include ions like sodium, potassium, chloride, and calcium. These osmolytes play a crucial role in stabilizing cellular structures, maintaining ion balance, and supporting metabolic and enzymatic functions under stress (Alpert 2006; Mann et al. 2019; Munns et al. 2020). Osmotic adjustment is achieved through the active transport of ions, synthesis of osmolytes, and sequestration of harmful ions into vacuoles. This comprehensive approach helps maintain cell turgor and overall stress resilience.

The impact of salinity on plants mirrors the effects of drought stress. High soil salinity diminishes a plant's ability to take up water due to an osmotic effect, which also impairs the absorption of crucial nutrients like potassium (K^+), calcium (Ca^{2+}), and nitrate (NO_3^-) while leading to the toxic buildup of sodium (Na^+) and chloride (Cl^-) in the plant's upper parts. To counteract these effects, plants must effectively regulate osmotic balance to maintain turgor pressure-vital for cell structure and function. Excessive salt in the soil reduces water potential, making water uptake challenging and resulting in cellular dehydration as the plant struggles to maintain its osmotic potential. In response, plants accumulate small molecular compounds such as glycine betaine and proline to enhance their water retention capacity (Zhang et al. 2020). As drought conditions worsen, proline levels in plants increase significantly, though this rate of increase slows over time. Initially, soluble protein content declines but later rises sharply once soil moisture falls below 31.38%, while soluble sugars continue to rise (Kishor et al. 2005). Osmotic stress also stimulates the production of lipid signals like sphingolipids and phosphatidic acid, which are integral to the plant's stress response (Hou et al. 2016; Zhang et al. 2023). Furthermore, stress conditions prompt the synthesis of osmoregulatory proteins that lower the osmotic potential within cells, helping to prevent dehydration and improve the plant's resilience to osmotic stress.

Osmotic stress triggers a swift response from the SnRK2 protein kinase family, which is crucial for enhancing plant tolerance to such stress (Maszkowska et al. 2021). During salt stress, early osmotic signals play a significant role in improving a plant's resilience to high salinity and drought by reducing stomatal openings, thereby conserving water (Lizana et al. 2006). Additionally, these signals help regulate stomatal movement in response to fluctuating light conditions, optimizing water usage by enabling quicker closure of stomata when light intensity transitions from high to low (Zhang et al. 2022). To mitigate ion toxicity, plants adeptly manage sodium (Na^+) and chloride (Cl^-) ions by sequestering them in vesicles and incorporating them into organic solutes. Nonetheless, plants exposed to high salinity may need to adapt their cell wall structures or water-salt transport mechanisms to achieve more energy-efficient osmoregulation (Zhang et al. 2023).

12.2.3 Oxidative Stress

Under abiotic stress conditions, plants often face an excessive buildup of reactive oxygen species (ROS), which can cause considerable cellular damage if not effectively managed. ROS, including singlet oxygen ($^1O2$), superoxide anion (O2−),

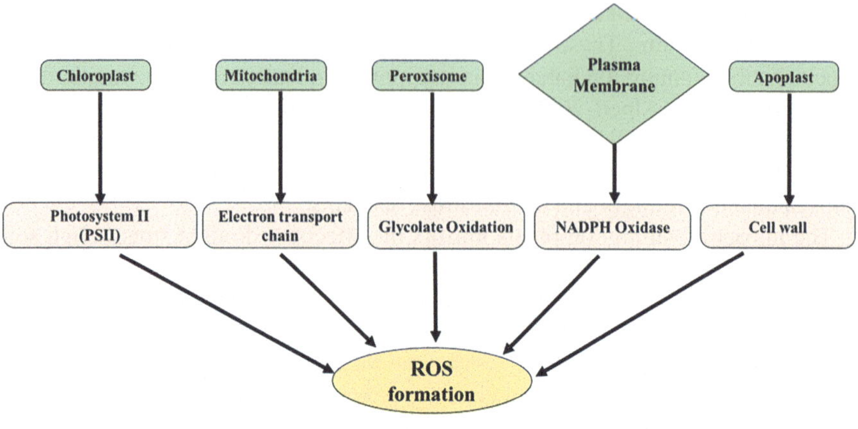

Fig. 12.2 The main sites of reactive oxygen species (ROS) production within cells include chloroplasts, mitochondria, peroxisomes, the apoplast, and plasma membranes. ROS are primarily generated through the electron transport processes in chloroplasts, mitochondria, and the plasma membrane, as well as through various metabolic pathways distributed across different cellular compartments. (Adapted from Sharma et al. 2012)

hydrogen peroxide (H_2O_2), and the highly reactive hydroxyl radical (OH), are reactive molecules derived from oxygen (Foyer et al. 1997). These species are produced naturally in various cellular compartments such as chloroplasts, mitochondria, plasma membranes, and microbodies like peroxisomes and microsomes (see Fig. 12.2). While ROS are a normal byproduct of aerobic metabolism, their levels can surge significantly due to disturbances in cellular homeostasis triggered by environmental stresses like drought, salinity, and metal toxicity (Shah et al. 2016; Mittler 2002; Sharma and Dubey 2005; Sharma and Dubey 2007; Maheshwari and Dubey 2009; Mann et al. 2024a). When ROS levels surpass the cell's defense mechanisms, the plant is said to be experiencing "oxidative stress."

Excessive accumulation of reactive oxygen species (ROS) under stress conditions can lead to oxidative stress, a condition where the equilibrium between ROS generation and the plant's antioxidant defenses is disrupted. This imbalance causes cellular damage through lipid peroxidation, protein oxidation, nucleic acid injury, enzyme inhibition, and the initiation of programmed cell death pathways (Sharma et al. 2012; Hasanuzzaman et al. 2020). Although ROS are often associated with harmful effects, they also serve as vital signaling molecules, acting as second messengers in various cellular processes, including the plant's response to abiotic stresses (Yan et al. 2007). The dual role of ROS—either as agents of damage or as signaling entities-hinges on the precise balance between their production and scavenging mechanisms (Dumanović et al. 2021).

Under normal conditions, there is a balance between ROS production and their removal by antioxidant systems, allowing ROS to act as signaling molecules without causing damage (Fig. 12.3). However, abiotic stresses can disrupt this equilibrium, leading to an overproduction of ROS and resulting in oxidative damage. In

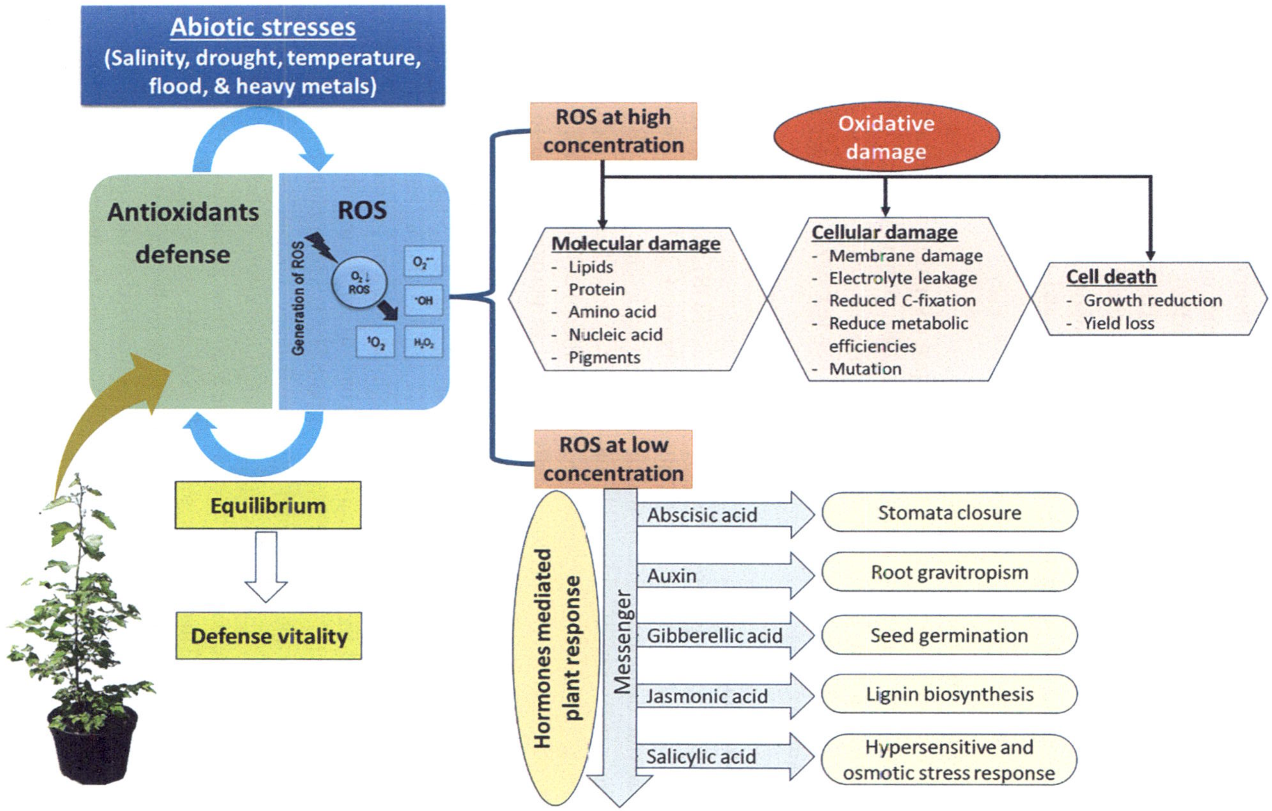

Fig. 12.3 Reactive oxygen species (ROS) such as singlet oxygen ($^1O_2$), superoxide anions (O_2^-), hydrogen peroxide (H_2O_2), and hydroxyl radicals (OH) are produced in response to environmental stress in plants. These ROS act as secondary messengers in plant hormone signaling, aiding in stress response and adaptation. However, an excess of ROS can lead to oxidative damage and cell death if not controlled. The plant's antioxidant defense system, including both enzymatic and nonenzymatic components, is essential for detoxifying ROS and mitigating oxidative stress. (Adapted from Sharma et al. 2012; Hasanuzzaman et al. 2020)

response, the plant's antioxidant defense system is crucial for neutralizing excess ROS and restoring balance, thereby protecting cellular integrity and supporting survival under stress. This defense system includes both enzymatic and nonenzymatic components. Key enzymatic antioxidants are superoxide dismutase (SOD), catalase (CAT), and guaiacol peroxidase (GPX), along with enzymes in the ascorbate-glutathione (AsA-GSH) cycle such as ascorbate peroxidase (APX), monodehydroascorbate reductase (MDHAR), dehydroascorbate reductase (DHAR), and glutathione reductase (GR) (Noctor and Foyer 1998). Nonenzymatic antioxidants, including ascorbate (AsA), glutathione (GSH), carotenoids, tocopherols, and phenolic compounds, are also essential for scavenging surplus ROS and mitigating oxidative stress.

Components of the Antioxidant Defense System

To counteract oxidative damage, plants have developed a sophisticated antioxidant defense system that integrates both enzymatic and nonenzymatic components (Hasanuzzaman et al. 2020). This multifaceted system involves various types of antioxidants working together to neutralize excess reactive oxygen species (ROS) and mitigate oxidative stress.

Enzymatic Antioxidants

play a crucial role in scavenging reactive oxygen species (ROS) produced in various subcellular compartments under stress conditions (Mishra et al. 2023). Key enzymes include superoxide dismutase (SOD), which initiates the defense against ROS by catalysing the conversion of superoxide radicals (O_2^-) into hydrogen peroxide (H_2O_2) and oxygen (O_2), thereby establishing the first line of defense against oxidative damage. Catalase (CAT), predominantly located in peroxisomes, detoxifies H_2O_2 efficiently by converting it into water and oxygen, a process essential for managing the byproducts of metabolic processes and protecting cells from oxidative stress (Mishra et al. 2023). Furthermore, peroxidases (POD), such as guaiacol peroxidase (GPX) and ascorbate peroxidase (APX), utilize reducing equivalents to convert H_2O_2 into water. APX, found in various cellular locations including chloroplasts, mitochondria, and the cytosol, uses ascorbate as an electron donor in the ascorbate-glutathione cycle, while GPX reduces both H_2O_2 and organic hydroperoxides, protecting membrane lipids from oxidative damage. Glutathione reductase (GR) is critical for regenerating reduced glutathione (GSH) from its oxidized form (GSSG), maintaining necessary GSH levels for the ascorbate-glutathione cycle and other redox processes. This process relies on NADPH as an electron donor. Monodehydroascorbate reductase (MDHAR) and dehydroascorbate reductase (DHAR) are vital for regenerating ascorbate from its oxidized forms, serving as key players in the ascorbate-glutathione cycle, which is crucial for detoxifying H_2O_2 and sustaining antioxidant defenses. Collectively, these enzymes form a comprehensive antioxidant defense system that safeguards plants from oxidative damage and ensures cellular health under stress conditions (Mittler 2002; Hasanuzzaman et al. 2020; Dumanović et al. 2021). In addition to enzymatic components, non-enzymatic antioxidants are equally crucial in regulating plant growth, development, and stress responses. Ascorbic acid (AsA), or vitamin C, acts as a primary low-molecular-weight antioxidant that scavenges ROS directly and functions as a key component of the ascorbate-glutathione cycle, maintaining oxidative balance. Glutathione (GSH), synthesized in the cytosol and chloroplasts, is essential for detoxifying ROS and regenerating ascorbate within the same cycle. Carotenoids, including β-carotene, lutein, and zeaxanthin, protect chlorophyll and cellular structures by quenching singlet oxygen and dissipating excess energy as heat. Tocopherols, particularly α-tocopherol (vitamin E), safeguard membrane lipids from peroxidation by scavenging lipid peroxyl radicals and stabilizing cell membranes. Phenolic compounds and flavonoids, as secondary metabolites, offer additional protection by scavenging ROS, chelating metal ions, and inhibiting ROS-producing enzymes in the vacuole and apoplast (Hasanuzzaman et al. 2020; Dumanović et al. 2021). The ascorbate-glutathione cycle exemplifies the plant's antioxidant defense system, involving enzymes such as APX, GR, MDHAR, and DHAR, as well as non-enzymatic antioxidants like ascorbate and glutathione. This cycle detoxifies hydrogen peroxide (H_2O_2) and regenerates antioxidants, ensuring ongoing protection against oxidative stress. Maintaining redox homeostasis through NADPH, which provides the reducing power for antioxidant regeneration, is fundamental for sustaining cellular functions and managing stress responses (Mittler 2002; Dumanović et al. 2021).

Role of the Antioxidant Defense System Under Abiotic Stress
Under stressful conditions, plants upregulate their antioxidant defense systems to curb excessive reactive oxygen species (ROS) accumulation and mitigate oxidative damage. The equilibrium between ROS production and antioxidant activity is pivotal in determining the extent of oxidative stress and the plant's capacity to endure and adapt to adverse environments. Increased antioxidant activity often correlates with enhanced stress tolerance in plants. Controlled ROS levels can trigger stress-responsive genes and pathways, while robust antioxidant systems manage ROS concentrations to avert cellular damage (Zhang et al. 2018; Szymańska et al. 2019).

The upregulation of key antioxidant enzymes such as superoxide dismutase (SOD), catalase (CAT), and peroxidase (POD) is essential for detoxifying ROS and safeguarding cells from oxidative stress (Zhang et al. 2018; Szymańska et al. 2019). Concurrently, heat shock proteins (HSPs) and other molecular chaperones stabilize and refold denatured proteins, and proteases degrade damaged proteins, all contributing to cellular protection. Additionally, modifications in cell wall composition enhance cell rigidity, which helps shield against mechanical damage induced by stress.

Changes in primary and secondary metabolic processes, including photosynthesis, respiration, and the synthesis of protective compounds, also play a crucial role in helping plants manage environmental stress (Shinozaki and Yamaguchi-Shinozaki 2000; Wan et al. 2002; Zhu 2002; Zhu 2016). Furthermore, research indicates that maintaining a high antioxidant capacity is closely linked to improved tolerance to abiotic stresses (Zaefyzadeh et al. 2009; Chauhan et al. 2023). Efficient ROS management not only prevents cellular damage but also bolsters the plant's resilience to challenging environmental conditions.

12.2.4 Stress Perception and Signaling Pathways

ROS Signaling

Reactive oxygen species (ROS), such as superoxide anions, hydrogen peroxide (H_2O_2), hydroxyl radicals, and singlet oxygen, are pivotal in how plants manage various abiotic stresses (Sharma et al. 2012; Hasanuzzaman et al. 2020; Qi et al. 2018). These molecules are produced in organelles like chloroplasts, mitochondria, and peroxisomes and serve crucial roles in stress signaling pathways (Sharma et al. 2012; Hasanuzzaman et al. 2020; Qi et al. 2018). They can trigger specific calcium influxes and electrical signals that modulate the plant's response to stress.

Among the ROS, hydrogen peroxide (H_2O_2) is particularly significant as a secondary messenger in multiple plant hormone signaling pathways. It interacts with hormones such as abscisic acid (ABA), salicylic acid (SA), jasmonic acid (JA), and ethylene to regulate processes like stomatal closure, defense mechanisms, and the activation of stress-responsive genes (Sharma et al. 2012). By influencing these hormonal pathways, ROS help orchestrate the plant's protective responses to stress, and in severe cases, can even initiate programmed cell death to prevent further damage.

Hormonal Signaling

Plants navigate environmental stresses like high salinity and drought through an intricate hormonal signaling network. Central to this adaptive response are abscisic acid (ABA) and ethylene, two crucial hormones that orchestrate the plant's strategy to cope with these adverse conditions. ABA primarily manages water stress and enhances drought resistance, while ethylene coordinates various stress responses, including growth adjustments and defense mechanisms. Together, these hormones ensure a well-regulated and effective response to environmental challenges.

Abscisic Acid (ABA)

Abscisic acid (ABA) stands out as a pivotal hormone in how plants manage abiotic stresses, especially drought and high salinity. Its influence extends across key processes such as stomatal closure, gene regulation, and osmotic adjustment, all aimed at preserving water and ensuring cellular stability (Kavar et al. 2007; Fukao et al. 2011; Syed et al. 2015). When faced with stress, plants swiftly ramp up ABA production, initiating a cascade of protective responses. This hormone effectively orchestrates a series of adaptive mechanisms that bolster the plant's resilience and maintain equilibrium under challenging conditions.

- **Activation of SnRKs:** Elevated levels of abscisic acid (ABA) trigger the activation of the SnRK (Sucrose Nonfermenting1-Related Protein Kinase) family, with a particular focus on SnRK2s. These kinases are crucial for maintaining ionic balance, managing water regulation (osmoregulation), modulating reactive oxygen species (ROS) production, and driving gene expression. SnRK2s also play a vital role in stomatal closure by influencing ion channels within guard cells, thereby reducing water loss during periods of stress.
- **ABA-Dependent Signaling:** The signaling pathways driven by ABA are fundamental for plant adaptation to drought and saline conditions. Key players in these pathways include ABA receptors such as RCAR (Regulatory Component of ABA Receptor), PYR1 (Pyrabactin Resistance 1), and PYL (PYR1-Like) proteins, which together comprise the PYL receptor family (Zhang et al. 2020). Upon stress-induced ABA synthesis, ABA binds to the hydrophobic pocket of the PYL START domain. This binding induces a conformational shift that promotes interaction with protein phosphatase type 2C (PP2C). The resulting inhibition of PP2C allows SnRK2 kinases to become active, thus regulating the expression of genes associated with stress responses (Zhang et al. 2020).
- **Energy Management:** During stress conditions, ABA aids in energy management by facilitating the breakdown of starch into sugars. These sugars act as osmolytes, helping to maintain cellular balance and stability. Moreover, ABA collaborates with calcium ion (Ca^{2+}) signaling to bolster the plant's defense mechanisms. This partnership enhances the regulation of ROS, preventing potential cellular damage and supporting the plant's resilience under stress.

Ethylene

Ethylene plays a complex and sometimes debated role in plant stress responses, particularly under salt stress. Its impact on growth and development is shaped by interactions with other hormones and signaling pathways (Drew et al. 1979; Visser et al. 1997; Riyazuddin et al. 2020). In response to salt stress, plants often ramp up ethylene production, which can enhance their tolerance through several mechanisms:

- **Ionic Balance Regulation:** Ethylene contributes to maintaining a stable ionic balance by modulating the levels of sodium (Na^+) and potassium (K^+). This adjustment helps alleviate the detrimental effects of salt stress on plant cells.
- **Management of Reactive Oxygen Species (ROS):** Ethylene influences ROS levels by promoting antioxidant defenses, thereby shielding plant cells from oxidative damage. Additionally, ethylene boosts the assimilation of vital nutrients like nitrates and sulfates, further aiding in stress management.

The role of ethylene in salt stress is not uniform across all plant species. For instance, in Arabidopsis, ethylene has a positive effect on salt tolerance, with its production and signaling pathways being integral to an effective stress response. Conversely, in rice, ethylene-related genes exhibit a contrasting effect; suppressing these genes can improve salt tolerance, whereas their overexpression tends to increase sensitivity to salt (Tao et al. 2015; Cebrián et al. 2021). This species-specific variation highlights the nuanced role ethylene plays in stress adaptation.

Other Hormones Salicylic Acid (SA), Jasmonic Acid (JA), and Gibberellins (GAs) also play significant roles in plant responses to stress by influencing gene expression and metabolic adjustments (Clarke et al. 2009; Zhang and Wang 2011; Dubois et al. 2018).

- **Gibberellins (GAs):** Plants exhibiting reduced levels of GAs or impaired GA signaling pathways often show heightened resistance to salt stress. In rice, for example, enhancing the degradation of GA-related genes has been linked to improved salt tolerance. This suggests that manipulating GA levels can be a strategic approach to bolster stress resilience.
- **Salicylic Acid (SA):** SA contributes to the regulation of plant responses under salt stress by interacting with other hormones such as ABA, ethylene, and GA. This interaction helps in maintaining ionic balance and managing osmotic stress in saline environments.

Mitogen-Activated Protein Kinase (MAPK) Cascades MAPK cascades are essential for transmitting stress signals from the cell membrane to the nucleus. These cascades involve a sequence of phosphorylation events that activate specific transcription factors, leading to the expression of genes responsive to stress (Hamel et al. 2006; Gao et al. 2008). By integrating diverse stress signals, MAPK pathways enable plants to mount effective responses to a range of environmental challenges.

12.2.5 Stress-Responsive Genes and Regulation Under Abiotic Stress

The regulation of abiotic stress responses in plants is a layered process involving transcriptional, translational, posttranslational, and epigenetic mechanisms. At the transcriptional level, plants adjust gene expression to manage stress, while translational regulation controls protein synthesis to respond efficiently. Posttranslational modifications fine-tune protein activity and stability, offering rapid adjustments. Epigenetic changes, like DNA methylation, enable plants to "remember" stress, improving future responses. Together, these mechanisms help plants adapt and survive in challenging environments.

Transcription Factors (TFs)

Transcription factors (TFs) are pivotal in regulating stress-responsive genes, adjusting plant gene expression in response to environmental challenges. Key transcription factors involved in abiotic stress responses include DREB, NAC, MYB, and bZIP proteins. Activated by upstream signaling pathways, these TFs bind to specific DNA sequences, either triggering or suppressing the expression of stress-related genes.

- **DREB Proteins**: As part of the AP2/ERF family, DREB transcription factors are central to managing stress from drought, salinity, and cold. By binding to dehydration-responsive elements (DREs) in target gene promoters, DREB TFs activate genes responsible for osmotic regulation, antioxidant protection, and other defensive measures essential for stress adaptation (Fowler and Thomashow 2002; Yamasaki et al. 2004).
- **NAC Proteins**: The NAC transcription factor family governs responses to drought, salinity, and pathogens. These proteins regulate genes involved in water retention, antioxidant production, and secondary metabolite biosynthesis, all of which enhance the plant's defense mechanisms against environmental stress (Sun et al. 2013; Srivastava and Verma 2017).
- **MYB Proteins**: MYB TFs regulate pathways linked to flavonoid biosynthesis, ROS scavenging, and secondary metabolism. They are crucial for managing drought and cold stress, influencing the expression of genes that support plant resilience under adverse conditions (Lee and Young 2013; Rienzo and Casamassimi 2016).
- **bZIP Proteins**: Basic leucine zipper (bZIP) transcription factors play a significant role in abscisic acid (ABA) signaling pathways, which are key in the plant's response to drought and salinity. By binding to ABA-responsive elements (ABREs) in gene promoters, bZIP TFs regulate the transcription of genes necessary for stress survival (Singer et al. 2016).

Each of these transcription factor families orchestrates distinct yet interconnected genetic responses, fine-tuning the plant's ability to endure and adapt to hostile environmental conditions.

Transcriptional Regulation

Transcriptional regulation plays a crucial role in enabling plants to adapt to abiotic stress by controlling the expression of genes vital for stress tolerance. Stress signals activate specific transcription factors (TFs) like DREB, AREB, and NAC, which bind to the promoters of stress-responsive genes, adjusting their expression. This process leads to the production of essential molecules such as osmolytes, heat shock proteins (HSPs), antioxidants, and other protective compounds (Zhang et al. 2023). Interestingly, various abiotic stresses, including cold, hypertonic environments, and high salinity, can activate overlapping transcriptional responses. These shared responses involve the upregulation of genes associated with MAPK cascades, calcium signaling, ROS regulation, programmed cell death (apoptosis), and protein degradation, all of which play key roles in stress survival and resilience (Zhang et al. 2023).

Promoters and Cis-Acting Elements Cis-acting elements, which are specialized DNA sequences in gene promoters, serve as binding sites for transcription factors and are essential for the precise control of gene expression under abiotic stress conditions. These elements play a critical role in fine-tuning plant responses to environmental changes. Common stress-responsive cis-acting elements include:

- **ABRE (ABA-Responsive Element)**: Central to ABA-driven stress responses, particularly in the context of drought and salinity stress.
- **DRE (Dehydration-Responsive Element)**: A key regulator of genes responding to dehydration, cold, and high salinity.
- **HSE (Heat Shock Element)**: Crucial for the activation of heat shock proteins during heat stress (Zhang 2003; Espasandin et al. 2014).

Together, these cis-acting elements and their interacting transcription factors form an intricate regulatory network that equips plants with the ability to adapt to a range of abiotic stresses, ensuring their survival in challenging environments.

Mechanisms of Transcriptional Regulation

The regulation of gene expression in plants under environmental stress is a highly complex and multifaceted process involving transcriptional control, epigenetic modifications, and the action of noncoding RNAs (Babu et al. 2004; Gomez-Mena et al. 2005). When plants are exposed to stressors like drought, salinity, or extreme temperatures, they activate intricate signaling pathways that lead to the activation or suppression of specific transcription factors (TFs). These TFs then move to the nucleus, where they bind to cis-acting elements in the promoters of stress-responsive genes, either triggering or repressing their transcription (Chandler et al. 2009).

Once bound to DNA, these TFs recruit co-factors that can alter chromatin structure, modifying its accessibility to the transcriptional machinery. This dynamic interaction determines whether RNA synthesis is promoted or suppressed, based on the type of stress and environmental conditions the plant is facing. Epigenetic mechanisms, such as DNA methylation, play a significant role in this regulation by modifying the chromatin landscape. DNA methylation, typically occurring on cytosine residues, can silence gene expression by either blocking transcription factor binding or recruiting proteins that condense chromatin, making it inaccessible for transcription (Aukerman and Sakai 2003; Lee et al. 2008).

Histone modifications, such as acetylation, methylation, and phosphorylation, further shape chromatin structure and influence gene expression. For example, histone acetylation loosens chromatin, thereby facilitating gene transcription, while other modifications can either activate or repress gene expression depending on the context.

Noncoding RNAs, particularly microRNAs (miRNAs) and small interfering RNAs (siRNAs), add another layer of regulation. miRNAs can suppress gene expression by targeting mRNAs for degradation or inhibiting their translation. siRNAs, on the other hand, guide the RNA-induced silencing complex (RISC) to complementary mRNAs, resulting in mRNA degradation and reduced gene expression (Hurtado et al. 2006; Wang et al. 2009).

Together, these mechanisms allow plants to fine-tune their response to environmental stress, ensuring that gene expression is tightly regulated for survival and adaptation in ever-changing conditions.

Examples of Transcriptional Regulation in Abiotic Stress Tolerance

Transcriptional regulation plays a critical role in bolstering plant tolerance to a wide array of abiotic stresses by orchestrating the activation of specific genes. During drought stress, for example, transcription factors like DREB2A bind to dehydration-responsive element (DRE) sequences, triggering the expression of genes involved in osmoprotection. These genes include those encoding late embryogenesis abundant (LEA) proteins and dehydrins, which help plants maintain cellular stability under water-deficit conditions (Agarwal et al. 2006; Augustine et al. 2015). Another key player in drought tolerance is the MYB96 transcription factor, which enhances plant resilience by regulating the biosynthesis of cuticular wax, effectively minimizing water loss through transpiration (Lee et al. 2019).

Under salinity stress, transcription factors like bZIP60 play an essential role in the unfolded protein response (UPR) pathway, which helps maintain protein stability and homeostasis in high-salinity environments (Iwata et al. 2008). Meanwhile, NAC67 is responsible for regulating genes that oversee ion transport and osmolyte biosynthesis, enabling plants to better manage ionic stress associated with excessive salt (Gonzalez 2016; Javed et al. 2020).

Temperature stress activates another set of transcriptional responses. Heat shock factors (HSFs), for example, trigger the production of heat shock proteins (HSPs), which function as molecular chaperones. These HSPs play a protective role by stabilizing and refolding denatured proteins, thus safeguarding cellular functions during extreme heat (Zhang et al. 2015; Guo et al. 2016). In cold stress conditions, the C-repeat Binding Factor (CBF) family of transcription factors binds to C-repeat/dehydration-responsive elements, initiating the expression of genes that facilitate cold acclimation. These genes often encode antifreeze proteins that help plants cope with freezing temperatures (Guy et al. 2008).

Through these finely tuned transcriptional networks, plants are able to activate a wide variety of stress-responsive genes, enabling them to adapt and survive in challenging environmental conditions.

Translational Regulation

Translational regulation is another crucial mechanism that enables plants to adapt to abiotic stresses by adjusting protein synthesis according to environmental challenges. During heat stress, canonical protein translation is downregulated, reallocating resources toward the production of heat shock proteins (HSPs). These HSPs serve as molecular chaperones, protecting cellular proteins from heat-induced denaturation and aggregation, thus ensuring protein stability under elevated temperatures (Zhang et al. 2023).

In cold stress scenarios, plants enhance their translational machinery to maintain stress resilience. The ribosomal biogenesis factor REIL2 is upregulated, which improves cold tolerance by promoting the processing of ribosomal RNA and stabilizing CBF protein levels, essential for cold acclimation (Zhang et al. 2023). Additionally, within chloroplasts, cold stress triggers an increased interaction between the RNA-binding protein RBD1 and 23S ribosomal RNA, a key step in the synthesis of chloroplast proteins that are vital for acclimating to lower temperatures (Zhang et al. 2023). Through these targeted translational adjustments, plants are able to efficiently allocate resources and fine-tune protein production, ensuring that key protective proteins are synthesized during environmental stresses, thereby enhancing their survival and adaptability.

Post-translational Regulation

Abiotic stress significantly impacts posttranslational modifications (PTMs), which play a key role in controlling the localization, stability, and function of proteins critical to stress adaptation. PTMs, such as phosphorylation, allow for the rapid adjustment of signaling proteins, enabling plants to quickly respond to shifting environmental conditions (Zhang et al. 2023). These modifications are not limited to signaling proteins; they also regulate the function of nonsignaling proteins crucial for stress resistance.

For instance, during osmotic stress and in response to abscisic acid (ABA) signaling, SnRK2 kinases become activated. These kinases phosphorylate a variety of target proteins, including transcription factors, transporters, and enzymes, which are integral to the plant's defense mechanisms against stress (Zhang et al. 2023). Through these targeted modifications, plants can rapidly adjust their physiological and metabolic processes, enhancing their resilience to stress.

Epigenetic Regulation

Plants demonstrate extraordinary adaptability to abiotic stresses through epigenetic mechanisms, such as DNA methylation and histone modifications (Chinnusamy et al. 2013). DNA methylation can either silence or adjust the expression of stress-responsive genes, allowing plants to precisely modulate their reaction to environmental stresses. Similarly, histone modifications like acetylation and methylation reshape chromatin architecture, affecting the accessibility of genes and their expression. These epigenetic tools offer plants the flexibility to fine-tune their stress responses, enabling them to better cope with and survive in hostile environmental conditions (Chinnusamy et al. 2013; Dahiya et al. 2023; Kumar et al. 2023). By dynamically adjusting gene regulation, epigenetic changes serve as a powerful mechanism for optimizing resilience against abiotic stresses.

12.2.6 Water Use Efficiency

Water use efficiency (WUE) is a vital physiological trait that measures how efficiently a plant utilizes water to generate biomass or yield. Its significance becomes particularly evident under abiotic stresses like drought, where water is scarce. Improving WUE is a crucial approach for boosting agricultural productivity and promoting sustainability, especially in arid and semi-arid regions where water resources are limited (Siahpoosh and Dehghanian 2012; Nagore et al. 2017). The concept of WUE encompasses several factors, including the balance between water absorption and loss, and its influence on growth and crop output. Understanding the mechanisms that govern WUE, such as stomatal regulation, photosynthetic efficiency, and root architecture, can inform strategies to enhance this trait in plants. Various approaches, including breeding, genetic modification, and agronomic practices, aim to improve WUE, offering potential solutions for mitigating the impacts of water scarcity on agriculture.

Types and Measurement of WUE

Water use efficiency (WUE) is a pivotal trait in determining plant survival and productivity, especially in environments with constrained water resources. WUE is divided into various categories, each shedding light on different dimensions of water utilization:

- **Intrinsic Water Use Efficiency (iWUE)**: This metric represents the ratio of photosynthetic carbon fixation to stomatal conductance. It reflects a plant's ability to assimilate carbon dioxide efficiently relative to its water loss through stomata (Condon et al. 2002; Song et al. 2015).
- **Instantaneous Water Use Efficiency (WUEi)**: This measures the ratio of photosynthesis to transpiration rates at a given moment, providing a real-time snapshot of a plant's water efficiency (Condon et al. 2002; Song et al. 2015).
- **Whole-Plant Water Use Efficiency**: This integrates the total biomass produced relative to the total water consumed over the plant's growing season, accounting for all physiological processes involved in plant growth (Condon et al. 2002).
- **Agronomic Water Use Efficiency (WUEa)**: This is defined as the ratio of crop yield to the total water used, including both natural rainfall and irrigation. It offers a practical evaluation of water efficiency in agricultural settings (Condon et al. 2002; Builes et al. 2011; Song et al. 2015).

These varied measures of WUE provide comprehensive insights into how plants and crops utilize water, helping to develop strategies for improving water management and enhancing productivity in water-limited environments.

Significance of WUE

Water use efficiency (WUE) is crucial for enhancing drought resilience and promoting sustainable agricultural practices. High WUE allows plants to maintain growth and yield even under limited water conditions, boosting their ability to survive and thrive during droughts. By optimizing WUE, we can reduce the need for irrigation, conserve valuable water resources, and support agriculture in regions facing water scarcity. Moreover, improving WUE is vital for adapting to the growing challenges of climate change, which is increasing both the frequency and intensity of droughts.

Enhancing WUE not only helps in mitigating the impacts of these climate-induced stressors but also plays a key role in ensuring food security by supporting more resilient and sustainable farming systems (Cowan 1982; Leakey et al. 2019). Several mechanisms influence water use efficiency (WUE) in plants, enhancing their ability to optimize water use under stress. Stomatal regulation ensures a balance between CO_2 uptake and water loss, improving WUE. C_4 and CAM photosynthetic pathways exhibit superior WUE compared to C_3 plants due to more efficient CO_2 fixation (Syvertsen et al. 1995; Terashima and Ono 2002). Adaptations in leaf morphology, like thicker cuticles, and root architecture, such as deeper roots, enhance water uptake and reduce loss. Biochemical factors like osmoprotectants help maintain photosynthetic efficiency, contributing to improved WUE under drought conditions (Soolanayakanahally et al. 2009; Tarara et al. 2011).

Examples and Applications

Improvements in water use efficiency (WUE) are demonstrated across diverse plant species and agricultural practices. Advances in breeding and genetic engineering have yielded drought-resistant crop varieties, such as certain maize and wheat cultivars, which showcase high WUE and better performance under water-stressed conditions. C4 and CAM plants, including maize, sorghum, and agave, inherently possess higher WUE, making them particularly suited for arid environments (Terashima and Ono 2002; Snyder et al. 2003; Tholen et al. 2008).

Furthermore, biotechnological innovations have significantly contributed to enhancing WUE. The transgenic overexpression of DREB transcription factors and the introduction of genes responsible for osmoprotectant synthesis have resulted in crops with improved WUE and drought resilience. These advancements offer promising solutions for agriculture facing the challenges of water scarcity (Soolanayakanahally et al. 2009; Tarara et al. 2011).

12.3 Biotechnological Approaches to Enhance Stress Tolerance

12.3.1 Genetic Engineering

Abiotic stresses like heat, drought, and salinity pose substantial challenges to the productivity of horticultural crops. Traditional breeding methods have often struggled to effectively tackle these stresses due to their complex, polygenic nature, which is frequently mapped to quantitative trait loci (QTLs). To address these challenges, it's crucial to identify candidate genes or QTLs associated with broad-spectrum tolerance to these environmental stresses.

These stresses disrupt crop growth by affecting both vegetative and reproductive stages, leading to a range of physiological, biochemical, and molecular changes that compromise cellular functions. Key issues include osmotic imbalance, ion distribution disturbances, and oxidative stress, which can result in protein denaturation. In response, plants employ a range of mechanisms involving cell signaling, transcriptional regulation, and the production of stress-related proteins, antioxidants, and osmotic solutes. These responses help maintain cellular homeostasis and repair damage (Parmar et al. 2017). Stress-sensitive plants often struggle to produce these protective compounds, leaving them more vulnerable.

Genetic engineering presents a promising solution for enhancing stress resilience by introducing genes responsible for stress protection into susceptible plants. This includes genes involved in synthesizing osmolytes (such as mannitol, glycine betaine, and proline), regulating ion and water transport (like aquaporins and ion transporters), and mediating transcriptional and signaling processes (such as MAPK and DREB1). Techniques like Agrobacterium tumefaciens-mediated gene transfer, gene gun methods, RNA Interference (RNAi), and CRISPR-Cas9-based gene editing enable the targeted introduction or silencing of these genes to boost stress resistance (Parmar et al. 2017, Mann et al, 2024b)

12.3.2 CRISPR/Cas9 and Other Gene-Editing Tools

CRISPR/Cas9 stands out as a highly precise genome-editing tool with the potential to revolutionize plant resilience against abiotic stresses like drought, heat, and salinity. This technology enables the targeted knockout of genes that hinder stress responses, thus bolstering plant resilience. Additionally, CRISPR/Cas9 can be used to activate or suppress genes involved in stress tolerance. For instance, it can enhance the synthesis of osmoprotectants or antioxidants, thereby improving a plant's ability to manage oxidative stress. Moreover, CRISPR/Cas9 facilitates the introduction of new genes or the modification of existing ones to enhance traits such as water-use efficiency, ion transport, or root architecture, which are crucial for surviving harsh environmental conditions (Razzaq et al. 2022).

RNA interference (RNAi) is another valuable technique for managing plant stress responses. By silencing stress-sensitive genes that lead to excessive production of reactive oxygen species (ROS), RNAi can mitigate oxidative damage under stress conditions. Both CRISPR/Cas9 and RNAi have demonstrated success in various crops, improving their tolerance to abiotic stresses and contributing to the development of crops that are better suited to unpredictable and extreme climate conditions (Tiwari and Rajam 2022). These innovative technologies are instrumental in advancing sustainable agriculture and ensuring food security in a changing world.

12.3.3 Identification of Quantitative Trait Loci (QTLs) and Marker-Assisted Breeding Programs

Abiotic stress tolerance in forest trees often involves complex, quantitative traits governed by multiple genes, making their analysis challenging. Quantitative trait locus (QTL) mapping addresses this complexity by associating single-gene markers with quantitative traits. The process starts with constructing pedigrees to maximize trait segregation and then using statistical analyses to link phenotypes with specific marker genotypes. Due to the high genetic diversity and extended life cycles of forest trees, QTL mapping typically uses outbred models rather than inbred lines. These models capitalize on the genetic variability of trees with contrasting phenotypes, using progeny from outcrosses as mapping populations to uncover marker-trait associations.

Once these associations are identified, they can be applied in marker-assisted selection (MAS) to streamline the selection and propagation of trees with desirable traits. This approach accelerates the breeding process, enhances genetic gains, and fosters the development of superior tree varieties better suited to environmental stresses. For instance, QTL mapping and MAS have been effectively used in species like *Sequoiadendron giganteum* and *Sequoia sempervirens* to enhance drought tolerance (De La Torre et al. 2022).

12.4 Conclusion

Enhancing plant resilience to abiotic stresses is pivotal as global environmental changes escalate challenges like salinity, drought, and extreme temperatures. Understanding how plants adapt and survive under these stresses is crucial for advancing both natural ecosystems and agricultural practices. Translating insights from model plants to diverse species and environments presents a significant challenge, as stress responses can vary widely.

A central aspect of stress response involves reactive oxygen species (ROS). While high ROS levels can cause severe damage to photosynthetic systems, especially by impairing PSII, moderate ROS production is essential for activating antioxidative defenses and signaling pathways. Balancing ROS production and scavenging is key to mitigating oxidative damage and enhancing stress tolerance.

Plants utilize a range of receptors and signaling molecules—such as ion channels, signaling proteins, phytohormones, and mitogen-activated protein kinases (MAPKs)—to detect and respond to stress. These molecular cascades activate regulatory systems that manage physiological and biochemical responses, improving photosynthetic efficiency and overall resilience under stress conditions (Chauhan et al. 2023).

Recent advances in biotechnology, such as CRISPR/Cas9 and other gene-editing technologies, offer promising methods to enhance stress tolerance by precisely modifying stress-responsive genes. However, ensuring these approaches are effective and safe in real-world conditions, where multiple stressors interact unpredictably, remains a challenge. Multi-omics approaches—integrating genomics, transcriptomics, proteomics, and metabolomics—are crucial for understanding the complex interplay of stress responses. These tools can uncover new genes and pathways involved in stress tolerance and reveal how various molecular and physiological processes are coordinated. Addressing these challenges requires a multidisciplinary approach. Collaboration among plant scientists, ecologists, agronomists, and policymakers will be essential to develop strategies that not only boost plant resilience but also maintain ecosystem health and ensure food security in an increasingly uncertain climate. Continued research into plant survival mechanisms will drive innovative solutions to the pressing environmental challenges we face.

References

Adler PB, Salguero-Gómez R, Compagnoni A, Hsu JS, Ray-Mukherjee J, Mbeau-Ache C, Franco M (2014) Functional traits explain variation in plant life history strategies. PNAS 111:740–745

Agarwal PK, Agarwal P, Reddy MK, Sopory SK (2006) Role of DREB transcription factors in abiotic and biotic stress tolerance in plants. Plant Cell Rep 25:1263–1274

Alpert P (2006) Constraints of tolerance: why are desiccation-tolerant organisms so small or rare? J Exp Biol 209:1575–1584

Anderegg WRL, Hicke JA, Fisher RA, Allen CD, Aukema J, Bentz B, Hood S, Lichstein JW, Macalady AK, McDowell N, Pan Y, Raffa K, Sala A, Shaw JD, Stephenson NL, Tague C, Zeppel M (2015) Tree mortality from drought, insects, and their interactions in a changing climate. New Phytol 208:674–683

Augustine SM, Ashwin Narayan J, Syamaladevi DP, Appunu C, Chakravarthi M, Ravichandran V, Tuteja N, Subramonian N (2015) Overexpression of EaDREB2 and pyramiding of EaDREB2 with the pea DNA helicase gene (PDH45) enhance drought and salinity tolerance in sugarcane (*Saccharum* spp. hybrid). Plant Cell Rep 34:247–263

Aukerman MJ, Sakai H (2003) Regulation of flowering time and floral organ identity by a MicroRNA and its APETALA2-like target genes. Plant Cell 15:2730–2741

Babu MM, Luscombe NM, Aravind L, Gerstein M, Teichmann SA (2004) Structure and evolution of transcriptional regulatory networks. Curr Opin Struct Biol 14:283–291

Banyal R, Rajkumar, Kumar M, Yadav RK, Dagar JC (2017) Agroforestry for Rehabilitation and Sustenance of Saline Ecologies. In: Dagar JC, Tewari VP (eds) Agroforestry: Anecdotal to Modern Science. Springer, Singapore, pp 413–454

Banyal R, Yadav RK, Sheoran P, Bhardwaj AK, Rajkumar PK, Tolia R (2018) Managing saline soils of indo-Gangetic plains with eucalyptus and Melia based agroforestry systems. Indian J Ecol 45(4):841–849

Banyal R, Saini V, Saini N, Rajkumar, Kumar M, Kuldeep (2023) Tree adaptive and reclamation mechanisms in managing waterlogged saline habitats: prospectus and impediments. In: Kumar A, Dhansu P, Mann A (eds) Salinity and drought tolerance in plants. Springer Nature, Singapore, pp 555–583

Bitocchi E, Rau D, Bellucci E, Rodriguez M, Murgia ML, Gioia T, Santo D, Nanni L, Attene G, Papa R (2017) Beans (Phaseolus ssp.) as a model for understanding crop evolution. Front Plant Sci 8

Brodribb TJ (2009) Xylem hydraulic physiology: the functional backbone of terrestrial plant productivity. Plant Sci 177:245–251

Brodribb TJ (2017) Progressing from 'functional' to mechanistic traits. New Phytol 215:9–11

Brodribb TJ, Cochard H (2009) Hydraulic failure defines the recovery and point of death in water-stressed conifers. Plant Physiol 149:575–584

Builes VHR, Porch TG, Harmsen EW (2011) Genotypic differences in water use efficiency of common bean under drought stress. Agron J 103:1206–1215

Cebrián G, Iglesias-Moya J, García A, Martínez J, Romero J, Regalado JJ, Martínez C, Valenzuela JL, Jamilena M (2021) Involvement of ethylene receptors in the salt tolerance response of Cucurbita pepo. Hortic Res 8:1–14

Chandler JW, Cole M, Flier A, Werr W (2009) BIM1, a bHLH protein involved in brassinosteroid signalling, controls *Arabidopsis* embryonic patterning via interaction with DORNROSCHEN and DORNROSCHEN-LIKE. Plant Mol Biol 69:57–68

Chauhan J, Prathibha M, Singh P, Choyal P, Mishra UN, Saha D, Kumar R, Anuragi H, Pandey S, Bose B, Mehta B, Dey P, Dwivedi KK, Gupta NK, Singhal RK (2023) Plant photosynthesis under abiotic stresses: damages, adaptive, and signaling mechanisms. Plant Stress 10:100296

Chinnusamy V, Dalal M, Zhu J-K (2013) Epigenetic regulation of abiotic stress responses in plants. In: Plant abiotic stress. Wiley, pp 203–229

Choat B, Jansen S, Brodribb TJ, Cochard H, Delzon S, Bhaskar R, Bucci SJ, Feild TS, Gleason SM, Hacke UG, Jacobsen AL, Lens F, Maherali H, Martínez-Vilalta J, Mayr S, Mencuccini M, Mitchell PJ, Nardini A, Pittermann J, Pratt RB, Sperry JS, Westoby M, Wright IJ, Zanne AE (2012) Global convergence in the vulnerability of forests to drought. Nature 491:752–755

Clarke SM, Cristescu SM, Miersch O, Harren FJM, Wasternack C, Mur LAJ (2009) Jasmonates act with salicylic acid to confer basal thermotolerance in *Arabidopsis thaliana*. New Phytol 182:175–187

Condon AG, Richards RA, Rebetzke GJ, Farquhar GD (2002) Improving intrinsic water-use efficiency and crop yield. Crop Sci 42:122–131

Cowan IR (1982) Regulation of water use in relation to carbon gain in higher plants. In: Lange OL, Nobel PS, Osmond CB, Ziegler H (eds) Physiological plant ecology II: water relations and carbon assimilation. Springer, Berlin, Heidelberg, pp 589–613

Dahiya A, Kumar A, Kumar A, Sanwal SK, Mann A, Kulshreshtha N (2023) Current status and future perspectives of epigenetic gene regulation for salt tolerance in wheat. In: Kumar A, Dhansu P, Mann A (eds) Salinity and drought tolerance in plants: Physiological perspectives. Springer Nature, Singapore, pp 333–345

De La Torre AR, Sekhwal MK, Puiu D, Salzberg SL, Scott AD, Allen B, Neale DB, Chin ARO, Buckley TN (2022) Genome-wide association identifies candidate genes for drought tolerance in coast redwood and giant sequoia. Plant J 109:7–22

Dong Q, Wallrad L, Almutairi BO, Kudla J (2022) Ca^{2+} signaling in plant responses to abiotic stresses. J Integr Plant Biol 64:287–300

Drew MC, Jackson MB, Giffard S (1979) Ethylene-promoted adventitious rooting and development of cortical air spaces (aerenchyma) in roots may be adaptive responses to flooding in *Zea mays* L. Planta 147:83–88

Dubois M, Van den Broeck L, Inzé D (2018) The pivotal role of ethylene in plant growth. Trends Plant Sci 23:311–323

Dumanović J, Nepovimova E, Natić M, Kuča K, Jaćević V (2021) The significance of reactive oxygen species and antioxidant defense system in plants: a concise overview. Front Plant Sci 11:552969

Espasandin FD, Maiale SJ, Calzadilla P, Ruiz OA, Sansberro PA (2014) Transcriptional regulation of 9-cis-epoxycarotenoid dioxygenase (NCED) gene by putrescine accumulation positively modulates ABA synthesis and drought tolerance in Lotus tenuis plants. Plant Physiol Biochem 76:29–35

Fowler S, Thomashow MF (2002) *Arabidopsis* transcriptome profiling indicates that multiple regulatory pathways are activated during cold acclimation in addition to the CBF cold response pathway. Plant Cell 14:1675–1690

Foyer CH, Lopez-Delgado H, Dat JF, Scott IM (1997) Hydrogen peroxide- and glutathione-associated mechanisms of acclimatory stress tolerance and signalling. Physiol Plant 100:241–254

Fukao T, Yeung E, Bailey-Serres J (2011) The submergence tolerance regulator SUB1A mediates crosstalk between submergence and drought tolerance in rice. Plant Cell 23:412–427

Gao M, Liu J, Bi D, Zhang Z, Cheng F, Chen S, Zhang Y (2008) MEKK1, MKK1/MKK2 and MPK4 function together in a mitogen-activated protein kinase cascade to regulate innate immunity in plants. Cell Res 18:1190–1198

Giarola V, Bartels D (2015) What can we learn from the transcriptome of the resurrection plant Craterostigma plantagineum? Planta 242:427–434

Gómez-Mena C, de Folter S, Costa MMR, Angenent GC, Sablowski R (2005) Transcriptional program controlled by the floral homeotic gene AGAMOUS during early organogenesis. Development 132:429–438

Gong Z, Xiong L, Shi H, Yang S, Herrera-Estrella LR, Xu G, Chao D-Y, Li J, Wang P-Y, Qin F, Li J, Ding Y, Shi Y, Wang Y, Yang Y, Guo Y, Zhu J-K (2020) Plant abiotic stress response and nutrient use efficiency. Sci China Life Sci 63:635–674

Gonzalez DH (2016) Chapter 1 – introduction to transcription factor structure and function. In: Gonzalez DH (ed) Plant transcription factors. Academic Press, Boston, pp 3–11

Greenway H, Munns R (1980) Mechanisms of salt tolerance in non-halophytes. Annu Rev Plant Physiol 31:149–190

Grift TE, Novais J, Bohn M (2011) High-throughput phenotyping technology for maize roots. Biosyst Eng 110:40–48

Guo M, Liu J-H, Ma X, Luo D-X, Gong Z-H, Lu M-H (2016) The plant heat stress transcription factors (HSFs): structure, regulation, and function in response to abiotic stresses. Front Plant Sci 7:114

Guy C, Kaplan F, Kopka J, Selbig J, Hincha DK (2008) Metabolomics of temperature stress. Physiol Plant 132:220–235

Hamel L-P, Nicole M-C, Sritubtim S, Morency M-J, Ellis M, Ehlting J, Beaudoin N, Barbazuk B, Klessig D, Lee J, Martin G, Mundy J, Ohashi Y, Scheel D, Sheen J, Xing T, Zhang S, Seguin A, Ellis BE (2006) Ancient signals: comparative genomics of plant MAPK and MAPKK gene families. Trends Plant Sci 11:192–198

Hasanuzzaman M, Bhuyan MHMB, Zulfiqar F, Raza A, Mohsin SM, Mahmud JA, Fujita M, Fotopoulos V (2020) Reactive oxygen species and antioxidant defense in plants under abiotic stress: revisiting the crucial role of a universal defense regulator. Antioxidants 9:681

Himabindu Y, Chakradhar T, Reddy MC, Kanygin A, Redding KE, Chandrasekhar T (2016) Salt-tolerant genes from halophytes are potential key players of salt tolerance in glycophytes. Environ Exp Bot 124:39–63

Hou Q, Ufer G, Bartels D (2016) Lipid signalling in plant responses to abiotic stress. Plant Cell Environ 39:1029–1048

Hurtado L, Farrona S, Reyes JC (2006) The putative SWI/SNF complex subunit BRAHMA activates flower homeotic genes in *Arabidopsis thaliana*. Plant Mol Biol 62:291–304

IPCC (2007) Climate change 2007: synthesis report. In: Core Writing Team, Pachauri RK, Reisinger A (eds) Contribution of working groups I, II and III to the fourth assessment report of the intergovernmental panel on climate change. IPCC, Geneva, p 104

Iwata Y, Fedoroff NV, Koizumi N (2008) *Arabidopsis* bZIP60 is a proteolysis-activated transcription factor involved in the endoplasmic reticulum stress response. Plant Cell 20:3107–3121

Javed T, Shabbir R, Ali A, Afzal I, Zaheer U, Gao S-J (2020) Transcription factors in plant stress responses: challenges and potential for sugarcane improvement. Plants (Basel) 9:491

Kavar T, Meglic V, Rozman L (2007) Diversity of Slovenian maize (Zea mays) populations by Hbr (MITE) markers and morphological traits. Genetika 43:1190–1197

Kishor PBK, Sangam S, Amrutha RN, Laxmi PS, Naidu KR, Rao KRSS, Rao S, Reddy KJ, Theriappan P, Sreenivasulu N (2005) Regulation of proline biosynthesis, degradation, uptake and transport in higher plants: its implications in plant growth and abiotic stress tolerance. Curr Sci 88:424–438

Koyro H-W, Geissler N, Seenivasan R, Huchzermeyer B (2016) Plant stress physiology: physiological and biochemical strategies allowing plants/crops to thrive under ionic stress. In: Pessarakli M (ed) Handbook of plant and crop stress, 3rd edn. CRC Press, Taylor & Francis Group, Boca Raton, pp 1081–1124

Kumar M (2024) Structural and compositional effects on tree-water relation. Universität Würzburg. https://opus.bibliothek.uni-wuerzburg.de/32624 (16 August 2024, date last accessed)

Kumar M, Waite P-A, Paligi SS, Schuldt B (2022) Influence of juvenile growth on xylem safety and efficiency in three temperate tree species. Forests 13:909

Kumar P, Kumar P, Suniti K, U., Avni, Mann, A. (2023) Transcriptional regulatory network involved in drought and salt stress response in rice. In: Kumar A, Dhansu P, Mann A (eds) Salinity and drought tolerance in plants. Springer, Singapore. https://doi.org/10.1007/978-981-99-4669-3_13

Leakey ADB, Ferguson JN, Pignon CP, Wu A, Jin Z, Hammer GL, Lobell DB (2019) Water use efficiency as a constraint and target for improving the resilience and productivity of C3 and C4 crops. Annu Rev Plant Biol 70:781–808

Lee TI, Young RA (2013) Transcriptional regulation and its Misregulation in disease. Cell 152:1237–1251

Lee J, Oh M, Park H, Lee I (2008) SOC1 translocated to the nucleus by interaction with AGL24 directly regulates leafy. Plant J 55:832–843

Lee HG, Park M-E, Park BY, Kim HU, Seo PJ (2019) The *Arabidopsis* MYB96 transcription factor mediates ABA-dependent triacylglycerol accumulation in vegetative tissues under drought stress conditions. Plants (Basel) 8:296

Lizana C, Wentworth M, Martinez JP, Villegas D, Meneses R, Murchie EH, Pastenes C, Lercari B, Vernieri P, Horton P, Pinto M (2006) Differential adaptation of two varieties of common bean to abiotic stress: I. Effects of drought on yield and photosynthesis. J Exp Bot 57:685–697

Maheshwari R, Dubey RS (2009) Nickel-induced oxidative stress and the role of antioxidant defence in rice seedlings. Plant Growth Regul 59:37–49

Mann A, Kumar N, Lata C, Kumar A, Kumar A, Meena BL (2019) Functional annotation of differentially expressed genes under salt stress in *Dichanthium annulatum*. Plant Physiol Rep 24:104–111

Mann A, Kumar A, Sanwal SK, Sharma PC (2020) Sustainable production of pulses under saline lands in India. In: Legume crops-prospects, production and uses. IntechOpen. https://doi.org/10.5772/intechopen.90304

Mann A, Lata C, Kumar N, Kumar A, Kumar A, Sheoran P (2023) Halophytes as new model plant species for salt tolerance strategies. Front Plant Sci 14:1137211

Mann A, Kumar N, Lata C, Kumar A, Meena BL, Kumar A (2024a) Physiological and differential gene expression reveals a trade-off between antioxidant capacity and salt tolerance in halophytes *Urochondra setulosa* and *Dichanthium annulatum*. Plant Growth Regul 102:555–570

Mann A, Poonam R, Priti C, Sujata Y, Noyonika K, Avni D, Nitish RP, Ashwani K, Arvind K, Satish KS (2024b) Genome editing technologies for enhancing plant resilience to biotic and abiotic stresses-brief review. J Soil Sal Water Qual 16(2):180–193

Maszkowska J, Szymańska KP, Kasztelan A, Krzywińska E, Sztatelman O, Dobrowolska G (2021) The multifaceted regulation of SnRK2 kinases. Cells 10:2180

Mishra N, Jiang C, Chen L, Paul A, Chatterjee A, Shen G (2023) Achieving abiotic stress tolerance in plants through antioxidative defense mechanisms. Front Plant Sci 14

Mittler R (2002) Oxidative stress, antioxidants and stress tolerance. Trends Plant Sci 7:405–410

Montanarella L, Pennock D, Mckenzie N et al (2015) The status of the world's soil resources (Technical summary). (FAO, 2015)

Munns R, Tester M (2008) Mechanisms of salinity tolerance. Annu Rev Plant Biol 59:651–681

Munns R, Passioura JB, Colmer TD, Byrt CS (2020) Osmotic adjustment and energy limitations to plant growth in saline soil. New Phytol 225:1091–1096

Nagore ML, Della Maggiora A, Andrade FH, Echarte L (2017) Water use efficiency for grain yield in an old and two more recent maize hybrids. Field Crop Res 214:185–193

Noctor G, Foyer CH (1998) ASCORBATE AND GLUTATHIONE: Keeping active oxygen under control. Annu Rev Plant Physiol Plant Mol Biol 49:249–279

Oh D-H, Barkla BJ, Vera-Estrella R, Pantoja O, Lee S-Y, Bohnert HJ, Dassanayake M (2015) Cell type-specific responses to salinity – the epidermal bladder cell transcriptome of mesembryanthemum crystallinum. New Phytol 207:627–644

Pampurova S, Van Dijck P (2014) The desiccation tolerant secrets of *Selaginella lepidophylla*: what we have learned so far? Plant Physiol Biochem 80:285–290

Parmar N, Singh KH, Sharma D, Singh L, Kumar P, Nanjundan J, Khan YJ, Chauhan DK, Thakur AK (2017) Genetic engineering strategies for biotic and abiotic stress tolerance and quality enhancement in horticultural crops: a comprehensive review. 3 Biotech 7:239

Praveen A, Dubey S, Singh S, Sharma VK (2023) Abiotic stress tolerance in plants: a fascinating action of defense mechanisms. 3 Biotech 13:102

Qi J, Song C-P, Wang B, Zhou J, Kangasjärvi J, Zhu J-K, Gong Z (2018) Reactive oxygen species signaling and stomatal movement in plant responses to drought stress and pathogen attack. J Integr Plant Biol 60:805–826

Radha B, Sunitha NC, Sah RP et al (2023) Physiological and molecular implications of multiple abiotic stresses on yield and quality of rice. Front Plant Sci 13

Razzaq MK, Akhter M, Ahmad RM, Cheema KL, Hina A, Karikari B, Raza G, Xing G, Gai J, Khurshid M (2022) CRISPR-Cas9 based stress tolerance: new hope for abiotic stress tolerance in chickpea (Cicer arietinum). Mol Biol Rep 49:8977–8985

Rensing SA, Goffinet B, Meyberg R, Wu S-Z, Bezanilla M (2020) The Moss Physcomitrium (Physcomitrella) patens: a model organism for non-seed plants. Plant Cell 32:1361–1376

Rienzo M, Casamassimi A (2016) Integrator complex and transcription regulation: recent findings and pathophysiology. Biochim Biophys Acta 1859:1269–1280

Ríos-Meléndez S, Valadez-Hernández E, Delgadillo C, Luna-Guevara ML, Martínez-Núñez MA, Sánchez-Pérez M, Martínez-Y-Pérez JL, Arroyo-Becerra A, Cárdenas L, Bibbins-Martínez M, Maldonado-Mendoza IE, Villalobos-López MA (2021) Pseudocrossidium replicatum (Taylor) R.H. Zander is a fully desiccation-tolerant moss that expresses an inducible molecular mechanism in response to severe abiotic stress. Plant Mol Biol 107:387–404

Riyazuddin R, Verma R, Singh K, Nisha N, Keisham M, Bhati KK, Kim ST, Gupta R (2020) Ethylene: a master regulator of salinity stress tolerance in plants. Biomol Ther 10:959

Rothan C, Diouf I, Causse M (2019) Trait discovery and editing in tomato. Plant J 97:73–90

Sah SK, Reddy KR, Li J (2016) Abscisic acid and abiotic stress tolerance in crop plants. Front Plant Sci 7

Salt DE, Blaylock M, Kumar NP, Dushenkov V, Ensley BD, Chet I, Raskin I (1995) Phytoremediation: a novel strategy for the removal of toxic metals from the environment using plants. Biotechnology (N Y) 13:468–474

Sharma SS, Dietz K-J (2006) The significance of amino acids and amino acid-derived molecules in plant responses and adaptation to heavy metal stress. J Exp Bot 57:711–726

Sharma P, Dubey RS (2005) Drought induces oxidative stress and enhances the activities of antioxidant enzymes in growing rice seedlings. Plant Growth Regul 46:209–221

Sharma P, Dubey RS (2007) Involvement of oxidative stress and role of antioxidative defense system in growing rice seedlings exposed to toxic concentrations of aluminum. Plant Cell Rep 26:2027–2038

Sharma P, Jha AB, Dubey RS, Pessarakli M (2012) Reactive oxygen species, oxidative damage, and Antioxidative defense mechanism in plants under stressful conditions. J Bot 2012:1–26

Sheoran P, Basak N, Kumar A, Yadav RK, Singh R, Sharma R, Kumar S, Singh RK, Sharma PC (2021) Ameliorants and salt tolerant varieties improve rice-wheat production in soils undergoing sodification with alkali water irrigation in Indo–Gangetic Plains of India. Agric Water Manag 243:106492

Shinozaki K, Yamaguchi-Shinozaki K (2000) Molecular responses to dehydration and low temperature: differences and cross talk between two stress signaling pathways. Curr Opin Plant Biol 3:217–223

Siahpoosh MR, Dehghanian E (2012) Water use efficiency, transpiration efficiency, and uptake efficiency of wheat during drought. Agron J 104:1238–1243

Sinclair TR, Muchow RC (2001) System analysis of plant traits to increase grain yield on limited water supplies. Agron J 93:263–270

Singer SD, Zou J, Weselake RJ (2016) Abiotic factors influence plant storage lipid accumulation and composition. Plant Sci 243:1–9

Snyder KA, Richards JH, Donovan LA (2003) Night-time conductance in C3 and C4 species: do plants lose water at night? J Exp Bot 54:861–865

Song L, Zhu J, Yan Q, Li M, Yu G (2015) Comparison of intrinsic water use efficiency between different aged Pinus sylvestris var. mongolica wide windbreaks in semiarid sandy land of northern China. Agrofor Syst 89:477–489

Soolanayakanahally RY, Guy RD, Silim SN, Drewes EC, Schroeder WR (2009) Enhanced assimilation rate and water use efficiency with latitude through increased photosynthetic capac-

ity and internal conductance in balsam poplar (Populus balsamifera L.). Plant Cell Environ 32:1821–1832
Sperry JS, Tyree MT (1988) Mechanism of water stress-induced xylem embolism. Plant Physiol 88:581–587
Srivastava V, Verma PK (2017) The plant LIM proteins: unlocking the hidden attractions. Planta 246:365–375
Sun Y, Niu Y, Xu J, Li Y, Luo H, Zhu Y, Liu M, Wu Q, Song J, Sun C, Chen S (2013) Discovery of WRKY transcription factors through transcriptome analysis and characterization of a novel methyl jasmonate-inducible PqWRKY1 gene from Panax quinquefolius. Plant Cell Tissue Organ Cult 114:269–277
Syed NH, Prince SJ, Mutava RN, Patil G, Li S, Chen W, Babu V, Joshi T, Khan S, Nguyen HT (2015) Core clock, SUB1, and ABAR genes mediate flooding and drought responses via alternative splicing in soybean. J Exp Bot 66:7129–7149
Syvertsen JP, Lloyd J, McCONCHIE C, Kriedemann PE, Farquhar GD (1995) On the relationship between leaf anatomy and CO2 diffusion through the mesophyll of hypostomatous leaves. Plant Cell Environ 18:149–157
Szymańska KP, Polkowska-Kowalczyk L, Lichocka M, Maszkowska J, Dobrowolska G (2019) SNF1-related protein kinases SnRK2.4 and SnRK2.10 modulate ROS homeostasis in plant response to salt stress. Int J Mol Sci 20:143
Tao J-J, Chen H-W, Ma B, Zhang W-K, Chen S-Y, Zhang J-S (2015) The role of ethylene in plants under salinity stress. Front. Plant Sci 6
Tarara JM, Pe AJEP, Keller M, Schreiner RP, Smithyman RP (2011) Net carbon exchange in grapevine canopies responds rapidly to timing and extent of regulated deficit irrigation. Funct Plant Biol 38:386–400
Terashima I, Ono K (2002) Effects of HgCl2 on CO2 dependence of leaf photosynthesis: evidence indicating involvement of Aquaporins in CO2 diffusion across the plasma membrane. Plant Cell Physiol 43:70–78
Tholen D, Boom C, Noguchi K, Ueda S, Katase T, Terashima I (2008) The chloroplast avoidance response decreases internal conductance to CO2 diffusion in *Arabidopsis thaliana* leaves. Plant Cell Environ 31:1688–1700
Tiwari R, Rajam MV (2022) RNA- and miRNA-interference to enhance abiotic stress tolerance in plants. J Plant Biochem Biotechnol 31:689–704
Tyree MT (2003) Hydraulic limits on tree performance: transpiration, carbon gain and growth of trees. Trees 17:95–100
Villalobos-López MA, Arroyo-Becerra A, Quintero-Jiménez A, Iturriaga G (2022) Biotechnological advances to improve abiotic stress tolerance in crops. Int J Mol Sci 23:12053
Visser EJW, Nabben RHM, Blom CWPM, Voesenek LACJ (1997) Elongation by primary lateral roots and adventitious roots during conditions of hypoxia and high ethylene concentrations. Plant Cell Environ 20:647–653
Waite P-A, Kumar M, Link RM, Schuldt B (2024) Coordinated hydraulic traits influence the two phases of time to hydraulic failure in five temperate tree species differing in stomatal stringencyKumar. Tree Physiol 44:tpae038
Wan J, Dunning MF, Bent AF (2002) Probing plant-pathogen interactions and downstream defense signaling using DNA microarrays. Funct Integr Genomics 2:259–273
Wang J-W, Czech B, Weigel D (2009) miR156-regulated SPL transcription factors define an endogenous flowering pathway in *Arabidopsis thaliana*. Cell 138:738–749
Yadav RK, Singh SP, Lal D, Kumar A (2007) Fodder production and soil health with conjunctive use of saline and good quality water in Ustipsamments of a semi-arid region. Land Degrad Dev 18(2):153–161
Yamasaki K, Kigawa T, Inoue M, Tateno M, Yamasaki T, Yabuki T, Aoki M, Seki E, Matsuda T, Nunokawa E, Ishizuka Y, Terada T, Shirouzu M, Osanai T, Tanaka A, Seki M, Shinozaki K, Yokoyama S (2004) A novel zinc-binding motif revealed by solution structures of DNA-binding domains of *Arabidopsis* SBP-family transcription factors. J Mol Biol 337:49–63

Yan J, Tsuichihara N, Etoh T, Iwai S (2007) Reactive oxygen species and nitric oxide are involved in ABA inhibition of stomatal opening. Plant Cell Environ 30:1320–1325

Yu J, Xu F, Wei Z, Zhang X, Chen T, Pu L (2020) Epigenomic landscape and epigenetic regulation in maize. Theor Appl Genet 133:1467–1489

Zaefyzadeh M, Quliyev R, Babayeva S, Abbasov M (2009) The effect of the interaction between genotypes and drought stress on the superoxide dismutase and chlorophyll content in durum wheat landraces. Turk J Biol 33:1–7

Zhang JZ (2003) Overexpression analysis of plant transcription factors. Curr Opin Plant Biol 6:430–440

Zhang S, Wang X (2011) Overexpression of *GASA5* increases the sensitivity of *Arabidopsis* to heat stress. J Plant Physiol 168:2093–2101

Zhang J, Li Y, Jia H-X, Li J-B, Huang J, Lu M-Z, Hu J-J (2015) The heat shock factor gene family in *Salix suchowensis*: a genome-wide survey and expression profiling during development and abiotic stresses. Front. Plant Sci 6

Zhang H, Zhang Y, Deng C, Deng S, Li N, Zhao C, Zhao R, Liang S, Chen S (2018) The *Arabidopsis* Ca^{2+}-dependent protein kinase CPK12 is involved in plant response to Salt stress. Int J Mol Sci 19:4062

Zhang H, Zhao Y, Zhu J-K (2020) Thriving under stress: how plants balance growth and the stress response. Dev Cell 55:529–543

Zhang H, Zhu J, Gong Z, Zhu J-K (2022) Abiotic stress responses in plants. Nat Rev Genet 23:104–119

Zhang Y, Xu J, Li R, Ge Y, Li Y, Li R (2023) Plants' response to abiotic stress: mechanisms and strategies. Int J Mol Sci 24:10915

Zhu J-K (2002) Salt and drought stress signal transduction in plants. Annu Rev Plant Biol 53:247–273

Zhu J-K (2016) Abiotic stress signaling and responses in plants. Cell 167:313–324

Homeostasis of Plant Metabolites in Mitigating Abiotic Stress Challenges

13

Palampreet Singh, Rashmi Mishra, Gurvarinder Kaur, Isha Madaan, and Geetika Sirhindi

Abstract

This chapter explores the critical role of metabolite homeostasis in mitigating the adverse effects of abiotic stresses on plants. Abiotic stresses such as drought, salt, temperature, and heavy metal toxicity cause an excess of reactive oxygen species (ROS) and other cytotoxic chemicals, disturbing cellular balance. Plants counter these stresses by maintaining a dynamic equilibrium of ROS and antioxidants, with major enzymatic (e.g. superoxide dismutase, catalase, ascorbate peroxidase) and non-enzymatic (e.g. ascorbate, glutathione) components playing critical roles. While osmolytes including proline, glycine betaine, and trehalose aid in osmotic adjustment and maintain cellular structures, the glyoxalase system detoxifies methylglyoxal (MG). These defenses are mediated by transcription factors such as WRKY, NAC, MYB, bZIP, and AP2/ERF, which are part of genetic and epigenetic regulatory networks that govern synthesis, transport, and compartmentalization of essential metabolites. Gaining an understanding of these systems can help improve agricultural sustainability and crop development in the face of changing environmental conditions by revealing important information about plant resilience and adaptation.

Keywords

Metabolite homeostasis · Abiotic stress · Reactive oxygen species · Antioxidant defense · Osmolytes

P. Singh · R. Mishra · G. Kaur · G. Sirhindi (✉)
Department of Botany, Punjabi University, Patiala, Punjab, India
e-mail: geetika@pbi.ac.in

I. Madaan
Department of Botany, Punjabi University, Patiala, Punjab, India

Government College of Education, Jalandhar, Punjab, India

A. Mann et al. (eds.), *Cutting Edge Technologies for Developing Future Crop Plants*,
https://doi.org/10.1007/978-981-96-2508-6_13

13.1 Introduction

As plants are sessile, maintaining metabolite homeostasis is a crucial approach for managing the challenges posed by abiotic stresses. Extreme temperatures, drought, salt, and heavy metal contamination are examples of abiotic stresses that upset the physiological balance in plant cells, causing an excess of reactive oxygen species (ROS) and other cytotoxic chemicals to be produced. These reactive species have the potential to harm proteins, lipids, carbohydrates, and nucleic acids, among other cellular constituents, threatening the plant's ability to survive.

In order to combat these challenges, plants have evolved highly developed defense mechanisms. The balance between the generation of ROS and the action of antioxidants that scavenge ROS is essential to these defenses. Through a complex interplay of enzymatic and non-enzymatic antioxidants, this dynamic balance is maintained. The conversion of reactive oxygen species (ROS) into less reactive molecules, facilitated by enzymes such as ascorbate peroxidase (APX), catalase (CAT), and superoxide dismutase (SOD), is essential for maintaining cellular integrity. Moreover, plants' capacity to detoxify ROS and reduce oxidative stress is greatly enhanced by non-enzymatic antioxidants such as glutathione (GSH) and ascorbate (AsA).

Another essential part of the plant's defensive mechanism is the glyoxalase system, which is made up of glyoxalase I (Gly I) and glyoxalase II (Gly II). Methylglyoxal (MG), a cytotoxic byproduct of cellular metabolism that builds up under extreme conditions, is detoxified by this mechanism. The glyoxalase system keeps cells in a state of equilibrium by transforming MG into less toxic compounds. Proline, glycine betaine (GB), and trehalose are examples of osmolytes and osmoprotectants that are essential for assisting plants in withstanding abiotic stresses. These substances stabilize proteins and membranes, aid in osmotic adjustment, and shield cellular structures. When a plant experiences stress, its reserves of osmolytes increase, which helps the plant hold onto water, keep up its turgor pressure and maintain its metabolic processes.

Intricate genetic and epigenetic networks govern the modulation of metabolite homeostasis during stress. The synthesis, transport, and compartmentalization of metabolites required for stress tolerance are ensured by these networks, which also regulate gene expression and enzyme activity. Important transcription factors that control these responses include AP2/ERF, WRKY, NAC, MYB, bZIP, and MYB. These transcription factors coordinate the expression of genes related to stress signaling pathways, osmolyte synthesis, and antioxidant defense. Developing ways to improve plant resilience requires an understanding of the mechanisms behind plant metabolite homeostasis in abiotic stress situations. The complex interactions between ROS and antioxidants, the function of the glyoxalase system, the importance of osmolytes, and the regulatory networks that control these reactions are all covered in detail in this chapter. Through investigating these aspects, we acquire knowledge about how plants endure and adjust to severe environmental circumstances, opening doors for developments in crop improvement and sustainable agriculture.

13.2 Metabolites in Abiotic Stress Mitigation

As part of regular cellular metabolism, living things produce reactive oxygen species (ROS), which are abundant in internal cell components such as mitochondria, cytosol, peroxisomes, and chloroplasts (Han et al. 2013; Nafees et al. 2019). Each plant cell maintains a dynamic balance between ROS and ROS-scavenging antioxidants. By generating a significant amount of ROS, abiotic stress disrupts this physiological equilibrium in favor of oxidative processes (Han et al. 2013). Under abiotic stress, insufficient energy consumption during photosynthetic processes lowers molecular oxygen and subsequently generates a lot of ROS, such as H_2O_2, O_2^{-}, $^1O_2$, OH•, and so forth (Hasanuzzaman et al. 2017a, b, c, d, e). ROS are highly reactive molecules that can break down and change the activities of a wide range of biological macromolecules, including proteins, lipids, carbohydrates, and nucleic acids (Soares et al. 2019; Mann et al. 2024). In addition, Methylglyoxal (MG), a cytotoxic molecule and reactive oxidant, spontaneously formed in a cell in modest quantity but under abiotic conditions, its synthesis increased and participated in producing oxidative stress. Similar to ROS, MG production is elevated under abiotic stress, which can damage the ultra-structural elements of cell and, cause mutation, and finally provoke programmed cell death (PCD). In addition to producing oxidative stress, ROS, and MG modulate certain antioxidants and their corresponding genes to influence acclimatization and defense responses through stress tolerance (Singh et al. 2019a, b). Overproduction of ROS and MG can also trigger a disruption in redox equilibrium, signaling the end of a plant's life cycle or cellular death (Han et al. 2013). Plant cells have a well-established glyoxalase system and antioxidant defense mechanism for scavenging harmful ROS and MG, respectively (Fig. 13.1). The antioxidant defense system consists of certain non-enzymatic components (AsA, Glutathione (GSH), alkaloids, α-tocopherol, non-protein amino acids, and

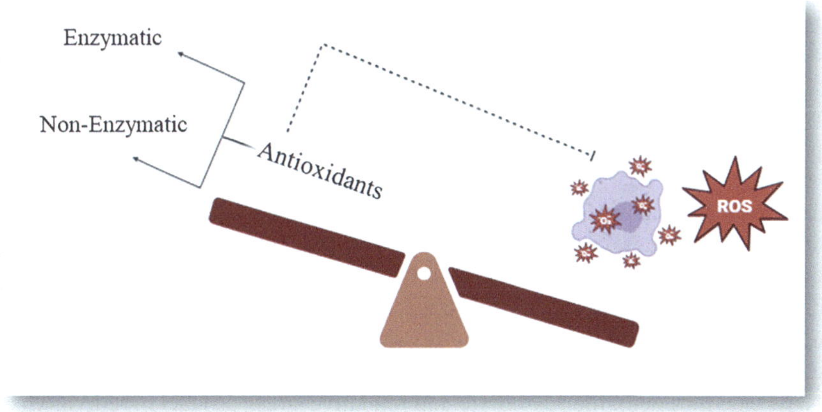

Fig. 13.1 Mitigation of stress-induced oxidative burst by antioxidants

phenolic compounds) and enzymatic components (SOD, CAT, APX, MDHAR, DHAR, glutathione reductase (GR), glutathione S-transferase (GST), and GPX (Hasanuzzaman et al. 2011). The AsA-GSH pool, which is part of the antioxidant defense system, uses four important enzymes, such as APX, MDHAR, DHAR, and GR, to scavenge ROS, hence reducing the stress effect (Latowski et al. 2010). SOD typically provides front-line protection against ROS in the antioxidant defense system by converting O2 − to H_2O_2. Subsequently, CAT and APX scavenge H_2O_2 to H_2O. Glutathione peroxidase and GST also scavenge H_2O_2 to H_2O with the help of GSH. Toxic MG is detoxified in the cell by glyoxalase system. Glutathione is an essential component of the AsA-GSH cycle and is also important for the MG detoxifying mechanism. Glyoxalase system is composed of two essential enzymes, Gly I and Gly II. GSH is used in the first phase of the MG detoxification process in the glyoxalase system to convert MG into S-D-lactoyl-glutathione. In the second step, S-D-lactoyl-glutathione is converted into D-Lactate, where GSH is recycled once more (Hasanuzzaman et al. 2017a, b, c, d, e). Moreover, GSH contributes to metal chelation. It boosts the amount of PC under stress condition, which builds a combination with metal and pushes into the cell vacuole as inert form (Hernández et al. 2017).

13.3 Ascorbate and Glutathione

In cellular organelles, plants have an antioxidant defense system that includes both enzymatic and non-enzymatic antioxidants that can partially scavenge various reactive oxygen species surpasses the antioxidant system's capacity to scavenge them. The antioxidant defense system includes the well-known non-enzymatic antioxidants ascorbate (AsA), glutathione (GSH), carotenoids, tocopherols, flavonoids, etc. (Gill and Tuteja 2010). Enzymatic antioxidant components that are widely recognized include ascorbate peroxidase (APX), glutathione reductase (GR), superoxide dismutase (SOD), catalase (CAT), glutathione peroxidase (GPX), glutathione S-transferase (GST), and peroxiredoxin (PRX) (Hasanuzzaman et al. 2017a, b, c, d, e). Of these, the AsA-GSH cycle is made up of AsA, GSH, APX, MDHAR, DHAR, and GR. Ascorbate is one of the most potent substrates for scavenging H_2O_2. Ascorbate maintains the decreased state of α-tocopherol. Ascorbate has been shown impact on the manufacture of zeaxanthin by preventing oxidative stress and dissipating excess light energy in the chloroplast's thylakoid membranes. Ascorbate sustains lower the status of prosthetic metal ions and maintains the activity of antioxidant enzymes (Hasanuzzaman et al. 2012). Glutathione affects several metabolic systems; it works as an antioxidant. Glutathione peroxidase and GST use glutathione supplement (GSH) as a substrate; GPX is in charge of detoxifying ROS, while GST is in charge of detoxifying xenobiotics (Hasanuzzaman et al. 2017a, b, c, d, e). The glyoxalase system comprised glyoxalase I (Gly I) and glyoxalase II (Gly II) enzymes detoxifies cytotoxic and oxidative stress producer methylglyoxal (MG), where Gly I utilizes GSH and after concluding MG detoxification, GSH is recycled (Hasanuzzaman et al. 2017a, b, c, d, e). The beneficial role of AsA-GSH cycle components has been shown in many plants that are impacted by abiotic stress. The presence of AsA and GSH has been observed to improve osmoregulation, plant water status and nutritional status, water use efficiency, photosynthetic activity, and the overall productivity of plant. In *O. sativa* under salt stress, where ROS production was high, Rahman et al. (2017) reported on the AsA-GSH cycle's significant involvement. Here, rice subjected to salt increased its content of reduced and oxidized GSH while containing less AsA due to increased levels of GR, APX, MDHAR, and DHAR actions against excessive ROS production. Salinity-induced oxidative stress was evident in the severe $O_2^-$and H_2O_2 overgeneration of *Vigna radiata* cultivated in saline conditions (Rahman et al. 2016). Salt-stressed *V. radiata* boosted GSH and GSSG contents along with decreasing AsA, whereas up-regulated the activity of all enzymatic antioxidants of AsA-GSH cycle and therefore responded with elevated ROS (Nahar et al. 2015). Recently, Singh et al. (2019a, b) reported the increased activity of enzymatic antioxidants, including APX, DHAR, and GR, with decreased AsA, GSH, and GSSG concentrations, because of salt-induced greater ROS formation in *Solanum lycopersicum*. Similarly, 150 mM salt-treated *S. lycopersicum* lowered AsA concentration, which might be employed in H_2O_2 detoxification, whereas improved GSH indicated its function in decreasing H_2O_2. Lou et al. (2018) demonstrated how *T. aestivum* responded to drought exposure. Hence, they discovered that the AsA-GSH cycle responded greatly with excess ROS formation by significant fluctuation of GSH/GSSG and AsA/DHA redox along with the constant increment of H_2O_2. Mahmud et al. (2017) demonstrated that due to Cr stress, the few components of AsA-GSH pathway increased their quantity or function in *B. juncea* L. cv. sBARI Sharisha-11. They reported 5 days duration of 0.15 mM and 0.3 mM K2CrO4 treatment lowered the concentration of AsA but did not affect the GSH content. Furthermore, the activities of APX and GR were elevated, and the activities of MDHAR and DHAR were decreased. The enhanced APX and GR activity might play a function in scavenging excess ROS. A similar elevation of APX and GR was also found in *B. napus* L. cv. BINA sharisha three due to Cd treatment (Hasanuzzaman et al. 2017a, b, c, d, e). *Ficus concinna* seedlings showed increased O_2 – production rate and H_2O_2 content after 48 hours of hexose transporters (HT) (35 °C and 40 °C) stress. AsA and GSH contents remained unaltered at 35 °C but decreased at 40 °C (Jin et al. 2015). The activity of APX, MDHAR, DHAR, and GR enzymes increased at 35 °C but then again lowered at 40 °C to the level of control plants. *V. radiata* seedlings showed reduced GSH content and MDHAR-DHAR activities under comparable heat stress conditions (40 °C, 48 hours) but increased APX-GR activities (Nahar et al. 2017). According to Sairam et al. (2011), waterlogging also decreased the rate at which H_2O_2 and O_2 were produced in susceptible cultivars, but had no effect on tolerant cultivars. However, the activity of the GR and APX enzymes increased in tolerant genotypes whereas it decreased in susceptible genotypes.

13.4 Osmolytes and Osmoprotectants Mitigating Stress

Osmolytes are solutes that are compatible with the cell and act as osmoprotectants, allowing the cell to maintain water levels without interfering with normal metabolism. The primary function of these organic metabolites is to control osmotic adjustment. These osmotic solutes, also known as osmoprotectants, assist organisms in enduring significant osmotic stress throughout their life cycle (Singh et al. 2015). These substances maintain the equilibrium of osmotic pressures between the cytosol and the surrounding cells (Wani et al. 2013). Additionally, by preventing the generation of ROS, they shield plants from oxidative damage. Osmolytes are neutral molecule that protect proteins and cell membranes from several stress that can affect cellular metabolism. Proline, sucrose, polyols, trehalose, GB, and alanine betaine are some of these metabolites. Osmoprotectants preserve the cellular activities of plants and are the primary approach used by plants to combat stress. Several studies have indicated the involvement of osmolytes in various types of stress, including drought (Anjum et al. 2017), salinity (Mann et al. 2019), osmotic (Conde et al. 2011), heavy metal, temperature, light, and pesticide stress (Ningthoujam et al. 2013). The decrease in water content and water potential of leaves is the most prevalent indication of drought stress in plants. It affects the physiological activities of plants like photosynthesis and causes death of the plants (Jaleel et al. 2008). Din et al. (2011) reported a sharp decrease in the quantity of chlorophyll in plants during exposure to drought stress, which attributed to the effect on enzymes involved to chlorophyll biosynthesis. As a defense against drought stress, plants have developed a number of defense mechanisms, one of which is the buildup of osmolytes, also known as osmoprotectants, in the plant (Tanveer et al. 2019; Anjum et al. 2017). When under water or drought stress, a variety of osmotically active molecules accumulate, including sugars, proline, GB, and organic acids. These molecules help to balance the water relations. The osmotic potential of the cell becomes extremely negative due to solute aggregation in the cell during water stress, which results in endosmosis of water into the cell and preserves the turgor of the cell. Proline stands out as the most significant osmolyte in drought-stressed plants. When blueberry plants were exposed to drought stress, spermidine, a polyamine, was given exogenously to increase plant resistance (Chen et al. 2017). Plants may resist drought stress mostly by osmotic adjustment and balancing the system of antioxidant defense that assists in scavenging ROS and providing stability to the cell membranes. Over one-third of Earth's landmass is affected by salt stress. This environmental restriction consists of two primary parts: an osmotic component and an ionic component. The former is linked to the soil's decreased external osmotic potential and the latter to an increase in ion build up that becomes dangerous at greater concentrations. Lipid peroxidation, nutritional imbalance, water and nutrient balance disturbance, disruption of enzyme function, and production of reactive oxygen species (ROS) are all consequences of salt stress that eventually harm the photosynthetic apparatus (Tang and Luo 2018). In order to withstand salinity, plants store low molecular weight osmolytes such proline, GB, and polyamines, which aid in preserving membrane stability. Proline accumulation is a defense mechanism against salt stress (Hasanuzzaman et al. 2013). Throughout the seasons, plants are exposed to a variety of temperature changes. Plants under high-temperature stress produce less chlorophyll synthesis (Reda and Mandoura 2011), which is a sign of stress for plastids (Li et al. 2010). Plants utilize various strategies to withstand temperature stress, one of which is the build-up of osmolytes or suitable solutes, which serve as protective agents. Exogenous administration of proline at low concentration enhanced tolerance toward cold stress (Jonytienė et al. 2012). Many studies have revealed the importance of GB in protecting plants under temperature stress (Kawakami et al. 2008). Due to improvement in technology, genes involved in synthesis of GB are now transferred in some non-accumulator plants that assist plants survive under stress. For instance, in *Oryza sativa*, the codA gene confers tolerance to cold stress, shielding the plant from ROS activity, enzyme damage, and membrane damage (Kathuria et al. 2009). Although PSII in the chloroplast is damaged by ROS generated during abiotic stress conditions, GB preserves protein activity and shields the plant from adverse environments. Heavy metal contamination has arisen as a widespread danger in addition to drought, temperature, and salinity stress (Kohli et al. 2018). Due to developments in urbanization and industrialization, the consumption of heavy metals has expanded at an alarming pace. Plants experience oxidative stress (Guo et al. 2017) as a result of heavy metal stress, which damages the cellular, biochemical, and physiological systems in plants (Sidhu et al. 2017). In *Coronopus didymus,* exposure to lead resulted in a drop in the concentration of leaf pigment, which may have been produced by a reduction in the synthesis of chlorophyll or its breakdown in response to heavy metal stress (Sidhu et al. 2017). Consequently, plants use a variety of strategies to resist heavy metal stress and detoxify heavy metals. Dhir et al. (2012) investigated osmolyte accumulation in *Salvinia natans* upon exposure to heavy metals. They detected higher levels of GB, sucrose, and mannitol in response to Cd, Cu, Ni, which is Zn, Pb, Fe, Mn, and Cr. Similar findings were made by Bhatti et al. (2013), who found that exogenous GB treatment improved wheat plants' resistance to heavy metal stress. They discovered that the application of GB increased biomass, photosynthetic pigments, root and shoot length and encouraged osmotic adjustments and ROS scavenging. Plant growth and development are regulated by a variety of environmental conditions. Plant metabolism is regulated by a number of factors, one of which is the intensity of sunlight. The changes in solar radiations alter the light responses of photosynthesis in thylakoid membranes (Kirchhoff et al. 2014). Numerous findings indicated that the stomatal density of mature leaves is influenced by light intensity. High light intensity, specifically in the UV area, impairs the photosynthesis process due to the overproduction of ROS, which damages the photosynthetic electron transport (Hideg et al. 2013). Wang et al. (2014) showed a higher accumulation of GB in response to high light stress in wheat, which aided in recovering chlorophyll content and the net photosynthetic rate in the plants via repairing PS II.

13.5 Regulatory Networks: Genetic and Epigenetic Control

In order maintain homeostasis, particularly in situations of abiotic stress, plants have complex regulatory networks that regulate gene expression and metabolic pathway activity. In order to regulate the synthesis and distribution of metabolites—which are critical for plant life and adaptation—these networks combine information from a variety of external stimuli. Maintaining plant metabolite homeostasis in situations of abiotic stress requires genetic regulation. It involves the control of gene expression, which is responsible for the transport, breakdown, and synthesis of metabolites necessary for stress tolerance. Osmoprotectants, which are essential for stress tolerance, are produced by plants and include proline, glycine betaine, and trehalose. Stress induces a common upregulation of genes that code for the enzymes needed for the manufacture of these substances. One such gene that is essential for proline production is P5CS (pyrroline-5-carboxylate synthase), which is increased in response to salt and drought stress. This gene aids in osmotic adjustment and protects cellular structures (Mladenov et al. 2021). Similar to this, salt stress induces the BADH (Betaine Aldehyde Dehydrogenase) gene, which is involved in the synthesis of glycine betaine and aids in membrane stability and osmoprotection (Khoshru et al. 2023). The TPS (Trehalose-6-phosphate synthase) gene controls the synthesis of trehalose and, when activated under stress, stabilizes proteins and membranes to offer protection (Mladenov et al. 2021). Superoxide Dismutase (SOD), Catalase (CAT), and ascorbate peroxidase (APX) genes are among the enzymes whose genes are increased in order to scavenge reactive oxygen species (ROS) and shield cellular constituents from oxidative damage. Superoxide radicals are transformed into hydrogen peroxide by SOD genes, and CAT and APX enzymes then further detoxify the hydrogen peroxide (Khoshru et al. 2023).

Hormonal regulation and signaling pathways are also involved in genetic control. The hormone abscisic acid (ABA) plays a pivotal role in abiotic stress responses. Under stress, genes related to ABA production and signaling are activated, resulting in osmoprotectant buildup and stomatal closure (Mladenov et al. 2021). According to Khoshru et al. (2023), genes encoding elements of the ethylene and jasmonic acid (JA) signaling pathways are also involved in stress responses. These pathways influence stress-responsive gene expression and metabolic changes.

Moreover, a further level of control over genetic responses to abiotic stress is provided by epigenetic alterations, which include DNA methylation, histone modifications, and chromatin remodeling. These modifications affect gene expression in response to stress (Mladenov et al. 2021). Stress treatments can also impact epigenetic mechanisms, which not only change the chromatin status of stress-responsive genes to regulate them at the transcriptional and posttranscriptional levels (Kim et al. 2015; Lamke and Baurle 2017; Luo and He 2019). Furthermore, stress memory is formed by epigenetic pathways and can be passed on to progeny (Friedrich et al. 2019). The types of epigenetic marks and their location on genes determine the effects of epigenetic mechanism-based regulation on gene expression. When heterochromatic markers like DNA methylation and H3K9me2 are found inside the promoter region, transcription is often inhibited, which has a repressive influence on downstream gene expression. Full-length transcript processing requires intragenic heterochromatic elements (Wang et al. 2013b; Lei et al. 2014; Duan et al. 2017a, b). Such a mechanism is of enormous significance to the dynamic regulation of stress responses, given the universality of transposable element (TE) insertion inside genes in plant genomes (Makarevitch et al. 2015), especially within stress-responsive genes. One of the defining characteristics of epigenetic mechanisms is DNA methylation at the 5′ position of cytosine (5mC). DNA methylation in plants takes place in various cytosine contexts, such as CG, CHG, and CHH (where H stands for A, T, or C). In the model plant species Arabidopsis, these contexts are catalyzed by Methyltransferase 1 (MET1), Chromomethylase 3 (CMT3), and Domains Rearranged Methyltransferase 2 (DRM2), respectively (Duan et al. 2018a, b; Zhang et al. 2018). Of these, asymmetric CHH methylation is a plant-specific de novo DNA methylation method that is facilitated by the 24 nt siRNA-dependent RNA-directed DNA methylation (RdDM) route (Law and Jacobsen 2010; Zhang et al. 2018). Several catalytic enzymes dynamically regulate epigenetic coding, including DNA methylation. DEMETER (DME), Repressor of Transcriptional Silencing 1 (ROS1)/Demeter-Like 1 (DML1), DML2, and DML3 in Arabidopsis are among the demethylases belonging to the DNA glycosylase family that have the ability to remove DNA methylation (Zhang et al. 2018; Liu and Lang 2019).

13.5.1 Histone Modification-Mediated DNA Methylation

DNA methylation and histone changes produce distinct epigenetic codes that control the expression of certain genes. Histone changes such as H3K9 and H3K27 dimethylation, for example, can cause DNA methylation, which in turn can lead to transcriptional gene silence. These alterations are facilitated by enzymes such as histone lysine methyltransferases (SUVH4/KYP) and DNA methyltransferases (CMT3). Additionally, permissive histone marks like H3K18 and H3K23 acetylation are necessary for DNA demethylation at particular locations. IDM1 is a histone acetyltransferase that binds to methylated DNA and catalyzes the acetylation of histone H3, which promotes a chromatin shape that is favorable to stress response and gene expression (Chinnusamy et al. 2014; Lindroth et al. 2004).

13.5.2 Stress Regulation of Genes for Histone Modification and RdDM

Genes involved in building and removing epigenetic marks are influenced by abiotic stressors in their expression. For instance, in rice, heavy metal stress downregulates certain genes, like MET1–2, but cold, salt, and drought stresses upregulate others, such OsAGO2 and OsCMT2 (Kapoor et al. 2008; Curtin et al. 2012). Histone modifications, chromatin remodeling, and DNA methylation are all affected by these changes in gene expression and are essential for the development of stress tolerance.

These genes are crucial for abiotic stress adaptation because mutants in them frequently exhibit reduced stress responses (Chinnusamy et al. 2014).

13.5.3 Gene Regulation Mediated by Stress-Induced Histone Modifications

The expression of genes that respond to abiotic stress is regulated by histone modifications. For example, genome-wide investigations conducted in Arabidopsis have demonstrated that dynamic regulation of histone methylation markers, including H3K4me, occurs during drought stress. Histone acetylation and methylation at particular locations are linked to the activation of ABA-responsive genes, and these alterations are important in ABA signaling pathways. On the other hand, lowering repressive markers such as H3K9 dimethylation can improve stress tolerance by increasing the expression of these genes (Wu and Chen 2010).

The inheritance of stress responses across generations is also influenced by epigenetic control. It is implied that plants have the ability to 'remember' previous stress exposures and adapt their responses in subsequent generations. Stable DNA methylation and histone modification patterns that are passed down to offspring mediate this transgenerational epigenetic memory (Mladenov et al. 2021).

Moreover, there are numerous levels of control involved in the intricate interaction between metabolite homeostasis and epigenetic regulation. For example, the effect of metabolism on the epigenetic regulation of stress responses in plants demonstrates how modifications to metabolism can affect epigenetic states, which in turn control genes that respond to stress. The methylation of DNA and histones is dependent on metabolites like folate and SAM (S-adenosyl methionine), which connects metabolic state to epigenetic control (Lu et al. 2023).

13.6 Key Transcription Factors

In plants, transcription factors (TFs) play a critical role in controlling gene expression in response to abiotic stresses and preserving metabolic balance. In order to help plants adapt to harsh environments, they attach to particular DNA sequences in the promoters of stress-responsive genes, modifying the expression of those genes (Fig. 13.2).

13.6.1 WRKY Transcription Factors

The vast family of WRKY TFs is well-known for its function in the responses of plants to stresses that are both biotic and abiotic. They control the expression of target genes by binding to W-box regions in their promoters. By modifying genes involved in antioxidant defense systems, WRKY TFs play a crucial role in controlling the plant's response to oxidative stress (Khoshru et al. 2023; Mann et al. 2021).

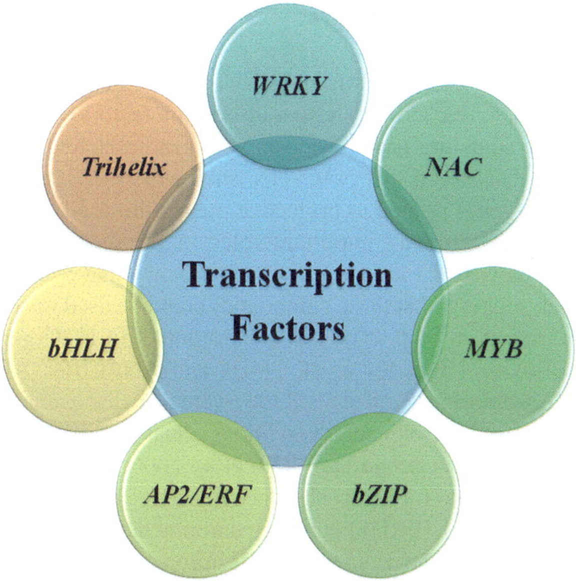

Fig. 13.2 Different transcription factors required for maintaining homeostasis of metabolites under stress

Various WRKY genes in cotton have been found, and their functions under different stress situations have been described. For example, through an ABA-dependent mechanism, tobacco overexpressing GhWRKY41 improved drought and salt tolerance. The expression of this gene improved plant survival in harsh environments by decreasing stomatal aperture and increasing the activity of stress-responsive genes. Furthermore, WRKY TFs have a role in the scavenging of reactive oxygen species (ROS), an important process for reducing oxidative damage in stressful environments. It has been demonstrated that genes such as GhWRKY6-like increase antioxidant enzyme activity, lowering ROS levels and shielding plants from oxidative stress. On the other hand, it has been discovered that certain WRKY genes, including GhWRKY17, raise susceptibility to salt and drought stressors by suppressing ROS scavenging mechanisms. Additionally, WRKY TFs interact with other signaling molecules that are essential for stress responses, like ethylene, salicylic acid (SA), and jasmonic acid (JA). Complex regulatory networks that precisely control the plant's reaction to different stresses are frequently the result of these interactions. For example, in Arabidopsis, overexpressing GarWRKY5 enhanced salt tolerance through controlling the JA, SA, and ROS scavenging pathways (Guo et al. 2019).

13.6.2 NAC Transcription Factors

NAC TFs, which include NAM, ATAF, and CUC, are an essential group that participates in responses to abiotic stress. These transcription factors control genes linked to stress-responsive signaling pathways, osmolyte synthesis, and cell wall biosynthesis. It has been discovered that NAC TFs, such as ANAC019, ANAC055, and ANAC072, activate genes related to protective metabolite production and osmotic adjustment, which in turn play important roles in responses to salt stress and drought. Research on NAC TF overexpression has shown that plants can withstand more stress, which suggests that these compounds could be used to create crops that are resistant to stress (Lan et al. 2017). A small number of NAC genes, including BnNAC from Brassica (Hegedus et al. 2003) and ANAC019, ANAC055, and ANAC072 from Arabidopsis (Tran et al. 2004), were also discovered to be involved in response to different environmental conditions. Guard cells are the primary site of SNAC1 activation during dehydration (Hu et al. 2006). In Arabidopsis, AtNAP and its homologs are crucial for leaf senescence (Guo and Gan 2006). Early responsive to dehydration stress 1 (ERD1) gene activation is dependent on TFs from the NAC family and ZF-HD, as determined by ERD1 promoter study (Tran et al. 2007). In Arabidopsis, XND1 is expressed in the xylem and linked to the ABA response, stress, and leaf senescence (Zhao et al. 2008). While GmNAC2, GmNAC3, and GmNAC4 were highly increased by osmotic stress, soybean 101 NAC domain-containing proteins that were found to be functionally non-redundant were implicated in response to abiotic stimuli and in cell death events (Pinheiro et al. 2009). While ABA, JA, and salt could also induce the soybean NACs GmNAC3 and GmNAC4, their reactions to cold vary. In tobacco leaves, GmNAC1, GmNAC5, and GmNAC6 were transiently expressed, which led to cell death and increased expression of senescence markers.

13.6.3 MYB Transcription Factors

The various roles that MYB TFs play in plant development and stress responses are well-known. They have a part in controlling the manufacture of secondary metabolites such as phenolic compounds and flavonoids, which are important for protecting plants from stress. It has been demonstrated that MYB TFs, including AtMYB2, AtMYB15, and AtMYB44, improve tolerance to salt and drought by regulating the expression of genes that respond to stress and pathways that are involved in ROS scavenging and ABA signaling (Sokouti 2024). Whether growth conditions are favorable or not, MYB TFs are crucial to numerous physiological processes, including secondary metabolism and growth conditions (Chen et al. 2006; Yanhui et al. 2006). In Arabidopsis, overexpression of MYB15 improves resistance to salt and drought (Ding et al. 2009). ABA sensitivity and osmotic tolerance are improved by increased expression levels of AtMYB2, AtMYC2, or both (Abe et al. 2003). Inducing ABA responsive stress genes, overexpression of 35S:AtMYC2, 35S:AtMYB2, and 35S:AtMYC2 + AtMYB2 in Arabidopsis resulted in an

ABA-hypersensitive phenotype with enhanced osmotic stress tolerance (Abe et al. 2003).

13.6.4 bZIP Transcription Factors

Basic leucine zipper (bZIP) transcription factors (TFs) play a crucial role in controlling how genes are expressed in response to abiotic stress conditions such as cold, salinity, and drought. They are important for the ABA signaling pathways that are necessary for stress adaptation. ABF (ABRE-binding factor), for instance, is one of the bZIP TFs that regulates the expression of genes that code for antioxidant defense and osmoprotectant production enzymes. Transgenic plants that overexpress ABF TFs have been linked to greater resistance to salinity and drought environments (Bhat et al. 2024).

13.6.5 AP2/ERF Transcription

Factor numerous genes that are responsive to stress are regulated by AP2/ERF (APETALA2/Ethylene Responsive Factor) TFs. It is well-known that these TFs affect the expression of genes connected to the ABA and ethylene signaling pathways, which are essential for plant stress reactions. It has been demonstrated that AP2/ERF TFs, such as DREB (Dehydration Responsive Element Binding), improve resistance to cold and drought stress by triggering the expression of genes associated with osmoprotectants and stress-related proteins (Lan et al. 2017).

13.6.6 bHLH Transcription Factors

Plant processes involving basic helix-loop-helix (bHLH) transcription factors include growth, development, and stress responses. They control the expression of genes related to stress responses and iron homeostasis. For instance, it has been demonstrated that bHLH TFs alter the expression of genes essential for preserving cellular homeostasis in the face of oxidative stress and iron deficiency (Carretero-Paulet et al. 2010).

13.6.7 Trihelix Transcription Factors

A less well-known but significant subset of transcription factors implicated in stress reactions are called trihelix TFs. These transcription factors control genes linked to signaling pathways for light and stress. For instance, it has been found that the trihelix TFs GmGT-2A and GmGT-2B in soybeans are involved in abiotic stress responses, namely in controlling gene expression in response to salt and drought stress (Xie et al. 2009).

13.7 Role of Key Enzymes in Plant Metabolite Homeostasis Under Abiotic Stress

In particular, when plants are under abiotic stress, key enzymes play a critical role in preserving metabolite homeostasis. These enzymes help plants adjust to environmental factors like salinity, drought, and high temperatures by facilitating a variety of biosynthetic and degradation pathways.

Reactive oxygen species (ROS), which are generally generated in response to abiotic stress, are critical to the management of antioxidant enzymes. These enzymes aid in the detoxification of ROS and protect the constituents of cells from oxidative damage. Superoxide dismutase (SOD), for instance, transforms superoxide radicals into hydrogen peroxide. This compound is subsequently further detoxified by additional antioxidant enzymes, including ascorbate peroxidase (APX) and catalase (CAT). These enzymes are necessary to keep the balance of reactive oxygen species and avoid cellular damage when under stress (Nadarajah 2021).

The synthesis of essential compounds that aid plants in overcoming stress is facilitated by biosynthetic enzymes. Proline is a crucial osmoprotectant that is produced in part by pyrroline-5-carboxylate synthase (P5CS). Glutamate is catalyzed to become pyrroline-5-carboxylate, which is a precursor to proline, by P5CS. Increased proline buildup results from P5CS overexpression under drought and salinity stress, which aids in osmotic adjustment and preserves cellular structures (Davies 2007). Another essential enzyme for the creation of trehalose, which stabilizes proteins and membranes under stress, is trehalose-6-phosphate synthase (TPS). Trehalose preserves cellular integrity, which helps shield plants from dehydration and other stress-related problems (Chen et al. 2022).

Enzymes that catalyze the breakdown of surplus or damaged metabolites are vital for maintaining cellular homeostasis and recycling vital constituents. The degradation of polyamines—small chemical molecules that are implicated in stress responses—is carried out by polyamine oxidase (PAO). When polyamines are oxidized by PAO, hydrogen peroxide ($H2O_2$) and other byproducts are produced. This process produces H_2O_2, a signaling molecule involved in stress reactions, in addition to regulating polyamine levels (Smith 2019). Autophagy, a stress-induced breakdown process that recycles damaged organelles and proteins, is mediated by autophagy-related (ATG) proteins. During extended stress, autophagy is essential for preserving cellular homeostasis and energy balance (Zhang and Li 2019).

Cytochrome P450 enzymes (CYPs) catalyze the metabolism of a wide variety of substrates, including hormones and secondary metabolites. These enzymes control the synthesis and breakdown of different chemicals, which has a major impact on how plants react to stress. For instance, CYPs play a key role in the biosynthesis and breakdown of phytohormones essential for stress responses, such as gibberellins and abscisic acid (ABA). CYPs aid in the adaptation of plants to shifting environmental conditions by controlling hormone levels (Jones and Brown 2020).

These enzymes help plants adapt to harsh environments by regulating the levels and activities of different metabolites, which ensures cellular survival and protection. They do this through their roles in antioxidant defense, biosynthesis pathways, degradation processes, and hormone metabolism.

13.8 Transport and Compartmentalization

Maintaining metabolite homeostasis in plants under abiotic stress involves not only the biosynthesis and degradation of metabolites but also their transport and compartmentalization within different cellular compartments. Membrane transporters and subcellular localization play crucial roles in these processes.

13.8.1 Role of Membrane Transporters

Membrane transporters are essential proteins that help molecules such as ions, carbohydrates, hormones, and amino acids pass through cell membranes more easily. These transporters play a crucial role in governing the intake and distribution of nutrients and metabolites, controlling osmotic balance, and preserving ion homeostasis.

The function of membrane transporters is frequently altered in plants to increase their tolerance to stress such salinity, drought, and variation in temperature. For instance, by eliminating excessive sodium ions from the cytoplasm, sodium-proton antiporters (NHX) play a critical role in preserving ion homeostasis and safeguarding cellular processes during salt stress (Saddhe and Manuka 2018). By enclosing sodium ions in vacuoles, these transporters minimize the detrimental effects of sodium ions on cellular metabolism (Zhu 2003).

Additionally essential to preserving ion balance in stressful situations are potassium (K+) transporters. Under saline conditions, potassium is necessary to maintain cellular turgor and enzyme activity. The high-affinity potassium transporter (HKT) family controls the absorption and distribution of potassium (Mäser et al. 2002). HKT transporters aid in plants' adaptation to high salinity conditions by regulating the distribution of potassium (Horie et al. 2009).

Besides ion transport, sugar transporters play a crucial role in osmotic control and energy supply. Sugars serve as osmoprotectants and energy sources, and their transportation is facilitated by sucrose transporters (SUTs) and hexose transporters (HTs). These transporters' expression is elevated in response to abiotic stress in order to preserve osmotic balance and guarantee a sufficient energy supply (Lalonde et al. 2004). For example, higher SUT activity during drought stress aids in sucrose redistribution to vital tissues, guaranteeing sustained metabolic activity and stress tolerance (Ayre 2011).

Additionally, phytohormones-important regulators of stress responses—are transported by membrane transporters. Certain transporters, such as ABC transporters and nitrate transporters, are responsible for the transportation of abscisic acid (ABA), a hormone essential to abiotic stress responses (Kang et al. 2010). These transporters make it possible for ABA to be quickly redistributed to various tissues, allowing for a coordinated reaction to stress signals (Kang et al. 2015).

Moreover, the transportation of organic acids and amino acids is essential for preserving metabolic adaptability in stressful situations. Proteins and stress-related metabolites are synthesized with the help of transporters including dicarboxylate transporters (DiTs) and amino acid permeases (AAPs), which assure the supply of nitrogen and carbon skeletons.s

13.8.2 Subcellular Localization (Vacuoles, Plastids, etc.)

Another essential strategy for preserving metabolite homeostasis in the face of abiotic stress is subcellular compartmentalization. Vacuoles, plastids, mitochondria, and the cytoplasm are examples of separate cellular compartments that play specific roles in the metabolism and storage of different substances.

For the storage of ions, metabolites, and osmoprotectants, vacuoles are essential. Vacuoles serve as storage spaces for extra ions and metabolites during stressful situations, shielding the cytoplasm from harmful consequences. The compartmentalization of Na + into vacuoles via NHX transporters helps to reduce salt stress. Furthermore, organic acids and secondary metabolites that are necessary for osmotic adjustment and stress tolerance are stored in vacuoles (Martinoia et al. 2007).

A large number of primary and secondary metabolites are synthesized in plastids, which includes chloroplasts, where photosynthesis occurs. Plasmids have a crucial role in the generation of stress-related metabolites such glutathione, proline, and ascorbate as well as antioxidants during abiotic stress. These metabolites aid in the scavenging of reactive oxygen species and the preservation of cellular structures (Gill and Tuteja 2010). The distribution of metabolic intermediates and inter-organelle communication are functions of the stromules, which are extensions of the plastid membrane (Hanson and Sattarzadeh 2011).

In addition to being necessary for the metabolism of energy, mitochondria are involved in the production of metabolites linked to stress. In response to stress, mitochondria modify their metabolic processes in order to maintain the production of ATP and to supply building blocks for the creation of defense molecules. Energy balance is preserved and oxidative stress is decreased by the dynamic fluctuations in mitochondrial membrane potential and the activation of alternate oxidase pathways (Millar et al. 2011).

The Golgi apparatus and endoplasmic reticulum (ER) are involved in the synthesis and modification of proteins. These organelles support appropriate folding and release of stress-responsive proteins, such as transporters and enzymes involved in the synthesis and breakdown of metabolites, during times of stress (Vitale and Boston 2008). The activation of several stress response pathways depends on calcium signaling, which is mediated by the ER (Kudla et al. 2010).

13.9 Conclusion

In conclusion, preserving plant metabolite homeostasis is essential for reducing the effects of abiotic stresses such as heavy metal pollution, drought, salt, and temperature stress. Plants accomplish this by maintaining a balance between reactive oxygen species (ROS) and antioxidants. This equilibrium involves both non-enzymatic antioxidants like glutathione (GSH) and ascorbate (AsA) and enzymatic antioxidants like superoxide dismutase (SOD), catalase (CAT), and ascorbate peroxidase (APX). Osmolytes such as proline, glycine betaine (GB), and trehalose aid in osmotic adjustment and cellular defense, whereas the glyoxalase system detoxifies toxic methylglyoxal (MG). These processes are controlled by networks of genetic and epigenetic modifications that involve transcription factors such AP2/ERF, WRKY, NAC, MYB, and bZIP. Gaining knowledge about these processes can help us increase agricultural sustainability and resilience by providing insights into how plants adapt to harsh environmental conditions and survive them.

References

Abe H, Urao T, Ito T, Seki M, Shinozaki K, Yamaguchi-Shinozaki K (2003) Arabidopsis AtMYC2 (bHLH) and AtMYB2 (MYB) function as transcriptional activators in abscisic acid signaling. Plant Cell 15(1):63–78

Anjum SA, Ashraf U, Tanveer M, Khan I, Hussain S, Shahzad B, Zohaib A, Abbas F, Saleem MF, Ali I, Wang LC (2017) Drought induced changes in growth, osmolyte accumulation and antioxidant metabolism of three maize hybrids. Front Plant Sci 8:69

Ayre BG (2011) Membrane-transport systems for sucrose in relation to whole-plant carbon partitioning. Mol Plant 4(3):377–394

Bhat BA, Mir RA, Mir WR, Hamdani SS, Mir MA (2024) Transcription factors-golden keys to modulate the plant metabolism to develop salinity tolerance. Plant Stress 11:100409

Bhatti KH, Anwar S, Nawaz K, Hussain K, Siddiqi E, Sharif R, Talat A, Khalid A (2013) Effect of exogenous application of glycinebetaine on wheat (Triticum aestivum L.) under heavy metal stress. Middle East J Sci Res 14:130–137

Carretero-Paulet L, Galstyan A, Roig-Villanova I, Martínez-García JF, Bilbao-Castro JR, Robertson DL (2010) Genome-wide classification and evolutionary analysis of the bHLH family of transcription factors in Arabidopsis, poplar, rice, moss, and algae. Plant Physiol 153(3):1398–1412

Chen Y, Fan X, Zhu L, Liu Z (2006) MYB transcription factors. Plant Sci 170(4):569–575

Chen X, Qiu L, Guo H, Wang Y, Yuan H, Yan D, Zheng B (2017) Spermidine induces physiological and biochemical changes in southern highbush blueberry under drought stress. Rev Bras Bot 40:841–851

Chen Y, Liu H, Zhang J (2022) Reprogramming of plant central metabolism in response to abiotic stresses. National Center for Biotechnology Information

Chinnusamy V, Dalal M, Zhu J (2014) Epigenetic regulation of abiotic stress responses in plants. In: Jenks MA, Hasegawa PM (eds) Plant abiotic stress, 2nd edn. Wiley, pp 210–212

Conde A, Silva P, Agasse A, Conde C, Gerós H (2011) Mannitol transport and mannitol dehydrogenase activities are coordinated in *Olea europaea* under salt and osmotic stresses. Plant Cell Physiol 52(10):1766–1775

Curtin SJ, Kantar MB, Yoon HW, Whaley AM, Schlueter JA, Stupar RM (2012) Co-expression of soybean Dicer-like genes in response to stress and development. Funct Integr Genomics 12(4):671–682

Davies WJ (2007) GH3-mediated auxin homeostasis links growth regulation with stress adaptation. J Biol Chem 282(13):10036–10046
Dhir B, Nasim SA, Samantary S, Srivastava S (2012) Assessment of osmolyte accumulation in heavy metal exposed *Salvinia natans*. Int J Bot 8(2):153–158
Din J Khan SU Ali I, Gurmani AR (2011) Physiological and agronomic response of canola varieties to drought stress
Ding Z, Li S, An X, Liu X, Qin H, Wang D (2009) Transgenic expression of MYB15 confers enhanced drought and salt tolerance in Arabidopsis thaliana. Biotechnol Lett 31(4):517–522
Duan CG, Wang CH, Guo HS (2017a) Epigenetic regulation of gene expression. J Genet Genomics 44(9):517–523
Duan CG, Wang X, Zhang L, Xiong X, Zhang Z, Tang K, Pan L, Hsu CCX, Tao H, Zhang WA, H. & Zhu, J. K. (2017b) A protein complex regulates RNA processing of intronic heterochromatin-containing genes in Arabidopsis. Proc Natl Acad Sci USA 114:E7377–E7384
Duan CG, Wang CH, Guo HS (2018a) Application of RNA silencing to plant disease resistance. Silence 9(1):32–39
Duan CG, Zhu JK, Cao X (2018b) Retrospective and perspective of plant epigenetics in China. J Genet Genomics 45(10):621–638
Friedrich T, Faivre L, Bäurle I, Schubert D (2019) Chromatin-based mechanisms of temperature memory in plants. Plant Cell Environ 42(3):762–770
Gill SS, Tuteja N (2010) Reactive oxygen species and antioxidant machinery in abiotic stress tolerance in crop plants. Plant Physiol Biochem 48(12):909–930
Guo Y, Gan S (2006) AtNAP, a NAC family transcription factor, has an important role in leaf senescence. Plant J 46(4):601–612
Guo H, Feng X, Hong C, Chen H, Zeng F, Zheng B, Jiang D (2017) Malate secretion from the root system is an important reason for higher resistance of *Miscanthus sacchariflorus* to cadmium. Physiol Plant 159(3):340–353
Guo Q, Zhao L, Fan X, Xu P, Xu Z, Zhang X, Meng S, Shen X (2019) Transcription factor GarWRKY5 is involved in salt stress response in diploid cotton species (Gossypium aridum L.). Int J Mol Sci 20(21): Article 5244
Han Y, Chaouch S, Mhamdi A, Queval G, Zechmann B, Noctor G (2013) Functional analysis of Arabidopsis mutants points to novel roles for glutathione in coupling H2O2 to activation of salicylic acid accumulation and signaling. Antioxid Redox Signal 18(18):2106–2121
Hanson MR, Sattarzadeh A (2011) Stromules: recent insights into a long neglected feature of plastid morphology and function. Plant Physiol 155(4):1486–1492
Hasanuzzaman M, Hossain MA, Fujita M (2011) Nitric oxide modulates antioxidant defense and the methylglyoxal detoxification system and reduces salinity-induced damage of wheat seedlings. Plant Biotechnol Rep 5(4):353–365
Hasanuzzaman M, Hossain MA, Teixeira da Silva JA, Fujita M (2012) Plant responses and tolerance to abiotic oxidative stress: Antioxidant defense is a key factor. In: Bandi V, Shanker AK, Shanker C, Mandapaka M (eds) Crop stress and its management: perspectives and strategies. Springer, pp 249–268
Hasanuzzaman M, Nahar K, Fujita M (2013) Plant response to salt stress and role of exogenous protectants to mitigate salt-induced damages. In: Ecophysiology and responses of plants under salt stress, pp 25–87
Hasanuzzaman M, Mahmud JA, Nahar K, Anee TI, Inafuku M, Oku H, Fujita M (2017a) Responses, adaptation, and ROS metabolism in plants exposed to waterlogging stress. In: Khan MIR, Khan NA (eds) Reactive oxygen species and antioxidant systems in plants: role and regulation under abiotic stress. Springer, pp 249–268
Hasanuzzaman M, Nahar K, Anee TI, Fujita M (2017b) Exogenous silicon attenuates cadmium-induced oxidative stress in *Brassica napus* L. by modulating AsA-GSH pathway and glyoxalase system. Front Plant Sci 8:1061
Hasanuzzaman M, Nahar K, Anee TI, Fujita M (2017c) Glutathione in plants: biosynthesis and physiological role in environmental stress tolerance. Physiol Mol Biol Plants 23(2):249–268

Hasanuzzaman M, Nahar K, Gill SS, Alharby HF, Razafindrabe BHN, Fujita M (2017d) Hydrogen peroxide pretreatment mitigates cadmium-induced oxidative stress in *Brassica napus* L.: an intrinsic study on antioxidant defense and glyoxalase systems. Front Plant Sci 8:115

Hasanuzzaman M, Nahar K, Hossain MS, Anee TI, Parvin K, Fujita M (2017e) Nitric oxide pretreatment enhances antioxidant defense and glyoxalase systems to confer PEG-induced oxidative stress in rapeseed. J Plant Interact 12(1):323–331

Hegedus D, Baldwin D, Kenaschuk E, Sharpe A, Parkin I (2003) NAC domain transcription factors. Can J Bot 81(2):109–121

Hernández JA, Barba-Espín G, Diaz-Vivancos P (2017) Glutathione-mediated biotic stress tolerance in plants. In: Hossain M, Mostofa M, Diaz-Vivancos P, Burritt D, Fujita M, Tran LS (eds) Glutathione in plant growth, development, and stress tolerance. Springer, pp 309–319

Hideg E, Jansen MA, Strid A (2013) UV-B exposure, ROS, and stress: inseparable companions or loosely linked associates? Trends Plant Sci 18(3):107–115

Horie T, Hauser F, Schroeder JI (2009) HKT transporter-mediated salinity resistance mechanisms in Arabidopsis and monocot crop plants. Trends Plant Sci 14(12):660–668

Hu H, Dai M, Yao J, Xiao B, Li X, Zhang Q, Xiong L (2006) Overexpressing a NAM, ATAF, and CUC (NAC) transcription factor enhances drought resistance and salt tolerance in rice. Proc Natl Acad Sci USA 103(35):12987–12992

Jaleel CA, Gopi R, Sankar B, Gomathinayagam M, Panneerselvam R (2008) Differential responses in water use efficiency in two varieties of *Catharanthus roseus* under drought stress. C R Biol 331(1):42–47

Jin C, Jiang M, Zhang J, Zhang S (2015) Heat stress-induced production of reactive oxygen species and the role of the antioxidant defense in Ficus concinna seedlings. Plant Physiol Biochem 96:57–63

Jonytienė V, Samuolienė G, Sirtautas R, Duchovskis P, Brazaitytė A (2012) Exogenous application of proline enhances tolerance of plants to cold stress. Acta Physiol Plant 34(4):1485–1493

Kang J, Hwang J-U, Lee M, Kim Y-Y, Assmann SM, Martinoia E (2010) PDR-type ABC transporter mediates cellular uptake of the phytohormone abscisic acid. Nature 464(7287):569–573

Kang J, Yoo H, Lee S, Kim SY (2015) Abscisic acid transporters in plants: Components and mechanisms. Plant Signa Behav 10(8):Article e1046609

Kapoor M, Arora R, Lama T, Nijhawan A, Khurana JP, Tyagi AK, Kapoor S (2008) Genome-wide identification, organization, and phylogenetic analysis of Dicer-like, Argonaute, and RNA-dependent RNA polymerase gene families and their expression analysis during reproductive development and stress in rice. Plant Physiol 146(1):165–181

Kathuria H, Giri J, Nataraja KN, Murata N, Udayakumar M, Tyagi AK (2009) Glycinebetaine-induced waterstress tolerance in codA-expressing transgenic indica rice is associated with up-regulation of several stress responsive genes. Plant Biotechnol J 7(5):512–526

Kawakami K, Sato Y, Yoshida M (2008) Genotypic differences in the accumulation of glycine-betaine and the mechanism of protection against abiotic stress in barley. Plant Cell Environ 31(6):1211–1218

Kim JM, To TK, Nishioka T (2015) Stress-induced DNA demethylation and the triggering of memories. Nat Rev Genet 16(9):593–605

Kirchhoff H, Sharpe RM, Herbstová M, Yarbrough R, Edwards GE (2014) Differential mobility of pigmentprotein complexes in granal and agranal thylakoid membranes of C3 and C4 plants. Plant Physiol 166(2):1571–1583

Kohli SK, Handa N, Sharma A, Bhardwaj R, Abd Allah EF, Alqarawi AA, Ahmad P (2018) Glycine betaine ameliorates plants under abiotic stresses: A review. Plant Physiol Biochem 123:14–23

Kudla J, Batistic O, Hashimoto K (2010) Calcium signals: The lead currency of plant information processing. The Plant Cell 22(3):541–563

Lalonde S, Wipf D, Frommer WB (2004) Transport mechanisms for organic forms of carbon and nitrogen between source and sink. Ann Rev Plant Biol 55:341–372

Lamke J, Bäurle I (2017) Epigenetic and chromatin-based mechanisms in environmental stress adaptation and stress memory in plants. Genome Biol 18(1):Article 124

Lan W, Zhang Y, Zhao T, Zhang H, Zhang S (2017) NAC transcription factors in plant abiotic stress response: An updated review of the functional mechanisms and applications in crop improvement. Front Plant Sci 8:Article 1905

Latowski D, Kruk J, Strzałka K (2010) Chloroplast isoprenoid quinones and their role in the antioxidant defense system. Photosynth Res 96(1):171–181

Lei M, Zhang H, Julian R, Tang K, Xie S (2014) DNA methylation regulates gene expression in plants. Plant Physiol 166(3):1172–1184

Li J, Pandeya D, Nath K, Zulfugarov IS, Yoo SC, Zhang H, Yoo JH, Cho SH, Koh HJ, Kim DS, Seo HS (2010) Zebra-necrosis, a thylakoid-bound protein, is critical for the photoprotection of developing chloroplasts during early leaf development. Plant J 62(4):713–725

Lindroth AM, Shultis D, Jasencakova Z, Fuchs J, Johnson L, Schubert D, Patnaik D, Pradhan S, Goodrich J, Schubert I, Jenuwein T, Khorasanizadeh S, Jacobsen SE (2004) Dual histone H3 methylation marks at lysines 9 and 27 required for interaction with chromomethylase 3. EMBO J 23(21):4286–4296

Liu R, Lang Z (2019) The mechanism and function of active DNA demethylation in plants. J Integr Plant Biol 61(1):33–49

Lou L, Li X, Chen J, Li Y, Tang Y, Lv J (2018) Photosynthetic and ascorbate-glutathione metabolism in the flag leaves as compared to spikes under drought stress of winter wheat (*Triticum aestivum* L.). PLoS One 13(3):e0194625

Lu Y, Bu Q, Chuan M, Cui X, Zhao Y, Zhou DX (2023) Metabolic regulation of the plant epigenome. Plant J 114(5):1001–1013

Luo X, He Y (2019) Experiencing winter for spring flowering – a molecular epigenetic perspective on vernalization. J Integr Plant Biol 61(1):33–49

Mahmud JA, Hasanuzzaman M, Nahar K, Rahman A, Hossain MS, Fujita M (2017) Maleic acid assisted improvement of metal chelation and antioxidant metabolism confers chromium tolerance in *Brassica juncea* L. Ecotoxicol Environ Saf 144:216–226

Makarevitch I, Waters AJ, West PT, Stitzer M, Hirsch CN, Ross-Ibarra J, Springer NM (2015) Transposable elements contribute to activation of maize genes in response to abiotic stress. PLoS Genet 11(1):e1004915

Mann A, Kumar A, Saha M, Lata C, Kumar A (2019) Stress induced changes in osmoprotectants, ionic relations, antioxidants activities and protein profiling characterize *Sporobolus marginatus* Hochst. ex A. Rich. salt tolerance mechanism. Indian J Exp Biol 57:672–679

Mann A, Kumar N, Kumar A, Lata C, Kumar A, Meena BL, Mishra D, Grover M, Gaba S, Parameswaran C, Mantri N (2021) de novo transcriptomic profiling of differentially expressed genes in grass halophyte *Urochondra setulosa* under high salinity. Sci Rep 11(1):5548

Mann A, Kumar N, Lata C, Kumar A, Meena BL, Kumar A (2024) Physiological and differential gene expression reveals a trade-off between antioxidant capacity and salt tolerance in halophytes *Urochondra setulosa* and *Dichanthium annulatum*. Plant Growth Regul 102:555–570

Martinoia E, Maeshima M, Neuhaus HE (2007) Vacuolar transporters and their essential role in plant metabolism. J Exp Bot 58(1):83–102

Mäser P, Gierth M, Schroeder JI (2002) Molecular mechanisms of potassium and sodium uptake in plants. Plant Soil 247(1):43–54

Millar AH, Mittova V, Kiddle G, Heazlewood JL, Bartoli CG, Theodoulou FL, Foyer CH (2011) Control of ascorbate synthesis by respiration and its implications for stress responses. Plant Physiol 153(2):572–585

Mladenov V, Fotopoulos V, Kaiserli E, Karalija E, Maury S, Baranek M, Segal NA, Testillano PS, Vassileva V, Pinto G, Nagel M (2021) Deciphering the epigenetic alphabet involved in transgenerational stress memory in crops. Int J Mol Sci 22(13):7118

Nadarajah K (2021) ROS homeostasis in abiotic stress tolerance in plants. Semantic Scholar

Nafees M, Fahad S, Shah F, Sha AM, Bukhari MA, Maryam A, Ahmed I, Ahmad S, Hussain S (2019) Reactive oxygen species signaling in plants. In: Hasanuzzaman M, Hakeem K, Nahar K, Alharby H (eds) Reactive oxygen species signaling in plants. Springer, pp 259–272

Nahar K, Hasanuzzaman M, Alam MM, Fujita M (2015) Roles of exogenous glutathione in antioxidant defense system and methylglyoxal detoxification during salt stress in mung bean. Biol Plant 59(4):745–756

Nahar K, Hasanuzzaman M, Alam M, Rahman A, Mahmud J-A, Suzuki T, Fujita M (2017) Insights into spermine-induced combined high temperature and drought tolerance in mung bean: Osmoregulation and roles of antioxidant and glyoxalase system. Protoplasma 254(2):445–460

Ningthoujam M, Habib K, Bano F, Zutshi S, Fatma T (2013) Exogenous osmolytes suppresses the toxic effects of malathion on *Anabaena variabilis*. Ecotoxicol Environ Saf 94:21–27

Pinheiro GL, Marques CS, Costa MD, Reis PA, Alves MS, Carvalho CM, Fietto LG, Fontes EP (2009) Complete inventory of soybean NAC transcription factors: sequence conservation and expression analysis uncover their distinct roles in stress response. Gene 444(1–2):10–23

Rahman A, Nahar K, Hasanuzzaman M, Fujita M (2016) Manganese-induced cadmium stress tolerance in rice seedlings: Coordinated action of antioxidant defense, glyoxalase system, and nutrient homeostasis. C R Biol 339(10):462–474

Rahman AK, Al Mahmud J, Hasanuzzaman M, Hossain MS, Fujita M (2017) Salt stress tolerance in rice: emerging role of exogenous phytoprotectants. In: Li Q (ed) Advances in international rice research. InTech, pp 139–174

Reda F, Mandoura HM (2011) Response of enzyme activities, photosynthetic pigments, proline to low or high temperature stressed wheat plant (*Triticum aestivum* L.) in the presence or absence of exogenous proline or cysteine. Biol Plant 55(2):335–340

Saddhe AA, Manuka R (2018) Plant sugars: homeostasis and transport under abiotic stress in plants. Semantic Scholar

Sairam RK, Dharmar K, Lekshmy S, Chinnusamy V (2011) Expression of antioxidant defense genes in mung bean (*Vigna radiata* L.) roots under water-logging is associated with hypoxia tolerance. Acta Physiol Plant 33(3):735–744

Sidhu GPS, Singh HP, Batish DR, Kohli RK (2017) Appraising the role of environment-friendly chelants in alleviating lead by *Coronopus didymus* from Pb-contaminated soils. Chemosphere 182:129–136

Singh M, Kumar J, Singh S, Singh VP, Prasad SM (2015) Roles of osmoprotectants in improving salinity and drought tolerance in plants: a review. Rev Environ Sci Biotechnol 14(3):407–426

Singh A, Kumar A, Yadav S, Singh IK (2019a) Reactive oxygen species-mediated signaling during abiotic stress. Plant Gene 18:100173

Singh M, Singh VP, Prasad SM (2019b) Nitrogen alleviates salinity toxicity in *Solanum lycopersicum* seedlings by regulating ROS homeostasis. Plant Physiol Biochem 141:466–476

Smith AM (2019) Polyamine catabolism in plant stress responses. Springer 111(3):719–728

Soares C, Carvalho MEA, Azevedo RA, Fidalgo F (2019) Plants facing oxidative challenges – a little help from the antioxidant networks. Environ Exp Bot 161:4–25

Sokouti B (2024) Molecular plant physiology for model plants under abiotic stress conditions. OBM Genetics 8(2):1–18

Tang W, Luo C (2018) Overexpression of zinc finger transcription factor ZAT6 enhances salt tolerance. Open Life Sci 13(1):431–445

Tanveer M, Shahzad B, Sharma A, Khan EA (2019) 24-Epibrassinolide application in plants: an implication for improving drought stress tolerance in plants. Plant Physiol Biochem 135:295–303

Tran LSP, Nakashima K, Sakuma Y, Simpson SD, Fujita Y, Maruyama K, Fujita M, Seki M, Shinozaki K, Yamaguchi-Shinozaki K (2004) Isolation and functional analysis of Arabidopsis stress-inducible NAC transcription factors that bind to a drought-responsive cis-element in the early responsive to dehydration stress 1 promoter. Plant Cell 16(9):2481–2498

Tran LSP, Quach TN, Guttikonda SK, Aldrich DL, Kumar R, Neelakandan A, Nguyen HT (2007) Comparative analysis of NAC transcription factor expression and function in soybean and Arabidopsis. Plant Mol Biol 63(6):787–805

Vitale A, Boston RS (2008) Endoplasmic reticulum quality control and the unfolded protein response: insights from plants. Traffic 9(10):1581–1588

Wang X, Li Q, Jin H, Liu X (2013b) Transposable elements and gene expression in plants. Genomics 102(4):240–245

Wang Y, Zhang H, Hou P, Su X, Zhao P, Zhao H, Liu S (2014) Foliar-applied salicylic acid alleviates heat and high light stress induced photoinhibition in wheat (*Triticum aestivum*) during the grain filling stage by modulating the *psbA* gene transcription and antioxidant defense. Plant Growth Regul 73(3):289–297

Wani SH, Singh NB, Haribhushan A, Mir JI (2013) Compatible solute engineering in plants for abiotic stress tolerance: role of glycine betaine. Curr Genomics 14(3):157–165. https://doi.org/10.2174/1389202911314030001

Wu K, Chen S (2010) Histone methylation and plant stress responses. J Exp Bot 61(3):791–799

Xie ZM, Zou HF, Lei G, Wei W, Zhou QY, Niu CF, Liao Y et al (2009) Soybean Trihelix transcription factors GmGT-2A and GmGT-2B improve plant tolerance to abiotic stresses in transgenic Arabidopsis. PLoS One 4(9):e6898

Yanhui C, Xiaoyuan Y, Kun H, Meihua L, Jigang L, Zhaofeng G, Xiaoxiao Z, Yunfei Z, Yunlong L, Xun L, Yiyu L, Li-Jia Q (2006) The MYB transcription factor superfamily of Arabidopsis: Expression analysis and phylogenetic comparison with the rice MYB family. Plant Mol Biol 60(1):107–124

Zhang X, Li H (2019) Autophagy and plant stress tolerance. Nat Rev Mol Cell Biol 20(5):285–298

Zhang H, Lang Z, Zhu JK (2018) Dynamics and function of DNA methylation in plants. Nat Rev Mol Cell Biol 19(8):489–506

Zhao C, Zhang Z, Xie S, Li M, Luo X (2008) XND1 transcription factor regulates xylem development and response to abiotic stress in Arabidopsis. Plant Cell 20(9):2693–2706

Zhu JK (2003) Regulation of ion homeostasis under salt stress. Curr Opin Plant Biol 6(5):441–445

Integrating Multi-omics Strategies to Enhance Crop Resilience in a Changing Climate

14

Gurpreet Kaur, Charu Lata, Anshul Sharma Manjul, Priti, and Naresh Kumar

Abstract

Climate change poses a significant threat to global food security, particularly for crop resilience. This chapter explores the complex and far-reaching influences of climate change on a wide range of crops, including vegetables, cereals and pulses. To address these challenges, a comprehensive and multidisciplinary approach is essential, and it offers a promising approach. Genomics, transcriptomics, metabolomics, proteomics, phenomics and ionomics provide a comprehensive understanding of the genetic, molecular and physiological mechanisms underlying crop responses to climate stress. By integrating these approaches, scientists can find key genes, pathways and metabolites involved in stress tolerance. This knowledge can be leveraged to develop climate-resilient crop varieties through breeding, genetic engineering and agronomic management practices. The chapter highlights the potential of multi-omics strategies in enhancing crop resilience by improving stress tolerance, yield stability and nutritional quality. Future research efforts should aim to expand the application of multi-omics techniques to a wider range of crop species and diverse climatic conditions. By integrating multi-omics data with field phenotyping and modelling, researchers can

G. Kaur
ICAR-Central Soil Salinity Research Institute, Karnal, Haryana, India

C. Lata · A. S. Manjul
ICAR-Indian Institute of Wheat and Barley Research, Regional Station, Shimla, Himachal Pradesh, India

Priti
K. L. Mehta Dayanand College for Women, Faridabad, Haryana, India

N. Kumar (✉)
Eternal University, Baru Sahib, Himachal Pradesh, India

A. Mann et al. (eds.), *Cutting Edge Technologies for Developing Future Crop Plants*,
https://doi.org/10.1007/978-981-96-2508-6_14

gain a more holistic understanding of how crops respond to climate change, ultimately facilitating the development of robust adaptation strategies.

Keywords

Omics · Climate change · Crops · Multi-omics · Food security

14.1 Introduction

Climate variability often destabilizes agroecosystems and disrupts the food supply chain (Semeraro et al. 2023). Because agriculture is highly dependent on prevailing climatic conditions, crops are frequently exposed to environmental and abiotic stresses resulting from these shifts (Kumar et al. 2018; Muluneh 2021). Crop plants may encounter multiple abiotic stresses at once, including salinity, low temperatures, drought, waterlogging, heavy metal exposure and temperature extremes, all of which severely affect their growth and development, leading to reduced agricultural productivity worldwide (Kar et al. 2024). Global warming, primarily caused by greenhouse gas emissions, has resulted in rising temperatures and more frequent droughts, posing additional risks to crop productivity. For example, wheat and maize yields have declined by 20–30% under these conditions (Muluneh 2021). While traditional plant breeding has successfully produced many high-yielding crop varieties, these approaches require long development and breeding cycles, making it difficult to quickly meet the increasing food supply demands in a changing climate. This time lag limits plant breeders' ability to respond rapidly to these emerging challenges.

Contemporary plant breeding prioritizes the development of high-yielding crop varieties that possess desired traits while requiring minimal external inputs throughout their growth cycle. This advancement has been achieved through the implementation of modern biotechnological techniques, which facilitate the introduction of genes with novel characteristics by overcoming the limitations imposed by species and genus (Giudice et al. 2021; Rashid and Stingl 2015). The ultimate goal is to expedite the development of crops that exhibit resilience to adverse environmental conditions inherited from their ancestors. However, this does not imply that conventional breeding is obsolete in agricultural practices. Indeed, the contemporary agricultural landscape is characterized by a concerted effort to maximize production from existing agricultural resources to meet the escalating demand for food in the face of a rapidly growing global population (Mrabet 2023). Achieving significant increases in agricultural yield while simultaneously enhancing crop resilience to adverse environmental conditions within limited land areas has proven to be beyond the capabilities of traditional breeding methods alone. While conventional breeding methods are traditional and often seen as less controversial, their effectiveness is hampered by several factors. These include the lengthy time required to achieve desired outcomes, the inability to precisely identify the introduction of specific

Table 14.1 Climate change-induced environmental stresses and the importance of multi-omics techniques in enhancing crop resilience

Abiotic stress	Effect on crops	Multi-omics approach	Role in crop improvement	References
Drought	Reduces water availability, causing stunted growth, wilting and reduced yield	*Genomics*: Identifying drought-resistant genes	Developing drought-resistant crop varieties	Paudel et al. (2024)
Heat stress	Leads to protein denaturation, reduced photosynthesis and impaired reproductive growth	*Proteomics*: Studying heat-tolerant proteins	Enhancing heat tolerance through protein-level modifications	Jan et al. (2023), Mondal et al. (2023)
Salinity	Causes ion toxicity, osmotic stress and nutrient imbalances	*Metabolomics*: Monitoring metabolic adjustments to salinity	Engineering crops with better ion homeostasis and metabolic regulation	Rajkumari et al. (2023), Salam et al. (2023)
Cold stress	Damages cell membranes and inhibits growth, flowering and seed set	*Transcriptomics and epigenomics*: Understanding gene expression changes under cold stress	Improving cold tolerance by altering epigenetic markers	Kumar and Rani (2023), Li et al. (2024)
Flooding/ waterlogging	Reduces oxygen availability to roots, leading to root damage and plant death	*Genomics and transcriptomics*: Identifying genes for waterlogging resistance	Developing varieties that tolerate hypoxic conditions	Owusu et al. (2023), Tyagi et al. (2023a, b)
Heavy metal toxicity	Causes toxicity, nutrient imbalances and disruption of plant physiological processes	*Transcriptomics, proteomics and metabolomics*: Tracking detoxification pathways	Engineering crops for better metal ion detoxification and tolerance	Sharma and Jha (2023), Yadav et al. (2023)

traits and the limited transfer of genetic variation between crossed species (Giudice et al. 2021). These constraints have accelerated the adoption of modern plant breeding and multi-omics approaches, which aim to tackle the urgent need for sustainable agricultural intensification amid rising global food demands and shifting climate conditions (Anand and Chaudhary 2024). Table 14.1 summarizes the general effects of climate-induced abiotic stresses and underscores the critical role of multi-omics approaches in advancing crop improvement. This chapter explores the challenges that climate change presents to crop production. Multi-omics analysis is presented as a potential solution for improving crop resilience and yield in changing environmental conditions.

14.2 Climate Change and Crop Yields

The agricultural systems across the globe are encountering unparalleled difficulties as the climate undergoes substantial transformations. It is anticipated that climate change will produce extensive repercussions on agriculture and food production, with certain regions exhibiting greater susceptibility compared to others (Singh et al. 2023). Whereas in regions like Sub-Saharan Africa and South Asia, where the problem of food insecurity is hard-hitting, the effects of climate change could exacerbate current challenges and paralyse such systems. One of the strongest pathways through which climate change affects food security is via impacts on agricultural yield.

Climate change can directly impact factors like plant growth, soil moisture and erosion, which are crucial for crop yields. Estimates suggest that by 2050, food demand will increase over fivefold in Africa and more than double in Asia, underscoring the immense strain that climate change is likely to place on agricultural systems in these regions (Mrabet 2023). Climate change is expected to reduce rainfall by 20% in Sub-Saharan Africa, potentially leading to arable land loss due to reduced soil moisture, increased aridity and salinity, and groundwater depletion (Omotoso et al. 2023). This would lead to a loss of agricultural land, reduced farmers' incomes and disruptions to food supply systems. South Asia faces similar trends, where rising temperatures and altered precipitation patterns are expected to adversely affect crop yields and intensify existing food insecurity. Understanding the impacts of climate change on crop productivity in these vulnerable regions is essential for developing policies and adaptation strategies to mitigate negative effects. Developing a better understanding of these complex relationships is a vital step in ensuring global food security in the face of a rapidly changing climate.

The production of vegetables, cereals and pulses, which are crucial components of the human diet, is particularly vulnerable to the effects of climate change. The changes in temperature, precipitation patterns and the frequency of extreme weather events can significantly disrupt the growth, yield and quality of these essential food crops. Recent studies have highlighted the complex and multifaceted nature of the impact of climate change on the agricultural sector (Kar et al. 2024; Kumar et al. 2018). The impacts of climate change on vegetable production are multifaceted and far-reaching. Elevated carbon dioxide levels, higher temperatures and altered precipitation patterns can directly affect plant growth, development and yield, leading to significant losses in vegetable production. Additionally, changes in the number of days suitable for plant growth, as well as the decreased availability of water and declines in soil quality, can pose significant challenges for vegetable cultivation. Furthermore, climate change can also indirectly impact vegetable production through the increased incidence of pests, diseases and the spread of invasive species, all of which can further threaten crop yields and quality.

Most studies dealing with climate change affecting wheat production focus on the expected rise in the level of carbon dioxide. The study on the potential adaptability of wheat to climate change in the Indo-Gangetic Plains, indicated that although global warming could favour wheat production in some areas, in those that are susceptible to extremely high temperatures, it would reduce the yields (Ortiz et al. 2008). They emphasized the need for heat-tolerant wheat genotypes to ensure such risks are reduced to some extent. In a study by Anwar et al. (2007), making use of the CropSyst model, simulated wheat yield potential under climate change in southeastern Australia. Their result showed that the mean wheat yield could decrease by up to about 25% because of increased atmospheric CO_2. Similarly, the CERES-wheat model was applied to various climate scenarios to estimate wheat yield under future climate conditions (Eitzinger et al. 2003). Findings by the said authors indicated that carbon dioxide had a major effect on the increase in crop yield within the study area. For Southern Australia, Luo et al., (2003) established the effects of climate change on wheat production in regard to the 2080s using the DSSAT 3.5 CERES-wheat models. They found that wheat yield is projected to show increments under all levels of carbon dioxide, but drier sites tend to be more suited for the production, which may find lower-quality wheat.

As a major global crop, the relationship between maize yield and climate change has been extensively studied. In a study, the CERES-Maize model with CCCM and GISS climate models to analyse rainfed maize yield and found that dry matter could increase by 1.4–2.1 t/ha with the CCCM model and by 3.5–5.6 t/ha with the GISS model (Cuculeanu et al. 2002). Conde et al. (2006) proposed strategies like construction of greenhouse, use of compost and drip irrigation to improve farmers' adaptation to climate change in Mexico, based on their rainfed maize analysis using the CERES-Maize model. The previous study employed the CERES-Maize model and the non-parametric Mann-Kendall test to predict sustainable maize production for smallholders in South Africa under various climate scenarios (Walker and Schulze 2006). Their results suggested that increase in the inorganic nitrogen and harvesting of rainwater could boost yield of crop in the long term. Similarly, Soler et al. (2007) investigated the effects of dates planting and weather conditions on production of maize in Brazil using the CERES-Maize model. They found that later planting dates could reduce average yield by 55% under rainfed conditions and 21% under irrigated conditions. Additionally, accurate yield forecasts could be made approximately 45 days before harvest.

The effects of climate change on crop yield vary regionally, with some areas experiencing increases and others decreases. These variations are influenced by latitude and irrigation practices. Crop yield can be enhanced through irrigation and increased precipitation during the growing season, but is more sensitive to precipitation than temperature. If future water availability declines, soils with high water-holding capacity can help reduce drought frequency and improve crop yield. Climate change may shorten the growing season, necessitating adjustments to planting dates for optimal crop production. Additionally, reduced crop rotation periods due to climate change require farmers to carefully consider crop diversities, date of sowing, density of crops and level of fertilization (Faye et al. 2023; Ghaffar et al. 2022). The positive impacts of climate change on agriculture include increased carbon dioxide concentrations, extended growing seasons in higher latitudes and mountainous areas and enhanced crop growth. However, negative effects include rising pest and disease incidence and soil degradation caused by temperature changes. These challenges have prompted scientists to develop crop varieties that are better suited to changing climates and degrading soils to ensure sufficient yields for the growing population. Quantifying climate uncertainty is a crucial factor in predicting crop yield variations under future climate scenarios.

14.3 Harnessing Omics for Crop Adaptation to Climate Change

14.3.1 Genomics Approach

The study of genomics focuses on the entire genetic composition of an organism including all its DNA sequences such as genes, regulatory components and non-coding segments. A key objective of genomics in crop science is to determine genetic variations linked to different characteristics which may include drought resistance, disease resistance or yield among others. This is as a result that Genomics-assisted breeding (GAB) has progressively turned into a building block for growing crops and livestock which can withstand changes in climate. In parallel the identifications of genes and alleles that are related to stress response and other important traits have been made easier by advancements in genome sequencing, assembly and annotation (Kamali and Singh 2023; Zenda et al. 2021a). Breeding systems have been streamlined through the integration of genomics and phenomics, which has led to more cost-effective and faster cultivar development cycles (Ndlovu 2020). GAB has been effective for enhancing tolerance to different stresses including heat, heat and salinity in major crops (Kole et al. 2015). Future prospects for improving crop resilience against climate change include high-throughput phenotyping, pan-genomics, CRISPR-Cas9 among others (McManus et al. 2020; Zenda et al. 2021a). These genomic approaches are vital for adapting agriculture to climate change and securing global food supplies.

Genomics enables the precise identification of specific genes and genetic markers linked to resilience against various climate-related stresses. Techniques such as quantitative trait loci mapping and GWAS are used for pinpoint genomic regions associated with stress tolerance. Advanced genomic technologies, including next-generation sequencing, have furthered the identification of stress-responsive genes, regulatory networks and genetic variations (Kamali and Singh 2023). Integrating transcriptomics, which uncovers gene expression changes under stress, with various omics data allows for manipulation of crop genomes to improve stress tolerance (Kamali and Singh 2023). These advancements have facilitated the development of stress-resilient crops using conventional breeding, genetic engineering and gene-editing techniques (Choudhary and Muthamilarasan 2022). By uncovering the molecular mechanisms and genetic mechanism behind stress tolerance, genomics and transcriptomics provide promising solutions for securing food production in the face of climate change (Kole et al. 2015). Once identified, these genomic regions can be targeted in breeding programmes to develop new crop varieties with enhanced resilience.

A key application of genomics in crop improvement is genomic selection, which utilizes DNA markers across the genome to predict the performance of individual plants in breeding programmes. DNA markers have revolutionized plant breeding by enabling more accurate and efficient selection of desirable traits. Genomic selection (GS), which leverages all marker data to predict plant performance, offers improved accuracy for complex traits compared to traditional marker-assisted selection (Jannink et al. 2010). This approach has accelerated crop improvement and reduced costs. Molecular markers, unaffected by environmental factors, are essential for QTL identification, genetic diversity assessment and genome mapping, with recent sequencing advancements further enhancing breeding programmes through EST-SSR and single-nucleotide polymorphism (SNP) markers (MirMohammadi Maibody and Golkar 2019). Marker-assisted selection has significantly shortened the time required to bring new crop varieties to market, empowering plant breeders globally (Hasan et al. 2021). By selecting plants with favourable genetic profiles, breeders can expedite the development of resilient crop varieties, particularly for complex traits such as drought tolerance, which are governed by multiple genes.

14.3.2 Transcriptomic Approach

Transcriptomics, which involves studying RNA transcripts produced by a genome, has become important method for understanding the expression of genes and regulation in crops. Unlike genomics, which provides a static map of the DNA sequence, transcriptomics offers a dynamic view of how genes are expressed under varying environmental conditions and developmental stages (Kumar et al. 2016; Marguerat and Bähler 2010). This dynamic approach is especially valuable in crop science, as understanding the molecular mechanisms behind characteristics, for example, stress tolerance, disease resistance and yield can greatly inform breeding programmes and genetic engineering initiatives.

The primary tool for transcriptomic analysis is RNA sequencing, which has emerged as a powerful tool for transcriptomic analysis in plant biology, enabling comprehensive gene expression profiling under various conditions. Studies have utilized RNA-seq to investigate plant responses to abiotic stresses such as drought and salinity. In wheat, drought stress during early reproductive stages significantly impacted gene expression, with 309 differentially expressed genes (DEGs) identified across developmental stages (Ma et al. 2017). Similarly, cucumber exhibited 2265 DEGs under drought stress, with genes related to sucrose biosynthesis and abscisic acid (ABA) signalling specifically induced (Wang et al. 2018). In chickpea, 2022 DEGs were recognized in salt stress, including those related to flowering time and pollen development (Kaashyap et al. 2022). Transcriptomic analysis of halophytic plants, *Dichanthium annulatum* (Mann et al. 2019) and *Urochondra setulosa* (Mann et al. 2021) identified DEGs specifically for high salt tolerance and oxidative stress. RNA-seq analysis of little millet under drought and salt stress revealed tissue-specific gene expression patterns and identified 61 common DEGs in both leaf and root tissues (Das et al. 2020). These studies demonstrate the utility of RNA-seq in elucidating plant stress responses.

Transcriptomic approaches have proven valuable in elucidating complex regulatory networks in crops under drought stress. Studies in rice, maize, sorghum, halophytes have revealed the upregulation of genes involved in osmotic adjustment, antioxidant defence and ABA signalling (Dugas et al. 2011; Miao et al. 2017; Mohanty et al. 2016; Mann et al. 2024). Maize drought-responsive genes are implicated in photosynthesis, stomatal regulation, hormone signalling, detoxification and sugar metabolism (Singh et al. 2019). A comparative transcriptome analysis of drought-resistant and drought-sensitive maize lines uncovered crucial genes associated with stress signalling, maintenance of redox balance, transcription factor regulation and the biosynthesis of amino acids (Zenda et al. 2019). In rice, transcriptome profiling has elucidated complex regulatory networks and signalling pathways during abiotic stresses (Kumar and Dash 2019). Across crops, abscisic acid (ABA) plays a central role in regulating stress responses through ABA-dependent and independent signalling pathways (Goharrizi et al. 2022). Proteomic studies have also revealed important protein modifications that contribute to stress tolerance. Integration of transcriptome data with genome-scale metabolic networks has identified key transcription factors, such as AP2/ERF, bZIP, MYB and NAC, that control important nodes in gene regulatory pathways (Mohanty et al. 2016). Co-expression analysis in maize grouped drought-response genes into modules, highlighting the crosstalk between stress and developmental signalling pathways (Miao et al. 2017). Systems biology approaches have facilitated the construction of network maps by integrating functional genomics data from both model plants and food crops, leading to the identification of commonly regulated signalling pathways and genes in response to drought (Jogaiah et al. 2013). These findings provide valuable information for developing drought-resistant crop varieties and improving breeding strategies (Goharrizi et al. 2022; Singh et al. 2019).

Additionally, transcriptomics is being combined with other omics approaches—such as genomics, proteomics and metabolomics—to offer a more comprehensive view of crop biology. This integrative approach helps to link changes in gene expression with alterations in protein levels and metabolic pathways, offering a more comprehensive view of how crops respond to environmental challenges. As transcriptomic technologies continue to evolve, new methodologies such as single-cell RNA sequencing (scRNA-seq) are being developed, offering even greater resolution and insights into gene expression at the single-cell level. These advances are expected to further enhance our understanding of crop resilience and drive the development of crops that are better equipped to withstand the stresses of climate change.

14.3.3 Metabolomics Approach

Metabolomics is the study of the complete set of metabolites within a biological sample. Metabolites are the end products of cellular processes and their levels can change rapidly in response to environmental conditions. Metabolomics is a powerful analytical tool for understanding plant responses to abiotic stresses and improving crop resilience (Carrera et al. 2021). This integration allows for an in-depth study of metabolic pathways and regulatory networks involved in stress tolerance, especially in key crops like wheat (Razzaq et al. 2021). Advanced technologies such as nuclear magnetic resonance (NMR) and mass spectrometry (MS) have significantly enhanced the identification and quantification of metabolites in both targeted and untargeted approaches (Katam et al. 2022). By profiling the metabolome, researchers can gain insights into the biochemical pathways that are activated or repressed during stress conditions. Metabolomics provides a functional readout of the cellular state, offering a direct link between gene expression and physiological responses.

Metabolomics is particularly valuable for climate change, as it helps identify biochemical markers associated with stress tolerance. For example, during drought conditions, certain osmoprotectants—such as proline, sugars and polyols—accumulate in plants to protect cells from dehydration. Metabolomic analysis can quantify these metabolites, providing insights into the metabolic adjustments that contribute to drought tolerance. Metabolomics also helps in understanding the trade-offs that occur during stress responses. For instance, while certain metabolic pathways may be upregulated to confer tolerance to a specific stress, this may come at the cost of growth or yield. By analysing the metabolome, researchers can identify these trade-offs and develop strategies to optimize stress tolerance without compromising productivity.

Several studies have demonstrated the power of metabolomics in improving crop resilience. Recent developments, including ion mobility spectrometry-MS, have improved the resolution of isomeric and isobaric metabolites (Mann et al. 2015; Katam et al. 2022). Metabolomics has applications in various areas of plant biology, including stress response mechanisms, spatial distribution of metabolites across different plant structures and single-cell analysis (Katam et al. 2022). Integration of metabolomics with other omics approaches and its application in speed breeding programmes offer promising avenues for crop improvement and ensuring food safety in the face of climate change (Razzaq et al. 2021). This technology has significant applications in crop improvement, including stress tolerance, phenotyping and biomarker identification (Alawiye and Babalola 2021; Razzaq et al. 2019). The implementation of advanced analytical methodologies, such as NMR, LC-MS and GC-MS, has facilitated the rapid advancement of metabolic profiling (Razzaq et al. 2019). Metabolomics plays a crucial role in studying plant stress responses by identifying stress-related compounds, signal transduction molecules and acclimation response components (Shulaev et al. 2008). Integration of metabolomics with other omics platforms enables a holistic view of plant stress responses and detailed pathway analysis (Shulaev et al. 2008). This approach has been applied to various crops, including maize, sunflower and wheat, to enhance growth, yield and pathogen resistance (Alawiye and Babalola 2021). Metabolomics-assisted breeding shows promise for efficient screening of crop yield and stress tolerance (Razzaq et al. 2019). Metabolomic analysis in wheat (*Triticum aestivum*) has revealed key metabolites linked to drought and heat tolerance, enabling the development of more resilient varieties. Similarly, in tomato (*Solanum lycopersicum*), metabolomics has been used to identify metabolic pathways that are activated in response to salinity stress, providing targets for breeding salt-tolerant varieties.

While each omics approach—genomics, transcriptomics and metabolomics—provides valuable insights into crop resilience, their integration offers a more holistic understanding of the complex interactions that govern stress responses. By combining data from multiple omics layers, researchers can identify key molecular networks and pathways that are critical for resilience. The combination of multi-omics data offerings several challenges, including data complexity, the need for advanced bioinformatics tools and the difficulty of interpreting large datasets. However, ongoing advancements in computational biology, machine learning and systems biology are helping to overcome these challenges. Future research will likely focus on developing more efficient methods for integrating omics data and applying these insights to the development of climate-resilient crops.

14.3.4 Proteomics

The comprehensive study of an organism's, tissue's, or cell's proteome throughout regular growth and development as well as in reaction to environmental changes is known as proteomics (Zenda et al. 2021b). Plant response to environmental changes includes changes in protein expression and post-translational modification of proteins to activate their defence system against the challenges. In response to environmental shifts, plants modify proteins post-translationally and change how they are expressed, triggering defence mechanisms to meet the challenges (Hashiguchi et al. 2010). Proteomics is a branch of molecular biology that uses a variety of techniques, broadly divided into gel-based and gel-free approaches. These techniques are frequently used in conjunction with mass spectrometry (MS) to identify, fractionate, analyse and treat proteins (Zenda et al. 2021b). Tropospheric ozone (O_3) is a major pollutant in the northern hemisphere that negatively impacts crop yields. Its levels are swiftly rising and climate models estimate that global average surface O_3 concentrations could increase by 20–25% by 2050. When soybean leaf tissue was subjected to high O_3 concentrations, proteomic profiling revealed notable changes. High O_3 levels were shown to cause 35 proteins to increase in abundance up to five times, 22 proteins to exhibit up to a fivefold increase in oxidation and 22 proteins to exhibit an increase in both oxidation and abundance. Numerous biological systems were impacted by these modifications, including the metabolism of carbon, photosynthesis, amino acid synthesis, biosynthesis of flavonoids and isoprenoid compounds, signalling, homeostasis and antioxidant pathways. According to the study, seasonal exposure to O_3 in soybeans changes the quantity and oxidation statuses of several redox-sensitive proteins (Galant et al. 2012).

Soybeans are produced in a variety of climate zones, but flooding stress—a problem made worse by climate change—significantly inhibits soybean development. Flooding causes the cell wall's reactive oxygen species (ROS)-scavenging enzymes to be downregulated, while the plasma membrane's signalling, stress response and antioxidative system-related proteins rise. The inhibition of proteins involved in mRNA processing and pre-ribosome formation results in the suppression of protein translation in the nucleus. There is a drop in proteins linked to protein folding in the endoplasmic reticulum and a reduction in proteins essential to the electron transport chain in the mitochondria. Post-translational modifications play a crucial role in regulating fundamental cellular processes involved in flood response, such as hormone signalling, energy metabolism and protein folding and degradation (Wang and Komatsu 2020).

Global climate change is worsening the conditions imposed by drought on wheat production. Comparative analysis of drought-tolerant and susceptible isolines revealed 41 differentially expressed proteins involved in hydrogen peroxide metabolism, ROS scavenging, photosynthesis, protein transport, cellular localization and oxidative stress response. Analysis of protein interactions and pathways revealed that transcription process, translation process, protein export, carbohydrate metabolism and photosynthesis are crucial processes involved in drought tolerance (Nouraei et al. 2024). Proteomic study sheds important light on the biochemical and molecular mechanisms regulating plant development and stress responses in various species. This information is crucial for breeding endeavours and may be used to find stress tolerance biomarkers, which will increase the yield of crops that are significant to the economy. Proteomics fills in the gaps between transcriptomics and metabolomics, functioning as a complimentary technique to genomics (Tan et al. 2017).

14.3.5 Phenomics

The study of phenomics is concerned with the observable characteristics of an organism that result from interactions between its environment and genome during the course of its life. The word "phenome" refers to the way that genes and environmental influences work together to form an organism's unique characteristics (Hoch et al. 2023). Crop phenomics studies how successfully crops adapt locally by examining variables like organ size, blooming period and seed size, which are crucial for important agronomic qualities like yield in particular environmental circumstances (Marsh et al. 2021). Plant phenotyping is an important technique in pre-breeding and breeding selection because it allows for the non-invasive evaluation of plant structure, function and environmental interactions. At different levels of plant organization, from individual organs to whole canopies, modern plant phenotyping techniques enable improved accuracy and precision in evaluating complex features linked to growth, yield and stress adaptability. High-throughput phenotyping (HTP) monitors vital aspects of plant responses to stress, including photosynthetic activity, growth conditions and total water content. It utilizes sensors to collect digital phenotypic data across the visible, near-infrared and induced fluorescence spectra (Janni et al. 2024).

Photosynthesis is a key factor in the yield and a number of important parameters that restrict photosynthesis are adversely impacted by climate change: mesophyll conductance, stomatal conductance, the Calvin–Benson cycle, biochemical capacity involving Rubisco, electron transport in thylakoid membranes, nonphotochemical quenching and carbon metabolism or fixation pathways. Therefore, in order to increase photosynthetic capacity and crop yield, future research should concentrate on methods to improve photoprotection, increase Rubisco's capacity to capture CO_2, optimize photorespiration and photosynthetic efficiency and create genetically engineered plants with greater tolerance to future climate challenges (Hussain et al. 2021). A decrease in leaf stomatal conductance and photosynthetic rate in context with climate change-related thermal stress has been documented in grapevine and this effect was variable with the genotype (Tzortzakis et al. 2020).

Climate change-related thermal stress has a direct impact on tomato extracts' complexity and flavonoid and phenolic content. Tomato species that are tolerant exhibited elevated levels of flavonoids and total phenolic content, along with heightened antioxidant activity (Alhaithloul et al. 2021). Similarly, climate change has a big influence on winemaking and a lot of domestic and foreign grapevine varietals have been researched for their resistance to heat and drought. In climate change-related droughts, there has been an increase in total phenols and antioxidant capacity. In conjunction with elevated superoxide dismutase activity, a rapid surge in hydrogen peroxide and lipid peroxidation was observed in the leaves. Nevertheless, the genetic makeup of the cultivar modulated these effects (Tzortzakis et al. 2020).

The architecture and anatomical features of root systems play a pivotal role in function of plant, involving the gaining of essential soil nutrients and water, as well as their subsequent translocation to aboveground tissues. Research in this area is essential for breeding climate-resilient crops (Calleja-Cabrera et al. 2020). Several root structural traits and anatomical features, like aerenchyma formation and cell wall modifications, can be targeted for improvement under specific stress conditions (Kaur et al. 2021 2022a). Numerous plant characteristics and metabolites are directly impacted by the diverse roles that microbiomes connected to plant roots play. Therefore, creating novel tactics to increase crop resilience to pathogens and environmental challenges requires a knowledge of the intricate relationships that exist between plants, the soil rhizosphere and microbiomes (Zenda et al. 2021b).

The time of blooming and phenology are critical for rice's climate change adaptation. Research estimated flowering-related characteristics such as cardinal temperatures, basic vegetative phase, photoperiod sensitivity and cold acclimation and a genome-wide association analysis utilizing single-nucleotide polymorphism (SNP) markers were also included in the investigation. These investigations revealed multiple genes engaged in the known blooming network, as well as genes implicated in adapting to heat stress and in epigenetic regulatory processes (Dingkuhn et al. 2017). High daytime temperatures and low night-time temperatures during key reproductive stages can cause spikelet sterility in rice and a study suggested that epigenetic regulation of acclimation could be important in indica rice genotypes adapted to cooler environments (Dingkuhn et al. 2017).

The northernmost part of the planet where field agriculture is feasible is comprised of the Nordic countries. Some of the natural obstacles that this area encounters are short growing seasons, long days and the requirement for crops that can withstand cold. These difficulties have been made more difficult by climate change. As a result, crops that are especially suited to the region's particular circumstances and potential future climatic scenarios are needed. Functional phenomics techniques that integrate plant physiology, bioinformatics and sophisticated high-throughput phenotyping are necessary to accomplish these objectives (Roitsch et al. 2022). Thus, integrating phenomics with other approaches can lead to the development of climate-resilient crops.

14.3.6 Ionomics

Ionomics is a high-throughput elemental profiling technique that investigates the fundamental processes of trace element composition and mineral nutrition in living things (Pita-Barbosa et al. 2019). Ionomics has made it possible to identify several genes that govern the ionome and their crucial function in controlling environmental adaptation by integrating genetics with high-throughput elemental analysis (Huang and Salt 2016; Mann et al. 2018; Janni et al. 2024).

Stresses brought on by climate change-significantly alter plant metabolism, especially with regard to the intake and transport of vital mineral elements. Abiotic stress-induced signalling pathways in plants use calcium (Ca^{2+}), a flexible secondary signalling messenger, in almost all of them (Calleja-Cabrera et al. 2020). Phosphorus (P) plays a vital role in reducing metal toxicity in crops by lowering metal concentrations and restricting metal mobility through the formation of stable metal-phosphate complexes (Maharajan et al. 2021). Similar to this, it has been demonstrated that other mineral nutrients, such as silicon (Si), magnesium (Mg), zinc (Zn), iron (Fe), selenium (Se) and sulphur (S), reduce the toxicity of heavy metals by boosting proline content, increasing antioxidant activity and increasing photosynthetic rates. Additionally, by competing with heavy metal uptake, these nutrients lessen the negative effects of heavy metals on biological processes (Ali et al. 2021). Because it promotes stomatal closure via the ethylene signalling pathway, potassium (K) is essential for plants because it increases photosynthetic rates (Sardans and Peñuelas 2021). According to guard cell ionomics profiling, stomatal closure is mediated by cadmium (Cd) and manganese (Mn), whereas stomatal opening is primarily mediated by zinc (Zn) and boron (B). Salinity stress typically results in an increased uptake of sodium (Na^+) ions by plants. This excess Na^+ can interfere with the uptake of other essential ions, such as nitrate (NO_3^-), potassium (K^+), calcium (Ca^{2+}) and magnesium (Mg^{2+}), leading to nutrient imbalances and reduced plant growth (Ali et al. 2021). Silicon (Si) is a beneficial element that significantly enhances crop tolerance to salinity stress. By promoting the accumulation and activation of a range of physiological and biochemical defence mechanisms, Si contributes to improved plant performance under saline conditions. For many plant species, silicon (Si) is nevertheless important, even though it is not considered an essential nutrient. Utilizing silicon can improve plant development and lessen a number of stressors, including those caused by nutrient imbalances, metal toxicity, heat, drought, salt and severe temperatures, as well as pressures associated with climate change. Even in high concentrations, silicon is harmless for plants and can delay the senescence of their leaves (Verma et al. 2020).

Seasonal variables have a bigger effect on the metabolism of Aleppo pine populations than soil conditions, according to a study that looked at seasonal ionomics and metabolic changes in mine tailings growing in a Mediterranean semi-arid environment. The availability of water in the topsoil layers may be related to this impact. If this finding is accurate, the Mediterranean region's anticipated drop in precipitation and rise in warmth as a result of global climate change may lead to prolonged summer-like patterns in pine needle metabolism. As such, this can have a negative impact on Aleppo pine population performance (Lopez-Orenes et al. 2018).

Without including salinity, ionomics research is insufficient since salinity has a detrimental effect on global agricultural productivity (Kaur et al. 2022b). Global soil salinity may increase more quickly as a result of climate change, especially in dry and semi-arid areas. A number of mitigation strategies can be used to address this problem, such as the use of subsurface drainage in wet saline areas, subsurface drainage in salt-tolerant plant genotypes, drip irrigation, agroforestry, climate-smart conservation agriculture, land-shaping techniques, agroforestry and the application of beneficial microorganisms. These techniques can enhance the biological and physicochemical characteristics of salt-affected soils worldwide, including microbial biomass carbon and enzyme activity, as well as bulk density, electrical conductivity, pH and available soil nutrients. This will increase soil health and increase crop productivity (Mukhopadhyay et al. 2021).

14.4 Conclusion and Future Perspectives

Climate change affects the growth cycle of crops, phonological development, productivity and quality of food. This temperature and carbon dioxide change, along with a rainfall pattern, seriously influences the agricultural crops and threatens their food security in the future. A deep understanding of how crop plants respond to different abiotic stresses, such as drought, salinity and extreme temperatures, is essential for developing crop varieties that can withstand the challenges posed by climate change. Genetic manipulation for the creation of new breeding to meet present demand for food and improve food security. All different multi-omics approaches types of plants and resistant cultivars has combined modern genetics with the traditional art of such as genomics, proteomics, transcriptomics, phenomics and ionomics are decisively important for determining the genetic basis of crop performance, improvement and stress tolerance mechanisms under different environmental conditions. Very soon, multi-omics approaches might identify candidate genes with their pathways, while the development of models is achieved that would predict traits important agronomically. Such information is to be used for enhancing the performance of crops through precision breeding in changing climates.

References

Alawiye TT, Babalola OO (2021) Metabolomics: current application and prospects in crop production. Biologia 76:227–239. https://doi.org/10.2478/s11756-020-00574-z

Alhaithloul HA, Galal FH, Seufi AM (2021) Effect of extreme temperature changes on phenolic, flavonoid contents and antioxidant activity of tomato seedlings (Solanum lycopersicum L.). PeerJ 9:e11193

Ali S, Tyagi A, Bae H (2021) Ionomic approaches for discovery of novel stress-resilient genes in plants. Int J Mol Sci 22:7182

Anand S, Chaudhary N (2024) Integrating multi-omics approaches for crop resilience under changing climatic conditions. In: Kesheri M, Kanchan S, Salisbury TB, Sinha RP (eds) Microbial omics in environment and health. Springer Nature, Singapore, pp 295–310. https://doi.org/10.1007/978-981-97-1769-9_11

Anwar MR, O'Leary G, McNeil D, Hossain H, Nelson R (2007) Climate change impact on rainfed wheat in South-Eastern Australia. Field Crop Res 104:139–147

Calleja-Cabrera J, Boter M, Oñate-Sánchez L, Pernas M (2020) Root growth adaptation to climate change in crops. Front Plant Sci 11. https://doi.org/10.3389/fpls.2020.00544

Carrera FP, Noceda C, Maridueña-Zavala MG, Cevallos-Cevallos JM (2021) Metabolomics, a powerful tool for understanding plant abiotic stress. Agronomy 11:824

Choudhary P, Muthamilarasan M (2022) Modulating physiological and transcriptional regulatory mechanisms for enhanced climate resilience in cereal crops. J Plant Physiol 278:153815

Conde C, Ferrer R, Orozco S (2006) Climate change and climate variability impacts on rainfed agricultural activities and possible adaptation measures. A Mexican case study. Atmósfera 19:181–194

Cuculeanu V, Tuinea P, Bălteanu D (2002) Climate change impacts in Romania: vulnerability and adaptation options. GeoJournal 57:203–209. https://doi.org/10.1023/B:GEJO.0000003613.15101.d9

Das RR, Pradhan S, Parida A (2020) De-novo transcriptome analysis unveils differentially expressed genes regulating drought and salt stress response in Panicum sumatrense. Sci Rep 10:21251

Dingkuhn M, Pasco R, Pasuquin JM, Damo J, Soulié J-C, Raboin L-M, Dusserre J, Sow A, Manneh B, Shrestha S (2017) Crop-model assisted phenomics and genome-wide association study for climate adaptation of indica rice. 2. Thermal stress and spikelet sterility. J Exp Bot 68:4389–4406

Dugas DV, Monaco MK, Olson A, Klein RR, Kumari S, Ware D, Klein PE (2011) Functional annotation of the transcriptome of Sorghum bicolor in response to osmotic stress and abscisic acid. BMC Genomics 12:514. https://doi.org/10.1186/1471-2164-12-514

Eitzinger J, Štastná M, Žalud Z, Dubrovský M (2003) A simulation study of the effect of soil water balance and water stress on winter wheat production under different climate change scenarios. Agric Water Manag 61:195–217. https://doi.org/10.1016/S0378-3774(03)00024-6

Faye B, Webber H, Gaiser T, Müller C, Zhang Y, Stella T, Latka C, Reckling M, Heckelei T, Helming K, Ewert F (2023) Climate change impacts on European arable crop yields: sensitivity to assumptions about rotations and residue management. Eur J Agron 142:126670. https://doi.org/10.1016/j.eja.2022.126670

Galant A, Koester RP, Ainsworth EA, Hicks LM, Jez JM (2012) From climate change to molecular response: redox proteomics of ozone-induced responses in soybean. New Phytol 194:220–229. https://doi.org/10.1111/j.1469-8137.2011.04037.x

Ghaffar A, Rahman MHU, Ahmed S, Haider G, Ahmad I, Khan MA, Afzaal M, Ahmed S, Fahad S, Hussain J (2022) Adaptations in cropping system and pattern for sustainable crops production under climate change scenarios. In: Improvement of plant production in the era of climate change. CRC Press

Giudice G, Moffa L, Varotto S, Cardone MF, Bergamini C, De Lorenzis G, Velasco R, Nerva L, Chitarra W (2021) Novel and emerging biotechnological crop protection approaches. Plant Biotechnol J 19:1495–1510. https://doi.org/10.1111/pbi.13605

Goharrizi KJ, Karami S, Hamblin MR, Momeni MM, Basaki T, Dehnavi MM, Nazari M (2022) Transcriptomic and proteomic profile approaches toward drought and salinity stresses

Hasan N, Choudhary S, Naaz N, Sharma N, Laskar RA (2021) Recent advancements in molecular marker-assisted selection and applications in plant breeding programmes. J Genet Eng Biotechnol 19:128

Hashiguchi A, Ahsan N, Komatsu S (2010) Proteomics application of crops in the context of climatic changes. Food Res Int Climate Change Food Sci 43:1803–1813. https://doi.org/10.1016/j.foodres.2009.07.033

Hoch L, Herdean A, Argyle PA, Ralph PJ (2023) High throughput phenomics for diatoms: challenges and solutions. Prog Oceanogr 216:103074

Huang X-Y, Salt DE (2016) Plant ionomics: from elemental profiling to environmental adaptation. Mol Plant 9:787–797

Hussain N, Ghaffar A, Zafar ZU, Javed M, Shah KH, Noreen S, Manzoor H, Iqbal M, Hassan IFZ, Bano H (2021) Identification of novel source of salt tolerance in local bread wheat germplasm using morpho-physiological and biochemical attributes. Sci Rep 11:1–12

Jan N, Rather AM-U-D, John R, Chaturvedi P, Ghatak A, Weckwerth W, Zargar SM, Mir RA, Khan MA, Mir RR (2023) Proteomics for abiotic stresses in legumes: present status and future directions. Crit Rev Biotechnol 43:171–190. https://doi.org/10.1080/07388551.2021.2025033

Janni M, Maestri E, Gullì M, Marmiroli M, Marmiroli N (2024) Plant responses to climate change, how global warming may impact on food security: a critical review. Front Plant Sci 14:1297569

Jannink J-L, Lorenz AJ, Iwata H (2010) Genomic selection in plant breeding: from theory to practice. Brief Funct Genomics 9:166–177

Jogaiah S, Govind SR, Tran L-SP (2013) Systems biology-based approaches toward understanding drought tolerance in food crops. Crit Rev Biotechnol 33:23–39. https://doi.org/10.3109/07388551.2012.659174

Kaashyap M, Ford R, Mann A, Varshney RK, Siddique KHM, Mantri N (2022) Comparative flower transcriptome network analysis reveals DEGs involved in chickpea reproductive success during salinity. Plan Theory 11:434. https://doi.org/10.3390/plants11030434

Kamali S, Singh A (2023) Genomic and transcriptomic approaches to developing abiotic stress-resilient crops. Agronomy 13:2903

Kar SK, Sharma A, Kar S, Dey A (2024) Impact on agricultural crop production under climate change scenario. In: Kumar AP (ed) Technological approaches for climate smart agriculture. Springer International Publishing, Cham, pp 109–132. https://doi.org/10.1007/978-3-031-52708-1_6

Katam R, Lin C, Grant K, Katam CS, Chen S (2022) Advances in plant metabolomics and its applications in stress and single-cell biology. Int J Mol Sci 23:6985

Kaur G, Sanwal SK, Sehrawat N, Kumar A, Kumar N, Mann A (2021) Assessing the effect of salinity stress on root and shoot physiology of chickpea genotypes using hydroponic technique. Indian J Genet Plant Breed 81:586–589

Kaur G, Sanwal SK, Sehrawat N, Kumar A, Kumar N, Mann A (2022a) Identification of salt tolerant chickpea genotypes based on yield and salinity indices. Legume Res Int J 45:1381–1387

Kaur G, Sanwal SK, Sehrawat N, Kumar A, Kumar N, Mann A (2022b) Getting to the roots of Cicer arietinum L.(chickpea) to study the effect of salinity on morpho-physiological, biochemical and molecular traits. Saudi J Biol Sci 29:103464

Kole C, Muthamilarasan M, Henry R, Edwards D, Sharma R, Abberton M, Batley J, Bentley A, Blakeney M, Bryant J (2015) Application of genomics-assisted breeding for generation of climate resilient crops: progress and prospects. Front Plant Sci 6:563

Kumar A, Dash PK (2019) Transcriptome analysis for abiotic stresses in rice (Oryza sativa L.). Transcript Anal:61–75

Kumar M, Rani K (2023) Epigenomics in stress tolerance of plants under the climate change. Mol Biol Rep 50:6201–6216. https://doi.org/10.1007/s11033-023-08539-6

Kumar D, Bansal G, Narang A, Basak T, Abbas T, Dash D (2016) Integrating transcriptome and proteome profiling: strategies and applications. Proteomics 16:2533–2544. https://doi.org/10.1002/pmic.201600140

Kumar P, Tokas J, Kumar N, Lal M, Singal HR (2018) Climate change consequences and its impact on agriculture and food security. Int J Chem Stud 6:124–133

Li Z, Wang W, Yu X, Zhao P, Li W, Zhang X, Peng M, Li S, Ruan M (2024) Integrated analysis of DNA methylome and transcriptome revealing epigenetic regulation of CRIR1-promoted cold tolerance. BMC Plant Biol 24:631. https://doi.org/10.1186/s12870-024-05285-0

Lopez-Orenes A, Bueso MC, Conesa H, Calderon AA, Ferrer MA (2018) Seasonal ionomic and metabolic changes in Aleppo pines growing on mine tailings under Mediterranean semi-arid climate. Sci Total Environ 637:625–635

Luo Q, Williams MA, Bellotti W, Bryan B (2003) Quantitative and visual assessments of climate change impacts on south Australian wheat production. Agric Syst 77:173–186

Ma J, Li R, Wang H, Li D, Wang X, Zhang Y, Zhen W, Duan H, Yan G, Li Y (2017) Transcriptomics analyses reveal wheat responses to drought stress during reproductive stages under field conditions. Front Plant Sci 8:592

Maharajan T, Ceasar SA, Krishna TPA, Ignacimuthu S (2021) Management of phosphorus nutrient amid climate change for sustainable agriculture. J Environ Qual 50:1303–1324. https://doi.org/10.1002/jeq2.20292

Mann A, Bishi SK, Mahatma MK, Kumar A (2015) Metabolomics and salt stress tolerance in plants. In: Wani SH, Hossain MA (eds) Managing salt tolerance in plants: molecular and genomic perspectives. CRC Press, pp 251–266

Mann A, Singh S, Kumar A, Kumar S, Kumar B (2018) Plant Ionomics: an important component of functional biology. In: Ramakrishna A, Gill SS (eds) Metabolic adaptations in plants during abiotic stress. CRC Press, pp 147–154

Mann A, Kumar N, Lata C et al (2019) Functional annotation of differentially expressed genes under salt stress in *Dichanthium annulatum*. Plant Physiol Rep 24:104–111

Mann A, Kumar N, Kumar A et al (2021) de novo transcriptomic profiling of differentially expressed genes in grass halophyte *Urochondra setulosa* under high salinity. Sci Rep 11:5548

Mann A, Kumar N, Lata C, Kumar A, Meena BL, Kumar A (2024) Physiological and differential gene expression reveals a trade-off between antioxidant capacity and salt tolerance in *Urochondra setulosa* and *Dichanthium annulatum*. Plant Growth Regul 102:555–570

Marguerat S, Bähler J (2010) RNA-seq: from technology to biology. Cell Mol Life Sci 67:569–579. https://doi.org/10.1007/s00018-009-0180-6

Marsh JI, Hu H, Gill M, Batley J, Edwards D (2021) Crop breeding for a changing climate: integrating phenomics and genomics with bioinformatics. Theor Appl Genet 134:1677–1690. https://doi.org/10.1007/s00122-021-03820-3

McManus CM, Rezende Paiva S, Faria D (2020) Genomics and climate change. Rev Sci Tech 39:481–490. https://doi.org/10.20506/rst.39.2.3099

Miao Z, Han Z, Zhang T, Chen S, Ma C (2017) A systems approach to a spatio-temporal understanding of the drought stress response in maize. Sci Rep 7:6590

MirMohammadi Maibody SAM, Golkar P (2019) Application of DNA molecular markers in plant breeding. Plant Genet Res 6:1–30

Mohanty B, Kitazumi A, Cheung CM, Lakshmanan M, de Los Reyes BG, Jang I-C, Lee D-Y (2016) Identification of candidate network hubs involved in metabolic adjustments of rice under drought stress by integrating transcriptome data and genome-scale metabolic network. Plant Sci 242:224–239

Mondal S, Karmakar S, Panda D, Pramanik K, Bose B, Singhal RK (2023) Crucial plant processes under heat stress and tolerance through heat shock proteins. Plant Stress 10:100227. https://doi.org/10.1016/j.stress.2023.100227

Mrabet R (2023) Sustainable agriculture for food and nutritional security. In: Sustainable agriculture and the environment. Elsevier, pp 25–90

Mukhopadhyay R, Sarkar B, Jat HS, Sharma PC, Bolan NS (2021) Soil salinity under climate change: challenges for sustainable agriculture and food security. J Environ Manag 280:111736

Mulunch MG (2021) Impact of climate change on biodiversity and food security: a global perspective – a review article. Agric & Food Secur 10:1–25. https://doi.org/10.1186/s40066-021-00318-5
Ndlovu N (2020) Application of genomics and phenomics in plant breeding for climate resilience. Asian Plant Res J 6:53–66
Nouraei S, Mia MS, Liu H, Turner NC, Khan JM, Yan G (2024) Proteomic analysis of near-isogenic lines reveals key biomarkers on wheat chromosome 4B conferring drought tolerance. Plant Genome 17:e20343. https://doi.org/10.1002/tpg2.20343
Omotoso AB, Letsoalo S, Olagunju KO, Tshwene CS, Omotayo AO (2023) Climate change and variability in sub-Saharan Africa: a systematic review of trends and impacts on agriculture. J Clean Prod 414:137487. https://doi.org/10.1016/j.jclepro.2023.137487
Ortiz R, Sayre KD, Govaerts B, Gupta R, Subbarao GV, Ban T, Hodson D, Dixon JM, Ortiz-Monasterio JI, Reynolds M (2008) Climate change: can wheat beat the heat? Agric Ecosyst Environ 126:46–58
Owusu AG, Lv Y-P, Liu M, Wu Y, Li C-L, Guo N, Li D-H, Gao J-S (2023) Transcriptomic and metabolomic analyses reveal the potential mechanism of waterlogging resistance in cotton (Gossypium hirsutum L.). Front Plant Sci 14. https://doi.org/10.3389/fpls.2023.1088537
Paudel P, Pandey MK, Subedi M, Paudel P, Kumar R (2024) Genomic approaches for improving drought tolerance in wheat (Triticum aestivum L.): a comprehensive review. Plant Archives 24(1):1289–1300
Pita-Barbosa A, Ricachenevsky FK, Flis PM (2019) One "OMICS" to integrate them all: ionomics as a result of plant genetics, physiology and evolution. Theor Exp Plant Physiol 31:71–89. https://doi.org/10.1007/s40626-019-00144-y
Rajkumari N, Chowrasia S, Nishad J, Ganie SA, Mondal TK (2023) Metabolomics-mediated elucidation of rice responses to salt stress. Planta 258:111. https://doi.org/10.1007/s00425-023-04258-1
Rashid M, Stingl U (2015) Contemporary molecular tools in microbial ecology and their application to advancing biotechnology. Biotechnol Adv 33:1755–1773
Razzaq A, Sadia B, Raza A, Khalid Hameed M, Saleem F (2019) Metabolomics: a way forward for crop improvement. Meta 9:303
Razzaq A, Guul W, Khan MS, Saleem F (2021) Metabolomics: a powerful tool to study the complexity of wheat metabolome. Protein Pept Lett 28:878–895
Roitsch T, Himanen K, Chawade A, Jaakola L, Nehe A, Alexandersson E (2022) Functional phenomics for improved climate resilience in Nordic agriculture. J Exp Bot 73:5111–5127
Salam U, Ullah S, Tang Z-H, Elateeq AA, Khan Y, Khan J, Khan A, Ali S (2023) Plant metabolomics: an overview of the role of primary and secondary metabolites against different environmental stress factors. Life 13:706. https://doi.org/10.3390/life13030706
Sardans J, Peñuelas J (2021) Potassium control of plant functions: ecological and agricultural implications. Plan Theory 10:419. https://doi.org/10.3390/plants10020419
Semeraro T, Scarano A, Leggieri A, Calisi A, De Caroli M (2023) Impact of climate change on agroecosystems and potential adaptation strategies. Land 12:1117. https://doi.org/10.3390/land12061117
Sharma D, Jha SR (2023) Chapter 15 – current understanding of genomics, transcriptomics, proteomics and metabolomics of plants upon heavy metal stress. In: Husen A, Ahmad A (eds) Genomics, transcriptomics, proteomics and metabolomics of crop plants, plant biology, sustainability and climate change. Academic Press, pp 327–338. https://doi.org/10.1016/B978-0-323-95989-6.00009-7
Shulaev V, Cortes D, Miller G, Mittler R (2008) Metabolomics for plant stress response. Physiol Plant 132:199–208. https://doi.org/10.1111/j.1399-3054.2007.01025.x
Singh N, Mittal S, Thirunavukkarasu N (2019) Effect of drought stress and utility of Transcriptomics in identification of drought tolerance mechanisms in maize. In: Rajpal VR, Sehgal D, Kumar A, Raina SN (eds) Genetic enhancement of crops for tolerance to abiotic stress: mechanisms and approaches, Vol. I, sustainable development and biodiversity. Springer International Publishing, Cham, pp 73–97. https://doi.org/10.1007/978-3-319-91956-0_4

Singh BK, Delgado-Baquerizo M, Egidi E, Guirado E, Leach JE, Liu H, Trivedi P (2023) Climate change impacts on plant pathogens, food security and paths forward. Nat Rev Microbiol 21:640–656. https://doi.org/10.1038/s41579-023-00900-7

Soler CMT, Sentelhas PC, Hoogenboom G (2007) Application of the CSM-CERES-maize model for planting date evaluation and yield forecasting for maize grown off-season in a subtropical environment. Eur J Agron 27:165–177. https://doi.org/10.1016/j.eja.2007.03.002

Tan BC, Lim YS, Lau S-E (2017) Proteomics in commercial crops: an overview. J Proteome 169:176–188

Tyagi A, Ali S, Park S, Bae H (2023a) Exploring the potential of multiomics and other integrative approaches for improving waterlogging tolerance in plants. Plan Theory 12:1544. https://doi.org/10.3390/plants12071544

Tyagi A, Sharma S, Srivastava H, Singh A, Kaila T, Ali S, Gaikwad AB, Singh NK, Gaikwad K (2023b) Transcriptome profiling of two contrasting pigeon pea (Cajanus cajan) genotypes in response to waterlogging stress. Front Genet 13. https://doi.org/10.3389/fgene.2022.1048476

Tzortzakis N, Chrysargyris A, Aziz A (2020) Adaptive response of a native Mediterranean grapevine cultivar upon short-term exposure to drought and heat stress in the context of climate change. Agronomy 10:249. https://doi.org/10.3390/agronomy10020249

Verma KK, Singh P, Song X-P, Malviya MK, Singh RK, Chen G-L, Solomon S, Li Y-R (2020) Mitigating climate change for sugarcane improvement: role of silicon in alleviating abiotic stresses. Sugar Tech 22:741–749. https://doi.org/10.1007/s12355-020-00831-0

Walker NJ, Schulze RE (2006) An assessment of sustainable maize production under different management and climate scenarios for smallholder agro-ecosystems in KwaZulu-Natal, South Africa. Phys Chemi Earth Parts A, B, C 31:995–1002

Wang X, Komatsu S (2020) Review: proteomic techniques for the development of flood-tolerant soybean. Int J Mol Sci 21:7497. https://doi.org/10.3390/ijms21207497

Wang M, Jiang B, Peng Q, Liu W, He X, Liang Z, Lin Y (2018) Transcriptome analyses in different cucumber cultivars provide novel insights into drought stress responses. Int J Mol Sci 19:2067

Yadav R, Singh G, Santal AR, Singh NP (2023) Omics approaches in effective selection and generation of potential plants for phytoremediation of heavy metal from contaminated resources. J Environ Manag 336:117730. https://doi.org/10.1016/j.jenvman.2023.117730

Zenda T, Liu S, Wang X, Liu G, Jin H, Dong A, Yang Y, Duan H (2019) Key maize drought-responsive genes and pathways revealed by comparative transcriptome and physiological analyses of contrasting inbred lines. Int J Mol Sci 20:1268

Zenda T, Liu S, Dong A, Duan H (2021a) Advances in cereal crop genomics for resilience under climate change. Life 11:502

Zenda T, Liu S, Dong A, Li J, Wang Y, Liu X, Wang N, Duan H (2021b) Omics-facilitated crop improvement for climate resilience and superior nutritive value. Front Plant Sci 12:774994

Advancing Sustainable Agriculture Through Plant–Microbial Interactions Amid Climate Change

15

Priyanka Chandra, Ram Kishor Fagodiya, Awtar Singh, Kailash Prajapat, Parul Sundha, Nirmalendu Basak, Kamlesh Verma, Gargi Sharma, Arvind Kumar Rai, Rajender Kumar Yadav, and Ashish Kumar Biswas

Abstract

The effects of climate change, such as increasing temperatures, altered precipitation patterns and intensities, and increased atmospheric CO_2 concentrations, have a direct influence on plant development. These effects also have an impact on transpiration rates, yield, and productivity of plants. Abiotic and biotic stressors in crops are the most common repercussions of these climate change-related circumstances. Numerous studies are being conducted to find ways to mitigate the climate change-related repercussions and adapt to its negative effects on agriculture. The development of heat- and drought-tolerant crops, adjusting or modifying agricultural growth schedules, prudent resource management techniques, etc. are the primary foci of these research programs. Nonetheless, agronomic management and crop improvement through breeding programs can somewhat lessen the effects of climate change. According to a number of recent research, beneficial microbes can help agricultural plants become more resilient through plant–microbial interaction and may be a more effective way to mitigate biotic and abiotic stresses. Through a variety of processes, these bacteria can colonise the rhizosphere and/or endorhizosphere of crop plants, hence promoting the growth of these plants. In order to cope up the detrimental effects of stresses brought on by climate change and to develop response mechanisms for climate

P. Chandra · R. K. Fagodiya (✉) · A. Singh · P. Sundha · N. Basak · K. Verma · G. Sharma
A. K. Rai · R. K. Yadav
Division of Soil and Crop Management, ICAR-Central Soil Salinity Research Institute, Karnal, Haryana, India

K. Prajapat
Division of Social Science Research, ICAR-Central Soil Salinity Research Institute, Karnal, Haryana, India

A. K. Biswas
ICAR–Indian Institute of Soil Science, Bhopal, Madhya Pradesh, India

A. Mann et al. (eds.), *Cutting Edge Technologies for Developing Future Crop Plants*,
https://doi.org/10.1007/978-981-96-2508-6_15

resilient agriculture, this book chapter will discuss the significance of these microorganisms and their potential to reduce stresses in crop plants.

Keywords

Climate change · Beneficial microorganisms · Abiotic stress · Biotic stress

15.1 Introduction

One of the main factors affecting human life and food security in the twenty-first century is climate change. The average global surface air temperature has increased by 1.53 °C since the Industrial Revolution (IPCC 2014). Global warming linked to climate change has been documented in nearly all manmade and natural systems across several nations. Climate change is primarily caused by a slow increase in the atmospheric concentration of greenhouse gases (GHGs), specifically CO_2, CH_4, N_2O, and fluorinated gases (F-gases), which are generated by a variety of human-induced activities (IPPC 2014; Kumar et al. 2020). Global warming brought on by rising atmospheric temperatures, variations in the amount and frequency of rainfall, an increase in CO_2 concentration in the atmosphere, and rising sea levels are all consequences of climate change that have a negative impact on food production and crops (Kumar and Gautam 2014; Fagodiya et al. 2020). These previously mentioned effects have a direct influence on plant growth by modifying transpiration and photosynthesis rates, which in turn impact yield and nutritional value. A rise in atmospheric CO_2 directly impacts plant growth by interfering with photosynthesis, which in turn alters root architecture and ultimately plant function. The main causes of this are decreased transpiration as a result of plants closing their stomata and increased net photosynthetic rates.

It has also been documented that high temperatures can cause spikelet and floret sterility, which prevents pollen grains from swelling and leads to poor thecae dehiscence and yield loss (Dubey et al. 2019). However, the shift in solar radiation causes respiration losses in addition to affecting crop growth, physiology and phenology due to the difference in night and day temperatures (Mahato 2014; Venkateswarlu and Shanker 2009). All of these factors contribute to a decrease in crop yield. Drought and an increase in salinity result from changes in rainfall and temperature because the salt in the saline soil rises to the surface. However, flooding is a result of heavy rains. These severe circumstances lead to physiological distress in plants, which lowers agricultural productivity. A result of the process of inundation brought on by sea level rise brought on by global warming is the loss of agricultural land. Additionally, this makes coastal groundwater and agricultural land more saline (Kumar et al. 2022). However, the main abiotic stressor in arid and semi-arid environments nowadays is drought or soil moisture stress, which results in low pH and soil salinity/alkalinity. Variations in atmospheric CO_2 facilitate the growth of diseases and pests (Kumar et al. 2018). Thus, climate change is a major contributing factor to biotic and abiotic stress that adversely affects agricultural growth and productivity.

15.2 Microbial Adaptation to Extreme Environmental Conditions

Microbes use a sophisticated multilevel regulatory gene system to help them adapt to stressful conditions (Rai et al. 2024). Microbes that live in harsh environments, such thermophiles and halophiles, have a variety of enzymes that aid in maintaining the stability of their membranes and allowing them to survive (Grover et al. 2011). Rhizospheric microbial populations that thrive in harsh environments like salt, drought, alkalinity, and acidity have been found to have potential uses in agriculture(Chandra et al. 2020, 2021c). These populations can also be employed as a preventative measure against the negative effects of climate change (Fig. 15.1) (Verma et al. 2017).

Extremophilic microbes have the capacity to endure in harsh environmental conditions, such as concentrations of salt between 0.80 and 4.5 molar (M) for halophiles, pH levels between 4 and 9 for acidophiles and alkaliphiles, and temperatures between 60 and 111 °C for thermophiles and −2 to −20 °C for psychophiles (Kumar et al. 2011). Small molecular weight osmoprotectants such as glutamate, trehalose, proline, proline betaine, glycine, beatine, and ectoine are produced by soil bacteria in response to these stressors and are involved in regulating their cytoplasmic osmolarity (Czech et al. 2018). Membrane-derived oligosaccharides are highly anionic polysaccharides that are found in significant amounts in the periplasm of bacterial cells. These polysaccharides are likely responsible for maintaining the turgor pressure within the periplasm of bacterial cells since they are unable to pass through porin proteins found in the cell wall (Csonka 1989).

Another way that bacteria adapt to stress is by the production of exopolysaccharides (EPS) (Yadav et al. 2018). By varying water potential, the EPS improves water retention, guards against hydric stress, and controls the transport of carbon sources within microbial cells. The unique cementing and water-holding qualities of EPS aid in the production and stabilisation of soil aggregates via the production of biofilms, EPS controls the flow of water and nutrients via plant roots (Costa et al. 2018).

Because soil bacteria have enough supply of ribosomes in their cells, they can synthesise proteins effectively even under stressful circumstances (Schimel et al. 2007). In reaction to increased temperatures, bacteria create heat shock proteins (HSPs), which bind to other proteins in non-native conformational states (Maleki et al. 2016). Small heat shock proteins (sHSP), proteases, and chaperons (GroEL, DnaK, DnaJ, GroES, ClpB, ClpA, and ClpX) made up these HSPs. While proteases break down permanently damaged proteins, chaperons are crucial for the correct folding of denatured proteins. Therefore, chaperones are essential for both the subsequent recovery of bacteria after heat stress as well as for their survival (Lund 2001). In a similar vein, it is critical to induce cryoprotective protein activities in microorganisms at low temperatures (Lorv et al. 2014).

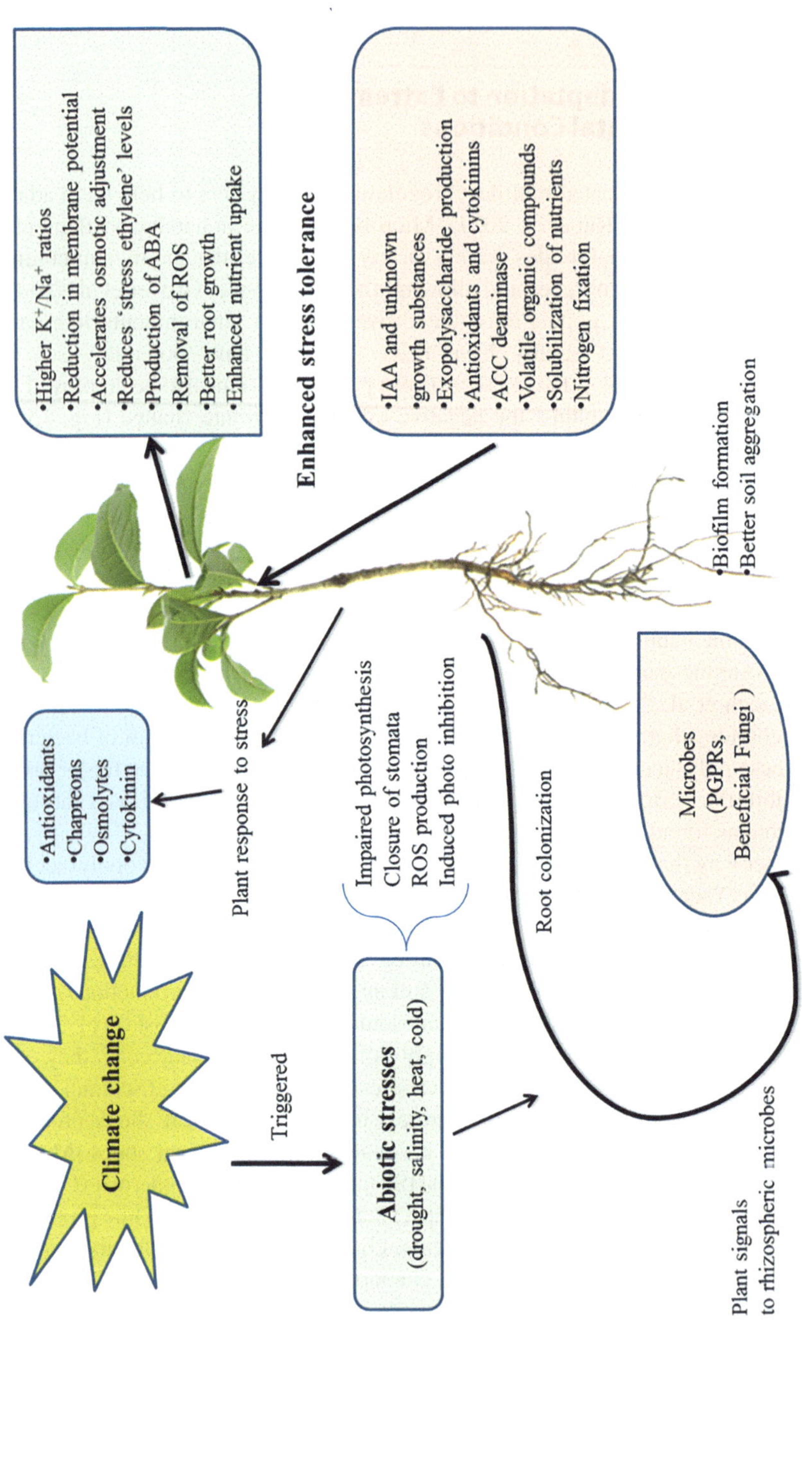

Fig. 15.1 Mechanisms of microbial abiotic stress management under the climate change

The semipermeable membrane of microbial cells prevents dehydration and cell death during drought conditions when water potentials decrease by collecting solutes that lower the internal water potential of the cells (Ngumbi and Kloepper 2016). Microbes use a variety of physiological strategies to safeguard their organelles and cell structures so they can survive in stressful environments. Among these processes are the production of spores and the buildup of suitable solutes (Berard et al. 2015). Complementary solute accumulation, such as trehalose, glycine, betaine, and proline, increases enzyme thermotolerance and prevents proteins from thermally denaturing, aiding in the preservation of membrane integrity (Berard et al. 2015).

Due to their ability to create endophytic symbiotic relationships and their evolution as intercellular root colonisers that communicate with other microorganisms in the rhizosphere, microbes also have the capacity to respond very well to biotic stressors (Verma et al. 2023). Plant growth is stimulated as a result of this mechanism, improving yield and disease defences (Olanrewaju et al. 2017). They may be able to shield plants from biotic stressors since they also release antagonistic substances like siderophore, hydrogen cyanide (HCN), and other enzymes that break down cell walls (Chandra et al. 2018, 2024a). Nonetheless, microorganisms' production of phytohormones links defence responses to plant development (Vejan et al. 2016). Additionally, microbes stimulate the development of genes in plants that aid in the management of plant diseases and their ability to withstand harsh environments (Kumar and Verma 2018). Extreme environments such as dry regions, acidic, saline, and sodic soils have been documented to harbour PGPRs such as *Rhizobium, Bradyrhizobium, Azotobacter, Azospirillum, Pseudomonas,* and *Bacillus* (Numan et al. 2018). The ability of bacteria to adapt could be important for managing stress happening due to climate change. Due to their unique ability to withstand harsh environments and their widespread and diverse character, microorganisms can serve as an excellent model for coping with stressful situations (Ahemad and Kibret 2014).

15.3 PGPR-Mediated Abiotic Stress Management

The climate change increased the risk of occurrence of different abiotic stresses like drought, chilling injury, salinity and high temperature and microbes has the potential to impart tolerance to plants against these stresses (Grover et al. 2011). Several bacteria with PGP potential have been reported to provide tolerance to the host plants under different abiotic stress environments (Chandra et al. 2021a, b) while some are listed in Table 15.1. Such microbes have the potential to alleviate the stresses in crops in climate change scenario thus expanding the spectrum of application of microorganisms.

Microbes have the ability to give plants resistance against many abiotic stresses (Chandra et al. 2024b, c), including as drought, chilling injury, salt, and high temperatures. The risk of these stresses occurring has increased due to climate change (Grover et al. 2011). Many bacteria have been shown to provide tolerance to the host plants under various abiotic stress environments, including *Rhizobium,*

Table 15.1 PGPRs demonstrating abiotic stress mitigation in several crops along with their mechanism

S. no	Root associated microbes	Type of abiotic stress	Mechanism of action	Crops	References
1	*Pseudomonas* spp.	Salt	ACC deaminase activity	*Brassica napus*	Akhgar et al. (2014)
2	*Rhizobium* spp.	Drought	EPS production and forming soil aggregation	*Helianthus annuus* L.	Alami et al. (2000)
3	*Pseudomonas* spp.	Heat	Induction of heat shock proteins; Improved plant biochemical status	*Sorghum bicolor*	Ali et al. (2009)
4	*Pseudomonas* spp.	Salt	ACC deaminase activity	*Solanum lycopersicum*	Ali et al. (2014)
5	*Azotobacter vinelandii*	Salt	Nitrogen fixation	*Triticum aestivum*	Aly et al. (2012)
6	*Pantoea agglomerans*	Drought	Rhizosphere soil aggregation through EPS	*Triticum aestivum*	Amellal et al. (1998)
7	*Pantoea agglomerans*	Salt	Exopolysaccharide production	*Triticum aestivum*	Amellal et al. (1999)
8	*Bacillus amylolequifaciens; Bacillus insolitus; Microbacterium* spp.;*P. syringae*	Salt	Inhibited influx of sodium due to formation of rhizosheath by exopolysaccharide	*Triticum aestivum*	Ashraf and Harris (2004)
9	*Dietzia natronolimnaea*	Salt	Modulation of abscisic acid signalling cascade	*Triticum aestivum*	Bharti et al. (2016)
10	*Bacillus safensis; Ochrobactrum pseudogregnonense*	Drought	Elevation of antioxidant responses	*Triticum aestivum*	Chakraborty et al. (2012)
11	*Pseudomonas putida*; *Enterobacter cloacae*; *P. putida*	Flooding	ACC deaminase activity	*Solanum lycopersicum*	Grichko and Glick (2001)
12	*Bacillus pumulis; Bacillus firmus*	Drought and salinity	Increase level of proline accumulation; ROS-scavenging enzymes activity	*Solanum tuberosum*	Gururani et al. (2012)
13	*Pseudomonades* spp.*; Bacillus lentus; Azospirillum brasilens*	Water	Improved antioxidant enzymes; photosynthetic pigments	*Ocimum basilicum*	Heidari and Golpayegani (2012)

(continued)

Table 15.1 (continued)

S. no	Root associated microbes	Type of abiotic stress	Mechanism of action	Crops	References
14	*Pseudomonas aeruginosa*	Zinc toxicity	Oxidative stress tolerance	*Triticum aestivum*	Islam et al. (2014)
15	*Pseudomonas pseudoalcaligenes; Bacillus pumilus*	Salt	Reduce lipid peroxidation; superoxide dismutase activity	*Oryza sativa*	Jha and Subramanian (2014)
16	*Pseudomonas fluorescens; Enterobacter hormaechei; Pseudomonas migulae*	Drought	ACC deaminase and EPS-producing activity	*Setaria italica* L.	Niu et al. (2018)
17	*Burkholderia phytofirmans*	Cold	Induced cell wall strengthening of mesophyll	*Arabidopsis thaliana*	Su et al. (2015)
18	*Paenibacillus polymyxa*	Drought	Induction of stress resistant gene	*Arabidopsis thaliana*	Timmusk and Wagner (1999)
19	*Bacillus megaterium; Bacillus subtilis; Azospirillum brasilense; Raoultella terrigena*	Cold	Production of antioxidant enzymes	*Triticum aestivum*; *Hordeum vulgare*	Turan et al. (2013)
20	*Pseudomonas fluorescens; P. fluorescens; P. putida*	Drought	ACC deaminase activity	*Pisum sativum*	Zahir et al. (2008)
21	*Pseudomonas fluorescens*	Water	Increase proline, total soluble sugars, catalase and peroxidase activity	*Zea mays* L.	Zarei et al. (2020)

Bacillus, Pseudomonas, Pantoea, Paenibacillus, Burkholderia, Achromobacter, Azospirillum, Microbacterium, Methylobacterium, Variovorax, and Enterobacter (Shrivastava and Kumar 2015). A selection of these bacteria is shown in Table 15.1. These bacteria may be able to reduce crop stressors under climate change scenarios, broadening the range of applications for microorganisms.

According to Yang et al. (2009), PGPR are also linked to plant immunity and are known to provide "induced systemic tolerance" (IST). It has been determined that some genes and the bacterial determinants of "induced systemic resistance" (ISR) activate plant activation pathways and cause IST in the host. The ISR and the IST-mediating pathways were demonstrated to be common in some instances while being very different in others. The foundation of a "eco-friendly stress-management strategy" is the possibility that the biochemical, molecular, and physical elicitation responses in plants caused by microbes may jointly offer defence against several biotic and abiotic stresses (Beneduzi et al. 2012; Van Loon 2007). Various factors underlie plants' ability to withstand stress induced by microorganisms (Choudhary et al. 2016). By producing more indole acetic acid (IAA), gibberellins, and other plant hormones, the PGPR increases resilience under stress conditions. This leads to increased root surface area, number of tips, and length, which improves growth and nutrient uptake (Vacheron et al. 2013). Furthermore, it has been discovered that PGPRs under salt stress enhance the growth of a number of crops, including oats, lettuce, tomato, and pepper (Rai et al. 2021). In these circumstances, the production of cytokinin and antioxidants by PGPRs leads to the accumulation of abscisic acid and the breakdown of reactive oxygen species (ROS) (Shrivastava and Kumar 2015). By inoculating *Arabidopsis thaliana* with *Paenibacillus polymyxa*, a drought-responsive gene called ERD15 was induced, hence conferring drought tolerance (Timmusk and Wagner 1999). When inoculated with *Azospirillum brasilense*, wheat showed improved "elastic adjustment" and better water status, leading to better nutritional status and grain production at harvest under drought stress (Kasim et al. 2013). The hormone produced by plants called ethylene regulates homoeostasis under stressful conditions, which inhibits the growth of roots and shoots in plants. But bacteria that produce 1-aminocyclopropane-1-carboxylate (ACC) deaminase sequester ACC from plants and use it for energy and nitrogen, which lessens the negative effects of ethylene. Accordingly, microorganisms that produce ACC deaminase help plants thrive and reduce stress (Olanrewaju et al. 2017).

In tomato and pepper, the PGPR strain of *Achromobacter piechaudii*, which produces ACC deaminase, confers induced systemic tolerance to salt stress and drought (Mayak et al. 2004). Plants that produce ACC deaminase are more resilient to drought stress because their roots grow larger and they take in more water and nutrients as a result. Additionally, microorganisms produce quaternary amines like glycine betaine, which give plants resistance against salinity, cold, and moisture shortage (Dimkpa et al. 2009). According to reports, the induction of ISR is responsible for the primed physiological condition of the plant inoculated with microbes. In fact, the priming phenomenon remains mostly enigmatic despite being linked to the buildup of dormant signalling proteins that undergo transduction upon activation in response to various stressors (Mhlongo et al. 2018). It has also been demonstrated that *Bacillus* sp. induces ISR, which helps to explain why crops have increased resistance to abiotic stressors (Chandra et al. 2018, 2023). *Arabidopsis thaliana* plants infected with *Erwinia carotovora*, *Paenibacillus polymyxa* causes variations in gene expression in diseased *A. thaliana* plants. Therefore, in order to counteract both biotic and abiotic challenges, plant genes implicated can be co-regulated by PGPRs (Timmusk and Wagner 1999). Osmotins are tiny defensive proteins secreted by bacteria that accumulate under conditions of salt stress or fungal infection. They respond to the phytohormones, abscisic acid and ethylene. *Vignu radiate* that was inoculated with *Bacillus cereus* has considerably elevated proline accumulation, POD, SOD, and CAT activities. Treatment plants were also found to have a high nutrient content (Faisal et al. 2016). Additionally, PGPR uses phytohormones to mobilise and/or solubilise nutrients and promote root development. But because they can solubilise phosphate, strains like *B. subtilis, Arthrobacter* sp., *B. aquimaris, B. cereus,* and have also been reported to improve nutrients in crops

(Yadav et al. 2017). In wheat under salinity and drought stress, PGPR enhanced N, P, and K uptake (Egamberdieva 2009). After being inoculated with Bacillus sp., the N, P, and K contents of lettuce increased by 5, 70, and 50%, respectively (Han and Lee 2005). Therefore, it can be said that applying PGPR strains to a plant can reduce stress and enhance its growth. Plant growth-promoting rhizobacteria that produce EPS can reduce the amount of Na^+ available in the rhizosphere for plant uptake, hence preventing salt stress in plants that are grown in saline conditions. The PGPR that generates exopolysaccharide is crucial to the development of rhizosheaths. PGPRs that produce EPS increase soil aggregation, which enhances soil structure. Higher water-holding capacity and water availability along with greater nutrient uptake were the outcomes. Moreover, EPS binds to soil-based Na^+. Plants experience less salt stress when Na + is bound by the hydroxyl, sulfhydryl, carboxyl, and phosphoryl functional groups of bacterial EPS (Yadav et al. 2018). When *P. chlororaphis* was injected into wheat, tomato, and cucumber grown in salinity, a rise in the amount of indole acetic acid was noted, along with an improvement in water conductivity (Egamberdieva 2011). The majority of rhizospheric microorganisms are capable of releasing cytokinins (CK), and when *Platycladus orientalis* was inoculated with PGPR strain *B. subtilis*, the resulting inoculum showed higher levels of CK in the shoots and greater drought resistance (Liu et al. 2013). According to Orhan (2016), IAA is produced by *Bacillus gibsonii, Staphylococcus succinus, Halomonas* sp., *Zhihengliuella* sp., and *Bacillus atrophaeus* it helps wheat develop under salt stress. Under various stressors (salt, drought, and heavy metals), the activity of enzymes that scavenge ROS, such as ascorbate peroxidase (APX) and catalase (CAT), dehydroascorbate reductase (DHAR), glutathione reductase (GR), and superoxide dismutase (SOD), was enhanced in potatoes inoculated with PGPR (Hahm et al. 2017). Furthermore, in salt-stressed okra, PGPR *Enterobacter* boosted the activities of SOD, APX, and CAT (Habib et al. 2016). According to a study, under saline conditions, pepper plants inoculated with PGPR have significant activity of the antioxidant enzymes glutathione peroxidase (GPX), CAT, and APX. When soybean plants were injected with PGPR strains, which lessened salt stress and enhanced development, an increase in proline buildup was observed. Proline concentration and total soluble sugar levels rise in PGPR-treated wheat plants, greatly enhancing their osmotolerance (Hahm et al. 2017).

15.4 PGPR-Mediated Biotic Stress Management

Because of its biological regulatory action in a climate change scenario, PGPR has the potential to function as a sustainable alternative. These processes include the activation of cellular components and the reinforcement of the cell wall as a result of the buildup of secondary metabolites. The negative effects of biotic stressors such as pathogens, viruses, nematodes, and insects can be lessened by inoculating plants with microbes (Fig. 15.2). By enhancing immunity against pathogens, these microbes help plants defend themselves from them (Kumar and Verma 2018).

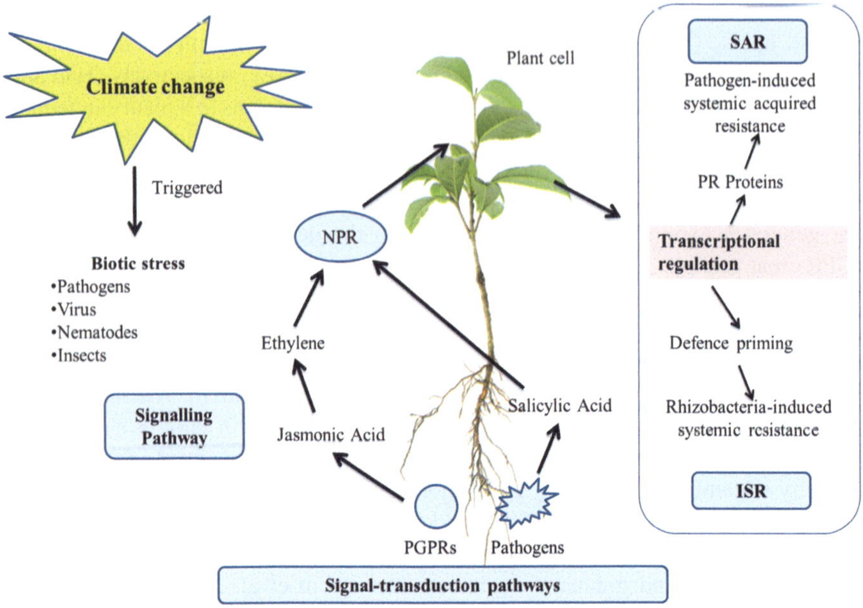

Fig. 15.2 Demonstration of PGPR-mediated biotic stress management

PGPRs can be used to stimulate plants, so inducing resistance to pathogens and carrying out the process of "immunisation" in plants. Plant–microbe relationships enhance a plant's ability to defend itself by preventing a wide range of diseases (Chandra et al. 2021b). Plant resistance is also induced as a result of PGPR colonisation of the roots. Plants are protected against various diseases by this response, which is referred to as induced systemic resistance (ISR) (Van Loon 2007). Numerous theories have been put out regarding the complex processes involved in the induction of ISR, including lipopolysaccharides from bacterial outer membranes, secondary metabolites such as siderophores, salicylic acid (SA), jasmonic acid (JA), and ethylene (ET) (Backer et al. 2018). Nonetheless, a number of reports indicate that SA-independent route superintended ISR. Plants that experience ISR produced by specific microorganisms also accumulate pathogenesis-related (PR) proteins, which in turn boost plant resistance to biotic stressors (Van Loon and Bakker 2006). Through the defence priming mechanism, PGPRs are also known to elicit resistance responses in plants. According to Pieterse et al. (2014), the physiological state known as the "primed state" is where plants acclimatise to respond more quickly to both biotic and abiotic stress. Many researches have documented that PGPRs during plant–microbe interactions induce ISR (van Loon and Bakker 2006; Rosier et al. 2018). The first instance of PGPR-mediated ISR enhancing plant defence responses by increasing phytoalexin in contrast to uninoculated control plants was seen in Fusarium oxysporum-infected Dianthus caryophyllus (Van Peer et al. 1991). ISR generated in *Phaseolus vulgaris* L. by *Bacillus pumilus* infected with *F. oxysporum*. This is caused by the appositions to quickly strengthen, obstructing the fungi's ability to enter the root cells. *Pseudomonas syringae* has been shown to increase ISR in *Arabidopsis* plants, and transcriptome analysis of the leaves showed increased expression of 81 genes that reflect the "primed state" of plants, in which they react more strongly to pathogen attack. Most of the genes whose expression was potentiated control either ET, JA, or both. It has been observed that in *Cucumis sativus* infected with *Pseudomonas syringae*, *Trichoderma asperellum* induces PR gene expression after primed on seeds of *Cucumis sativus* (Shoresh et al. 2005). In wheat, the PGPR strain *Serratia marcescens* increases salt tolerance and ISR while also controlling the severity of fungal infection-related diseases. Treatment of wheat plants with varying salt levels also affects antioxidant enzymes, such as peroxidase, catalase, and superoxide dismutase, which lessens the oxidative damage that salinity stress causes to plants. In addition to enhancing immunity through inducing ISR, PGPR also modifies the expression of defence genes in plants. Numerous PGPRs, including *Bacillus*, *Serratia* spp., *Burkholderia*, *Pseudomonas*, and *Streptomyces*, emit volatile chemicals that can also cause priming. It has been noted that microorganisms' volatile organic compounds (VOCs) prevent fungal sporulation as well as vegetative growth. *Bacillus subtilis* induced an ethylene-dependent signalling pathway independent of SA, JA, or the gene known as the nonexpressor of pathogenesis-related genes 1 (NPR1). *Bacillus subtilis* not only induces the signalling pathway but also releases secondary metabolites including butanediol and 3-hydroxy-2-butanone, which strengthen plant defences against pathogens (Rudrappa et al. 2010).

15.5 Conclusions

Due to alterations in environmental conditions, agriculture is particularly vulnerable in the event of climate change. It has been demonstrated that rhizospheric bacteria can lessen both biotic and abiotic stressors in host plants. One of the most crucial traits that bacteria have in order to reduce the effects of both biotic and abiotic stresses on their hosts is the ability to induce a systemic response in the plants. Therefore, the identification of the bacteria and their metabolites that offer protection against biotic and abiotic challenges, or against two or more abiotic stresses, should be the main focus of future research initiatives. Furthermore, thorough research should be done on the relationships between various plants and helpful microorganisms, or plant–bacterial interactions. In order to combat the negative effects of climate change, more research on the useful applications of these beneficial bacteria in agriculture is desperately needed.

References

Ahemad M, Kibret M (2014) Mechanisms and applications of plant growth promoting rhizobacteria: Current perspective. J King Saud Univ Sci 26(1):1–20

Akhgar AR, Arzanlou M, Bakker PAHM, Hamidpour M (2014) Characterization of 1-Aminocyclopropane-1-Carboxylate (ACC) Deaminase-Containing *Pseudomonas* spp. in the Rhizosphere of Salt-Stressed Canola. Pedosphere 24(4):461–468

Alami Y, Achouak W, Marol C, Heulin T (2000) Rhizosphere soil aggregation and plant growth promotion of sunflowers by an exopolysaccharide-producing *Rhizobium* sp. strain isolated from sunflower roots. Appl Environ Microbiol 66(8):3393–3398. https://doi.org/10.1128/AEM.66.8.3393-3398.2000

Ali SZ, Sandhya V, Grover M, Kishore N, Rao LV, Venkateswarlu B (2009) *Pseudomonas* sp. strain AKM-P6 enhances tolerance of sorghum seedlings to elevated temperatures. Biol Fertil Soils 46:45–55. https://doi.org/10.1007/s00374-009-0404-9

Ali S, Charles TC, Glick BR (2014) Amelioration of high salinity stress damage by plant growth promoting bacterial endophytes that contain ACC deaminase. Plant Physiol Biochem 80:160–167

Aly MM, Tork S, AlGarni S, Nawar L (2012) Production of lipase from genetically improved *Streptomyces exfoliates* LP10 isolated from oil-contaminated soil. Afr J Microbiol Res 6(6):1125–1137

Amellal N, Burtin G, Bartoli F, Heulin T (1998) Colonization of wheat roots by an exopolysaccharide-producing pantoea agglomerans strain and its effect on rhizosphere soil aggregation. Appl Environ Microbiol 64(10):3740–3747

Amellal N, Bartoli F, Villemin G, Talouizte A, Heulin T (1999) Effect of inoculation of EPS-producing *Pantoea agglomerans* on wheat rhizosphere soil aggregation. Plant Soil 211:93–101

Ashraf M, Harris PJC (2004) Potential biochemical indicators of salinity tolerance in plants. Plant Sci 166:3–16

Backer R, Rokem JS, Ilangumaran G, Lamont J, Praslickova D, Ricci E, Subramanian S, Smith DL (2018) Plant growth-promoting rhizobacteria: context, mechanisms of action and roadmap to commercialization of biostimulants for sustainable agriculture. Front Plant Sci 23(9):1473

Beneduzi A, Ambrosini A, Passaglia LM (2012) Plant growth-promoting rhizobacteria (PGPR): their potential as antagonists and biocontrol agents. Genet Mol Biol 35(4):1044–1051

Berard A, Sassi MB, Kaisermann A, Renault P (2015) Soil microbial community responses to heat wave components: drought and high temperature. Clim Res 66:243–264

Bharti N, Pandey SS, Barnawal D, Patel VK, Kalra A (2016) Plant growth promoting rhizobacteria *Dietzia natronolimnaea* modulates the expression of stress responsive genes providing protection of wheat from salinity stress. Sci Rep 6:34768. https://doi.org/10.1038/srep34768

Chakraborty U, Chakraborty B, Chakraborty AP, Dey P (2012) Water stress amelioration and plant growth promotion in wheat plants by osmotic stress tolerant bacteria. World J Microbiol Biotechnol 29:789. https://doi.org/10.1007/s11274-012-1234-8

Chandra P, Tripathi P, Chandra A (2018) Isolation and molecular characterization of plant growth-promoting Bacillus spp. and their impact on sugarcane (Saccharum spp. hybrids) growth and tolerance towards drought stress. Acta Physiol Plant 40:199. https://doi.org/10.1007/s11738-018-2770-0

Chandra P, Dhuli P, Verma P et al (2020) Culturable microbial diversity in the rhizosphere of different biotypes under variable salinity. Trop Ecol 61:291–300. https://doi.org/10.1007/s42965-020-00089-3

Chandra A, Chandra P, Tripathi P (2021a) Whole genome sequence insight of two plant growth-promoting bacteria (B. subtilis BS87 and B. megaterium BM89) isolated and characterized from sugarcane rhizosphere depicting better crop yield potentiality. Microbiol Res 247:126733. https://doi.org/10.1016/j.micres.2021.126733

Chandra P, Khobra R, Sundha P et al (2021b) Plant growth promoting Bacillus-based bio formulations improve wheat rhizosphere biological activity, nutrient uptake and growth of the plant. Acta Physiol Plant 43:139. https://doi.org/10.1007/s11738-021-03310-5

Chandra P, Wunnava A, Verma P et al (2021c) Strategies to mitigate the adverse effect of drought stress on crop plants – influences of soil bacteria: a review. Pedosphere 31:496–509. https://doi.org/10.1016/S1002-0160(20)60092-3

Chandra P, Yadav S, Kumar A, Sheoran P, Mann A (2023) Arbuscular mycorrhizal fungi–plant interaction for salinity and drought stress alleviation. In: Kumar A, Dhansu P, Mann A (eds) Salinity and drought tolerance in plants. Springer, Singapore. https://doi.org/10.1007/978-981-99-4669-3_19

Chandra P, Khobra R, Sundha P et al (2024a) Bacillus subtilis strain BS87 as a biocontrol agent against spot blotch disease: effect on growth, nutrient status, and antioxidant enzymes in wheat. Acta Physiol Plant 46:35. https://doi.org/10.1007/s11738-024-03657-5

Chandra P, Rai AK, Basak N et al (2024b) In vitro P – solubilization activity of halophilic fungi in salt – affected soils and their potential as bio – inoculants. Environ Eng Res 29:230760. https://doi.org/10.4491/eer.2023.760

Chandra P, Fagodiya RK, Rai AK, Singh R, Sheoran P, Prajapat K, Singh A, Verma K, Verma VK, Yadav RK, Biswas AK (2024c) Different tillage and residue management practices affect soil biological activities and microbial culturable diversity in rice-wheat cropping system under reclaimed sodic soils. J Ecol Eng 25(5):193–207

Choudhary DK, Kasotia A, Jain S, Vaishnav A, Kumari S, Sharma KP, Varma A (2016) Bacterial-mediated tolerance and resistance to plants under abiotic and biotic stresses. J Plant Growth Regul 35:276–300

Costa O, Raaijmakers JM, Kuramae EE (2018) Microbial extracellular polymeric substances: ecological function and impact on soil aggregation. Front Microbiol 9:1636

Csonka LN (1989) Physiological and genetic responses of bacteria to osmotic stress. Microbiol Rev 53(1):121–147

Czech L, Hermann L, Stoveken N, Richter AA, Hoppner A, Smits S, Bremer E (2018) Role of the extremolytes ectoine and hydroxyectoine as stress protectants and nutrients: genetics, phylogenomics, biochemistry and structural analysis. Genes 9(4):177

Dimkpa C, Weinand T, Asch F (2009) Plant-rhizobacteria interactions alleviate abiotic stress conditions. Plant Cell Environ 32(12):1682–1694

Dubey R, Pathak H, Singh S, Chakravarti B, Thakur AK, Fagodia RK (2019) Impact of sowing dates on terminal heat tolerance of different wheat (Triticum aestivum L.) cultivars. Natl Acad Sci Lett 42(6):445–449

Egamberdieva D (2009) Alleviation of salt stress by plant growth regulators and IAA producing bacteria in wheat. Acta Physiol Plant 31:861–864

Egamberdieva D (2011) Survival of Pseudomonas extremorientalis TSAU20 and P. chlororaphis TSAU13 in the rhizosphere of common bean (Phaseolus vulgaris) under saline

Fagodiya RK, Kumar A, Kumari S, Medhi K, Shabnam AA (2020) Role of nitrogen and its agricultural management in changing environment. In: Naeem M, Ansari A, Gill S (eds) Contaminants in agriculture. Springer, Cham. https://doi.org/10.1007/978-3-030-41552-5_12

Faisal I, Tahira Y, Shafaqat A, Basharat A, Sohail H, Weijun Z, Muhammad A (2016) Plant growth promoting bacteria confer salt tolerance in Vigna radiata by up-regulating antioxidant defense and biological soil fertility. Plant Growth Regulations 80:23–26

Grichko V, Glick B (2001) Amelioration of flooding stress by ACC deaminase-containing plant growth-promoting bacteria. Plant Physiol Biochem 39:11–17. https://doi.org/10.1016/S0981-9428(00)01212-2

Grover M, Ali SZ, Sandhya V, Rasul A, Venkateswarlu B (2011) Role of microorganisms in adaptation of agriculture crops to abiotic stresses. World J Microbiol Biotechnol 27(5):1231–1240

Gururani MA, Upadhyaya CP, Baskar V, Venkatesh J, Nookaraju A, Park SW (2012) Plant growth-promoting rhizobacteria enhance abiotic stress tolerance in *Solanum tuberosum* through inducing changes in the expression of ROS-scavenging enzymes and improved photosynthetic performance. J Plant Growth Regul 32:245–258

Habib SH, Kausar H, Saud HM (2016) Plant growth-promoting rhizobacteria enhance salinity stress tolerance in Okra through ROS-scavenging enzymes. Biomed Res Int 2016:6284547

Hahm MS, Son JS, Hwang YJ, Kwon DK, Ghim SY (2017) Alleviation of salt stress in pepper (Capsicum annum L.) plants by plant growth-promoting Rhizobacteria. J Microbiol Biotechnol 27(10):1790–1797

Han HS, Lee KD (2005) Plant growth promoting rhizobacteria effect on antioxidant status, photosynthesis, mineral uptake and growth of lettuce under soil salinity. Res J Agric Biol Sci 1(3):210–215

Heidari M, Golpayegani A (2012) Effects of water stress and inoculation with plant growth promoting rhizobacteria (PGPR) on antioxidant status and photosynthetic pigments in basil (Ocimum basilicum L.). J Saudi Soc Agric Sci 11:57–61. https://doi.org/10.1016/j.jssas.2011.09.001

IPPC (2014) IPCC (Intergorvernmental panel on climate change), synthesis report 5

Islam F, Yasmeen T, Ali Q, Ali S, Arif MS, Hussain S, Rizvi H (2014) Influence of *Pseudomonas aeruginosa* as PGPR on oxidative stress tolerance in wheat under Zn stress. Ecotoxicol Environ Saf 104:285–293

Jha Y, Subramanian RB (2014) PGPR regulate caspase-like activity, programmed cell death, and antioxidant enzyme activity in paddy under salinity. Physiol Mol Biol Plants 20(2):201–207

Kasim WA, Osman ME, Omar MN, El-Daim IAA, Bejai S, Meijer J (2013) Control of drought stress in wheat using plant-growth promoting bacteria. J Plant Growth Regul 32:122–130

Kumar R, Gautam HR (2014) Climate change and its impact on agricultural productivity in India. J Climatol Weather Forecast 2(1):109

Kumar A, Verma JP (2018) Does plant – microbe interaction confer stress tolerance in plants: a review? Microbiol Res 207:41–52

Kumar L, Awasthi G, Singh B (2011) Extremophiles: a novel source of industrially important enzymes. Biotechnology 10:121–135

Kumar P, Tokas J, Kumar N, Lal M, Singal HR (2018) Climate change consequences and its impact on agriculture and food security. Int J Chem Stud 6(6):124–133

Kumar A, Medhi K, Fagodiya RK, Subrahmanyam G, Mondal R, Raja P, Malyan SK, Gupta DK, Gupta CK, Pathak H (2020) Molecular and ecological perspectives of nitrous oxide producing microbial communities in agro-ecosystems. Rev Environ Sci Biotechnol 19:717–750. https://doi.org/10.1007/s11157-020-09554-w

Kumar R, Singh A, Bhardwaj AK, Kumar A, Yadav RK, Sharma PC (2022) Reclamation of salt-affected soils in India: progress, emerging challenges, and future strategies. Land Degrad Dev 1–12:2169. https://doi.org/10.1002/ldr.4320

Liu F, Xing S, Ma H, Du Z, Ma B (2013) Cytokinin producing, plant growth promoting rhizobacteria that conferresistance to drought stress in Platycladus orientalis container seedlings. Appl Microbiol Biotechnol 97(20):9155–9164

Lorv JS, Rose DR, Glick BR (2014) Bacterial ice crystal controlling proteins. Scientifica (Cairo) 2014:1–20

Lund PA (2001) Microbial molecular chaperones. Adv Microb Physiol 44:93–140

Mahato A (2014) Climate change and its impact on agriculture. Int J Sci Res Publ 4(4):1–6

Maleki F, Khosravi A, Nasser A, Taghinejad H, Azizian M (2016) Bacterial heat shock protein activity. J Clin Diagn Res 10(3):BE01–BE03

Mayak S, Tsipora T, Bernar GR (2004) Plant growth-promoting bacteria that confer resistance to water stress in tomatoes and peppers. Plant Sci 166(2):525–530

Mhlongo MI, Piater LA, Madala NE, Labuschagne N, Dubery IA (2018) The chemistry of plant–microbe interactions in the rhizosphere and the potential for metabolomics to reveal signaling related to defense priming and induced systemic resistance. Front Plant Sci 9:112

Ngumbi E, Kloepper J (2016) Bacterial-mediated drought tolerance: current and future prospects. Appl Soil Ecol 105:109–125

Niu X, Song L, Xiao Y, Ge W (2018) Drought-tolerant plant growth-promoting rhizobacteria associated with foxtail millet in a semi-arid agroecosystem and their potential in alleviating drought stress. Front Microbiol 8:2580. https://doi.org/10.3389/fmicb.2017.02580

Numan M, Bashir S, Khan Y, Mumtaz R, Shinwari ZK, Khan AL, Khan A, AL-Harrasi A (2018) Plant growth promoting bacteria as an alternative strategy for salt tolerance in plants: a review. Microbiol Res 209:21–32

Olanrewaju OS, Glick BR, Babalola OO (2017) Mechanisms of action of plant growth promoting bacteria. World J Microbiol Biotechnol 33(11):197

Orhan F (2016) Alleviation of salt stress by halotolerant and halophilic plant growth-promoting bacteria in wheat (Triticum aestivum). Braz J Microbiol 47:621

Pieterse CM, Zamioudis C, Berendsen RL, Weller DM, Van-Wees SC, Bakker PA (2014) Induced systemic resistance by beneficial microbes. Annu Rev Phytopathol 52:347–375

Rai AK, Dinkar A, Basak N et al (2021) Phosphorus nutrition of oats genotypes in acidic soils: exploiting responsive plant-microbe partnership. Appl Soil Ecol 167:104094. https://doi.org/10.1016/j.apsoil.2021.104094

Rai AK, Basak N, Chandra P et al (2024) Occurrence of Sphingomonas olei with elemental S oxidation capability in sodic soil: potential role in sodicity reclamation and plant growth promotion. Ecotoxicol Environ Saf 278:116396. https://doi.org/10.1016/j.ecoenv.2024.116396

Rosier A, Medeiros FHV, Bais HP (2018) Defining plant growth promoting rhizobacteria molecular and biochemical networks in beneficial plant-microbe interactions. Plant Soil 428:35–55

Rudrappa T, Biedrzycki ML, Kunjeti SG, Donofrio NM, Czymmek KJ, Pare PW, Bais HP (2010) The rhizobacterial elicitor acetoin induces systemic resistance in Arabidopsis thaliana. Commun Integrat Biol 3(2):130–138

Schimel J, Teri CB, Wallenstein M (2007) Microbial stress-response physiology and its implications for ecosystem function. Ecology 88(6):1386–1394

Shoresh M, Yedidia I, Chet I (2005) Involvement of jasmonic acid/ethylene signaling pathway in the systemic resistance induced in cucumber by Trichoderma asperellum T203. Phytopathology 95:76–84

Shrivastava P, Kumar R (2015) Soil salinity: a serious environmental issue and plant growth promoting bacteria as one of the tools for its alleviation. Saudi J Biol Sci 22(2):123–131

Su F, Jacquard C, Villaume S, Michel J, Rabenoelina F, Clément C, Barka EA, Dhondt-Cordelier S, Vaillant-Gaveau N (2015) *Burkholderia phytofirmans* PsJN reduces impact of freezing temperatures on photosynthesis in *Arabidopsis thaliana*. Front Plant Sci 6:810. https://doi.org/10.3389/fpls.2015.00810

Timmusk S, Wagner EG (1999) The plant-growth-promoting rhizo-bacterium Paenibacillus polymyxa induces changes in Arabidopsis thaliana gene expression: a possible connection between biotic and abiotic stress responses. Mol Plant-Microbe Interact 12(11):951–959

Turan M, Güllüce M, Çakmakçı R, Sahin F (2013) Effect of plant growth-promoting rhizobacteria strain on freezing injury and antioxidant enzyme activity of wheat and barley. J Plant Nutr 36:731. https://doi.org/10.1080/01904167.2012.754038

Vacheron J, Desbrosses G, Bouffaud ML, Touraine B, Moenne-Loccoz Y, Muller D, Legendre L, Wisniewski-Dye F, Prigent-Combaret C (2013) Plant growth-promoting rhizobacteria and root system functioning. Front Plant Sci 4:356

Van Loon LC (2007) Plant responses to plant growth-promoting rhizobacteria. Eur J Plant Pathol 119:243–254

Van-Loon LC, Bakker PAHM (2006) Induced systemic resistance as a mechanism of disease suppression by Rhizobacteria. In: PGPR: biocontrol and biofertilization, pp 39–66

Van-Peer R, Niemann GJ, Schippers B (1991) Induced resistance and phytoalexin accumulation in biological control of fusarium wilt of carnation by Pseudomonas sp. strain WCS417r. Phytopathology 81:728–734

Vejan P, Abdullah R, Khadiran T, Ismail S, Boyce NA (2016) Role of plant growth promoting rhizobacteria in agricultural sustainability- a review. Molecules 21(5):573

Venkateswarlu B, Shanker AK (2009) Climate change and agriculture: adaptation and mitigation strategies. Indian J Agron 54:226–230

Verma P, Yadav AN, Kumar V, Dhananjaya PS, Saxena A (2017) Beneficial plant-microbes interactions: biodiversity of microbes from diverse extreme environments and its impact for crop improvement. In: Plant-microbe interactions in agro-ecological, pp 543–580

Verma P, Chandra P, Rai AK et al (2023) Native rhizobacteria suppresses spot blotch disease, improves growth and yield of wheat under salt–affected soils. Plant Stress 10:100234. https://doi.org/10.1016/j.stress.2023.100234

Yadav AN, Verma P, Singh B, Chauahan VS, Suman A, Saxena AK (2017) Plant growth promoting bacteria: biodiversity and multifunctional attributes for sustainable agriculture. Adv Biotechnol Microbiol 5(5)

Yadav SN, Singh AK, Peter JK, Masih H, Benjamin JC, Singh DK, Chaudhary S, Ramteke PW, Ojha SK (2018) Study of exopolysaccharide containing pgprs on growth of okra plant under water stress conditions. Int J Curr Microbiol App Sci 7(11):3337–3374

Yang J, Kloepper JW, Ryu CM (2009) Rhizosphere bacteria help plants tolerate abiotic stress. Trends Plant Sci 14(1):1–4

Zahir ZA, Munir A, Asghar HN, Shaharoona B, Arshad M (2008) Effectiveness of rhizobacteria containing ACC deaminase for growth promotion of peas (*Pisum sativum*) under drought conditions. J Microbiol Biotechnol 18(5):958–963

Zarei T, Moradi A, Kazemeini SA et al (2020) The role of ACC deaminase producing bacteria in improving sweet corn (*Zea mays* L. var saccharata) productivity under limited availability of irrigation water. Sci Rep 10:20361. https://doi.org/10.1038/s41598-020-77305-6

CRISPR/Cas-Based Fungal Genome Engineering for Secondary Metabolite Production: Progress and Challenges

16

Noyonika Kaul, Poonam Ranga, Sujata Yadav, Priti Choudhary, Annu Dahiya, Deepti Rani Sehrawat, and Anita Mann

Abstract

Genome editing has evolved significantly, with early methods like zinc finger nucleases, homologous recombination, and TALENs revolutionizing molecular biology but facing challenges such as complexity, high error rates, and low efficiency. The advent of CRISPR/Cas technology, inspired by bacterial immune responses to viruses, represents a major breakthrough in the field. CRISPR/Cas offers precise, efficient targeting of specific DNA sequences with simpler design and broader applications. This system has been transformative in various organisms, including fungi, where it is used for gene knockout, gene insertion, and metabolic pathway engineering. Advances in fungal genetics have significantly improved natural product research by revealing hidden compounds and the processes that create them. With CRISPR/Cas technologies, scientists can now modify genes by adding, removing, or altering them, making it possible to manipulate gene clusters responsible for producing various compounds in a wide range of fungi. Most importantly, editing disease-specific genes or nutrient-responsive or certain metabolites necessary for plant growth can enhance the potential of fungi in agriculture. The simplicity, versatility, and efficiency of CRISPR/Cas make it a powerful tool for fungal genetics, enabling advancements in research, biotechnology, and industrial applications. This chapter explores the CRISPR/Cas system's mechanisms and its key applications in fungal biology.

Keywords

Biotic-stress · CRISPR · Mycotoxin · Gene-editing · Cas9 · Fungi · Metabolite

N. Kaul · P. Ranga · S. Yadav · P. Choudhary · A. Dahiya · A. Mann (✉)
Department of Crop Improvement, ICAR-Central Soil Salinity Research Institute, Karnal, Haryana, India

D. R. Sehrawat
Department of Anthropology, Panjab University, Chandigarh, India

A. Mann et al. (eds.), *Cutting Edge Technologies for Developing Future Crop Plants*,
https://doi.org/10.1007/978-981-96-2508-6_16

16.1 Introduction

Fungi play a very crucial role across various industries including agriculture, textiles, food production, pharmaceuticals, and bioremediation. In natural ecosystems, fungi are essential for nutrient cycling, particularly in decomposing organic matter and acting as biofertilizers. The latest developments in fungal molecular research have made it possible to sequence the entire genome and examine gene and protein expression arrays. These developments provide interesting pathways for providing new products such as mammalian protein molecules, medicines, pigments, and fatty acids utilizing fungal systems, favoring the welfare of humans. (Adrio and Demain 2003; Idnurm and Meyer 2014).

The ability to use fungal strains for protein and metabolite biosynthesis is based on parameters such as yields for production and compliance with regulations, which is particularly important in sectors like medicines and food manufacturing. This chapter explores these applications and considerations, highlighting the pivotal role of fungi in modern biotechnological advancements. Multiple fungal species are currently employed for large-scale biosynthesis of recombinant proteins and metabolites (Punt 2006). Fungi produce a wide range of synthetic metabolites, which include both useful substances with possible applications in medicines, agricultural chemicals, and cosmetics, as well as destructive toxins that may have detrimental impact people, plants, animals, and the natural environment (Erb and Kliebenstein 2020). For instance, phytotoxins produced by plant-pathogenic fungi contribute significantly to crop diseases, leading to substantial economic losses and environmental challenges. Additionally, mycotoxins, toxic compounds generated by specific fungal metabolism, pose serious threats to both livestock and human health (Fumagalli et al. 2021). Contrary to this, fungi are abundant sources of important secondary metabolites used in a variety of industries, including industry, agriculture, and medicine. These metabolites are synthesized through diverse pathways, utilizing primary metabolites as foundational components. Major molecular classes include polyketides, terpenoids, and non-ribosomal peptides.

In fungal genomes, the genes responsible for the biosynthesis of secondary metabolites are typically organized into clusters known as biosynthetic gene clusters (BGCs). These BGCs generally contain a set of genes that work together to produce complex molecules. The typical components of a fungal BGC include genes encoding core synthases or synthetases, which are responsible for the initial steps of secondary metabolite production, as well as biosynthetic tailoring enzymes that modify these core structures (Rokas et al. 2020). Additionally, BGCs often contain genes for regulatory proteins, transporters, and enzymes related to self-resistance, ensuring the effective production and protection of the organism from its own metabolites. The rapid advancement in sequencing technologies and genomic tools has dramatically increased the number of publicly available fungal genomes in recent years. This surge in genomic data has provided researchers with a rich resource for studying fungal BGCs. Annotated genomic information has become more accessible, significantly accelerating the identification and characterization of these clusters.

Automated genome mining tools have been instrumental in this progress. Tools like antiSMASH, MIBiG 2.0, and BiG-SCAPE allow researchers to efficiently scan fungal genomes for BGCs, predict their functions, and compare them across various species (Blin et al. 2021). These tools use sophisticated algorithms to identify BGCs based on conserved sequence motifs and structural features, facilitating large-scale analysis and cross-species comparisons. The integration of advanced sequencing technologies, comprehensive genomic databases, and powerful bioinformatics tools has not only enhanced our understanding of fungal SM biosynthesis but also paved the way for discovering new natural products. These discoveries have significant implications for medicine, agriculture, and industry, where fungal secondary metabolites are highly valued for their bioactive properties.

16.2 CRISPR/Cas9-Based Genome Editing in Fungi

The CRISPR/Cas9 technology has been successfully utilized to edit the genomes of various filamentous fungi and higher fungi, including members of the *Ascomycetes* and *Basidiomycetes* groups, as well as fungus-like eukaryotic microorganisms. The application of programmable site-specific nucleases for targeted genome engineering in these fungi has shown encouraging and promising outcomes (Table 16.1).

For instance, the CRISPR/Cas9 system has been effectively applied to several fungal species. Notable examples include *Aspergillus, Penicillium and Saccharomyces*. These advancements highlight the versatility and efficacy of CRISPR/Cas9 in fungal genome editing, paving the way for significant progress in fungal biology and biotechnology. The ability to precisely edit the genomes of these

Table 16.1 Gene-editing applications in fungal species: tools, promoters, and transformation methods

Fungal species	Gene editing tool	Promoters of sgRNA	Gene delivery	References
Alternaria alternata	CRISPR/Cas9	*gpdA*	PEG mediated protoplast transformation	Wenderoth et al. (2017)
Aspergillus fumigatus	CRISPR/Cas9	*SNR52*	Protoplast transformation	Fuller et al. (2015)
Aspergillus oryzae	PtFg TALEN	*glaA142*	PEG-mediated Protoplast transformation	Mizutani et al. (2017)
Candida albicans	CRISPR/Cas9	SNR52	Lithium acetate method	Vyas et al. (2015)
Neurospora crassa	CRISPR/Cas9	*Beta-tubulin*	–	Matsu-ura et al. (2015)
Penicillium chrysogenum	CRISPR/Cas9-RNP	*U6*	Protoplast transformation	Pohl et al. (2016)
Saccharomyces cerevisiae	CRISPR/Cas9	*TEF1*	LiAc/SS carrier DNA/PEG method	Jessop-Fabre et al. (2016)

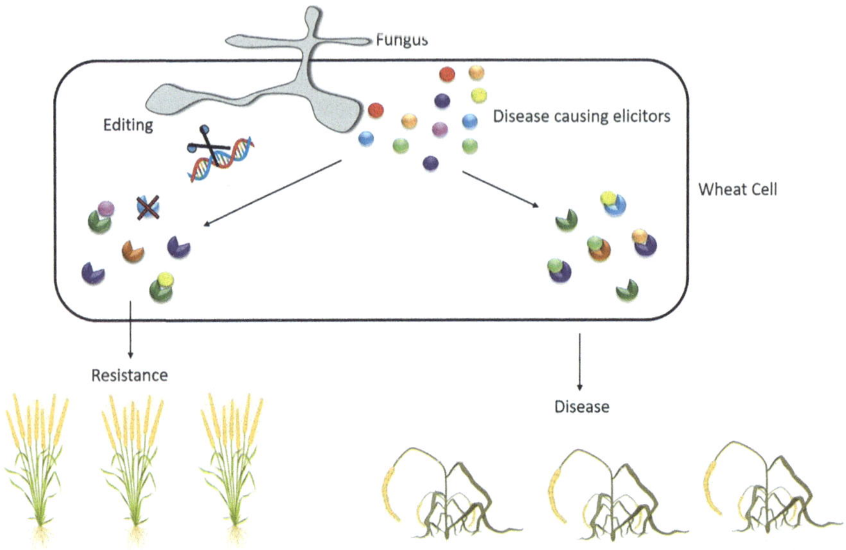

Fig. 16.1 Use of CRISPR to combat fungal pathogens thereby strengthening wheat crop's defense system

organisms opens up new possibilities for research and applications, including the improvement of fungal strains for industrial use, the study of gene function, and the development of new strategies for fungal disease control. The successful implementation of CRISPR/Cas9 in these diverse fungal species underscores its potential as a powerful tool for genome engineering in a wide range of fungal and fungus-like microorganisms. A brief workflow of fungal genome editing for disease resistance is depicted in Fig. 16.1.

16.3 Industrial Applications of CRISPR/Cas9 Methods in Fungi

Fungi play a pivotal role in various facets of human life, influencing our health, food production, and livestock. These organisms offer numerous benefits through mechanisms such as enhancing plant nutrition and protection, and generating enzymes, antibiotics, and therapeutic compounds. Known for their versatility, fungi produce a wealth of valuable natural properties (Zhong and Xiao 2009). Today, fungi are integral to the large-scale production of secondary metabolites, proteins, and organic acids. Fungal cell factories are essential for synthesizing these critical substances. However, to maximize the potential of fungi as sources of innovative molecules and as microbial cell factories, several challenges need to be addressed. Efforts must be made to enhance and optimize the use of fungi for producing a wide range of cell-mediated products (Meyer et al. 2016).

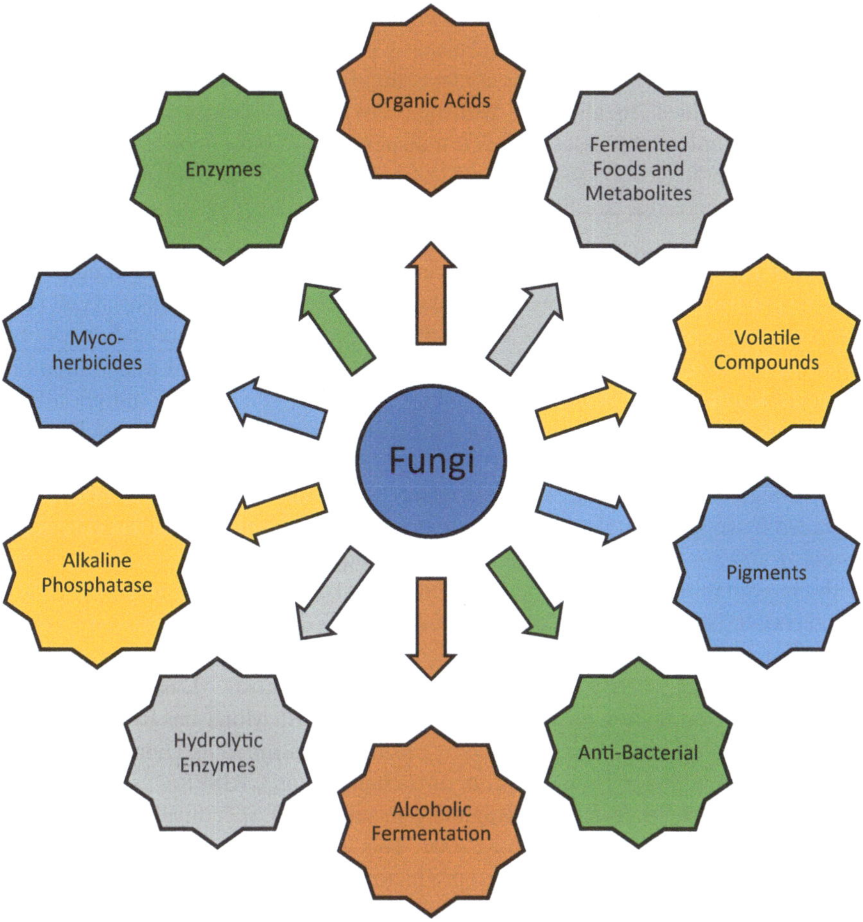

Fig. 16.2 Important metabolites and enzymes synthesized by various fungal species

Substances with industrial and pharmaceutical applications (Fig. 16.2). These include alkaloids, polysaccharides, steroids, and terpenes, many of which possess antibacterial and anti-tumor Overcoming these challenges is crucial for fully leveraging the capabilities of fungi, leading to significant advancements in industrial and pharmaceutical sectors. *Saccharomyces cerevisiae*, commonly known as baker's yeast, serves as a model organism extensively used in industrial applications. It is pivotal in the development of recombinant proteins, various food products, beverages, analytical tools, and biofuels (Stovicek et al. 2017). The CRISPR-Cas9 gene-editing technology has been employed to create auxotrophic markers through gene disruption, resulting in *S. cerevisiae* mutants that exhibit resistance to heat stress. This genetic modification has proven to be a sustainable solution for fermentation inhibitors, making it highly relevant in industries producing alcoholic beverages like wine and beer, as well as in the fermentation processes for organic acids, glucose, glycerol, and other substances (Salazar-Cerezo et al. 2020).

The use of *S. cerevisiae* in these contexts highlights its versatility and efficiency as an industrial host. Its ability to withstand environmental stresses and facilitate various biochemical processes makes it indispensable in both traditional and modern industrial practices. As research and technology continue to advance, the applications of *S. cerevisiae* are likely to expand, further solidifying its role in diverse industrial sectors.

Penicillium subrubescens, an ascomycete fungus, is rich in carbohydrate-active enzyme families that degrade plant biomass, making it a valuable industrial cell factory for enzyme production. The CRISPR/Cas9 system has been successfully established in *P. subrubescens* by developing protoplast generation and transformation methods, leading to the deletion of the *ku70* gene. This gene editing did not affect the fungus's growth or protein production, showcasing the potential of CRISPR/Cas9 for further strain improvement and gene functional characterization (Mondal et al. 2022). Cellulose, the most abundant biopolymer on Earth, is efficiently degraded by fungi through cellulase synthesis. Genetic manipulation has enhanced fungal strains for better cellulase production, though traditional methods are slow. With the availability of full-genome sequences for key filamentous fungi, CRISPR-Cas technology has become essential for efficiently developing improved strains. The main goal of current cellulase research is to develop fungal strains with enhanced cellulase activity and reduced production costs (Liu et al. 2020). Mycotoxin contamination poses serious health risks, with Monascus Red, produced by *Monascus purpureus*, being limited by citrinin contamination. Genetic suppression of citrinin in this polykaryotic fungus is challenging. CRISPR/Cas system was developed to delete large genomic fragments in such fungi, specifically targeting the 15-kb citrinin gene cluster in strain KL-001. This led to stable, citrinin-free mutants and a 2–5% increase in Monascus Red production, offering a robust solution for the food industry (Zhou et al. 2024). Traditional Chinese food therapies have inspired modern medicine, with *Monascus*, a fungus used for centuries in Chinese cuisine, being a notable example. *Monascus* produces valuable metabolites, including natural pigments and lipid-lowering substances. Recent research by Zhou in 2024 has advanced *Monascus* development at the lab scale, focusing on metabolite biosynthesis, strain modification, and fermentation optimization. It emphasized the potential for industrial-scale production through automated solid-state fermentation and the integration of synthetic biology with process intensification, aiming to enhance *Monascus* applications across various industries.

16.4 CRISPR-Cas9: A Novel Approach for Biological Control

Biological control has long been viewed as a viable alternative to pesticide-based pest management strategies, yet its global impact and adoption remain limited and inconsistent. The rapid advancement of molecular techniques, particularly those related to DNA, is transforming many life sciences fields. Molecular techniques are increasingly used alongside traditional morphological taxonomic methods for identifying and studying biological control agents, including microbes. A range of molecular techniques has been applied to investigate food-web interactions. Polymerase chain reaction (PCR) methods have largely replaced immunological approaches such as enzyme-linked immunosorbent assay (ELISA) and are now being overtaken by next-generation sequencing (NGS) techniques, which offer unprecedented power at rapidly decreasing costs. Molecular techniques also have the potential to manipulate biological control agents, a process that will accelerate with the advent of gene-editing tools like the CRISPR/Cas9 system (Mann et al. 2024). These gene-editing tools provide unparalleled capabilities to both elucidate and manipulate plant defense mechanisms, including those involving the attraction of natural enemies to attacked plants. Rapid technological advances will lead to the development of even more innovative pest management options, with their adoption likely limited primarily by regulatory challenges. Fungi are the culprits behind various plant diseases, such as smut, mildew, rot, and rust, which greatly diminish crop yields and lower the quality of produce globally. Additionally, some fungi generate mycotoxins-harmful secondary metabolites that pose serious health dangers to both humans and animals when ingested through contaminated food (Wang et al. 2014). To combat these challenges, researchers have developed several strategies to improve plant resistance to fungal infections. These strategies leverage current insights into the molecular processes involved in plant-pathogen interactions. Key genes that play a role in enhancing plant resistance to fungi have been identified, and these genes are now principal targets for genome editing using CRISPR/Cas9 technology. This approach holds significant promise for increasing crop resilience and ensuring food security. Marcio Vinicius in 2023 (Cortes et al. 2023) details the first successful marker-free genetic engineering of *Trichoderma harzianum* using a CRISPR/Cas9 system. By targeting the *albA* and *ku70* genes, researchers created mutant strains that confirmed the efficacy of the gene-editing system. Another finding was reported by the first successful use of CRISPR/Cas9 gene-editing in *Trichoderma harzianum*, a fungus used for biological pest control in agriculture. Researchers created mutants by targeting the pyr4 gene, which was confirmed through sequencing. These mutants showed reduced growth without uridine and resistance to 5-fluorotic acid, while retaining their mycoparasitic activity. The findings demonstrated that CRISPR/Cas9 can effectively edit genes in *T. harzianum* without compromising its biological control properties (Vieira et al. 2021). *Aspergillus flavus*, a producer of hepatocarcinogenic aflatoxin, poses significant health and trade issues. The study developed by (Chang 2024) included a dual CRISPR/Cas9 system with 60-nucleotide donor DNA to delete harmful gene clusters in *A. flavus*. It efficiently removed a 102-kb segment containing aflatoxin and cyclopiazonic acid gene clusters, and a 27-kb ustiloxin B gene cluster, with deletion efficiencies ranging from 66.6% to 85.6%. Larger chromosomal deletions (201 kb and 301 kb) were also achieved with efficiencies of 57.7–69.2%. This method generates marker-free, atoxigenic biocontrol strains suitable for field trials.

16.5 Challenges in Editing Fungal Genome

Despite the numerous advancements and promising applications of fungi in various industries, several significant challenges remain in comparison to genome editing in plants (Fig. 16.3). Ensuring fungal strains meet stringent regulatory standards, especially in pharmaceuticals and food production, is a major hurdle. Regulatory bodies require comprehensive safety assessments and adherence to guidelines, which can be resource-intensive and time-consuming. Mycotoxins, toxic secondary metabolites produced by certain fungi, pose serious health risks to humans and animals. Preventing and managing mycotoxin contamination in food and feed products is a critical challenge that requires constant monitoring and advanced mitigation strategies. Identifying and optimizing fungal strains for high-yield production of desired metabolites and proteins is complex. This process often involves extensive genetic manipulation and fermentation process optimization, which can be technically challenging and costly.

Maintaining consistent production yields in large-scale operations is difficult due to the variability in fungal growth conditions and genetic stability. Ensuring reliable and scalable production processes is essential for industrial applications. While CRISPR/Cas9 and other genome editing tools have revolutionized fungal biotechnology, their efficiency and specificity can vary among different fungal species. Developing robust genome editing protocols for diverse fungal strains is an ongoing challenge. The functional characterization of genes within biosynthetic gene clusters (BGCs) is still limited. Understanding the precise roles of individual genes and their interactions within BGCs is crucial for manipulating and optimizing secondary

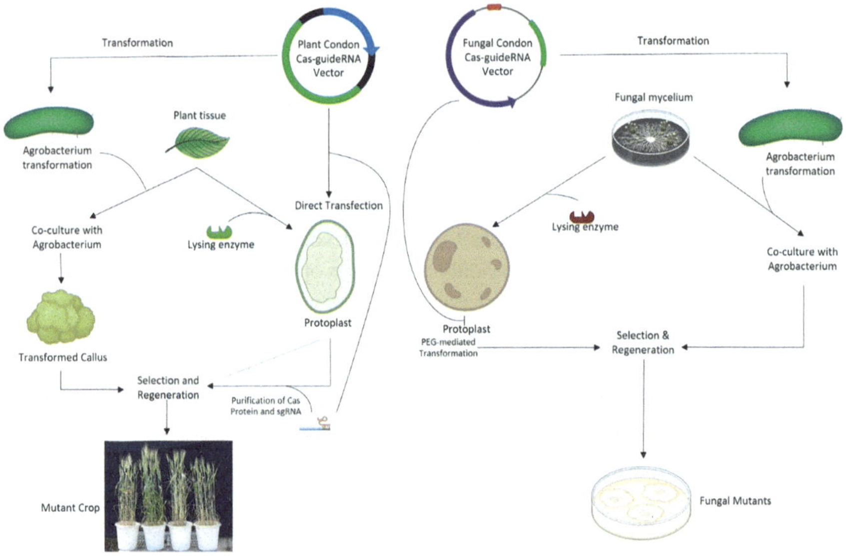

Fig. 16.3 CRISPR technology: differing approaches in plant and fungal biotechnology

metabolite production. The cost of production, including raw materials, energy, and labor, can be high for fungal biotechnology processes. Developing cost-effective production methods is vital for commercial viability. Industrial-scale fungal production can have environmental impacts, such as waste generation and resource consumption. Implementing sustainable practices and minimizing environmental footprints are essential for the long-term success of fungal biotechnologies. Biological control using fungi to manage plant diseases and pests requires consistent and reliable efficacy under varying environmental conditions. Ensuring the stability and performance of fungal biocontrol agents is a significant challenge. Pathogens and pests can develop resistance to fungal biocontrol agents over time. Monitoring and managing resistance development is crucial to maintaining the effectiveness of biological control strategies. Addressing these challenges requires a multidisciplinary approach, integrating advancements in molecular biology, genomics, bioinformatics, and industrial biotechnology. Collaborative efforts between academia, industry, and regulatory bodies are essential to overcome these obstacles and fully realize the potential of fungi in various sectors.

16.6 Conclusion and Future Prospects

Fungi are indispensable across numerous industries and natural ecosystems due to their diverse roles in nutrient cycling, bio-fertilization, and the production of valuable metabolites. Advances in fungal molecular biology, including genome sequencing and gene expression analysis, have unlocked new potential for fungal applications in biotechnology. The development of CRISPR/Cas9 and other gene-editing technologies has revolutionized our ability to manipulate fungal genomes, enhancing their utility in producing recombinant proteins, secondary metabolites, and other industrially relevant compounds. Despite their benefits, fungi also pose challenges, such as the production of phytotoxins and mycotoxins, which can adversely affect agriculture and health. However, with the integration of advanced genomic tools and bioinformatics, researchers are better equipped to explore fungal biosynthetic pathways, leading to the discovery of new natural products with significant applications in medicine, agriculture, and industry. The ongoing research and technological advancements promise to further harness the capabilities of fungi, ultimately contributing to sustainable and innovative solutions across various sectors.

References

Adrio JL, Demain AL (2003) Fungal biotechnology. Int Microbiol 6:191–199

Blin K, Shaw S, Kloosterman AM, Charlop-Powers Z, Van Wezel GP, Medema MH, Weber T (2021) antiSMASH 6.0: improving cluster detection and comparison capabilities. Nucleic Acids Res 49(W1):W29–W35

Chang PK (2024) Creating large chromosomal segment deletions in *aspergillus flavus* by a dual CRISPR/Cas9 system: deletion of gene clusters for production of aflatoxin, cyclopiazonic acid, and ustiloxin B. Fungal Genet Biol 170:103863

Cortes MVDCB, Barbosa ET, de Sousa Oliveira MI, Maciel LHR, Junior ML, Contesini FJ, de Filippi MCC, da Silva-Lobo VL (2023) *Trichoderma harzianum* marker-free strain construction based on efficient CRISPR/Cas9 recyclable system: a helpful tool for the study of biological control agents. Biol Control 184:105281

Erb M, Kliebenstein DJ (2020) Plant secondary metabolites as defenses, regulators, and primary metabolites: the blurred functional trichotomy. Plant Physiol 184(1):39–52

Fuller KK, Chen S, Loros JJ, Dunlap JC (2015) Development of the CRISPR/Cas9 system for targeted gene disruption in *Aspergillus fumigatus*. Eukaryot Cell 14(11):1073–1080

Fumagalli F, Ottoboni M, Pinotti L, Cheli F (2021) Integrated mycotoxin management system in the feed supply chain: innovative approaches. Toxins 13(8):572

Idnurm A, Meyer V (2014) Welcome to fungal biology and biotechnology. Fungal Biol Biotechnol 1:1–2

Jessop-Fabre MM, Jakočiūnas T, Stovicek V, Dai Z, Jensen MK, Keasling JD, Borodina I (2016) EasyClone-MarkerFree: a vector toolkit for marker-less integration of genes into *Saccharomyces cerevisiae* via CRISPR-Cas9. Biotechnol J 11(8):1110–1117

Liu W, An C, Shu X, Meng X, Yao Y, Zhang J, Chen F, Xiang H, Yang S, Gao X, Gao SS (2020) A dual-plasmid CRISPR/Cas system for mycotoxin elimination in polykaryotic industrial fungi. ACS Synth Biol 9(8):2087–2095

Mann A, Ranga P, Choudhary P, Yadav S, Kaul N, Dahiya A, Prakash NR, Kumar A, Kumar A, Sanwal SK (2024) Genome Editing Technologies for Enhancing Plant Resilience to Biotic and Abiotic Stresses-Brief Review. Journal of Soil Salinity and Water Quality 16(2):180–193

Matsu-Ura T, Baek M, Kwon J, Hong C (2015) Efficient gene editing in *Neurospora crassa* with CRISPR technology. Fungal Biol Biotechnol 2:1–7

Meyer V, Andersen MR, Brakhage AA, Braus GH, Caddick MX, Cairns TC, de Vries RP, Haarmann T, Hansen K, Hertz-Fowler C, Krappmann S (2016) Current challenges of research on filamentous fungi in relation to human welfare and a sustainable bio-economy: a white paper. Fungal Biol Biotechnol 3:1–17

Mizutani O, Arazoe T, Toshida K, Hayashi R, Ohsato S, Sakuma T, Yamamoto T, Kuwata S, Yamada O (2017) Detailed analysis of targeted gene mutations caused by the platinum-fungal TALENs in *aspergillus oryzae* RIB40 strain and a ligD disruptant. J Biosci Bioeng 123(3):287–293

Mondal S, Halder SK, Mondal KC (2022) Tailoring in fungi for next generation cellulase production with special reference to CRISPR/CAS system. Syst Microbiol Biomanuf 2:1–17

Pohl C, Kiel JA, Driessen AJ, Bovenberg RA, Nygard Y (2016) CRISPR/Cas9 based genome editing of *Penicillium chrysogenum*. ACS Synth Biol 5(7):754–764

Punt PJ (2006) Filamentous fungi as cell factories for protein production. Microb Cell Factories 5(Suppl 1):S32

Rokas A, Mead ME, Steenwyk JL, Raja HA, Oberlies NH (2020) Biosynthetic gene clusters and the evolution of fungal chemodiversity. Nat Prod Rep 37(7):868–878

Salazar-Cerezo S, Kun RS, De Vries RP, Garrigues S (2020) CRISPR/Cas9 technology enables the development of the filamentous ascomycete fungus *Penicillium subrubescens* as a new industrial enzyme producer. Enzym Microb Technol 133:109463

Stovicek V, Holkenbrink C, Borodina I (2017) CRISPR/Cas system for yeast genome engineering: advances and applications. FEMS Yeast Res 17(5):fox030

Vieira AA, Vianna GR, Carrijo J, Aragão FJ, Vieira PM (2021) Generation of *Trichoderma harzianum* with pyr4 auxotrophic marker by using the CRISPR/Cas9 system. Sci Rep 11(1):1085

Vyas VK, Barrasa MI, Fink GR (2015) A *Candida albicans* CRISPR system permits genetic engineering of essential genes and gene families. Sci Adv 1(3):e1500248

Wang Y, Cheng X, Shan Q, Zhang Y, Liu J, Gao C, Qiu JL (2014) Simultaneous editing of three homoeoalleles in hexaploid bread wheat confers heritable resistance to powdery mildew. Nat Biotechnol 32(9):947–951

Wenderoth M, Pinecker C, Voß B, Fischer R (2017) Establishment of CRISPR/Cas9 in *Alternaria alternata*. Fungal Genet Biol 101:55–60

Zhong JJ, Xiao JH (2009) Secondary metabolites from higher fungi: discovery, bioactivity, and bioproduction. In: Biotechnology in China I: from bioreaction to bioseparation and bioremediation, pp 79–150

Zhou J, Pan Q, Xue Y, Dong Y, Chen Y, Huang L, Zhang B, Liu ZQ, Zheng Y (2024) Synthetic biology for Monascus: from strain breeding to industrial production. Biotechnol J 19(7):2400180

Microbial Contributions to Crop Adaptation: Innovation for Climate-Resilient Agriculture

17

Sadhana Giri, Himani Panwar, Kajal Singh, and Mukesh Kumar

Abstract

The escalating impact of climate change on agriculture demands innovative strategies to enhance crop resilience and ensure global food security. Microorganisms, encompassing bacteria, fungi, and other soil-dwelling microbes, play a crucial role in aiding crop adaptation to environmental stresses. The multifaceted contributions of microbes to crop resilience and their ability to enhance nutrient uptake, improve water-use efficiency, and bolster resistance to both biotic and abiotic stresses. By forming symbiotic relationships with plants, these microbes facilitate critical functions such as nitrogen fixation, phosphate solubilization, and the production of growth-promoting hormones. Moreover, the potential of microbial communities to mitigate the adverse effects of climate-related stressors such as drought, salinity, and extreme temperatures offers a promising avenue for developing climate-resilient crops. This chapter explores the latest advancements in microbial biotechnology, including the use of microbial inoculants and the engineering of plant-microbe interactions, to enhance crop performance under changing environmental conditions. The integration of these microbial strategies into sustainable agricultural practices could revolutionize crop production, offering a pathway to achieve resilience in the face of climate uncertainty.

S. Giri
Department of Microbiology, Kanya Gurukul Campus, Gurukula Kangri (Deemed to be University), Haridwar, Uttarakhand, India

H. Panwar (✉) · K. Singh
Department of Botany and Microbiology, Gurukula Kangri (Deemed to be University), Haridwar, Uttarakhand, India

M. Kumar
Department of Microbiology, Chaudhary Charan Singh Haryana Agricultural University, Hisar, Haryana, India

A. Mann et al. (eds.), *Cutting Edge Technologies for Developing Future Crop Plants*,
https://doi.org/10.1007/978-981-96-2508-6_17

Keywords
Crop resilience · Soil microbes · Sustainable agriculture · Climate changes · Agricultural innovation

17.1 Introduction

Actions by humans, like the burning of fossil fuels, deforestation, and manufacturing activities, are important factors causing climate change, which is significantly disrupting global agricultural systems. The rise in greenhouse gases, such as nitrous oxide (N_2O), carbon dioxide (CO_2), and methane (CH_4), is contributing to rising global temperatures, altered precipitation patterns, and a growing frequency of extreme weather events. These climatic shifts are having substantial effects on crop productivity, livestock health, and overall food security. One of the primary consequences of climate change is the impact on crop yields. Changing temperature and precipitation patterns are already altering agricultural productivity in various regions. In many areas, increased temperatures are shortening the growing seasons of major crops, leading to lower yields. Studies have shown that staple crops like wheat, rice, and maize are seeing declining yields in regions affected by rising temperatures and unpredictable rainfall. Additionally, droughts and heat stress can hinder crop development, reduce pollination, and exacerbate pest infestations and disease outbreaks (Zhao et al. 2017). Although some regions may temporarily benefit from warmer temperatures or elevated CO_2 levels, these advantages are often outweighed by the broader negative impacts of climate change on agriculture.

Soil degradation is another serious consequence of climate change, driven by processes such as erosion, salinization, and the loss of organic matter. Heavy rainfall can erode topsoil, while extended droughts reduce soil moisture. Moreover, shifting climate patterns are impacting water availability, resulting in both excessive flooding and severe water shortages in many agricultural regions. These challenges put pressure on irrigation systems, particularly in areas where consistent water supplies are essential for crop cultivation (FAO 2020). The unpredictability of agricultural outputs due to climate change is exacerbating food insecurity, particularly in vulnerable regions. Declining crop yields, coupled with rising global food prices, are intensifying the risk of hunger and malnutrition. Smallholder farmers, especially in developing countries, are disproportionately affected as they often lack the financial and technical resources to adapt to these climatic changes (IPCC 2022). Considering these challenges, various adaptation and mitigation strategies are being explored to safeguard agriculture from the impacts of climate change.

Microorganisms are fundamental to sustaining soil health and fertility, performing crucial functions in nutrient cycling, organic matter decomposition, and soil structure stabilization. In agricultural systems, soil microbes are pivotal in breaking down organic materials and converting them into nitrogen, phosphorus, and potassium are examples of essential nutrients that are necessary for the most successful development of plants.

One of the key microbial processes in agriculture is biological nitrogen fixation. Phosphorus, another crucial nutrient for plants, is often in limited supply in soils due to its tendency to form insoluble compounds. Phosphate-solubilizing bacteria (PSB) address this issue by converting these insoluble phosphorus forms into soluble forms that plants can absorb. This microbial action not only improves soil fertility but also promotes sustainable farming practices by reducing the need for chemical fertilizers, which can have adverse effects on ecosystems when used excessively (Sharma et al. 2013; Singh and Reddy 2022). Thus, soil microorganisms are essential for enhancing soil fertility and supporting sustainable agricultural practices by improving nutrient availability and reducing reliance on chemical inputs. Their role in nutrient cycling and organic matter decomposition underscores their importance in maintaining healthy and productive soils.

The increasing awareness of climate change has made it imperative to effectively utilize the beneficial soil microorganisms and their interactions to improve soil fertility and health, increase agroecosystem resilience, and maximize crop productivity. This chapter focuses on the different tactics used by microbial diversity to survive in climate change and make the crops better adapt to the changing climatic conditions as illustrated in Fig. 17.1.

17.2 Soil Microbial Community

The soil microbiome, comprising a vast and varied assemblage of bacteria, fungi, and other microorganisms, is crucial for sustaining soil health and facilitating vital ecological functions. This complex microbial community is a significant reservoir of biodiversity, playing a pivotal role in processes such as nutrient cycling, organic matter decomposition, disease suppression, and plant vitality. Our knowledge of soil microbial communities has been completely transformed by recent developments in molecular techniques, such as high-throughput sequencing. These technologies have unveiled the intricate structure and diverse functional roles of soil microbes, offering deeper insights into how these microorganisms interact with their environment and contribute to ecosystem dynamics. The study of the soil microbiome is increasingly recognized as essential for promoting sustainable agricultural practices and advancing environmental conservation efforts.

17.2.1 Structure of the Soil Microbiome

The soil microbiome exhibits substantial variability based on numerous factors, including soil type, pH, moisture content, organic matter levels, and land management practices (Fierer et al. 2017). Typically, bacterial communities are the most prevalent in soil ecosystems, playing key roles in the decomposition of organic matter, nitrogen fixation, and nutrient cycling. With plant roots, fungi, especially mycorrhizal fungi, form mutualistic relationships that improve nutrient uptake and increase plant resistance to environmental stress (Yadav et al. 2024).

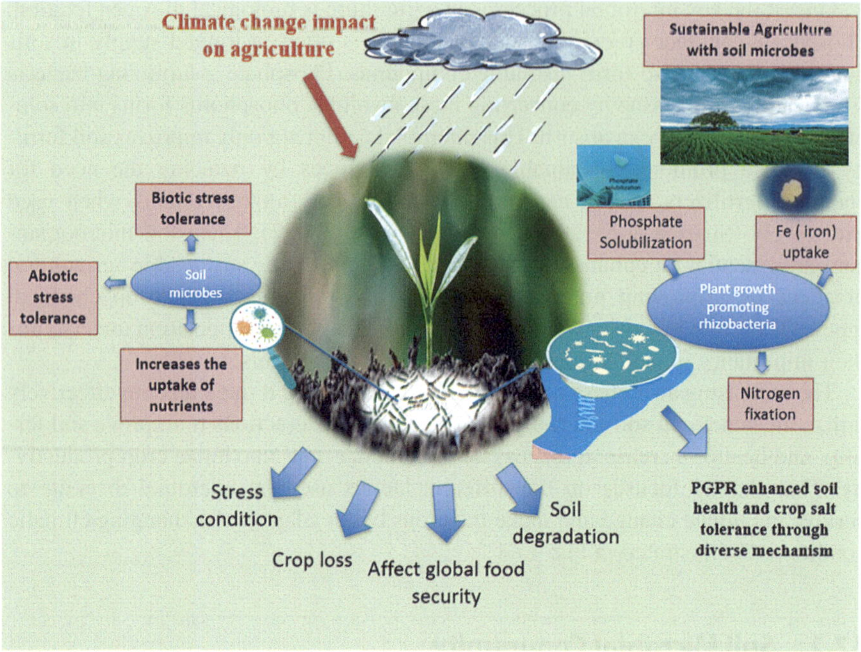

Fig. 17.1 Overview of the role of microbes in crop adaptation to various stress conditions

Emerging research underscores the significance of microbial diversity and community structure in sustaining ecosystem functions. Soil microbial communities are not uniform; rather, they comprise intricate networks where various taxa interact and affect soil processes (Banerjee et al. 2019).

17.2.2 Composition of Soil Microbial Communities

Soil microbial communities are vital to ecosystem functioning and are responsible for a wide range of processes that support plant growth, nutrient cycling, and carbon sequestration. These communities are made up of viruses, bacteria, fungi, archaea, and protozoa, with the majority of soils having a higher concentration of bacteria and fungi. Many elements, including soil type, climate, vegetation, and land management techniques, influence the diversity and make-up of soil microbial communities. The most prevalent and varied class of microorganisms in soil are called bacteria. They play essential roles in nutrient cycling, including nitrogen fixation, nitrification, and denitrification (Islam et al. 2020). Bacterial communities in soil are typically dominated by a few key phyla, including Proteobacteria, Acidobacteria, Actinobacteria, Bacteroidetes, and Firmicutes. However, within these phyla, there is a high level of taxonomic and functional diversity, with many species adapted to specific soil environments. Proteobacteria is one of the most metabolically diverse and contains many genera involved in nitrogen fixation, such as *Rhizobium* and *Azospirillum*. These bacteria form symbiotic relationships with legumes, contributing to soil fertility by converting atmospheric nitrogen into forms available to plants. Acidobacteria members of the phylum Acidobacteria are common in soils, particularly in acidic environments. These microorganisms participate in the breakdown of organic materials and play roles in carbon cycling. Recent studies have shown that Acidobacteria abundance increases in soils with high organic content and low pH (Kielak et al. 2016). Actinobacteria include important soil-dwelling bacteria such as Streptomyces, which are known for their ability to degrade complex organic compounds, including lignin and cellulose. Actinobacteria are also responsible for producing many of the antibiotics used in medicine, as well as playing critical roles in soil nutrient cycling and fungi are another dominant group supporting the cycling of nutrients and the breakdown of organic matter in soil microbial communities. Soil fungi are primarily classified into two functional groups: saprotrophic fungi, which decompose dead organic material, and mycorrhizal fungi, which live in symbiotic partnerships with plant roots.

17.2.3 Factors Influencing Soil Microbial Diversity

The organic matter content, soil pH, moisture content, and land management techniques are just a few of the environmental factors that affect soil microbial diversity. Among them, soil pH is a particularly critical determinant of microbial diversity. Studies have shown that bacterial diversity tends to peak in soils with neutral pH and decreases in both highly acidic and highly alkaline conditions (Liu et al. 2021). In contrast, fungal communities often demonstrate a greater tolerance for extreme pH levels, maintaining diversity across a broader pH range. The availability of organic matter is another key factor affecting soil microbial diversity and composition. Soils rich in organic content generally support more diverse microbial communities, as organic matter serves as an energy source for various microorganisms (Banerjee et al. 2019). Agricultural techniques that increase the amount of organic matter in the soil, like crop rotation, cover crops, and the application of organic amendments like manure or compost are associated with improved microbial diversity. Land management practices, including tillage, fertilization, and pesticide use, also significantly influence soil microbial communities. For example, no-till farming and the use of organic fertilizers have been found to promote higher microbial diversity compared to conventional agricultural practices, which often involve more intensive soil disturbance (Hartman et al. 2018; Thapa et al. 2021). Conversely, excessive application of synthetic fertilizers can diminish microbial diversity by favoring fast-growing, nitrogen-utilizing bacteria, while suppressing the growth of slower-developing microorganisms that play crucial roles in nutrient cycling and soil health.

17.3 Microbial Contributions to Nutrient Uptake in Crops

Plant health and growth depend on the soil having the appropriate concentration of various macronutrients, micronutrients, and trace elements (Awasthi et al. 2022). It is also essential that these nutrients exist in a biologically accessible form in order to affect plant growth and yield. For example, although the soil contains large amounts of phosphate and sulfate, plants only use the soluble ionic form of these nutrients, which is absorbed through a variety of processes (Saini and Sharma 2019).

According to Sahu et al. (2018), Microorganisms are essential to plants' ability to mobilize and absorb nutrients. Arbuscular mycorrhizal fungi (AMF) and plant growth-promoting rhizobacteria (PGPR) are two examples of soil microorganisms that have been demonstrated to have positive effects on plant nutrition, growth, and health. The equilibrium of the soil microbial community has been seriously disrupted by the overuse of pesticides in an attempt to boost agricultural productivity.

As a result, both environmental pollution and the natural geochemical cycle have been impacted. It was often suggested that they be used in order to reduce the amount of chemicals that are used in agriculture. When it comes to improving crop yield and quality and restoring soil nutrients for environmentally friendly and sustainable agriculture, microorganisms are a better bioinoculant than pesticides (Alori and Babalola 2018).

17.3.1 Nitrogen Fixation by Rhizobia and Other Symbiotic Bacteria

Bacteria have a wide range of advantageous direct and indirect effects on plant growth. Bacteria that stimulate plant growth have a variety of direct effects on plants, such as solubilizing mineral nutrients like potassium and phosphorus to increase their availability, fixing nitrogen, sequestering iron through the help plants deal with stress, siderophores are produced, plant growth regulators are made, and ACC deaminase is synthesized. Depletion of iron, the synthesis of antibiotic compounds, the synthesis of antifungal metabolites, and the production of extracellular enzymes that break down the cell walls of phytopathogens like fungi are examples of indirect effects (Shah et al. 2021).

By making macronutrients like N, P, and K as well as micronutrients like Fe more available, bacteria enhance plant growth and development. A significant part of nitrogen cycling and plant uptake in the soil is facilitated by bioinoculants (Shahwar et al. 2023). The use of nitrogenous fertilizers has led to nitrate pollution of water sources and eutrophication; hence, biological nitrogen fixation is crucial. Numerous types of bacteria, including cyanobacteria and archaea, may fix nitrogen. These bacteria are capable of both free-living and symbiotic coexistence. Phylogenetically diverse free-living nitrogen-fixing bacteria, including Acetobacter, Arthrobacter, Azospirillum, Azotobacter, Bacillus, Burkholderia, Klebsiella, and Pseudomonas, as well as members of the genus Rhizobium and its relatives belonging to the class Alphaproteobacteria, provide nitrogen for a wide range of important crops (Saini and Sharma 2019).

17.3.2 Phosphate Solubilization and Mineralization

Although large reservoirs of phosphorus (P) are found in agricultural soils, plants cannot access these reservoirs, necessitating the ongoing application of phosphorus fertilizer. P-solubilizing microorganisms found in the rhizosphere of plants are essential to the cycling of both organic and inorganic phosphorus, which leads to the fast commercial growth of inoculants (Sharma et al. 2013). It has been discovered that, depending on the situation, various bacteria are crucial to the mineralization of organic phosphates and the solubilization of inorganic phosphates. Numerous bacteria belonging to the genera Achromobacter, Agrobacterium, Bacillus, Burkholderia, Flavobacterium, Gluconacetobacter, Micrococcus, Pseudomonas, Rhizobium, and others have the ability to convert phosphates from their insoluble forms to soluble ones (Singh and Reddy 2022). They solubilize by a number of mechanisms, such as the production of inorganic acids, phosphatase enzyme release, organic acid synthesis, and other chelating agents. The production of organic acids is the main step in the solubilization of phosphate. More phosphate-solubilizing bacteria, which are metabolically more active than bacteria from other areas, are found in the rhizosphere (Wang et al. 2017). After phosphate-solubilizing rhizobacteria were inoculated, improved plant growth, organic matter, and yield of several crops and plants were observed (Rizvi et al. 2014).

17.3.3 Mycorrhizal Associations and Nutrient Acquisition

Many fungal groups, including those that stimulate plant growth, ectomycorrhizal fungi, and arbuscular fungi, are essential for enhancing plant growth and supplying nutrients through a range of mechanisms, such as the increased aboveground photosynthesis, the phosphorus-promoting hormone synthesis, and the solubilization of phosphorus. The soil microbial community is largely composed of arbuscular mycorrhizal (AM) fungi. They are found all over the world and are associated mutualistically with more than 80% of land plants' roots. Apart from their numerous other benefits, like increasing crop water availability, protecting crops from pests, increasing crop resilience to environmental stressors, and playing a vital part in the biogeochemical cycle of nutrients in the soil, AM fungi also help with the mobilization and uptake of carbon, phosphorus, nitrogen, and other vital minerals (Wahab et al. 2023, Chandra et al. 2023).

Plants associated with AM absorb three to five times more P through AM roots than do direct roots, making P the most studied nutrient that AM roots can absorb. Furthermore, in low-P soils, the roots of AM-infected plants with mycorrhizal associations have the ability to absorb and accumulate more P in comparison to the roots of plants lacking AM symbiosis (Nadeem et al. 2017). The multiple-meter-long mycelia of AM increase the surface area of absorption multiple times to obtain the same nutrients that nearby plant roots are attempting to obtain. According to Olsson et al. (2014), Hyphae are more easily absorbed into the soil due to their finer structure than plant roots. This increases the surface area available for absorption and encourages the growth and development of nearby plants. Additionally, it has been found that AM fungi facilitate better biological N fixing, aid in nitrogen absorption by plants, and improve phosphorus and nitrogen uptake (Tang et al. 2022).

17.4 Enhancing Abiotic Stress Tolerance Through Microbes

Abiotic stresses such as salinity, water logging, drought, and others have a significant impact on plant growth and agricultural productivity. Furthermore, changes in the ideal temperature, nutrient depletion, and heavy metal deposition all have an impact on plant growth and yield. Depending on the crop and the environment in which it is grown such as in semi-arid or arid regions or areas with typically high salinity, abiotic stress can cause a 20–50% loss in crop yield. It can also cause a decrease in the rate of aboveground biomass growth and impair crops' ability to photosynthesize due to reduced water uptake.

Abiotic Stress Management

Abiotic stresses are known to have a negative impact on plant growth and productivity as well as to cause morphological, physiological, biochemical, and molecular changes in plants. These changes frequently result in large-scale, frequently total crop failures. For instance, cold stress reduces the agricultural productivity of plants in hilly regions.

17.4.1 Drought Stress Tolerance

Crop yield and growth are restricted by drought stress, particularly in semi-arid areas. Plants produce more proline, an osmolyte, in response to a water shortage, which raises the osmotic potential of individual cells (Ghosh et al. 2022). Similar to this, exudates released by bacteria in the root zone also contain osmolytes that can work in concert with osmolytes produced by plants in response to stress, hence enhancing drought tolerance. Plants have been observed to experience drought when their proline levels are elevated (Nair et al. 2017). Plants develop a variety of defense mechanisms, such as mycorrhizal associations with soil fungi, to combat climate hazards and drought. *Glomus intraradices, G. claroideum*, and other AM symbioses impact the water dynamics of their host plants. *Glomus versiforme*, an arbuscular mycorrhizal fungus, and water stress have an effect on the reactive oxygen metabolism and antioxidant synthesis of citrus (*Citrus tangerine*). They demonstrated how AM symbiosis contributes to increases in the production of both enzymatic and non-enzymatic antioxidants, which in turn improves drought tolerance in AM plants (Wu et al. 2013).

17.4.2 Salinity Stress Tolerance

More than 20% of irrigated land globally experiences a decrease in crop production due to salinity, making it one of the most severe environmental stresses. Arbuscular mycorrhizal plants exhibit enhanced photosynthesis and water use efficiency when exposed to salt stress, and it has been demonstrated that AM fungi can enhance crop tolerance to salinity. AM fungi help to reduce the impact of salt stress on the host crops by improving nutrition, increasing the ratio of K^+/Na^+ in plant tissues, and improving osmotic adjustment by accumulating compatible solutes such as soluble sugars, glycine betaine, and proline (Porcel et al. 2012; Chandra et al. 2024). According to Latef and Chaoxing (2011), AM fungus by lessening the oxidative stress caused by salt, *Glomus mosseae* can protect tomato plants from salinity. They achieve this by mitigating the detrimental effects of salt on tomato plants, such as the loss of root colonization, development of surface area of leaves, quantity of chlorophyll, weight at harvest, and fruit production. Rhizosphere microorganisms mitigate salinity stress by preventing plant cell damage, restoring ion and osmotic balance, and allowing plants to grow again in the face of salt stress (Liu et al. 2022).

17.4.3 Heavy Metal Stress Tolerance

Heavy metals in the environment can be harmful to water sources, soils, and plants. The sustainability of food production is put at risk by certain metals, such as zinc, copper, iron, manganese, nickel, molybdenum, and cobalt, even though their prolonged presence in soils and high concentrations can negatively impact soil and water quality. These metals are essential micronutrients for plants and are necessary for their development as well as their growth. It has been demonstrated time and time again that AM fungi benefit plants under stress from heavy metals. According to Karimi et al. (2011), they can eliminate dangerous heavy metals, act as a sink to reduce metal concentrations near roots and prevent metals from reaching plants, protecting them from stress and metal toxicity.

17.4.4 Temperature Stress Tolerance

Extreme temperature variations bring plants under stress. In tropical and subtropical regions, high soil temperatures already pose a significant challenge to agricultural productivity. Increased temperatures have an impact on the timing of fruiting, plant water relations, and photosynthetic rate. Low temperatures in some regions restrict agricultural crop productivity (Etesami and Beattie 2017). To treat tomato plants with the isolates that showed significant tolerance to chilling, Subramanian et al. (2016) isolated 40 psychrotrophic bacterial isolates from the genera *Arthrobacter, Flavimonas, Flavobacterium, Massilia, Pedobacter,* and *Pseudomonas.* The isolates were observed to reduce membrane damage, activate antioxidant enzymes, and cause proline synthesis in leaf tissue when exposed to chilling temperature conditions (15 °C).

17.5 Microbes in Biotic Stress Management

Agricultural productivity is often hampered by biotic stressors, such as pests and diseases, which lead to substantial crop loss. According to Naamala and Smith (2020), crop diseases and pests have emerged because of increases in temperature and fluctuations in the amount of precipitation in specific regions due to climatic changes across the globe. Technologies like genetic engineering and plant breeding have been used to meet these demands. Plant breeding is a labor-intensive process, and created cultivars may introduce new pathogens and diseases, despite reports of favorable outcomes. By providing agricultural plants with disease resistance against a variety of pathogenic organisms through specialized chemical secretions, microbes contribute significantly to agricultural sustainability by lowering the stress that pathogens offer to plants.

PGPR may be utilized as a biocontrol agent for safeguarding plants against bacteria, viruses, fungi, and insects, and other disease-causing microorganisms (El-Saadony et al. 2022). They have greater advantages for the environment and humans than synthetic pesticides, break down faster in soil, and are less likely to promote pathogen resistance. They have numerous advantages over chemical insecticides. Utilizing seeds coated with biocontrol agents may help prevent disease in major crops including rice, wheat, and maize, according to previous research. Rhizosphere bacteria use an array of strategies to suppress plant root pathogens: they can act indirectly by competing with the pathogen for nutrients and space on the root, they can act directly by causing resistance in the host plant or by producing biosurfactants, antibiotics, siderophores that sequester iron, or enzymes that break down the cell wall of pathogens. The amount of ethylene produced by pathogen-infected plants can be modulated by plant growth-promoting bacteria that contain the enzyme ACC deaminase, reducing the damage by the pathogen (Nascimento et al. 2016; Saini and Sharma 2019).

According to Beneduzi et al. (2012), Rhizobacteria produce siderophores that inhibit the growth of fungal infections and prevent fungal growth from utilizing iron. *Burkholderia cepacia* or its siderophore could potentially be utilized as a biocontrol agent against *Colletotrichum gloeosporioides* (Santos-Villalobos et al. 2012). Cyclic lipopeptides produced by *Pseudomonas* spp. are important for controlling pathogens such as *Pythium ultimum*, *Rhizoctonia* spp., *Phytophthora infestans*, and *R. solani* (Iqbal et al. 2023). Additionally, it has been proposed that PGPR produces defensive enzymes including 1-aminocyclopropane1-carboxylate (ACC) deaminase, which lessens biotic stress (Shahid et al. 2023). The ability of species belonging to the genera *Bacillus, Pseudomonas, Burkholderia, Alcaligenes, Rhizobium, Aeromonas,* and *Streptomyces* to suppress plant pathogens has been demonstrated (Shah et al. 2021). It has been determined that these species are biocontrol agents. These biological agents produce some secondary metabolites that have a variety of activities and possess the capacity to effectively combat plant diseases. Among these are a range of biocontrol agents that yield hydrogen cyanide (HCN), an efficient chemical for a wide range of plant diseases. Many rhizobacteria produce HCN, commonly referred to as prussic acid, a volatile, wide-spectrum secondary metabolite that is vital to the biological control of a variety of pathogenic microbes in soil. For instance, the secretion of HCN by bacterial strains has been related to the suppression of diseases such as sunflower charcoal rot, which is caused by *Macrophomina phaseolina*, respectively (Reetha et al. 2014).

The influence of AMF on the reduction of soil-borne diseases has primarily been assessed in research on fungal pathogens such as *Phytophthora, Aphanomyces, Fusarium, Verticillium,* and others that cause root rots and lesions (Kalamulla et al. 2022) and nematodes that cause galls. Arbuscular mycorrhizal (AM) fungus can protect host plants from certain root diseases by suppressing their growth and improving plant health and nutrition. Disease suppression is also facilitated by several additional mechanisms, such as the synthesis of phytohormones, siderophores, defensive plant compounds, defense-related gene expression, and an increase in the colonization of rhizobacteria that promote plant growth (Saini and Sharma 2019).

17.6 Microbial Inoculation for Crop Improvement

Redesigning agricultural methods and implementing technology that allows crops to adapt to these challenges is necessary to combat climate change. Microbial inoculants may be an invaluable tool in redesigning and developing of novel agricultural practices to accomplish the necessary goals (Fig. 17.2). The challenge of creating microbial inoculants to help crops combat the negative consequences of climate change still exists though (Fadiji et al. 2022). Microbial inoculant formulations with one or more strains of microorganisms are available on the global market. These formulations may not always have the desired effect on the crop, despite the sector's enormous expansion. This is due to the fact that agro-environmental factors such as farming practices, soil environment, climatic conditions, and water availability deeply impact the microorganisms that comprise the formulations. According to this concept, it is critical to offer solutions that fulfill farmers' needs, particularly those related to the product's efficacy (Kavadia et al. 2020).

To mitigate the negative impacts of climate change, it is imperative to create artificial microbial consortia that possess specialized roles in assisting plants to adapt to factors influencing agricultural productivity. Wallenstein (2017) stated that inoculation can be used to create a plant's rhizosphere, ultimately helping plants cope with both abiotic and biotic stresses. Next-generation inoculants should also establish strong relationships with the native rhizosphere microbial community structure in a way that will support plant growth, as plants do not grow in isolation but rather in the presence of other organisms and a range of pedo-climatic circumstances (de Vries and Wallenstein 2017). Targeting the drought, salt, or heat tolerance of particular crops will be the primary use of habitat-specific and functionally relevant synthetic microbial communities in the framework of climate-smart agriculture (Saad et al. 2020).

Microbial inoculation is generally required as part of agricultural practice due to topsoil depletion, infertile soil, stunted growth of plants, low yield, and inadequate variety of native bacteria. Utilizing plant-microbe interactions will promote a

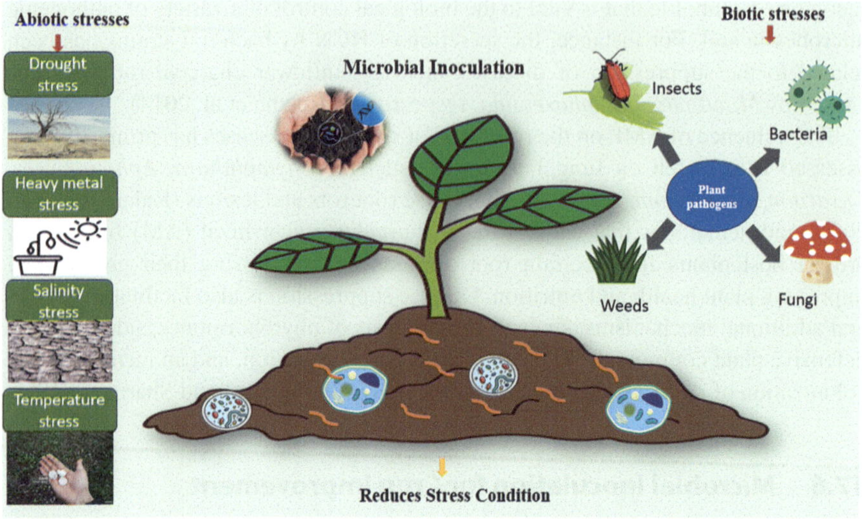

Fig. 17.2 Representation of the use of microbial inoculation to tackle stress conditions

sustainable economy to achieve economic stability, in addition to aiding in the mitigation of climate change. It results in the creation of plant cultivars with the capacity to flourish in the face of climate change and increased CO_2 levels (Saini and Sharma 2019). Beneficial soil microbes can also support a variety of ecosystem activities and processes while lowering reliance on artificial fertilizers and pesticides, which can have a significant positive socioeconomic impact on the world economy. Hence, by utilizing beneficial microbes to regulate the soil environment, it is possible to enhance nutrient availability for the production of high-yielding, high-quality crops; shield crops from pests, diseases, and weeds; and control other elements that restrict output, the provision of ecosystem services, and resilience to stresses like droughts (Saini and Sharma 2019).

Numerous soil microorganisms, such as plant growth-promoting rhizobacteria (PGPR), have been recognized as an environmentally benign means of efficiently encouraging plant development while dealing with both abiotic and biotic stress conditions (Vocciante et al. 2022). Furthermore, a variety of microbial inoculants have been shown to lessen the toxicity level of heavy metals in plants, contribute to nitrogen fixation, help in the transformation of nutrients, and produce siderophores that aid in the ability of the plant to absorb enough iron, synthesize phytohormones such as Auxin, and offer protection against a variety of plant pathogens (Mohanty and Swain 2018).

Effective nutrient cycling and plant growth in cultivated soils are significantly influenced by the activity of soil microorganisms (Yadav et al. 2021). For commercial applications, these PGPRs are being developed as effective biofertilizers. To maintain production and lower the chemical burden, more natural microorganisms can be used as biofertilizers, biopesticides, biofungicides, and other products.

Furthermore, using farm yard manure improves the soil microbial activity's resistance and resilience to heat stress; however, providing fertilizer that has only nitrogen in it, may reduce the soil's capacity to withstand and recover from heat stress (Kumar et al. 2014).

17.7 Conclusion

Microbial contributions to crop adaptation present a transformative opportunity for building climate-resilient agricultural systems. Throughout this exploration, we have seen how diverse microbial communities, ranging from soil bacteria and fungi to plant-associated symbionts, play vital roles in enhancing nutrient uptake, mitigating abiotic stressors, and defending crops against pathogens. These beneficial interactions between plants and microbes offer an innovative, eco-friendly approach to addressing the pressing challenges posed by climate change. The integration of microbial technologies into agricultural practices not only improves crop resilience but also promotes sustainability by reducing reliance on synthetic fertilizers and chemical pesticides. By leveraging microbial inoculants, genetic engineering, and the precise manipulation of plant-microbe interactions, we can develop more robust crops capable of thriving in increasingly unpredictable environments. Furthermore, the application of these technologies on a larger scale could revolutionize farming practices, enabling farmers to cultivate healthier, higher-yielding crops even under adverse conditions. However, while the potential is vast, significant challenges remain, including understanding the complex dynamics of soil microbiomes, ensuring the consistent performance of microbial inoculants, and scaling these innovations for widespread adoption. Addressing these obstacles will require interdisciplinary research, collaborative partnerships, and supportive policy frameworks. The path to climate-resilient agriculture lies in embracing the power of microbial contributions. As research and technological advancements in microbial biotechnology continue to evolve, we move closer to securing global food production and sustainability in an era of climate uncertainty.

References

Alori ET, Babalola OO (2018) Microbial inoculants for improving crop quality and human health in Africa. Front Microbiol 9:2213

Awasthi S, Chauhan R, Srivastava S (2022) The importance of beneficial and essential trace and ultratrace elements in plant nutrition, growth, and stress tolerance. In: Plant nutrition and food security in the era of climate change. Academic Press, pp 27–46

Banerjee S, Walder F, Büchi L, Meyer M, Held AY, Gattinger A, van der Heijden MG (2019) Agricultural intensification reduces microbial network complexity and the abundance of keystone taxa in roots. ISME J 13(7):1722–1736

Beneduzi A, Ambrosini A, Lucianae MPP (2012) Plant growth-promoting rhizobacteria (PGPR): their potential as antagonists and biocontrol agents. Genet Mol Biol 35(4):1044–1051

Chandra P, Yadav S, Kumar A, Sheoran P, Mann A (2023) Arbuscular Mycorrhizal fungi–plant interaction for salinity and drought stress alleviation. In: Kumar A, Dhansu P,

Mann A (eds) Salinity and drought tolerance in plants. Springer, Singapore. https://doi.org/10.1007/978-981-99-4669-3_19

Chandra P, Fagodiya RK, Rai AK, Singh R, Sheoran P, Prajapat K, Singh A, Verma K, Verma VK, Yadav RK, Biswas AK (2024) Different tillage and residue management practices affect soil biological activities and microbial culturable diversity in rice-wheat cropping system under reclaimed sodic soils. J Ecol Eng 25(5):193–207

de Los Santos-Villalobos S, Barrera-Galicia GC, Miranda-Salcedo MA, Peña-Cabriales JJ (2012) Burkholderia cepacian XXVI siderophore with biocontrol capacity against Colletotrichum gloeosporioides. World J Microbiol Biotechnol 28(8):2615–2623

de Vries FT, Wallenstein MD (2017) Below-ground connections underlying above-ground food production: a framework for optimising ecological connections in the rhizosphere. J Ecol 105(4):913–920

El-Saadony MT, Saad AM, Soliman SM, Salem HM, Ahmed AI, Mahmood M et al (2022) Plant growth-promoting microorganisms as biocontrol agents of plant diseases: mechanisms, challenges and future perspectives. Front Plant Sci 13:923880

Etesami H, Beattie GA (2017) Plant-microbe interactions in adaptation of agricultural crops to abiotic stress conditions. Probiot Plant Health:163–200

Fadiji AE, Babalola OO, Santoyo G, Perazzolli M (2022) The potential role of microbial biostimulants in the amelioration of climate change-associated abiotic stresses on crops. Front Microbiol 12:829099

FAO (2020) The impact of climate change on water resources and agriculture. FAO Water Reports

Fierer N, Bradford MA, Jackson RB (2017) Toward an ecological classification of soil bacteria. Ecology 88(6):1354–1364

Ghosh UK, Islam MN, Siddiqui MN, Cao X, Khan MAR (2022) Proline, a multifaceted signalling molecule in plant responses to abiotic stress: understanding the physiological mechanisms. Plant Biol 24(2):227–239

Hartman K, van der Heijden MG, Wittwer RA, Banerjee S, Walder F, Schlaeppi K (2018) Cropping practices manipulate abundance patterns of root and soil microbiome members paving the way to smart farming. Microbiome 6(1):1–14

IPCC (2022). Climate change 2022: impacts, adaptation, and vulnerability. Contribution of working group II to the sixth assessment report of the intergovernmental panel on climate change

Iqbal O, Li C, Lodhi AM (2023) Antagonistic pseudomonas: alternative to chemical fungicides for the management of phytopathogens. In: Biofungicides: eco-safety and future trends. CRC Press, pp 216–246

Islam W, Noman A, Naveed H, Huang Z, Chen HY (2020) Role of environmental factors in shaping the soil microbiome. Environ Sci Pollut Res 27:41225–41247

Kalamulla R, Karunarathna SC, Asad S, Tibpromma S, Stephenson SL, Promputtha I, Yapa N (2022) Biocontrol potential of arbuscular mycorrhizal fungi on soil-borne pathogens in rice and other crops. benefits 3:5

Karimi A, Khodaverdiloo H, Sepehri M, Sadaghiani MR (2011) Arbuscular mycorrhizal fungi and heavy metal contaminated soils. Afr J Microbiol Res 5(13):1571–1576

Kavadia A, Omirou M, Fasoula D, Ioannides IM (2020) The importance of microbial inoculants in a climate-changing agriculture in eastern Mediterranean region. Atmos 11(10):1136

Kielak AM, Barreto CC, Kowalchuk GA, Van Veen JA, Kuramae EE (2016) The ecology of Acidobacteria: moving beyond genes and genomes. Front Microbiol 7:744

Kumar S, Patra AK, Singh D, Purakayastha TJ (2014) Long-term chemical fertilization along with farmyard manure enhances resistance and resilience of soil microbial activity against heat stress. J Agron Crop Sci 200(2):156–162

Latef AAHA, Chaoxing H (2011) Effect of arbuscular mycorrhizal fungi on growth, mineral nutrition, antioxidant enzymes activity and fruit yield of tomato grown under salinity stress. Sci Hortic 127(3):228–233

Liu J, Sui Y, Yu Z, Shi Y, Chu H, Jin J (2021) Soil pH determines fungal diversity along an elevation gradient in a forest ecosystem. Front Microbiol 12:648195

Liu Y, Xun W, Chen L, Xu Z, Zhang N, Feng H et al (2022) Rhizosphere microbes enhance plant salt tolerance: toward crop production in saline soil. Comput Struct Biotechnol J 20:6543–6655
Mohanty S, Swain CK (2018) Role of microbes in climate smart agriculture. In: Microorganisms for green revolution: volume 2: microbes for sustainable agro-ecosystem, pp 129–140
Nadeem SM, Khan MY, Waqas MR, Binyamin R, Akhtar S, Zahir ZA (2017) Arbuscular mycorrhizas: An overview. Arbuscular mycorrhizas and stress tolerance of plants 1–24
Naamala J, Smith DL (2020) Relevance of plant growth promoting microorganisms and their derived compounds, in the face of climate change. Agronomy 10(8):1179
Nair SS, Sahu PK, Brahmaprakash GP (2017) Microbial inoculants for agriculture under changing climate
Nascimento F, Rossi MJ, Glick B (2016) Role of ACC deaminase in stress control of leguminous plants. In: Subramaniam G et al (eds) Plant growth promoting Actinobacteria. Springer Science, Business Media, Singapore, pp 179–192
Olsson O, Olsson PA, Hammer EC (2014) Phosphorus and carbon availability regulate structural composition and complexity of AM fungal mycelium. Mycorrhiza 24:443–451
Porcel R, Aroca R, Ruiz-Lozano JM (2012) Salinity stress alleviation using arbuscular mycorrhizal fungi. A review. Agron Sustain Dev 32:181–200
Reetha AK, Pavani SL, Mohan S (2014) Hydrogen cyanide production ability by bacterial antagonist and their antibiotics inhibition potential on Macrophomina phaseolina (Tassi.) Goid. Int J Curr Microbiol Appl Sci 3:172–178
Rizvi A, Khan MS, Ahmad E (2014) Inoculation impact of phosphate-solubilizing microorganisms on growth and development of vegetable crops. In: Phosphate solubilizing microorganisms: principles and application of Microphos technology, pp 287–297
Saad MM, Eida AA, Hirt H (2020) Tailoring plant-associated microbial inoculants in agriculture: a roadmap for successful application. J Exp Bot 71(13):3878–3901
Sahu A, Bhattacharjya S, Mandal A, Thakur JK, Atoliya N, Sahu N et al (2018) Microbes: a sustainable approach for enhancing nutrient availability in agricultural soils. In: Role of Rhizospheric microbes in soil: volume 2: nutrient management and crop improvement, pp 47–75
Saini R, Sharma S (2019) Climate resilient microbes in sustainable crop production. In: Contaminants in agriculture and environment: health risks and remediation, vol 1, p 264
Shah A, Nazari M, Antar M, Msimbira LA, Naamala J, Lyu D et al (2021) PGPR in agriculture: a sustainable approach to increasing climate change resilience. Front Sustain Food Syst 5:667546
Shahid M, Singh UB, Khan MS, Singh P, Kumar R, Singh RN et al (2023) Bacterial ACC deaminase: insights into enzymology, biochemistry, genetics, and potential role in amelioration of environmental stress in crop plants. Front Microbiol 14:1132770
Shahwar D, Mushtaq Z, Mushtaq H, Alqarawi AA, Park Y, Alshahrani TS, Faizan S (2023) Role of microbial inoculants as bio fertilizers for improving crop productivity: a review. Heliyon 9(6):e16134
Sharma M, Singh S, Kumar V (2013) Role of phosphate-solubilizing bacteria in enhancing phosphorus availability in soil. J Soil Sci Plant Nutr 13(2):315–326
Singh RS, Reddy MS (2022) Phosphate-solubilizing microorganisms for sustainable agriculture: current status and future perspectives. Microbiol Res 241:126611
Subramanian P, Kim K, Krishnamoorthy R, Mageswari A, Selvakumar G, Sa T (2016) Cold stress tolerance in psychrotolerant soil bacteria and their conferred chilling resistance in tomato (Solanum lycopersicum mill.) under low temperatures. PLoS One 11(8):e0161592
Tang H, Hassan MU, Feng L, Nawaz M, Shah AN, Qari SH et al (2022) The critical role of arbuscular mycorrhizal fungi to improve drought tolerance and nitrogen use efficiency in crops. Front Plant Sci 13:919166
Thapa VR, Marschner P, Wang J, Ma X (2021) Effect of organic and conventional management on soil microbial diversity and activity in subtropical tea plantations. Appl Soil Ecol 159:103845
Vocciante M, Grifoni M, Fusini D, Petruzzelli G, Franchi E (2022) The role of plant growth-promoting rhizobacteria (PGPR) in mitigating plant's environmental stresses. Appl Sci 12(3):1231

Wahab A, Muhammad M, Munir A, Abdi G, Zaman W, Ayaz A et al (2023) Role of arbuscular mycorrhizal fungi in regulating growth, enhancing productivity, and potentially influencing ecosystems under abiotic and biotic stresses. Plan Theory 12(17):3102

Wallenstein MD (2017) Managing and manipulating the rhizosphere microbiome for plant health: a systems approach. Rhizosphere 3:230–232

Wang Z, Xu G, Ma P, Lin Y, Yang X, Cao C (2017) Isolation and characterization of a phosphorus-solubilizing bacterium from rhizosphere soils and its colonization of chinese cabbage (*Brassica campestris* ssp. *chinensis*). Front Microbiol 8:1270

Wu Q-S, Srivastava AK, Zou Y-N (2013) AMF-induced tolerance to drought stress in citrus: a review. Sci Hortic 164:77–87

Yadav AN, Kour D, Kaur T, Devi R, Yadav A, Dikilitas M et al (2021) Biodiversity, and biotechnological contribution of beneficial soil microbiomes for nutrient cycling, plant growth improvement and nutrient uptake. Biocatal Agric Biotechnol 33:102009

Yadav R, Kumar A, Saharan BS, Gera R, Rana A, Parshad J (2024) Arbuscular mycorrhizal fungi as biostimulant for plant growth and abiotic stress amelioration. In: Microbial biostimulants for plant growth and abiotic stress amelioration. Elsevier, pp 81–95

Zhao C, Liu B, Piao S (2017) Temperature increase reduces global yields of major crops in four climate zones. Nat Clim Chang 7(3):333–339

Effect of Climate Change on Seed Quality Development

18

Gajender Yadav, Satender Kumar, Govind Prasad,
Manish Yadav, and Rajender Kumar Yadav

Abstract

The impact of climate change on seed quality development is an emerging area of concern within agricultural and ecological research. This chapter includes the effects of various climate factors including temperature, rainfall, drought stress, elevated CO_2, and other climate change agents on seed quality development, from germination to genetic composition. Elevated temperatures can accelerate seed development but may lead to early maturation, altered biochemical composition, and reduced viability. Variations in rainfall influence soil moisture levels, critical for seed germination, seedling establishment, and seed longevity. Prolonged drought conditions can induce dormancy or result in poor seed set and induce salinity in arid regions affecting overall seed quality. Elevated CO_2 levels can enhance photosynthesis and crop growth yet may alter seed biochemical composition, forced maturity, and nutrient contents in the seed. Other climate-related factors, such as increased incidence of extreme weather events and changes in seasonal patterns, further complicate seed development processes. The interplay of these factors can lead to significant modifications in the genetic makeup of the seed. Understanding these complex interactions is crucial for developing strategies to mitigate the adverse effects of climate change on seed quality, ensuring sustainable agricultural practices and food security in the face of a changing global climate.

Keywords

Seed quality development · Climate change · Viability · Longevity

G. Yadav (✉) · S. Kumar · G. Prasad · M. Yadav · R. K. Yadav
ICAR-Central Soil Salinity Research Institute, Karnal, Haryana, India

A. Mann et al. (eds.), *Cutting Edge Technologies for Developing Future Crop Plants*,
https://doi.org/10.1007/978-981-96-2508-6_18

18.1 Introduction

Global agriculture is significantly and widely impacted by climate change. Climate patterns have changed as a result of the increase in global temperatures. The atmosphere, ocean, cryosphere, and biosphere have all undergone extensive and rapid changes. This has resulted in numerous negative effects, losses, and harm to both people and the environment (Yadav et al. 2024a). These changes include an increase in the frequency of extreme weather events, erratic precipitation, and longer warm seasons. The output of agricultural products is directly impacted by these changes, which also increase the frequency of extreme weather events such as storms, floods, and droughts. Moreover, melting glaciers, rising sea levels, and altered ecosystems have all resulted from climate change, putting strain on the world's agricultural sector (Upadhyay 2020). Climate change has numerous effects on agriculture, including altered temperature and precipitation patterns, altered dynamics of pests and diseases, altered atmospheric carbon dioxide levels, and altered crop nutritional quality (Mpandeli et al. 2019). Scientists estimated that the earth will have warmed by 3.5 to over 8 °C by the year 2100 (Hawkins et al. 2022). Periodically, there will be a rise in simultaneous losses in important food-producing regions due to increased and perhaps contemporaneous climate extremes. Plants exist in tiny forms, and their seeds provide a window into the health and nutritional quality of those plants in the future. The seed is a fertilized, matured ovule of a flowering plant containing an embryo or rudimentary plant. The cereal grain is an indehiscent, small, and one-seeded fruit, in which the entire seed coat is closely integrated with the pericarp (Borczak et al. 2023). Seed development is a series of processes, from flowering, through ovule fertilization to seed maturation. A good seed is expected to develop into a potentially independent plant in the seedbed. Seed quality ultimately refers to the value of the seed to the farmer to establish a subsequent vigorous, uniform crop. Seed quality is a much broader concept than just seed purity (the absence of weeds, other species seeds foreign bodies, etc.), seed health (the absence of seed-borne disease), and viability (Xhemali et al. 2024). The quality of a seed is a combination of all physical, biological, pathological, and genetic attributes that contribute to the final yield of a crop established from the seed (Mor et al. 2008; Doggalli et al. 2024). Seed viability is nevertheless an important measure of seed quality. A special concern of seed producers is those environmental factors prevailing before harvest that affect subsequent viability (Dadlani et al. 2023). Therefore, seed quality cannot be seen in isolation and should be integrated within overall studies of crops in response to the environment. Seed growth in the seed-filling period is determined by field environmental factors, i.e., temperature, rainfall, and light. These field environmental factors may also determine seed quality by influencing seed quality development (Maity et al. 2016).

A seed's ability to grow and develop is mostly hampered by changes in environmental variables. Predicting the date of germination is based on how temperature affects plant growth and development (Maity et al. 2023). Seasonal variations, temperature, light intensity, soil moisture content, and seed size all have an impact on seed germination. Numerous factors determine the quality of seeds, including germination, physical and genetic purity, viability, vigour, and health, as well as morphological traits including weight and size (Maity et al. 2016). Seed quality refers to a seed's total planting value, which includes a high standard for genetic purity, physical purity, germination, vigour, health, and moisture, as well as dimensions, weight, colour, etc. As a crucial component of most farming systems, seed quality has a significant impact on livestock and human food and nutritional needs, particularly in light of the difficulties that climate change presents for sustainable agriculture. The seed size, nutrient concentration in the seed, and germination percentage are the main factors that determine the quality of a seed. Climate change effects can also influence pollinator behaviour, affect crop suitability zones, promote the spread of novel seed-borne diseases, and reduce seed yields, storability, and longevity. High temperatures and CO_2 can change the physiological processes involved in flowering. The reproductive stage is extremely sensitive to environmental stressors and can eventually fail seed set and timely development (Tushabe et al. 2023). While high temperatures accelerate anthesis and shorten crop duration, lower temperatures during the crop growth stage may lengthen crop duration. Climate change's effects on the environment and crop productivity have garnered a lot of attention lately; little research has been done on how it will affect the development of high-quality seeds, which are the essential component of food production. As the main planting material for crops, seeds are also responsible for introducing the majority of new agricultural technologies to the field.

18.2 Effect of Climate Change on Seed Quality

Climate change has a significant impact on various aspects of agriculture, including seed quality. Changes in temperature, carbon dioxide (CO_2) levels, precipitation patterns, and the frequency of extreme weather events can affect the physiological and genetic characteristics of seeds. These climate change agents influence seed quality parameters such as viability, germination rate, vigour, and dormancy (Table 18.1).

18.2.1 Effect of Temperature Change

Researches indicate that elevated temperatures during seed development can significantly impact the seed quality of crops. High temperatures can accelerate seed maturation, leading to premature and underdeveloped seeds (Reed et al. 2022; Macova et al. 2021). The effects of temperature stress on seed germination vary among species, but seed vigour is consistently reduced by high temperatures both before and after physiological maturity (Hampton et al. 2012). Temperature stress can alter gene expression in developing seeds, affecting endosperm development and seed size. The identification of heat-sensitive genes, such as OsMADS87 in rice, offers potential targets for improving thermal resilience in crops (Chen et al. 2016). These findings underscore the need for adaptive strategies to maintain seed quality and

Table 18.1 Effect of different climate change agents on seed quality parameters

Factor	Crop	Impact on seed quality	References
Temperature	Wheat	Reduced germination	Cheng et al. (2009), Sanyal et al. (2022)
		Reduced seed vigour and longevity	Sanyal et al. (2022), Joshi et al. (2016)
	Rice	Delayed germination	Akman (2009), Liu et al. (2019), Suriyasak et al. (2020)
		Reduced radicle length	Iloh et al. (2014)
	Barley	Reduced germination and increased water uptake rate	Khan (1977)
		Increased germination	Sakai et al. (2022), Roy et al. (2024)
	Field pea	Reduced seed setting, seed germination, and test weight	Lamichaney et al. (2021)
CO_2	Cabbage, corn, bean, wheat	Seed germination not affected at 3–4% CO_2 concentration in soil	He et al. (2016)
		Inhibited seed germination of all at >30% CO_2 concentration in soil	
	Rice	Seed germination not affected at 580 ppm	Chen et al. (2015)
		Reduced germination at 720 ppm; seed vigour at 610 ppm	Lamichaney et al. (2019)
		Increased seed germination and seed vigour at 800 ppm	Abzar et al. (2017)
		Grain quality traits not affected	Hu et al. (2022)
UV radiation	Wheat	UV-C increased seed germination, vigour, and biochemical contents	Semenov et al. (2020)
		UV-B increased seedling biometrics	Tertyshnaya et al. (2017)
	Rice	UV-C increased the speed of seed germination and germination percentage	Mutum et al. (2024)
Drought	Various crops	Poor seed development	Pinedo-Taco et al. (2022)
	Rice	Accumulation of proline, poor seed set and weight	Panja et al. (2022)
Salinity	Mustard	Decreased oil, protein, and crude fibre contents	Neha et al. (2022)
	Maize, soybean	Lower germination count	Pinedo-Taco et al. (2022)
	Pepper	Negative impact on seed weight, germination percentage, and higher proline content	Shams et al. (2023)
Rainfall	Wheat	Germination, longevity, seed development, and seed weight	Yadav and Ellis (2016), Ellis and Yadav (2016), Yadav and Ellis (2018)

crop productivity in the face of climate change. For *Brassica napus*, increased temperatures result in reduced fertilization rates, higher abortion rates, and pre-harvest sprouting, ultimately decreasing seed yield. Temperature stress also alters hormone levels, particularly auxin and ABA, affecting embryo development and seed dormancy (Macova et al. 2022). Furthermore, elevated temperatures can reduce seed vigour, impacting germination speed and seedling growth (Reed et al. 2022). These findings highlight the need for developing heat-resilient crop varieties to ensure food security in a changing climate.

Recent studies indicate that increased frequency and intensity of heat waves can significantly impact seed germination and plant establishment, particularly in vulnerable ecosystems. Heat waves can prolong germination duration for extreme events (Alvarez et al. 2023) and shift germination phenology from spring to autumn in alpine plants, while heat waves can increase germination in some species, coupling with drought conditions generally decreases germination rates (Orsenigo et al. 2015). These changes may have major implications for future plant population size and structure. Heat stress during seed development can reduce seed size, number, and vigour in various crops (Reed et al. 2022). In India, where heat waves have become more frequent due to climate change, governments have implemented heat action plans to mitigate impacts on health, productivity, and agriculture (Pillai et al. 2023).

Climate change creates many barriers to the ongoing production of high-quality seeds. Although temperature is usually the main factor influencing this process, increased temperatures affect pollen sterility, germination, and crop phenology, which in turn reduces seed set (Singh et al. 2013). Lower temperatures during the crop growth stage lengthen the crop, whereas higher temperatures accelerate anthesis and crop maturity. Because of the rising temperatures, potential seed yields are predicted to decrease in many tropical and subtropical regions. Elevated temperatures promote evaporation, causing soil aridity and influencing the rate of growth and reproduction (Bucheli et al. 2022). In addition, abnormally high temperatures have the potential to negatively impact the ability to absorb nutrients and form grains by hindering the photosynthesis ability. Elevations in temperature have the potential to alter the distribution of species by impacting seed germination, growth, and fitness, leading to significant ecological and economic ramifications. Certain physiological and biochemical processes, such as the need for oxygen during seed imbibition, are impacted by high temperatures (Nascimento et al. 2008). Moisture, hormones, and enzymatic activity are the three key temperature factors that have the greatest effects on seed germination. Elevated temperature has different effects on seed germination in different species. Each species has an ideal temperature range, and germination cannot take place below or beyond these ranges. For example, research revealed that short bursts of high temperature during and shortly before crop flowering restrict the number of seeds in several cereal crops (Reed et al. 2022). Without genetic improvement, effective adaptation, special management, and CO_2 fertilization, it has been calculated that a one-degree increase in the global mean temperature will reduce the output of wheat, rice, and maize globally by 6.0, 3.2, and 7.4%, respectively (Zhao et al. 2017). Furthermore, because elevated temperatures can impact the production of proteins and starch, wheat quality may also be impacted, leading to a reduction in overall quality and weakened gluten strength (Yanagi 2024).

It has been found that in Arabidopsis, high temperatures cause an increase in abscisic acid (ABA) and reactive oxygen species buildup, which causes seeds to go dormant and restricts seed germination (Toh et al. 2008). High temperature stress causes the production of heat shock proteins (HSPs), which are connected to dormancy. Under heat stress, soybeans' total free amino acid content and total protein concentration decreased nutritionally, while their oil concentration increased significantly (Janni et al. 2020). Suriyasak et al. (2020) observed that seeds from plants subjected to elevated temperatures during the grain-filling stage sprouted considerably later than seeds from plants that were not under stress. Delay persisted even after dormancy release treatments indicating that the primary seed dormancy identified during grain filling was not the cause. When compared to the control, heat stress during grain filling greatly increased seed germination. It was discovered that after imbibition, genes linked to ABA catabolism, GA biosynthesis, NADPH oxidase, and ABA biosynthesis were considerably increased in heat-stressed seeds of barley (Sakai et al. 2022). High temperatures during the reproductive period can cause abnormalities such as reduced pollen number, pollen tube growth, pollen germination, loss of pollen viability, shrunken pollen, poor fertilization, empty pollen grains, embryo abortion, seed abortion, and forced maturity. While climatic variations at harvest and in the post-harvest phases may not necessarily affect the seed output, they can reduce seed morphology and biochemistry, which will ultimately lower its planting value and field performance. Extremely low temperatures can also have an impact on wheat. For example, low temperatures can cause freezing and frost, which can directly injure mature plants and seedlings. They can also cause slow development and delayed reproductive processes, which can have an impact on production and quality (Yanagi 2024). The climatic factors and their effect on seed quality are illustrated in Fig. 18.1.

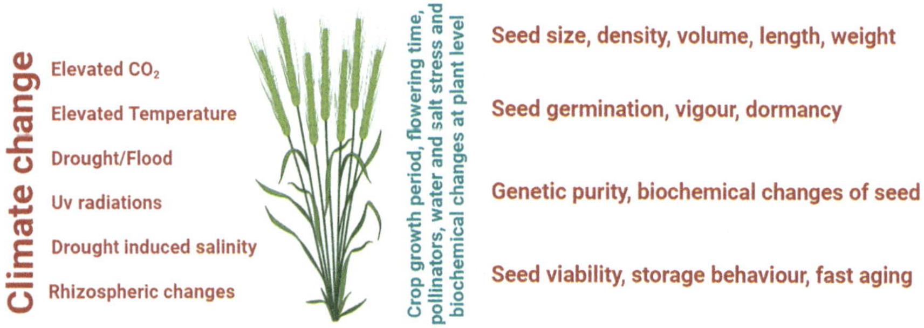

Fig. 18.1 Effect of climate change agents on seed quality

18.2.2 Effect of Carbon Dioxide Concentration

The atmospheric CO_2 concentrations can rise to 540 and 960 ppm, for the lowest and maximum examined emission scenarios respectively, until 2100. By the end of the century, there is a 1% yearly increase in the concentration of CO_2 in the global atmosphere, according to various climate models of the IPCC (Nelson and Nelson 2024). Increased CO_2 is anticipated to have an impact on the physiological state and biochemical makeup of the seeds, which dictate their propensity to germinate, as well as their size, shape, colour, uniformity, and health (Marty and BassiriRad 2014). Increases in atmospheric CO_2 are predicted to increase seed weight, since assimilates are more readily available, although a review of the literature showed that crops' responses to CO_2 varied widely. Studies have demonstrated that higher CO_2 has increased seed weight due to increased assimilate availability (Lamichaney and Maity 2021). In a comprehensive review work, Marty and BassiriRad (2014) found a positive association between the reactions of seed mass to CO_2 and germination performance.

In addition to increasing photosynthesis and plant biomass, higher CO_2 concentrations significantly change the nutritional value of C_3 plants such as rice and wheat (Ebi et al. 2021). Elevated CO_2 levels act as a carbon source, accelerating wheat growth by encouraging photosynthesis. Wheat protein content may decrease even if the elevated CO_2 concentration can promote the synthesis of starch. This could have an impact on the wheat gluten strength and other quality attributes. Second, the reproductive phases of wheat are impacted by elevated CO_2 concentrations. Higher CO_2 concentrations may reduce the wheat reproductive cycle, allowing it to quickly reach the heading and grain-filling phases. This could cause wheat plants to reach the reproductive growth stage earlier, which would have an impact on the growth of spikes and the creation of grains (Yanagi 2024). A physiological and biochemical change in the seeds, decreased metabolic activity, reserve accumulation, and decreased activity of hydrolytic enzymes (protease and amylase) are the main causes of the decrease in rate, speed, and percent germination of seed upon CO_2 exposure. On the other hand, dormancy break, enhanced gibberellic acid and ethylene production, increased rate of water intake and oxygen availability, and enhanced activity of antioxidant enzymes that scavenge radicals are the reasons for the favourable effect of elevated CO_2 on seed germination (Lamichaney and Maity 2021).

18.2.3 Changes in Precipitation Patterns

Excessive rainfall during seed development can significantly impact crop yield and quality. Studies on buckwheat showed that waterlogging and flooding, especially during early growth stages, reduced seed yield by decreasing flower numbers (Sugimoto and Sato 2000). In wheat, simulated rainfall during seed development temporarily decreased seed longevity, but this effect was largely reversible if conditions improved before harvest (Ellis and Yadav 2016, Yadav and Ellis 2018). Process-based crop models are being developed to simulate the complex effects of

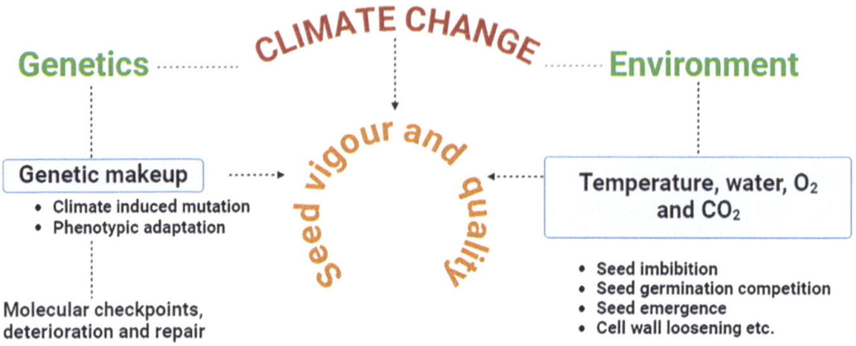

Fig. 18.2 Climate change and seed quality interactions

excessive rainfall, including waterlogging, submergence, lodging, and increased pest and disease pressure (Kim et al. 2024). While these models show promise, they require more data for calibration and evaluation. Interestingly, water stress in sweet corn, particularly during silking, also negatively affected seed yield and quality and was associated with increased fungal infections (Revilla et al. 2021). These findings highlight the importance of optimal water management for seed production. Figure 18.2 shows the interaction of various climatic factors with seed quality.

An increased risk of mouldy beans was also linked to excessive rainfall during the late growth season. Humid tropical circumstances encourage the formation of mould and other fungi, which supports the findings made here that heavy rainfall towards the end of the growing season raises the likelihood of coffee bean defects, particularly mouldy beans (Poltronieri and Rossi 2016). Rainfall at the ripening stage decreased the quality and longevity of the seeds during storage in addition to causing pre-harvest sprouting (Baek and Chung 2014). In the vegetative and reproductive stages, excessive rainfall creates flooding-like conditions, while insufficient rainfall creates drought-like conditions that have a significant impact on disease infestation, seedling growth, germination, and crop physiology. Second, floods can cause waterlogging in wheat fields, which would directly harm the plants mechanically. Overwatering can cause plants to lodge and break, disrupting natural growth and seed development. Furthermore, floods have the potential to remove a lot of nutrients from the soil, which could lead to nutrient deficits in wheat plants and consequently impair production and quality. Wheat may become nitrogen deficient as a result of the soil's diluted nitrogen content following floods (Yanagi 2024).

The impact of drought on plant nutrition varies among genotypes, suggesting potential for breeding drought-tolerant varieties. Drought stress significantly impacts plant growth, seed development, and nutrient dynamics. It reduces seed size, yield, and harvest index in crops such as soybeans and common beans (Smith et al. 2019). While drought decreases nutrient concentrations in leaves, its effect on seed nutrient content is less consistent. Some studies report increased mineral nutrient concentrations in drought-stressed soybean seeds, while others find no significant changes in common bean seeds (Samarah et al. 2004). Drought impairs nutrient uptake and transport from roots to shoots by reducing transpiration rates and altering membrane permeability. Understanding these interactions is crucial for developing strategies to mitigate drought effects on crop production and nutritional quality (da Silva et al. 2011). Drought also decreases photosynthetic rates and chlorophyll content during seed development (Sehgal et al. 2018).

18.3 Mitigation Strategies

Climate change poses significant challenges to agricultural productivity and seed development. Implementing adaptation strategies can mitigate these effects and ensure food security. Water-efficient irrigation systems can reduce water usage while improving yields and quality of produce (Henry and Krutz 2016; Jyolsna et al. 2024; Yadav et al. 2024b). Optimizing planting schedules, adjusting sowing dates, and selecting appropriate cultivars can help adapt to changing climatic conditions and ensure the quality of grains (Hampton et al. 2016; Sheoran et al. 2022). Soil management practices, such as conservation techniques to reduce moisture loss, are crucial in drier environments to secure seed quality losses due to water stress conditions (Rosales et al. 2023). An optimization-simulation framework can balance seed and food production with resource efficiency, potentially saving water and fertilizer while increasing the nutritional quality of seeds (Chen et al. 2023).

Adaptation to climate change can be understood as a continuous process of integrating current risk management techniques into practice and lowering the possible risk from climate change impacts (Singh et al. 2020, 2022). In order to adapt to climate change, we must ensure the availability of water and nutrients, minimize the harm caused by extreme weather events (heat waves and heavy rain), and modify the gene pool of cultivars to the new growing circumstances. Numerous biological, economic, and social elements will affect the adaptability of agricultural production systems. Seed nutritional makeup can be changed through breeding, crop management techniques, and/or post-harvest methods. Post-harvest solutions include enhanced processing, storage, and preservation as well as manual fortification. Pre-harvest options include biofortification by breeding (traditional trait selection and genetic engineering) and agronomic approaches (Pena-Rosas et al. 2019). The worldwide breeding goal has been to generate crop cultivars that are high-yielding, resistant to disease and pests, and have less variation in the gene pool. Crop management strategies, including irrigation scheduling, ideal plant density, integrated fertilizer management, precise pesticide application, and routine crop scouting, must be customized for the unique environmental pressures that local cropping systems face. Delayed sowing may be a smart management tactic to lessen the impact of heat stress on the quantity and quality of seeds (Marcinkowski and Piniewski 2018).

In high-value crops, seed coating with a thermostable polymer that can include the necessary insecticides and growth supplements (such as hormones or plant growth regulators) is a viable solution to deal with unfavourable field circumstances. As a result, breeding occurs more quickly, allowing animals to adjust to the effects of climate change more quickly. The detrimental effects of excessive rainfall and flood occurrences can be mitigated by agricultural practices such as improving drainage systems, selecting cultivars with more flexibility, and modifying planting dates to avoid periods of peak rainfall. Agricultural practices must modify their pest and disease management tactics in response to variations in temperature and humidity. To prevent pests from invading crops, this may entail implementing biological management techniques, introducing natural predators, and using insecticides sparingly.

18.4 Conclusions

Climate change is real and has already impacted every facet of life for humans on our planet. A significant prerequisite for global agriculture is the delivery of higher quality seeds to both maintain and increase food production. To counteract climatic change challenges, we must find new locations for seed production and creative ways to improve the quality of seed. The negative effects of climate change on the production of food grains have previously been recognized, but less attention has been paid to how these effects would affect the world supply and production of high-quality seeds. Enhancing the worldwide coordination of agricultural research is a substantial opportunity that will boost our capacity to create crop and cropping systems that are more adaptable to climate change. This involves improving agricultural resilience by increasing output and quality through genetic engineering, digital agriculture, and modern agricultural technologies. Supporting agricultural technology research and implementation and integrating technological innovation into national agricultural development objectives are crucial when it comes to policymaking. The seed sector must also devise plans to feed the world's 9 billion inhabitants by the year 2050.

References

Abzar A, Said MNM, Ahmad WJW et al (2017) Elevated CO_2 concentration enhance germination, seedling growth and vigour of rice. Ecol Environ Conservat 23:2058

Akman Z (2009) Comparison of high temperature tolerance in maize, rice and sorghum seeds by plant growth regulators. J Anim Vet Adv 8:358–361

Alvarez PR, Krix DW, Offord CA et al (2023) Impact of extreme heatwaves and life-history traits on seed germination responses in Cumberland plain woodland native plant species. Aust J Bot 71:395–405

Baek JS, Chung NJ (2014) Influence of rainfall during the ripening stage on pre-harvest sprouting, seed quality, and longevity of rice (*Oryza sativa* L.). Korean J Crop Sci 59(4):406–412

Borczak B, Marqués JMC, Paredes-Lopez O et al (2023) Structure and composition of Latin-American crops. In: Latin-American seeds. CRC Press, New York, pp 119–159

Bucheli J, Dalhaus T, Finger R (2022) Temperature effects on crop yields in heat index insurance. Food Policy 107:102214

Chen C, Jiang Q, Ziska L et al (2015) Seed vigour of contrasting rice cultivars in response to elevated carbon dioxide. Field Crop Res 178:63–68

Chen C, Begcy K, Liu K et al (2016) Heat stress yields a unique MADS box transcription factor in determining seed size and thermal sensitivity. Plant Physiol 171:606–622

Chen S, Liu W, Yan Z et al (2023) Adaptation strategy can ensure seed and food production with improving water and nitrogen use efficiency under climate change. Earth's Future 11:e2022EF002879

Cheng WH, Chiang MH, Hwang SG et al (2009) Antagonism between abscisic acid and ethylene in Arabidopsis acts in parallel with the reciprocal regulation of their metabolism and signaling pathways. Plant Mol Biol 71:61–80

da Silva EC, Nogueira RJMC, da Silva MA et al (2011) Drought stress and plant nutrition. Plant stress 5:32–41

Dadlani M, Gupta A, Sinha SN, Kavali R (2023) Seed storage and packaging. In: Seed science and technology. Springer Nature Singapore, Singapore, pp 239–266

Doggalli G, Deshpande SK, Patil BR et al (2024) Assessing genetic variability in cowpea yield components and seed quality parameters; a comprehensive review. Asian J Soil Sci Plant Nutr 10:489–495. https://doi.org/10.9734/ajsspn/2024/v10i2307

Ebi KL, Anderson CL, Hess JJ et al (2021) Nutritional quality of crops in a high CO_2 world: an agenda for research and technology development. Environ Res Lett 16:064045. https://doi.org/10.1088/1748-9326/abfcfa

Ellis RH, Yadav G (2016) Effect of simulated rainfall during wheat seed development and maturation on subsequent seed longevity is reversible. Seed Sci Res 26:67–76. https://doi.org/10.1017/s0960258515000392

Hampton JG, Boelt B, Rolston MP, Chastain TG (2012) Effects of elevated CO2 and temperature on seed quality. J Agric Sci 151:154–162. https://doi.org/10.1017/s0021859612000263

Hampton J, Conner A, Boelt B et al (2016) Climate change: seed production and options for adaptation. Agriculture 6:33. https://doi.org/10.3390/agriculture6030033

Hawkins P, Geza W, Mabhaudhi T et al (2022) Dietary and agricultural adaptations to drought among smallholder farmers in South Africa: a qualitative study. Weather Climate Extremes 35:100413. https://doi.org/10.1016/j.wace.2022.100413

He W, Moonis M, Chung H, Yoo G (2016) Effects of high soil CO2 concentrations on seed germination and soil microbial activities. Int J Greenhouse Gas Control 53:117–126. https://doi.org/10.1016/j.ijggc.2016.07.023

Henry WB, Krutz LJ (2016) Water in agriculture: improving corn production practices to minimize climate risk and optimize profitability. Curr Clim Chang Rep 2:49–54. https://doi.org/10.1007/s40641-016-0035-9

Hu S, Li T, Wang Y et al (2022) Effects of free air CO_2 enrichment (FACE) on grain yield and quality of hybrid rice. Field Crop Res 306:109237. https://doi.org/10.1016/j.fcr.2023.109237

Iloh AC, Omatta G, Ogbadu GH, Onyenekwe PC (2014) Effects of elevated temperature on seed germination and seedling growth on three cereal crops in Nigeria. Sci Res Essays 9:806–813. https://doi.org/10.5897/sre2014.5968

Janni M, Gullì M, Maestri E et al (2020) Molecular and genetic bases of heat stress responses in crop plants and breeding for increased resilience and productivity. J Exp Bot 71:3780–3802. https://doi.org/10.1093/jxb/eraa034

Joshi MA, Faridullah S, Atul K (2016) Effect of heat stress on crop phenology, yield and seed quality attributes of wheat (*Triticum aestivum* L.). J Agrometeorol 18:206–215. https://doi.org/10.54386/jam.v18i2.937

Jyolsna T, Vashisht BB, Yadav M et al (2024) Field and simulation studies on yield, water and nitrogen dynamics and use efficiency in rice-wheat crops in sequence. Field Crop Res 311:109366. https://doi.org/10.1016/j.fcr.2024.109366

Khan RA (1977) Effect of high-temperature stress on the growth and seed characteristics of barley and cotton. In: Genetic diversity in plants. Springer, Boston, pp 319–324

Kim Y-U, Webber H, Adiku SGK et al (2024) Mechanisms and modelling approaches for excessive rainfall stress on cereals: waterlogging, submergence, lodging, pests and diseases. Agric For Meteorol 344:109819. https://doi.org/10.1016/j.agrformet.2023.109819

Lamichaney A, Maity A (2021) Implications of rising atmospheric carbon dioxide concentration on seed quality. Int J Biometeorol 65:805–812. https://doi.org/10.1007/s00484-020-02073-x

Lamichaney A, Kumar Swain D, Biswal P et al (2019) Elevated atmospheric carbon–dioxide affects seed vigour of rice (Oryza sativa L.). Environ Exp Bot 157:171–176. https://doi.org/10.1016/j.envexpbot.2018.10.011

Lamichaney A, Parihar AK, Hazra KK et al (2021) Untangling the influence of heat stress on crop phenology, seed set, seed weight, and germination in field pea (Pisum sativum L.). Front Plant Sci 12. https://doi.org/10.3389/fpls.2021.635868

Liu J, Hasanuzzaman M, Wen H et al (2019) High temperature and drought stress cause abscisic acid and reactive oxygen species accumulation and suppress seed germination growth in rice. Protoplasma 256:1217–1227. https://doi.org/10.1007/s00709-019-01354-6

Macova K, Prabhullachandran U, Spyroglou I, et al (2021) Effects of long-term high-temperature stress on reproductive growth and seed development in development in Brassica napus. https://doi.org/10.1101/2021.03.11.434971

Macova K, Prabhullachandran U, Stefková M et al (2022) Long-term high-temperature stress impacts on embryo and seed development in Brassica napus. Front Plant Sci 13. https://doi.org/10.3389/fpls.2022.844292

Maity A, Natarajan N, Vijay D et al (2016) Influence of metal nanoparticles (NPs) on germination and yield of oat (Avena sativa) and berseem (Trifolium alexandrinum). Proc Natl Acad Sci India Sect B: Biol Sci 88:595–607. https://doi.org/10.1007/s40011-016-0796-x

Maity A, Paul D, Lamichaney A et al (2023) Climate change impacts on seed production and quality: current knowledge, implications, and mitigation strategies. Seed Sci Technol 51:65–96. https://doi.org/10.15258/sst.2023.51.1.07

Marcinkowski P, Piniewski M (2018) Effect of climate change on sowing and harvest dates of spring barley and maize in Poland. Int Agrophys 32:265–271. https://doi.org/10.1515/intag-2017-0015

Marty C, BassiriRad H (2014) Seed germination and rising atmospheric CO_2 concentration: a meta-analysis of parental and direct effects. New Phytol 202:401–414. https://doi.org/10.1111/nph.12691

Mor VS, Deswal DP, Mann A, Dahiya BS, Beniwal BS (2008) Characterization of marigold (Tagetes spp.) genotypes using SDS-PAGE and RAPD markers. Seed Sci Technol 36(3):757–766

Mpandeli S, Nhamo L, Moeletsi M et al (2019) Assessing climate change and adaptive capacity at local scale using observed and remotely sensed data. Weather Climate Extremes 26:100240. https://doi.org/10.1016/j.wace.2019.100240

Mutum B, Bera K, Chaudhury KM, Dutta P (2024) Impact of priming with UV radiation on seed germination and seedling growth of Chakhao Rice cultivar. Int J Plant Soil Sci 36:63–70. https://doi.org/10.9734/ijpss/2024/v36i24364

Nascimento WM, Vieira JV, Silva GO et al (2008) Carrot seed germination at high temperature: effect of genotype and association with ethylene production. HortScience 43:1538–1543. https://doi.org/10.21273/hortsci.43.5.1538

Neha YG, Yadav RK et al (2022) Salt removal through residue-filled cut-Soiler simulated preferential shallow subsurface drainage improves yield, quality and plant water relations of mustard (Brassica juncea L.). Sustain For 14:4146. https://doi.org/10.3390/su14074146

Nelson M, Nelson DB (2024) Decoupling CO_2 from climate change. Int J Geosci 15:246–269. https://doi.org/10.4236/ijg.2024.153015

Orsenigo S, Abeli T, Rossi G et al (2015) Effects of autumn and spring heat waves on seed germination of High Mountain plants. PLoS One 10:e0133626. https://doi.org/10.1371/journal.pone.0133626

Panja S, Gupta AD, Dey N (2022) Impact of drought stress on grains filling in rice and its management: a review. Agric Rev. https://doi.org/10.18805/ag.r-2369

Pena-Rosas JP, Mithra P, Unnikrishnan B et al (2019) Fortification of rice with vitamins and minerals for addressing micronutrient malnutrition. Cochrane Database Syst Rev 2019. https://doi.org/10.1002/14651858.cd009902.pub2

Pillai AV, Singh C, Patil S, et al. (2023) How is India Adapting to Heatwaves? An Assessment of Heat Action Plans with Insights for Transformative Climate Action. https://doi.org/10.24943/hiah03.2023

Pinedo-Taco R, Figueroa-Serrudo C, Alvarado-Huamán L (2022) Water stress, heat, and salinity in the physiological quality of the seeds. In: Seed biology updates. IntechOpen

Poltronieri P, Rossi F (2016) Challenges in specialty coffee processing and quality assurance. Challenges 7:19. https://doi.org/10.3390/challe7020019

Reed RC, Bradford KJ, Khanday I (2022) Seed germination and vigor: ensuring crop sustainability in a changing climate. Heredity 128:450–459. https://doi.org/10.1038/s41437-022-00497-2

Revilla P, Anibas CM, Tracy WF (2021) Sweet corn research around the world 2015–2020. Agronomy 11:534. https://doi.org/10.3390/agronomy11030534

Rosales F, Casanova PL, Flota BC et al (2023) Irrigated agriculture as an adaptation strategy against climate change: a review. J Agric Sci 15:56. https://doi.org/10.5539/jas.v15n6p56

Roy D, Biswakarma N, Ghosh T et al (2024) Climate change on seeds physiology. In: Advances in global change research. Springer Nature Singapore, Singapore, pp 347–382

Sakai Y, Suriyasak C, Inoue M et al (2022) Heat stress during grain filling regulates seed germination through alterations of DNA methylation in barley (*Hordeum vulgare* L.). Plant Mol Biol 110:325–332. https://doi.org/10.1007/s11103-022-01278-5

Samarah N, Mullen R, Cianzio S (2004) Size distribution and mineral nutrients of soybean seeds in response to drought stress. J Plant Nutr 27:815–835. https://doi.org/10.1081/pln-120030673

Sanyal A, Joshi MA, Tomar BS (2022) Effect of post-anthesis high temperature on seed longevity and vigour in wheat (*Triticum aestivum* L.). Curr Sci 123:87–92. https://doi.org/10.18520/cs/v123/i1/87-92

Sehgal A, Sita K, Bhandari K et al (2018) Influence of drought and heat stress, applied independently or in combination during seed development, on qualitative and quantitative aspects of seeds of lentil (*Lens culinaris*) genotypes, differing in drought sensitivity. Plant Cell Environ 42:198–211. https://doi.org/10.1111/pce.13328

Semenov A, Korotkova I, Sakhno T et al (2020) Effect of UV-C radiation on basic indices of growth process of winter wheat (Triticum aestivum L.) seeds in pre-sowing treatment. Acta Agric Slov 116. https://doi.org/10.14720/aas.2020.116.1.1563

Shams M, Yildirim E, Khadivi A et al (2023) Differences in seed germination, plant growth, and proline content of some pepper (Capsicum annuum L.) genotypes under salinity stress. J Crop Health 76:309–316. https://doi.org/10.1007/s10343-023-00962-w

Sheoran P, Sharma R, Kumar A, Singh RK, Barman A, Prajapat K, Kumar S, Sharma PC (2022) Climate resilient integrated soil–crop management (CRISCM) for salt affected wheat Agri–food production systems. Sci Total Environ 837:155843

Singh RP, Prasad PVV, Reddy KR (2013) Impacts of changing climate and climate variability on seed production and seed industry. In: Advances in agronomy. Elsevier, pp 49–110

Singh RK, Singh A, Kumar S, Sheoran P, Sharma DK, Stringer LC, Quinn CH, Kumar A, Singh D (2020) Perceived climate variability and compounding stressors: implications for risks to livelihoods of smallholder Indian farmers. Environ Manag 66:826–844

Singh RK, Singh A, Kumar S, Sheoran P, Jat HS, Sharma PC, Sharma DK, Hazarika BN, Bhowmik SN, Sureja AK, Bhardwaj R (2022) Experimental co-production of knowledge to adapt to environmental change in northern India. Environ Sci Pol 136:357–368

Smith MR, Veneklaas E, Polania J et al (2019) Field drought conditions impact yield but not nutritional quality of the seed in common bean (*Phaseolus vulgaris* L.). PLoS One 14:e0217099. https://doi.org/10.1371/journal.pone.0217099

Sugimoto H, Sato T (2000) Effects of excessive soil moisture at different growth stages on seed yield of summer buckwheat. Jpn J Crop Sci 69:189–193. https://doi.org/10.1626/jcs.69.189

Suriyasak C, Oyama Y, Ishida T et al (2020) Mechanism of delayed seed germination caused by high temperature during grain filling in rice (Oryza sativa L.). Sci Rep 10. https://doi.org/10.1038/s41598-020-74281-9

Tertyshnaya YV, Levina NS, Elizarova OV (2017) Impact of ultraviolet radiation on germination and growth processes of wheat seeds. Agric Mach Technol:31–36. https://doi.org/10.22314/2073-7599-2017-2-31-36

Toh S, Imamura A, Watanabe A et al (2008) High temperature-induced abscisic acid biosynthesis and its role in the inhibition of gibberellin action in Arabidopsis seeds. Plant Physiol 146:1368–1385. https://doi.org/10.1104/pp.107.113738

Tushabe D, Altmann F, Koehler E et al (2023) Negative effects of high-temperature stress on gametophyte performance and their consequences for seed reproduction in wild plants. Environ Exp Bot 216:105532

Upadhyay RK (2020) Markers for global climate change and its impact on social, biological and ecological systems: a review. Am J Clim Chang 09:159–203. https://doi.org/10.4236/ajcc.2020.93012

Xhemali B, Giovanardi D, Biondi E, Stefani E (2024) Tomato and pepper seeds as pathways for the dissemination of Phytopathogenic bacteria: a constant challenge for the seed industry and the sustainability of crop production. Sustain For 16:1808. https://doi.org/10.3390/su16051808

Yadav G, Ellis RH (2016) Development of ability to germinate and of longevity in air-dry storage in wheat seed crops subjected to rain shelter or simulated supplementary rainfall. Seed Sci Res 26:332–341. https://doi.org/10.1017/s0960258516000209

Yadav G, Ellis RH (2018) Longevity of 285 seed lots of wheat in hermetic storage compared with independent estimates from the seed viability equation. Seed Sci Technol 46:341–347. https://doi.org/10.15258/sst.2018.46.2.1429

Yadav SL, Kumar D, Yadav M et al (2024a) Environmental resource management with reference to global climate change. In: Environmental nexus for resource management. CRC Press, Boca Raton, pp 330–350

Yadav M, Vashisht BB, Jalota SK et al (2024b) Improving water efficiencies in rural agriculture for sustainability of water resources: a review. Water Resour Manag 38:3505–3526. https://doi.org/10.1007/s11269-024-03836-6

Yanagi M (2024) Climate change impacts on wheat production: reviewing challenges and adaptation strategies. Adv Resou Res 4(1):89–107

Zhao C, Liu B, Piao S et al (2017) Temperature increase reduces global yields of major crops in four independent estimates. Proc Natl Acad Sci 114:9326–9331. https://doi.org/10.1073/pnas.1701762114

Part IV

Emerging Crops Resilience to Climate Change

Recent Advances Towards Abiotic Stresses Tolerance and Improvement in Barnyard Millet: A Climate-Resilient Crop for Food Security

19

Anish Sehrawat, Deepti Rani Sehrawat, Anita Mann, and Anita Rani Sehrawat

Abstract

Various abiotic stresses are increasing environmental pressure due to the combined impacts of climate change, along with competitive use of land generating uncertainties about the capacity to grow sufficient food. Millets offer alternative solutions for food production during climate change. Barnyard millet is the fastest growing among the millets and is an important food source for fragile regions due to its quick growth, resistance to various abiotic stresses, and capacity to flourish in unfavourable conditions. This crop can be utilized as a model for stress tolerance in crop improvement or as a substitute crop in changing climatic conditions. 'Omic' approaches are pivotal in identifying novel genes and mechanisms responsible for abiotic stress tolerance, which can be leveraged to enhance the stress resilience of staple crops while maintaining yield. Integrating conventional farming methods with cutting-edge technologies can serve for the development of better stress-resilient crops that can help meet current food demands.

Keywords

Abiotic stress · Barnyard millet · Climate resilience · Crop improvement · Food security

A. Sehrawat · A. R. Sehrawat (✉)
Department of Botany, Maharshi Dayanand University, Rohtak, Haryana, India
e-mail: anitasehrawat@mdurohtak.ac.in

D. R. Sehrawat
Department of Anthropology, Panjab University, Chandigarh, Punjab, India

A. Mann
ICAR-Central Soil Salinity Research Institute, Karnal, Haryana, India

A. Mann et al. (eds.), *Cutting Edge Technologies for Developing Future Crop Plants*,
https://doi.org/10.1007/978-981-96-2508-6_19

19.1 Introduction

The importance of small millets in the agricultural production system is rising due to the changing climate circumstances, including frequent dry spells as well as inadequate availability of water for cultivation. About 40% of the population in Africa, particularly those living in the sub-Saharan nations, live in regions with less rainfall, but the figure is on the rise and will likely double by 2050 (Willett et al. 2019). South Asia and Africa seem to be the regions most affected by changing climates, especially global warming, according to reports on the event (Nelson et al. 2009; Numan et al. 2021; Warren et al. 2018). Water scarcity, contamination of soil, elevated soil, and water salinization are the result of the combined impacts of climate change (Khedr et al. 2021). Underutilized crops that are not grown and traded globally, such as sorghums (Hadebe et al. 2017) and millets (Revoredo-Giha et al. 2022; Kumar et al. 2023a, b), are often adapted to arid and semi-arid regions, offering alternative solutions for food production during climate change (Prasad 2020; Rosero et al. 2020; Muthamilarasan et al. 2019; Chivenge et al. 2015; Benitez-Alfonso et al. 2023). The interplay between climate change and the growing socio-economic burden on our natural resources under increasing environmental pressure along with competitive use of land generates uncertainties about the capacity to grow sufficient food to meet the world's rapidly increasing population. Food insecurity and malnutrition hazards are further exposed to the poor by price volatility, rising commodity costs, and increasingly inelastic demand in the rich world (OECD-FAO 2011). The present global agriculture and food systems have to be changed, according to the United Nations Sustainable Development Goal 2 (United Nations General Assembly 2015), which aims to 'end hunger, achieve food security and improved nutrition and promote sustainable agriculture' (Cappelli et al. 2021, Cappelli and Cini 2021). The world's agricultural output has been severely impacted by factors such as diminishing agrobiodiversity, limited land, insufficient water, and destruction of the environment (Li and Siddique 2020; Mann et al. 2020).

In order to solve the problems mentioned above, agricultural production needs to shift to something more sustainable as it is not viable at the moment (FAO, IFAD 2020). Integrating conventional farming methods with cutting-edge technologies could help meet current food demands while protecting resources for future generations (Davis et al. 2016; Singh and Singh 2017). This would also maintain the agricultural system's sustainability and improve agricultural production (National Institute of Food and Agriculture 2021; Cappelli et al. 2021; Cappelli and Cini 2021). Considering the worldwide circumstances, the Food and Agriculture Organization of the United Nations (FAO) designated 2023 the International Year of Millets and highlighted millets as an essential component of attaining sustainable development goals (SDGs) (Antony Ceasar and Maharajan 2022). The diminishing variety of crops grown for agricultural purposes highlights serious questions about how sensitive present agricultural practices are to the effects of climate change. It is quite likely that as a result of climate change and simultaneous weather extremes, some of these significant species may become extinct, and some unknown or forgotten crops will become our staple crops (Muthamilarasan and Prasad 2020;

Vetriventhan et al. 2020). Less global warming potential (GWP), higher resistance to abiotic and biotic stresses, shorter growing seasons, higher water and nutrient efficiency, and adaptability to various cropping systems under both irrigated and dry farming conditions are some of the advantages of small millets (Sanyal et al. 2021). With the onset of climate change that could cause food chaos in the future, cultivation of minor millets as 'catch crops' is favoured by historical versatility, resilience to adverse environment conditions, long and easy storage, resistant to pests or diseases, and almost no required maintenance, which is advantageous for hilly and terrace cultivation although little input during growth is necessary (FAO and ICRISAT 1996). Traditional varieties frequently perform better than improved kinds in the absence of transcending inputs. Traditional drought-resistant crops—sometimes enhanced via thoughtful farmer selection—offer resilience and stability in the face of climate change (Florin et al. 2016). Barnyard millet is the fastest growing among the millets (Sood et al. 2015) and can take 45–60 days, depending on the growth environment and accession, to complete the lifespan from seed to seed (Padulosi et al. 2009; Denton 1987). Barnyard millet is an important food source for fragile regions due to its quick growth, resistance to drought, and capacity to flourish in unfavourable conditions (De Wet et al. 1983; Kono et al. 1987; Majid et al. 1989; Sood et al. 2015). The genus *Echinochloa*, including the barnyard millet, comprises almost 250 annual and perennial species under the *Poaceae* family and the *Panicoideae* sub-family (Clayton and Renvoize 2006). These species are broadly dispersed throughout the world's warmer as well as temperate climates (Bajwa et al. 2015). According to their taxonomic and evolutionary relationships, 35 species have been defined so far through morphological, cytological, and molecular marker analyses (Yabuno 1966, 1987; Yamaguchi et al. 2005; Yuichiro et al. 1999). Japanese barnyard millet was cultivated in some regions of Japan where the weather, soil, and facilities for irrigation were not favourable to the rice crop. In years when a cool climate severely damaged the rice harvest, barnyard millet, which is resilient to cold weather, helped fight hunger (Yabuno 1987). In India, barnyard millet is mainly grown in hilly regions of Tamil Nadu and Uttarakhand up to 2300 metres above mean sea level (MSL), primarily during the kharif season (Kaur and Sharma 2020). Indian barnyard millet is also known as billion-dollar grass (Farrell 2011; USDA NCRS 2002).

Numerous millets have seen notable improvements and the introduction of new cultivars as a result of traditional breeding techniques such as landrace selection, mutation breeding, hybridization for delivering revolutionary variation followed by pedigree selection, and utilization of hybrid robustness (Serba and Yadav 2016; Serba et al. 2020; Vetriventhan et al. 2020). Millets are gradually gaining popularity as a food crop worldwide, but their yield still remains low despite their recent captivating cuisine and health advantages, as well as their ability to adapt to harsh environmental circumstances (Taylor 2019; Dias-Martins et al. 2018; Alavi et al. 2019). Since these crops can be utilized as a model for stress tolerance in crop improvement or as a substitute crop in places affected by drought, it is assumed that they will ensure food security in the future (Plaza-Wuthrich et al. 2016). In this chapter, we attempt to elucidate the different tolerance mechanisms of barnyard millet towards various abiotic stresses and try to propose a comprehensive idea about the improvement of abiotic stress tolerance through the integration of conventional breeding with available knowledge of modern genetic or genomic resources of millets. Variable adaptability and higher abiotic stress tolerance mechanism of barnyard millet can serve for the development of better stress-resilient crops to support food and nutritional policy for the increasing population of the world.

19.2 Effect of Drought Stress

Severe drought stress significantly affects barnyard millet growth and physiology at any stage of their life cycle. Different tolerance levels among *Echinochloa* species are shown by their varying responses to osmotic or drought stress. During the anthesis stage, the chlorophyll stability index, soluble protein content, and relative water content show a decreasing trend (Divya et al. 2019). The elongation zone length, crown root elongation rate, and photosynthetic rate all decrease, owing mostly to lower stomatal conductance (Matsuura et al. 2000). Root length density in deep soil layers decreases by 74% as a result of this stress, which also causes a 46% reduction in transpiration rate, therefore lowers leaf area and shoot dry weight (Zegada-Lizarazu and Iijima 2005). Additionally, this stress results in reduced height, biomass, and seed production in jungle rice (*E. colona*) (Chauhan and Johnson 2010). The root-to-shoot ratio decreases in American and Korean *E. oryzicola*, respectively (Park et al. 2016). Water stress during the anthesis stage causes a reduction in the number of reproductive tillers and yield loss due to the reduced number and fertility of surviving spikelets (Royo et al. 1999). Grain yields are reduced due to a decrease in seed number per ear and seed weight, attributed to stress effects on pollination, floret abortion, and cytokinin reduction, which lead to fewer endosperm cells in seeds (Bradford 1994; Ashok et al. 2018). Drought stress also impairs the adaptation of quinclorac-resistant *E. crusgalli* by limiting germination and development, as well as decreasing GST activity (Wu et al. 2019). The stress causes decreased midday leaf water potential and wilting symptoms such as leaf rolling and senescence of the lower leaves in *E. cruss-galli var. formosensis* and *E. oryzicola* (Yamasue et al. 1989a, b).

19.3 Effect of Waterlogging

Waterlogging stress significantly reduces water use efficiency in plants (Zegada-Lizarazu and Iijima 2005). Plants may suffer from anoxia or hypoxia in waterlogged circumstances due to the tremendously poor ability of oxygen to transfer through water (Jackson et al. 1985). This might result in the accumulation of phytotoxic chemicals (Ernst 1990; Kreuzwieser et al. 2004; Lamers et al. 1998; Ponnamperuma 1984) and reduced nutrient availability (Ejiri and Shiono 2019; Laanbroek 1990). Under hypoxic conditions, a reduced concentration of nitrogen in leaves and stems was observed along with a reduced root growth rate, while sodium concentration increased throughout the plant (Matsuura et al. 2022). Waterlogging stress has a significant impact on barnyard millet's growth and development. The species tends to show reduced root biomass and a decline in root elongation, which hampers the plant's ability to absorb nutrients effectively. This stress condition causes a decrease in photosynthetic activity due to chlorophyll degradation and impaired gas exchange. Barnyard millet's tolerance to waterlogging is generally low, resulting in stunted growth, chlorosis, and premature leaf senescence under prolonged waterlogged conditions. Additionally, waterlogging can lead to an increase in ethylene production in barnyard millet, which negatively affects root and shoot growth. This stress condition also disrupts carbohydrate metabolism, leading to a reduced energy supply for growth and development. The reduction in oxygen availability under waterlogged conditions forces barnyard millet to switch from aerobic to anaerobic respiration, which is less efficient and produces toxic by-products such as ethanol and lactic acid.

19.4 Effect of Salinity

Salinity stress is characterized by ion toxicity and osmotic stress in plants, which manifest as oxidative stress (Kim et al. 2004; Abogadallah 2010; Makrana et al. 2023; Kesh et al. 2022; Mann et al. 2023). The effect of salt on six Italian populations of *E. crus-galli* was studied at germination and seedling growth stage, under nine different concentrations of NaCl, ranging from 0 to 400 mM. Up to a saline concentration of 250 mM, resistant and sensitive populations showed stable seed germination rates of more than 60% and 80%, respectively. However, a gradual increase in salt concentration above 250 mM NaCl led to a reduction in seed germination to 13% of the population at 400 mM NaCl (Serra et al. 2018). Seed germination rates of about 68% in *E. crus-galli* (Sadeghloo et al. 2013) and more than 60% in *E. glabrescens* (Opena et al. 2014) were observed at 225 mM NaCl and 250 mM NaCl, respectively. Chauhan and Johnson in 2009 observed complete inhibition of *E. colona* (L.) Link seeds under a salt concentration of 200 mM. There was a gradual decrease in germination for both *E. glabrescens* and *E. colona* as the concentration of salt increased (Opena et al. 2014; Chauhan and Johnson 2009). Germination percentage, germination rate, and shoot and root length gradually decreased with increased NaCl stress in *Echinochloa frumentacea* (Parkash et al. 2018). Deterioration of ion balance in seeds due to induced water and osmotic stress under saline conditions delayed and reduced seed germination. Accumulation of putrescine, biosynthesis of polyamines, changes in unsaturated fatty acids, and increased lipid peroxidation, along with altered membrane properties and structure, were observed following salt treatment (Yun et al. 2004). A gradual decrease in germination percentage, shoot and root lengths, fresh and dry weight, relative shoot and root lengths, relative germination percentage, and vigour index was observed as salt concentration increased. Shoot length was more affected than root length at 150 mM NaCl due to reduced nutrient and metabolite supply to the shoots, as nutrient absorption was competed by ions. The salt damage index was 84% at 50 mM, 96% at mM, and 99% at 150 mM and 200 mM NaCl concentrations, respectively, based on germination percentage (Arthi et al. 2019). Variable reductions in germination among different genotypes were observed under salt stress, with declined shoot, root, and seedling lengths, along with decreased relative water capacity and increased relative salt injury rates, as salt concentration increased (Williams et al. 2019). When compared to the control, the biomass of *E. crus-galli* was dramatically reduced by severe saline stress. Salt stress caused a considerable drop in relative water content (RWC) below control levels. Chlorophyll a, b, and total chlorophyll concentrations were not significantly reduced compared to the control under mild salt stress but considerably dropped under severe salt stress. Chloroplast ultrastructure showed slightly abnormal shapes of chloroplasts with inflamed peroxisomes and mitochondria. When compared to mild saline stress, severe saline stress had a considerable impact on *E. crus-galli* (Islam et al. 2018). According to Rahman and Ungar (1990), *E. crus-galli* seeds could germinate up to 1.5% NaCl, but they could only withstand somewhere 20 days at a concentration of nearly 1% NaCl right after emergence. Compared to the germination and emergence stages, *E. crus-galli* exhibits reduced tolerance during the early seedling stage (Serra et al. 2018). Plant height was less affected than weight, with greater reductions observed in shoot and root weight compared to their lengths as salt levels increased (Wilson and Read 2006). Emergence was more affected by salinity than germination, with notable decreases in emergence at higher salt rates. The concentration of NaCl was entitled to decrease emergence by 50% varied between 123.58 mM and 287.76 mM. The loss in seedling weight suggests that a falling value for the chlorophyll a/b ratio may reflect a slowed-down rate of photosynthesis caused by salinity-related stress (Fogliatto et al. 2021). NaCl treatment inhibited the growth of the *E. crus-galli* shoots, resulting in a decrease in the fresh weight of the entire shoot. The symptoms were not severe at 120 mM NaCl concentration, as seen by growth inhibition and a 27% drop in photosynthetic energy conversion efficiency (FPSII) in *E. crus-galli* seedlings (Sawada et al. 2008). *E. stagnina* was shown to exhibit a decrease in tiller length, tiller count, leaf area, and number of leaves over the salinity stress ranging from 0 to 300 mM NaCl. At high salinity levels of around 400 mM NaCl, *E. stagnina* was unable to grow. There was a reduction in water use efficiency and stomatal conductance when salinity rose above a 100 mM NaCl threshold to 300 mM NaCl concentrations. Reductions in photosynthesis and transpiration rates were observed, with a drastic effect on photosynthesis compared to transpiration (Khedr et al. 2021). The growth of *E. oryzicola* was retarded, with a decreased photosynthetic rate and significantly increased Na content. Sodium content in leaves was lower than in roots of *E. oryzicola* (Kim et al. 2004). Rubisco and chlorophyll content reductions appeared to be reasons for the wild type's reduced photosynthetic activity (Abogadallah et al. 2010a). To restrict 50% of the shoot biomass of jungle rice and barnyard grass, respectively, the electrical conductivity (EC) needed was 13 dS/m and 15 dS/m (Chauhan et al. 2013). At lower concentrations, both Na2SO4 and NaCl may encourage the growth of *E. frumentacea*, but at greater concentrations, they can inhibit it, with NaCl having a larger inhibitory impact than Na2SO4 (Lu et al. 2021).

19.5 Effect of Temperature

In comparison to low light intensity, high light intensity causes chlorosis to occur under salt stress earlier. This is because of an increase in reactive oxygen species (ROS), especially hydrogen peroxide under high light intensity (Giacomelli et al. 2007). This is further supported by the increased H_2O_2 amounts in chlorotic plants (Abogadallah 2010). Chilling conditions lead to decreased carbon uptake and activities of C4 enzymes. Specifically, NADP+-malate dehydrogenase and pyruvate P ~ dikinase steps of the C4 pathway are probably linked to the decrease in CO2 uptake in C4 plants in cold environments (Potvin et al. 1986).

E. crus-galli exhibits a purely plastic response to environmental conditions in terms of seed production. When grown at high temperatures, the plants produce seeds with lower average mass, variance, and germinability. Conversely, cold maturation results in the production of larger seeds compared to warm maturation, with the latter seeds being deprived of reserve constituents. Seeds matured at high temperatures contain higher concentrations of potassium (K+) and manganese (Mn_2+) but lower concentrations of starch and protein (Charest and Potvin 1993). After being exposed to high light (HL), the photochemical quenching (qP) in the leaves of *E. crus-galli* somewhat reduces. On the other hand, following photo-inhibitory light treatment, the rate of photosynthesis quickly approaches control plant levels. A larger pool of antheraxanthin and zeaxanthin (A + Z) and enhanced non-radiative energy (NPQ) dissipation are correlated with a decrease in photochemical quenching in the plants under evaluation (Romanowska et al. 2017).

19.6 Effect of Metals

At a soil concentration of 250 mg/kg of cadmium (Cd), a germination percentage of 65% was observed in barnyard millet (*Echinochloa frumentacea*) cultivar CO2. The seedling vigour index is negatively influenced, leading to a drop in germination percentage. The growth of root-shoot length, fresh, and dry weight is affected as increased Cd concentration decreases germination percentage and morpho-physiological growth parameters compared to control (Revathy and Shanker 2023; Abe et al. 2011). When *Echinochloa* species are exposed to $MgCl_2$ or $MgSO_4$, their growth is limited to 33–42% comparable to control plants, showing symptoms resembling calcium deficiency. Excess $MgCl_2$ greatly decreases the content of calcium in the plant body, significantly decreases the Ca content in shoots, and increases the contents of Mg and K in roots and Cl and K in shoots, with a slight increase in the Mg content in shoots of *E. oryzicola* (Kobayashi et al. 2005).

19.7 Effect of Other Independently or Combined Stresses

Echinochloa species are well-characterized for crop mimicry and are among the most dangerous herbicide-resistant weeds globally (Norsworthy et al. 2014; Guo et al. 2018). Herbicide effectiveness is influenced by factors such as plant size and soil moisture, with water stress decreasing the absorption and translocation of herbicides (Ahmadi et al. 1980). Increased quinclorac resistance in *E. crusgalli* may be related to the heightened metabolic activity of glutathione S-transferase (GST), which may reduce adaptability to drought (Wu et al. 2019). 2,4-D's phytotoxic action has been reduced on saline-stressed *E. crusgalli* plants, particularly when mild saline +2,4-D treatment is applied. This means that more weed control activities are required along with longer application periods or larger herbicide application rates, for weeds affected by salt (Islam et al. 2018). When subjected to multiple stresses, plants activate specific and unique responses (Pandey et al. 2015). For instance, mild salinity treatment increases tolerance significantly, primarily because of increased antioxidant enzyme activity, dynamic regulation of ABA and IAA, and a better-suited metabolism with more metabolite accumulation that is associated with osmotic adjustment and signalling. Despite the administration of 2,4-D, saline stress reduces electrolyte leakage (EL) and lipid peroxidation and promotes comparatively better development in *E. crusgalli* by boosting the plant's overall antioxidative capacity (Islam et al. 2018).

Salt stress drastically inhibits germination and GST activity of the quinclorac-resistant population of *E. crusgalli*, with higher resistance correlating with lower germination rates. The plant height and fresh weight of herbicide-resistant *E. crusgalli* under salt stress are lesser than those of herbicide-sensitive *E. crusgalli*, suggesting that herbicide-resistant *E. crusgalli* is more susceptible to adapt to salt environments than herbicide-sensitive biotypes. This represents the fitness cost that herbicide-resistant *E. crusgalli* must bear in order to adapt to unfavourable stress conditions (Wu et al. 2022a, b, c). A reasonably strong salt stress led to the formation of the chlorotic phenotype, which became even worse with high light intensity (Abogadallah et al. 2010). *Echinochloa* species are tolerant to a variety of abiotic stressors in both their farmed and wild forms.

19.8 Drought Tolerance

In addition to variety improvement, it is critical to prioritize the evaluation and conservation of varied genetic resources, the understanding of stress tolerance mechanisms, and the conservation of genetic resources in order to increase crop yield in areas susceptible to stress (Matsuura et al. 2022). It is believed that during evolution, natural hybridization and chromosomal doubling allowed *Echinochloa* species to expand their habitats from flooded areas to dry upland regions. This allowed *E. crusgalli* to adapt to a wider range of habitats than *E. oryzicola* (Park et al. 2016). A significant minor millet crop, barnyard millet, is ideally suited to regions with low to moderate rainfall (500–700 mm) because of its fast maturity (Sampath et al.

1989). Among different species, barnyard grass is the most drought and salinity-stress-resistant, followed by alfalfa and sorghum (Shayghan and Sedghi 2013).

Germination is a crucial phenological stage in a plant's life cycle, ensuring successful establishment and final yield (Zare et al. 2006). Enhancing the seed germination rate can result in better seedling establishment, especially under salinity and drought stress conditions (He et al. 2002). Barnyard millet cultivars CO2 and MDU1 showed 100% and 94% seed germination under mild stress conditions, respectively. The CO2 cultivar exhibited maximum shoot, root, and seedling lengths along with higher germination percent, under mild to moderate drought stress, while the MDU1 cultivar had the highest root-to-shoot ratio under high drought stress (Francis et al. 2021). When exposed to osmotic stress, *E. crus-galli* var. praticola's root-to-shoot ratio increased, demonstrating its ability to withstand high stress levels by sustaining root growth and minimizing shoot growth inhibition, enabling adaptation to dry upland conditions (Park et al. 2016). Root growth is an energy-consuming mechanism that has a clear connection to the accessibility of shoot assimilates (Eliasson 1968; Brower 1977; Aguirrezabal et al. 1994). Mean root length, photosynthetic rate, assimilation rate, and the plant's ability to withstand prolonged water flow are all linked to the barnyard millet plant's ability to maintain a relative growth rate of root length under low osmotic potential. By doing so, the specific root length of lateral roots is increased, and the relative growth rate of the root in dry weight is maintained. Good water uptake under stress is possible in barnyard millet since it experiences osmotic stress without affecting the plant's resistance to water flow through it or its mean root length (Matsuura et al. 2000). Barnyard millet outperformed other small millets in shoot and root length under mild and high stress conditions (Francis et al. 2021).

Particularly, in regions where millet is grown, the efficient use of deep-water resources is important. When topsoil is subjected to drought stress, deuterium labelling of deep soil water reveals that finger millet, Job's tears, and barnyard millet enhance water extraction from deep layers (Zegada-Lizarazu and Iijima 2005). Barnyard millet significantly increases water uptake from deep soil under topsoil drought stress, with well-adapted deep roots to anaerobic conditions. This enhances water uptake efficiency (4.5 times) and reduces transpiration rate under drought compared to well-watered controls. Barnyard millet's higher drought resistance is reflected in its sustained leaf area, a critical factor for water use efficiency (WUE) in drought-prone environments (Zegada-Lizarazu and Iijima 2005; Passioura 1996). High chlorophyll stability index percentages help plants withstand stress through better chlorophyll availability. Under stress, higher chlorophyll stability index percentages lead to achieving maximal productivity, higher dry matter production, and an elevated rate of photosynthetic activity (Madhan Mohan et al. 2000). The highest chlorophyll stability index of 85.58% was observed by genotype ESL-94, with ESD-83 following closely behind at 80.88%, under drought conditions at the anthesis stage (Divya et al. 2019). Barnyard millet can improve its resilience to stress by re-establishing the equilibrium between transpiration and photosynthesis, resulting in a robust water usage efficiency (Khedr et al. 2021). Nitrate reductase enzyme activity, coordinated with photosynthesis and carbon skeleton availability, is also important for drought tolerance (Zhao et al. 2017). The genotypes ESLG-94 and ESD-83 recorded maximum nitrate reductase activity, indicating no detrimental effect on nitrate reductase activity and photosynthetic efficiency in drought-tolerant genotypes (Divya et al. 2019).

Barnyard grass exhibits a higher photosynthesis rate and carboxylation efficiency than rice and weedy rice in both instances of water scarcity and well-watering conditions (Agostinetto et al. 2022). Drought during the reproductive stage significantly decreases leaf area index (LAI) in small millets, but barnyard millet recorded the highest LAI compared to finger millet and little millet at reproductive and maturity stages. Barnyard millet showed better recovery after re-watering, with a lesser reduction in LAI compared to other millets, enabling higher solar radiation harvest and sustained net assimilation rate and leaf area duration under stress (Senthil et al. 2018). Barnyard millet performed better under drought at the reproductive stage when weighed against other millets such as little millet and finger millet, in terms of more productive tillers, ear heads, and lesser reduction in ear head length, ear head weight, and 1000-grain weight (Ashok et al. 2018). It recorded higher nitrate reductase activity compared to finger millet and little millet during reproductive and maturity stages (Senthil et al. 2018). H_2O_2 content in barnyard millet under water deficit appears to be more connected to cell signalling than oxidative damage, with no increase in thiobarbituric acid reactive substances (TBARS) or protein reduction (Agostinetto et al. 2022). Barnyard millet variety VL172 in the Kumaon region of the Himalayas responds better to drought by accumulating higher levels of proline and ascorbic acid and increased antioxidant enzymes activity, such as APX, GPX, SOD, and CAT under stress (Singh et al. 2021).

Biomass and seeds yielded were comparable at 50–100% soil field capacity in jungle rice plants, indicating high water use efficiency in barnyard millet species. Compared to plants at 100% field capacity, plants at 25% field capacity still obtained 48% of seeds and 42% of biomass (Chauhan and Johnson 2010). Barnyard millet recorded the highest grain and straw yield under drought conditions compared to finger millet and little millet. It showed greater recovery after re-watering, with a lesser reduction in ear heads per plant and yield, demonstrating resilience under water stress (Ashok et al. 2018). Pink-pigmented facultative methylotrophic bacteria (PPFMs) positively correlated with yield increases in barnyard millet. Exogenous PPFM application alleviates drought stress effects, improves germination, growth, development, quality, and yield, and prevents chlorophyll breakdown under drought, delaying senescence in kudiraivali. The highest plant growth and yield parameters were recorded with seed imbibition and foliar spray of Methylobacterium at 30 and 60 DAS (Balaji et al. 2019; Poorniammal et al. 2020; Poorniammal et al. 2021).

19.9 Tolerance to Waterlogging

Yabuno (1975) states that all Japanese wild species of *Echinochloa* are annual and can be categorized into three groups: lowland (*Echinochloa crus-galli var. crus-galli*), upland fields (*Echinochloa crusgalli var. praticola*), and those growing only in paddy fields (*Echinochloa crus-galli var. oryzicola* and *Echinochloa crus-galli var. formosensis*; Matsuura et al. 2022). The worldwide distribution of *E. oryzicola* is extensive with large intraspecific variation and distinct survival methods prevalent in other nations (Michael 1983; Yabuno 1966). *E. oryzicola* has adapted to flooded rice in Japan through traits such as phenotypic plasticity of plant height, photoperiodic sensitivity in heading, shattering habit, dormancy, mimicry, and anaerobic germinability (Yamasue 2001). It was suggested through phylogenetic analysis that the hexaploid species *Echinochloa crus-galli* was derived from a tetraploid *Echinochloa oryzicola* (Aoki and Yamaguchi 2008, 2009). While the *Echinochloa oryzicola* habitat is restricted to regions with water logging, *Echinochloa crus-galli* has varieties that can grow in both wet and dry habitats (Rao et al. 2007; Tanesaka et al. 2010; Yamasue 2001). Originating from *Echinochloa crus-galli var. crus-galli*, Japanese millet exhibits resistance to waterlogging (Matsuura et al. 2022; Matsuura et al. 2016; Kono et al. 1987). *E. utilis* is reported to be more hypoxia tolerant than *S. italica* and *B. ramosa*. In hypoxia, it grows its roots more quickly and has root hairs close to the tip, aiding it to absorb nutrients (Nye 1966, Gahoonia et al. 2001). The smaller proportion of stele in hypoxia-tolerant millets such as *E. utilis* helps suppress oxygen consumption. These millets also propagate inherent aerenchyma in the same way as wetland plants, with root hairs near the root tip, indicating hypoxia tolerance due to oxygen supply to root tips through lysigenous aerenchyma along adventitious roots (Matsuura et al. 2022). A few features of millet species that can withstand waterlogging include smaller stele that use less oxygen and higher nutrient absorption through root hairs, as seen in *E. utilis* (Matsuura et al. 2016). Barnyard millet induces cortical sclerenchyma in nodal roots, which filters out excess soil water, protects against microbes, stops roots from leaking oxygen, and gives cortical aerenchyma physical support (Galamay et al. 1991).

Echinochloa species, including some wetland plants, form a constitutive radial oxygen loss (ROL) barrier, helping them to respond to variable environmental stresses such as salt drought and waterlogging stress (Manzur et al. 2015; McDonald et al. 2001, 2002; Visser et al. 2000). More than 85% of the root exodermis cells in non-waterlogged areas have well-developed Casparian strips and suberin lamellae, which form a radial oxygen leaking barrier. Phytotoxic chemicals from the soil cannot pass through this barrier. While *E.* crus-*galli* may have varied its habitat by obtaining such a barrier at the exodermis, *E. oryzicola* lacks a constitutive ROL barrier in an aerated environment (Ejiri and Shiono 2019). Hexokinase (HXK) activity allows *Echinochloa* species to germinate and flourish in flood conditions, hypothesized to maximize hexose entry into glycolysis for generation of ATPs anaerobically (Fox et al. 1998). Aldehyde dehydrogenase (ALDH) may detoxify acetaldehyde formed through alcoholic fermentation during anaerobic germination, as seen in *E. crusgalli var. formosensis* (tolerant) and *praticola* (intolerant) (Fukao et al. 2003).

19.10 Tolerance to Salt Stress

Plant responses to salt stress involve complex gene networks and metabolic processes that vary depending on the plant's inherent tolerance, the severity, and duration of the stress exposure (Hasegawa et al. 2000; Zhu 2002; Parul et al. 2023). On the other hand, sensitive reactions may cause the plant to age more quickly and ultimately die (Abogadallah 2010). In barnyard millet, high salt tolerance has been observed up to 200 mM NaCl, surpassing rice cultivars such as FL 478 known for their tolerance of up to 100 mM (Arthi et al. 2019). Even at an EC of 24 dS/m, barnyard grass maintained 4% biomass production compared to controls, whereas rice did not survive at this level of salinity (Chauhan et al. 2013). *Echinochloa stagnina* adapts to salinity mainly by minimizing water loss by decreased leaf area or stomatal closure and increased antioxidant enzyme activity to detoxify reactive oxygen species (ROS) generated under stress (Khedr et al. 2021). *Echinochloa colona* from salt-affected fields exhibits higher Na + and K+ accumulation in roots, stimulating extensive root growth to maintain water balance (Al-Sherif 2007; Bernstein et al. 1995; Ejiri and Shiono 2019). Barnyard millet shows germination even at 150 mM and 200 mM NaCl concentrations, indicating inherent salt tolerance of certain genotypes (Arthi et al. 2019). However, germination percentages sharply decline at 400 mM NaCl, underscoring variable tolerance among different genotypes (Sadeghloo et al. 2013; Serra et al. 2018). The reduction in germination, root, and shoot growth under salt stress varies significantly among genotypes, with osmotic adjustment potentially playing a role in maintaining relative water content (RWC) under moderate salinity levels (Williams et al. 2019).

Echinochloa frumentacea and *E. crusgalli* display contrasting root morphologies and responses to salt stress, highlighting their distinct adaptive strategies. In comparison to *E. crusgalli*, *E. frumentacea* has longer primary roots, fewer root hairs, and a lower local branching density. Significantly, under salt stress circumstances, the root volume of *E. frumentacea* is 1.43 times greater than that of *E. crusgalli*, suggesting a selective utilization of photosynthetic products for roots. This differential allocation strategy is further underscored by the varying effects of different salt concentrations on the growth of radial root and biomass across both species (Zhang et al. 2024).

For increasing salt stress levels, *E. frumentacea* adapts by gradually thickening its roots, increasing root hair density, and enhancing root complexity. *E. crusgalli*, on the other hand, uses an approach intended for growing roots longer while preserving a consistent root thickness and producing densely packed longer root hairs. This dichotomy is evident in the study's findings: under moderate salt stress (100 mmol L-1), *E. frumentacea* shows significant increases in root length, surface area, tip number, fork number, and crossing number, whereas higher salt concentrations (220 and 340 mmol L-1) lead to decreases in these parameters. Conversely, *E. crusgalli* exhibits increased total surface area and length of roots, specifically under severe salt stress (340 mmol L-1), highlighting its capacity for lateral root growth in response to high salinity. High salt concentrations notably promote root thickening in *E. crusgalli* without affecting the diameter of *E. frumentacea's* roots significantly. This suggests that the *E. crus-galli* roots' lateral development strategy is particularly beneficial under high salt conditions. Conversely, under low salt stress, *E. frumentacea's* robust root system offers advantages in water and nutrient acquisition over *E. crus-galli,* underscoring its higher growth capacity in such conditions. The root hair dynamics further differentiate these species: Salt stress has little effect on *Echinochloa crus-galli's* ability to retain long, thick root hairs, whereas *E. frumentacea* shows a gradual increase in root hair density with rising salt concentrations. These adaptations reflect their respective capabilities to compete for water and nutrients in saline environments, with *E. crus-galli's* more complex root system potentially offering a wider range and efficiency in resource acquisition under stress. In summary, the contrasting root morphologies and adaptive responses of *E. crus-galli* and *E. frumentacea* to salt stress highlight their ecological flexibility and potential utility in managing saline soils. These insights underscore the importance of understanding species-specific adaptations in agricultural and ecological contexts (Zhang et al. 2024).

Echinochloa crus-galli exhibits robust tolerance to salt stress, evident even at 120 mM NaCl, surpassing the sensitivity observed in rice (Sawada et al. 2008). This resilience is underscored by stable chlorophyll fluorescence (FPSII) in leaf blades under NaCl treatment, indicative of maintained photosynthetic pigment levels, including chlorophyll a, chlorophyll b, and carotenoids, which are known markers of salt tolerance in plants (Stepien and Johnson 2009; Fogliatto et al. 2021; Ashraf and Harris 2013; Acosta-Motos et al. 2015). Barnyard grass mutants, characterized by elevated chlorophyll, Rubisco, and photosynthetic rates compared to wild types, highlight the role of these components in enhancing photosynthetic efficiency under saline conditions although no similar trends were observed in PEPC contents (Abogadallah et al. 2010a). In *Echinochloa stagnina,* differential responses to salinity include higher proline content across all salt treatments, indicating a consistent adaptive strategy under stress ($P < 0.05$) (Khedr et al. 2021). Catalase activity and soluble sugars peak at treatments of 100 mmol·L-1 Na2SO4 and 75 mmol·L-1 NaCl, respectively, reflecting pivotal roles in osmotic adjustment and oxidative stress management (Lu et al. 2021). Conversely, a negative correlation between soluble and insoluble sugars in *E. stagnina* suggests a trade-off where the accumulation of soluble sugars occurs at the cost of structural polysaccharides under stress conditions (Khedr et al. 2021).

Echinochloa oryzicola demonstrates effective sodium ion (Na+) regulation under salt stress conditions, maintaining high potassium ion (K+) content and a consistently higher K+/Na + ratio compared to rice. This ability is crucial for minimizing oxidative stress, as lower Na + accumulation in *E. oryzicola* suggests reduced oxidative damage relative to rice (Nguyen et al. 2005). Under high salinity environments, plants require a constant level of an elevated cytosolic K+/Na + ratio in order to survive, supporting the adaptive strategy observed in *E. oryzicola* (Zhu et al. 1998; Apse et al. 1999). Despite its lower antioxidative capacity compared to rice, *E. oryzicola* exhibits greater salt tolerance, emphasizing the role of osmotic adjustment mechanisms facilitated by proline and other osmoprotectants (Nguyen et al. 2005). Furthermore, *Echinochloa crus-galli,* noted for its superior salt tolerance compared to rice, accumulates higher levels of proline and glycine betaine under saline conditions, essential for osmotic adjustment rather than Na + exclusion (Fogliatto et al. 2021; Kim et al. 1999 and 2000). This species' ability to selectively uptake potassium over sodium contributes significantly to its resilience in saline environments (Wilson and Read 2006). Despite suggestions that Na + accumulation may be slightly higher in *E. crus-galli* than in rice, the plant's robust response toward salinity is attributed to its effective management of oxidative stress and maintenance of cellular integrity (Kim et al. 2000). Studies on barnyard grass mutants underscore the importance of ROS scavenging mechanisms in enhancing salt tolerance. These mutants exhibit reduced lipid peroxidation and improved activities of antioxidant enzymes such as superoxide dismutase (SOD), catalase (CAT), and peroxidase (POD) under salt stress, mitigating oxidative damage to cellular components (Abogadallah et al. 2010a). The suppression of reactive oxygen species accumulation correlates with improved growth and survival rates in salt-stressed barnyard grass populations (Abogadallah et al. 2010a, b). The adaptive strategies of salt-tolerant grass species, including *E. oryzicola* and *E. crus-galli*, highlight the complex interplay between ion regulation, osmotic adjustment, and ROS scavenging mechanisms in enhancing plant resilience under saline conditions (Sawada et al. 2008). These results add to a deeper knowledge of how different grass species manage salt stress, providing insights into potential agricultural applications for improving crop tolerance to environmental challenges.

In barnyard grass, under salt stress conditions, ascorbate peroxidase, peroxidase, and glutathione reductase exhibit differential regulation between green and chlorotic leaves. Higher activities of these enzymes in green leaves indicate their crucial role in maintaining ROS balance and preventing oxidative damage. Conversely, chlorotic leaves with reduced enzyme activities suffer from ROS overaccumulation, which correlates with chlorophyll loss, slowed growth, and increased oxidative stress (Abogadallah et al. 2010b). The decline in POD, GR, and APX activities in chlorotic leaves may result from excessive H_2O_2 production, damaging these enzymes and impairing their regulatory functions. In barnyard grass mutants (e.g., fows B19 and B21), high catalase (CAT) activity induced by elevated H_2O_2 levels in chlorotic plants does not compensate for the loss of POD, GR, and APX activities (Abogadallah et al. 2010b). This imbalance leads to oxidative stress despite increased CAT levels, highlighting the crucial role of POD, GR, and APX in maintaining ROS at physiological levels for stress tolerance.

Echinochloa crusgalli similarly relies on POD, APX, and GR for ROS regulation under salt stress. Loss of their activities leads to ROS build-up, even with elevated CAT activity, resulting in oxidative stress that reduces plant fitness. Lower CAT activities in salt-tolerant mutants reflect effective ROS control by POD, APX, and GR, demonstrating their essential role in stress adaptation (Abogadallah et al. 2010b). Interestingly, the lowest superoxide dismutase (SOD) activity in the most salt-tolerant mutant suggests that excessive SOD activity may not be necessary under these conditions, further emphasizing the specific roles of different antioxidative enzymes in stress tolerance mechanisms. These findings underscore the critical importance of antioxidative enzymes such as POD, SOD, APX, CAT, and GR in the salt tolerance of barnyard grass and *Echinochloa crusgalli*, elucidating their roles in ROS regulation and oxidative stress management under adverse environmental conditions.

Echinochloa stagnina exhibits significant capabilities in reducing soil salinity over extended periods. In laboratory-controlled soil columns, *E. stagnina* effectively decreased soil salinity (SS) by 65–87% and significantly reduced electrical conductivity (EC), with more pronounced effects observed compared to controls (Helalia et al. 1992). Field studies corroborate these findings, where *E. stagnina* grown in saline Vertisols and saline-sodic clay soils demonstrated substantial reductions in soil salinity within relatively short periods. Specifically, reductions of SS by 72–81% in Vertisols and significant EC reductions from 27.6 to 4.3 dS/m in saline-sodic clay soils were reported (Helalia et al. 1992; Ado et al. 2016). The mechanisms underlying *E. stagnina's* soil salinity reduction involve its root system enhancing soil structure by increasing macro-porosity (6–9%), facilitating salt leaching (Ado et al. 2019). Additionally, salt accumulation within the plant biomass contributes to overall soil salinity reduction (Ado et al. 2016, Barbiero et al. 2001, Zhao et al. 2013). These properties position *E. stagnina* as a beneficial fodder crop capable of improving salt-affected lands and enhancing agricultural productivity through diversified cropping systems (Ado et al. 2022).

Similarly, *Echinochloa frumentacea* shows promise for cultivation in moderately and severely salinized areas, where it has been observed to decrease soil bulk density and increase soil porosity and capillary porosity while also reducing total soil salt and alkalinity levels (Yupeng et al. 2020). These characteristics underscore its potential for mitigating soil salinity issues and supporting sustainable agricultural practices in challenging environments. In barnyard millet, certain accessions exhibit high germination percentages (>70%) under severe salt stress conditions (250 mM), highlighting their potential as tolerant varieties (Williams et al. 2019). Enhanced antioxidant responses in these millet accessions are associated with reduced oxidative damage and improved salinity tolerance (El-Shabrawi et al. 2010). However, the exact mechanisms governing seed germination under salt stress remain incompletely understood despite extensive research attention on salt stress modes and mechanisms during vegetative growth (Abogadallah and Quick 2009). Overall, these studies underscore the potential of *Echinochloa* species and barnyard millet in ameliorating soil salinity, enhancing agricultural sustainability, and advancing our understanding of plant responses to salt stress.

19.11 Tolerance to Temperature Stress

Barnyard grass (*Echinochloa crus-galli*) demonstrates significant adaptive strategies to cope with temperature stress and environmental variability. Under future climate scenarios with elevated CO_2 and temperature, barnyard grass exhibits remarkable increases in total dry mass and leaf area, highlighting its potential competitiveness (Miskelyte et al. 2017). This adaptive capacity is further underscored by the species' ability to cultivate and develop seeds in various levels of soil moisture, contributing to its weedy nature in diverse environments (Chauhan and Johnson 2010). Regarding physiological responses, barnyard grass maintains robust photosynthetic activity even under challenging conditions. Despite high light stress, the species shows resilience in maintaining photochemical efficiency, attributed to effective strategies such as PSII protein phosphorylation and a high pool of protective carotenoids (Rinta et al. 1996). This resilience is crucial for optimizing CO2 assimilation rates and ensuring sustained photosynthetic performance (Romanowska et al. 2017). Moreover, barnyard grass exhibits distinct biochemical adaptations, including the utilization of malate in mitochondrial respiration, which supports high ATP production necessary for metabolic activities and stress tolerance (Romanowska et al. 2017). Comparative studies highlight barnyard grass's unique chloroplast characteristics, such as lower grana development in bundle sheath chloroplasts compared to mesophyll chloroplasts, which likely contribute to its efficient light harvesting and energy utilization (Wasilewska-Dębowska et al. 2024). These adaptations reflect the species' evolutionary strategy to thrive under variable environmental conditions, ensuring its ecological success and resilience in the face of climate change challenges. In conclusion, barnyard grass exemplifies a comprehensive suite of adaptive traits—from physiological adjustments in photosynthetic machinery to biochemical strategies in energy metabolism—that equip it to thrive under temperature stress and fluctuating environmental conditions. These insights underscore the ecological versatility and adaptive potential of *Echinochloa crus-galli* in responding to future climate scenarios.

The efficient activity of the NDH complex in *Echinochloa crus-galli's* chloroplasts is responsible for its remarkable resilience to variations in light intensity and quality. Bundle sheath chloroplasts in barnyard grass exhibit significantly higher NDH complex activity, facilitating enhanced cyclic electron transport and ATP production compared to mesophyll chloroplasts, under varying light conditions (Wasilewska-Dębowska et al. 2024). This mechanism plays a crucial role in mitigating oxidative stress by levelling the redox balance of the plastoquinone pool and protecting the photosynthetic apparatus from light-induced damage. Furthermore, barnyard millet (*Echinochloa frumentacea*) demonstrates superior water use efficiency compared to C3 crops such as wheat, soybean, and sunflower, particularly under conditions of varying vapour pressure deficit (VPD). Despite increased transpiration rates with rising VPD, barnyard millet maintains constant photosynthesis rates, resulting in more than double the water use efficiency of C3 crops (Rawson et al. 1977). This efficiency is supported by higher stomatal and boundary layer resistances, along with lower mesophyll resistance, which collectively optimizes carbon dioxide gas exchange and minimizes water loss in barnyard millet. These adaptive traits underscore the ecological and agricultural potential of both *Echinochloa* species, highlighting their resilience to environmental stresses and their capacity to thrive in diverse climatic conditions. The efficient utilization of light and water resources not only enhances their ecological competitiveness but also positions them as promising candidates for sustainable agricultural practices in the face of climate change challenges.

19.12 Metal and Herbicides Stress Tolerance

Echinochloa colona L. exhibits notable tolerance to heavy metals chromium (Cr) and nickel (Ni) through callus cultures derived from various explants. Calli cultivated on Murashige and Skoog medium accompanied by 1.5 mg/L Cr, 2.0 mg/L Ni, or a combination thereof showed vigorous growth, indicating their capacity to withstand metal stress. SDS-PAGE analysis revealed specific proteins in tolerant calli, distinct from controls, suggesting a biochemical basis for metal tolerance (Murashige and Skoog 1962; Samantaray et al. 2001). Successful plant regeneration from somatic embryos further confirmed the viability and normal growth of these metal-tolerant plantlets. Similarly, *Echinochloa polystachya* demonstrates hyperaccumulation capabilities for cadmium (Cd), making it a promising candidate for phytoextraction from polluted tropical sites (Solís-Domínguez et al. 2007). In contrast, *Echinochloa pyramidalis* (Lam) Hutch serves as an effective macrophyte in filtering physical, organic, and mineral pollutants. Studies with EP-F filters showed significant reductions in turbidity, suspended matter, chemical oxygen demand (COD), ammonium (NH4+), nitrate (NO3-), and phosphate (PO43-) ions after 90 days, underscoring its purifying performance (Mouko et al. 2024). Regarding Cd stress, higher concentrations adversely affect the relative shoot and root growth indices in *Echinochloa* species although initial stages of leaf development maintain high relative water content before declining as leaves mature (Revathy and Shanker 2023). Contrastingly, rice demonstrates greater tolerance to excess magnesium (Mg) due to potential calcium (Ca) deficiency interactions, with Mg levels in *Echinochloa oryzicola* plateauing beyond 10 mM MgCl2 concentrations (Kobayashi et al. 2005). These findings highlight the diverse adaptive strategies and ecological roles of *Echinochloa* species in responding to and mitigating metal and herbicide stresses, underscoring their potential for environmental bioremediation and agricultural applications under challenging conditions.

A combination of environmental stress factors, including variable temperatures, drought, and waterlogging, poses significant challenges to crop production, particularly in rainfed agriculture systems (Khanthavong et al. 2022). These combined stresses often result in greater crop failure compared to individual stressors due to their cumulative impact. Barnyard grass (*Echinochloa crus-galli*) exhibits distinct phenotypic traits that differentiate it from rice, making it difficult to control in agricultural fields. The occurrence of numerous glutathione S-transferase (GST) and cytochrome P450 (CYP) genes in *E. crus-galli* potentially enhances its detoxification capabilities, contributing to its weediness (Yu and Powles 2014). Allelopathic interactions between rice and barnyard grass are being explored for breeding improved cultivars in the future, highlighting potential strategies for weed management (Guo et al. 2018). Moreover, environmental stress factors influence crop yield and growth, as well as herbicide activity and the effectiveness of chemical-based weed management approaches (Islam et al. 2018). Plants activate specific stress responses when exposed to multiple stresses simultaneously, enhancing their tolerance and survival strategies (Pandey et al. 2015; Foyer et al. 2016). For instance, *E. crus-galli* exhibits increased resistance to quinclorac herbicide under salt stress conditions, albeit with lower GST activity (Wu et al. 2022a, b, c). Stomatal conductance is crucial for maintaining photosynthesis and regulating leaf temperature through evapotranspiration, critical for biomass production and crop resilience to varying soil water status and temperature fluctuations (Khanthavong et al. 2022). The growth stimulation of *E. crus-galli* under saline and 2,4-D herbicide treatments correlates with elevated antioxidant activities and regulated hormone levels, such as indole-3-acetic acid (IAA) and abscisic acid (ABA) (Islam et al. 2018). These physiological responses contribute to its resilience under combined stress conditions.

The induction of endogenous IAA under salt stress further enhances *E. crus-galli's* salt tolerance compared to other crop plants (Hakim et al. 2013). Studies also indicate interactions between herbicide treatments, soil moisture regimes, and enzymatic activities such as ascorbate peroxidase and superoxide dismutase, which influence plant responses to herbicide stress (Marchezan et al. 2017). In conclusion, understanding these multifaceted interactions and adaptive mechanisms of *Echinochloa* species provides insights into their ecological roles, weed management strategies, and potential applications in sustainable agriculture under challenging environmental conditions.

19.13 Conventional Breeding Strategies for Stress-Tolerant Barnyard Millet

Barnyard millet (*Echinochloa* spp.) has evolved mechanisms to thrive under diverse environmental stresses, including variable climatic conditions and soil salinity. This adaptability is reflected in its wide genetic diversity, with approximately 8000 accessions conserved globally (Renganathan et al. 2020). Characterizing these accessions helps identify specific traits that contribute to stress tolerance, essential for further genetic improvement efforts (Trivedi et al. 2017). Studies on barnyard millet have highlighted specific stress tolerance mechanisms, such as enzymatic activities and physiological responses. For instance, catalase activity in certain accessions such as IC340999, peroxidase activity in IC548613, and superoxide dismutase activity in IC279703 indicate their role in stress mitigation (Trivedi et al. 2017). These enzymatic activities are crucial for detoxifying reactive oxygen species (ROS) generated under stress conditions, thereby maintaining cellular homeostasis. Genotypic studies have identified correlations between morpho-phenological traits and stress tolerance indicators such as seed vigour index, germination percentage, and seedling length under saline conditions (Dhanalakshmi et al. 2019; Subramanian et al. 2019).

This information aids in selecting superior genotypes, such as MDU 1 and BAR 119, which exhibit robust tolerance across different growth stages (Subramanian et al. 2020). In regions such as South Korea, barnyard millet is cultivated on reclaimed lands with moderate salinity, producing nutritious fodder despite salt concentrations of up to 0.5% (Cho Kwang-Min et al. 2013a). However, the plant's ability to accumulate salts in its tissues, as observed in stem, leaves, and roots, underscores its adaptation strategy under saline stress (Cho Kwang-Min et al.

2013b). Drought stress studies reveal barnyard millet's resilience, with higher root length stress tolerance index and plant height stress tolerance index compared to other millet varieties such as foxtail millet and kodo millet (Francis et al. 2021). Physiologically efficient genotypes such as ESLG-94 and ESD-83 maintain higher yields under drought conditions, suggesting potential metabolomic and molecular mechanisms for drought tolerance (Divya et al. 2019). Moreover, barnyard millet shows promising salt tolerance, with accessions such as ACM161, ACM295, ACM335, and IEc167 exhibiting varying degrees of tolerance to high salt concentrations (Williams et al. 2019). These accessions, mainly derived from local landraces, are valuable for further breeding programmes aimed at enhancing salt tolerance in agricultural crops. Leveraging the genetic diversity and adaptive traits of barnyard millet through traditional breeding approaches offers substantial potential for developing stress-tolerant cultivars. This approach not only enhances resilience against climate variability but also contributes to sustainable agriculture practices worldwide.

19.14 Hybrids

Hybridization plays a pivotal role in advancing genetic studies and breeding strategies, particularly for mapping quantitative trait loci (QTL) and conducting marker-assisted back-crossing. When hybridizing between wild and cultivated forms of barnyard grass species such as *E. colona/E. frumentacea* and *E. crus-galli/E. esculenta*, normal meiotic behaviour with 27 bivalents results in reproductively fit offspring. However, interspecific hybrids such as *E. crus-galli/E. colona* and *E. esculenta/E. frumentacea* exhibit unequal meiotic behaviour, leading to reproductively infertile offspring (Yabuno 1966, 1984). Despite sterility, these hybrids have the potential for genetic enhancement and stress tolerance improvement due to their combined traits (Sood et al. 2014). In efforts to enhance stress tolerance, Samantaray et al. (2001) successfully developed chromium and nickel-tolerant cell lines from *E. colona*, highlighting the potential of hybridization in creating novel traits for environmental adaptation and agricultural resilience. This approach not only expands the genetic diversity available for breeding programmes but also underscores the importance of hybridization in exploring and exploiting the inherent variability within barnyard grass species to meet future agricultural challenges.

19.15 Mutants

Mutational breeding represents a crucial approach for generating novel variation and expanding the genetic diversity of polyploid species such as *Echinochloa*. In the case of barnyard grass, stable waxy mutant lines were developed upon irradiation of the low amylase landrace 'Nogehie', illustrating the potential of induced mutations to introduce desirable traits (Hoshino et al. 2010). Gamma irradiation has also been instrumental in increasing genetic variance for important agronomic traits such as grain yield, plant height, tiller number, and head length in barnyard grass (Mehta et al. 2005). While hybrids between early maturing *E. frumentacea* and high-yielding *E. esculenta* are sterile, wild ancestors of these species offer fertile hybrids, underscoring their value as genetic reservoirs for crop improvement (Mandelbaum et al. 1995). Moreover, mutational breeding has proven effective in developing salt-tolerant genotypes of barnyard grass. For instance, mutagenesis with sodium azide yielded three salt-tolerant genotypes capable of germinating and growing under high salt conditions (Abogadallah and Quick 2009). These mutants not only exhibited tolerance to osmotic and ionic stresses induced by salt but also showed enhanced antioxidant activity, crucial for mitigating the effects of reactive oxygen species under stress (Abogadallah and Quick 2009). Combining these mutants through crossing offers the potential for developing genotypes with comprehensive salt tolerance, encompassing both germination and vegetative growth stages (Abogadallah et al. 2010a). This integrated approach highlights the importance of mutational breeding in expanding the adaptive potential and resilience of barnyard grass in challenging environments.

19.16 Existing 'Omic' Strategies for Improving Stress Tolerance in Barnyard Millet

Recent advancements in phenotyping and genomic resources have unveiled the untapped potential of orphan crops such as barnyard millet, emphasizing their innate ability to cope with various stresses. These 'omic' approaches are pivotal in identifying novel genes and mechanisms responsible for abiotic stress tolerance, which can be leveraged to enhance the stress resilience of staple crops while maintaining yield, crucial for global food security under changing climatic conditions (Kumar et al. 2023a, b). Despite their inherent resilience compared to cereals, barnyard millets require additional efforts to further enhance their stress tolerance and yield in fragile lands (Muthamilarasan and Prasad 2021). Sequencing technologies have a critical role in unravelling the genetic blueprints of various *Echinochloa* species. The NCBI database hosts sequences of several *Echinochloa* species, including *E. crus-galli, E. oryzicola, E. phyllopogon, E. colona, E. pyramidalis, E. frumentacea, E. esculenta,* and *E. oryzoides*. Notably, the cultivated species *E. frumentacea* and *E. esculenta* are relatively underrepresented in current nucleotide sequence databases. These genomic insights provide a foundation for exploring hidden tolerant genes in orphan crops such as barnyard millet, facilitating targeted breeding efforts to enhance stress tolerance and ensure sustainable food production in the face of climate variability and environmental challenges.

19.17 Marker-Assisted Genotyping

Building linkage maps, investigating genetic diversity, and developing marker-assisted crop selection techniques are all made possible by molecular markers (Tomar 2010). Random amplified polymorphic DNA (RAPD) markers, as highlighted by Hilu (1994), have been instrumental in distinguishing between wild ancestors and domesticated species in the genus *Echinochloa*. Meanwhile, the amplified fragment length polymorphism (AFLP) marker system, noted by, surpasses RAPD by revealing a more intricate genetic landscape of *Echinochloa* species, because it can produce a greater number of alleles per primer (Danquah et al. 2002, Tabacchi et al. 2009). Insertion or deletion (InDel) markers targeting genes such as ITS, rbcL, and matK have significantly enriched DNA barcoding efforts across *Echinochloa* species, including *E. crus-galli*, *E. colona*, and *E. oryzicola* (Lee et al. 2014). These markers not only delineate species boundaries but also illuminate genetic relationships, contributing to agricultural sustainability and bolstering climate-resilient cropping. Twenty-two EST-SSR markers developed through comprehensive transcriptomic analyses have proven potent in studying the herbicide resistance in *E. crus-galli* variants (Li et al. 2013; Yang et al. 2013). Babu et al. (2017) underscored the utility of these markers, demonstrating their efficacy in genetic studies. The versatility of molecular markers extends to various genotyping applications, including phylogenetic relationships, cross-transferability, and comparative genomic mapping of crops (Cordeiro et al. 2001; Varshney et al. 2005; Xie et al. 2010; Yu et al. 2004). Moreover, the cross-genera transferability of foxtail millet markers suggests potential synteny with barnyard millet, opening avenues for leveraging genomic insights across related species (Kumari et al. 2013; Yadav et al. 2014). These insights underscore the ongoing challenges and opportunities in leveraging molecular tools to enhance barnyard millet's resilience and productivity under diverse environmental conditions.

19.18 Gene-Associated Adaptive Traits

Whole genome sequencing (WGS) holds immense promise for finding the essential genes and metabolic processes that determine agricultural traits with considerable economic importance in crops such as barnyard millet. RNA-Seq, complementing WGS, provides insights into gene expression patterns and transcript structures, vital for understanding non-coding regions (Kumar et al. 2023a, b). Despite its potential, barnyard millet research faces challenges due to its complex genome and limited funding, hindering the application of modern biotechnological approaches (Kumari et al. 2013; Pandey et al. 2013). To effectively utilize the genetic potential of barnyard millet, efforts must be focused on characterizing germplasm resources and locating trait-specific donors and QTLs. Comparative genomics with foxtail millet, which shares genomic similarities, offers a pathway for leveraging existing molecular insights (Lata et al. 2013; Muthamilarasan et al. 2014, Venkata Suresh et al. 2013). The chloroplast genomes of *Echinochloa* species, including barnyard millet, contain 111–131 genes, varying in protein-coding, tRNA, and rRNA genes. Genetic polymorphisms among species stem from differences in gene copy numbers, influencing genome size and phenotypic traits (Kumar et al. 2023a, b). Despite challenges, recent advancements in chloroplast genome sequencing provide a foundation for further genetic exploration (Suryawanshi et al. 2016; Ye et al. 2014).

The utilization of RFLP, SSRs, and SNPs has expedited QTL mapping and linkage map construction in barnyard millet (Fukao et al. 2004). Jayakodi et al. (2019b), exemplify this approach, demonstrating its efficacy in identifying markers linked to key agro-morphological traits. However, the absence of a comprehensive genetic linkage map for barnyard millet remains a notable gap. Pioneering studies by Ishikawa et al. (2013) and Renganathan et al. (2019) have contributed significantly to barnyard millet mapping efforts. They utilized BSA and EST-SSR markers to explore the genetic basis of pigmentation traits, highlighting specific SSR markers such as BMESSR 39. These insights are pivotal for advancing marker-assisted selection strategies in barnyard millet. Stress tolerance and genetic diversity in the core collection of barnyard millet have been investigated through notable research using SNP markers. About 10,816 SNPs were found using a genotype-by-sequencing method among 65 biotypes of *E. colona* and 8217 SNPs among 22 biotypes of *E. crus-galli*, revealing distinct clusters within each species. They highlighted these findings, underscoring the complexity posed by the hexaploid genome ($2n = 6\times = 54$) of barnyard millet (Wallace et al. 2015). *E. phyllopogon* has evolved resistance to cyhalofop-butyl through enhanced herbicide metabolism and reduced absorption, mechanisms crucial for weed management (Ruiz-Santaella et al. 2006). Elevated CO2 levels stimulate carbon assimilation in MR-*E. colona*, facilitating the synthesis of protective compounds against herbicidal inhibition of fatty acid synthesis (Refatti et al. 2019). Roma-Burgos et al. (2018) and Rouse et al. (2018) demonstrated that multiple resistant genotypes exhibit constitutive upregulation of stress tolerance genes and hypersensitivity to herbicides such as propanil, enhancing their competitive ability under stress. These adaptive mechanisms are further bolstered by temperature increases, favouring *Echinochloa* growth and potentially reducing herbicidal efficacy (Refatti et al. 2019). Regulatory elements play a crucial part in enhancing tolerance against abiotic stress (Zafar et al. 2020). While genes linked to stress tolerance have been identified in millets, functional annotations remain a critical gap (Renganathan et al. 2020). The genomic resources of *Echinochloa* species encompass a wide range of genes intricate in pathways crucial for flavonoid biosynthesis, C4 pathways, photosynthesis (including NADH-plastoquinone oxidoreductase, PS I and PS II components, and ATP synthase), sh4 related to non-shattering grain traits, micronutrient transport (e.g., nicotianamine synthase 1 and 2 and Fe_2C transport protein 2-like protein), waxy grain traits, DNA/RNA polymerases, resistance against flooding, and herbicides stress. These genetic insights pave the way for understanding the adaptive mechanisms and potential applications in crop improvement (Jayakodi et al. 2019b; Yamasue et al. 1989a, b).

19.19 Transcriptomic and Proteomic Approach

Successful genome editing or transformation efforts in barnyard millet are constrained by limited transformation systems and regeneration protocols. Gupta et al. in 2001 explored transformation methods in barnyard millet, highlighting the efficacy of the uq1 promoter in driving gene expression, unlike other promoters tested (CaMV35S, Act1, RbcS, and Ft). This pioneering work laid the foundation for stable transformation methods in millets, facilitating the regeneration of up to 2266 plants per 100 seed calli of *E. crus-galli* (Gupta et al. 2001). In addition to molecular markers, successful regeneration protocols have been documented for *E. crus-galli* (Gupta et al. 2001), *E. colona* (Samantaray et al. 2001), and *E. frumentacea* (Bobkov 2005; Rajak et al. 2018). RNA-Seq has revolutionized gene expression profiling, surpassing microarrays in efficiency and cost-effectiveness (Weber 2015). Molecular studies have identified glutathione S-transferases (GSTs) as pivotal in conferring saline-alkaline tolerance in *Echinochloa* species, shedding light on mechanisms previously obscured by limited genomic data in barnyard millet (Wu et al. 2022a, b, c). These advancements underscore the evolving landscape of molecular tools and regeneration techniques, crucial for enhancing barnyard millet's agronomic traits and stress resilience through biotechnological interventions.

Significant progress has been achieved in knowing complex characteristics linked to adaptations and invasiveness in *Echinochloa* species, such as homeobox gene function, photosynthetic dynamics, herbicide resistance, and flood responses. Deep insights have been obtained using RNA-sequencing technologies, which have tremendously aided this exploration (Gao et al. 2018; Guo et al. 2017; Nah et al. 2015; Pan et al. 2022; Rangani et al. 2022; Xu et al. 2015; Yang et al. 2013). After a thorough investigation, a total of 97,065 transcripts were found in the transcriptome of the domestic variety CO2 of *E. frumentacea*. This study identified critical genes associated with critical features such as drought tolerance and accumulation of zinc (Zn) and iron (Fe). These findings underscore the utility of transcriptomic approaches in elucidating genetic mechanisms underlying agronomically significant traits (Jayakodi et al. 2019a). The higher ploidy level and enormous genomic size of *Echinochloa* species have resulted in restricted genomic resources. However, recent studies have highlighted the importance of gene expression modifications in signalling components, which play a pivotal part in the adaptation of *Echinochloa*. The RLK-encoding genes family revealing conserved genomic regions with foxtail millet has been identified as major contributors to adaptive diversity, demonstrating a notable difference in environmental tolerance even among *E. crus-galli* species (Nah et al. 2015). They identified ERF homologs potentially involved in submergence tolerance, highlighting their role in signalling pathways during flooding as well as in response to various abiotic stresses. Recently, Wu et al. (2022a, b, c) undertook comprehensive genome sequencing efforts covering two weedy accessions (*E. oryzicola* and *E. crus-galli*) and a domesticated version of *E. Colona var. frumentacea* (PI463783). Their meticulous de novo assembly efforts resulted in complete genome establishment for respective species, providing a foundational resource for further genetic and genomic studies.

Indian barnyard millet exhibits potential survival under drought stress through mechanisms such as high photosynthetic capacity and rapid growth (Jayakodi et al. 2019b; Tadele 2016). Transcriptomic analyses have revealed that transcripts unique to Indian barnyard millet are abundant in biological processes associated with photosynthetic activity, such as electron transport during photosynthetic activity, light harvesting in photosystems I and II, and chlorophyll-binding. This was supported by qRT-PCR data, which revealed that under drought stress circumstances, Indian barnyard millet expressed these genes more than *E. crus-galli* (Jayakodi et al. 2019b). Comparative transcriptome data between *E. crus-galli* and *E. frumentacea* have highlighted significant differences in gene expression related to photosystem components such as apoprotein A1 of photosystem I P700 chlorophyll a, chlorophyll apoprotein CP43 of photosystem II, and D1 protein of PS II P680 reaction centre. *E. frumentacea* generally exhibits higher expression levels of these genes compared to *E. crus-galli*, reflecting potential differences in drought tolerance mechanisms (Yamasue et al. 1989a, b).

Noteworthy contributions from China have significantly advanced our understanding of *Echinochloa* species, particularly through genome sequencing efforts. Chu-Yu et al. (2020) employed PacBio technology to sequence the allohexaploid genome of *E. crus-galli*, revealing a comprehensive assembly that encompasses 103,853 protein-coding genes. This genomic resource has been pivotal in elucidating the genetic underpinnings of *E. crus-galli's* invasiveness and adaptation attributes in agricultural contexts within China (Chu-Yu et al. 2020). In addition to *E. crus-galli, E. glabrescens* has also been subject to detailed genomic scrutiny. The genome of *E. glabrescens* is about the size of the maize genome and is about 14 times larger than that of *Arabidopsis thaliana*. It features a mixed C4 metabolic pathway that uses both cytosolic PEPCK decarboxylases and chloroplast NADP-ME (Masumoto et al. 2010). This species' adaptation to waterlogged growth habitats includes unique metabolic pathways involving ammonium assimilation, facilitated by chloroplast-localized PEPC, which is suppressed in contrast to C_3-crop rice (Masumoto et al. 2010). Further comparative transcriptomic analyses involving rice, maize, and *E. glabrescens* have identified 1610 genes, including 117 transcription factors, crucial for C_4 photosynthesis (Covshoff et al. 2016; Masumoto et al. 2010). This research highlights 132 TFs under selection, potentially beneficial for rice-breeding programmes aimed at enhancing adaptability and productivity under diverse environmental conditions (Covshoff et al. 2016). These genomic insights underscore the diversity and adaptive potential of *Echinochloa* species, paving the way for future studies aimed at harnessing these genetic resources for crop improvement and environmental resilience.

This study marks a significant advancement in barnyard millet research by identifying candidate reference genes crucial for normalizing gene expression data under various stress conditions using RT-qPCR. α-TUB and UBC5 were identified as the most reliable drought stress reference genes, while GAPDH and α-TUB proved optimal for salinity conditions. GAPDH emerged as the top choice, followed by APRT, for low-temperature stress, whereas RP II and EF-1α were the most reliable reference genes during heat stress (Subramanian et al. 2020). The abiotic stress tolerance-related functional genomics and metabolomics research utilizing barnyard millet is further enhanced by these results, which guarantee the validity of gene expression studies. Molecular markers exhibiting above 55% cross-genera transferability, derived from both non-millet and millet species, have significantly augmented genomic resources in barnyard millet. This cross-species synteny analysis has revealed substantial similarities between barnyard millet and rice in traits such as spikelet characteristics, panicle length, yield-related traits, amylose content, leaf senescence, and resistance to pests such as blast (Babu et al. 2018). These insights offer promising avenues for enhancing stress-resilient characters in barnyard millet by leveraging information from related genera. Together, these studies underscore the potential of barnyard millet as a robust model for studying and improving stress resilience, supported by comprehensive genomic and transcriptomic resources and insights into cross-species trait conservation.

19.20 Future Prospects

The study represents a significant leap forward in barnyard millet research by identifying critical reference genes essential for normalizing gene expression data across various stress conditions using RT-qPCR (Subramanian et al. 2020). These findings not only ensure the reliability of gene expression studies but also bolster the utility of barnyard millet in advancing functional genomics and metabolomics research focused on abiotic stress tolerance. Furthermore, transcriptome sequencing of both cultivated and wild barnyard millet has enriched our comprehension of the genetic mechanisms governing the accumulation of micronutrients and tolerance toward drought stress (Jayakodi et al. 2019a). Molecular markers derived from different species, with more than 55% cross-genera transferability, have significantly expanded the genomic resources available for barnyard millet. Together, these studies underscore barnyard millet's potential as a robust model for investigating and enhancing stress resilience, supported by comprehensive genomic and transcriptomic resources, as well as insights into the conservation of traits across different species. This research holds promise for advancing agricultural sustainability and food security under challenging environmental conditions.

The transfer of a large chromosomal region (68 kb) comprising five stress-related protein-coding genes from Panicum to wild Hordeum species has been highlighted by recent, notable discoveries (Mahelka et al. 2021; Verhage 2021). These findings challenge previous assumptions about the rarity of horizontal gene transfer (HT) among plants, particularly from *Panicoideae* to *Triticeae* within grasses (Mahelka et al. 2021). It is plausible that future studies may uncover more horizontal transfer events along this unidirectional and biased pathway, potentially enhancing *Triticeae's* ability to adapt to environmental stressors (Wu et al. 2022a). The expanding collection of species with documented transcriptomes offers a robust platform to investigate the distribution and function of chloroplast-targeted PEPC across diverse phylogenetic contexts. Future research should delve into how rising temperatures impact these metabolic interactions and consequently influence weed responses to herbicides (Refatti et al. 2019). Additionally, NF-Y transcription gene families, well-studied in model plants such as rice, Arabidopsis, and foxtail millet (Feng et al. 2015), can hold promise for genetic enhancement programmes in small millets such as barnyard millet. A crucial chance for additional research in this field is the availability of completely annotated and draft genome sequences for a few minor millets (Rani et al. 2023). These advancements underscore the potential for leveraging genomic insights to enhance agricultural resilience and sustainability across diverse millet species.

References

Abe T, Fukami M, Ogasawara M (2011) Effect of hymexazole (3-hydroxy-5-methylisoxazole) on cadmium stress and accumulation in Japanese millet (*Echinochloa frumentacea* link). J Pestic Sci 36(1):48–52

Abogadallah GM (2010) Insights into the significance of antioxidative defense under salt stress. Plant Signal Behav 5(4):369–374

Abogadallah GM, Quick WP (2009) Vegetative salt tolerance of barnyard grass mutants selected for salt tolerant germination. Acta Physiol Plant 31:815–824

Abogadallah GM, Serag MM, El-Katouny TM, Quick WP (2010a) Salt tolerance at germination and vegetative growth involves different mechanisms in barnyard grass (*Echinochloa crusgalli L.)* mutants. Plant Growth Regul 60:1–12

Abogadallah GM, Serag MM, Quick WP (2010b) Fine and coarse regulation of reactive oxygen species in the salt tolerant mutants of barnyard grass and their wild-type parents under salt stress. Physiol Plant 138(1):60–73

Acosta-Motos JR, Diaz-Vivancos P, Álvarez S, Fernández-García N, Sanchez-Blanco MJ, Hernández JA (2015) Physiological and biochemical mechanisms of the ornamental *Eugenia myrtifolia* L. plants for coping with NaCl stress and recovery. Planta 242:829–846

Ado MN, Guero Y, Michot D, Soubeiga B, Kiesse TS, Walter C (2016) Phytodesalinization of irrigated saline Vertisols in The Niger Valley by *Echinochloa stagnina*. Agric Water Manag 177:229–240

Ado MN, Michot D, Guero Y, Hallaire V, Dan Lamso N, Dutin G, Walter C (2019) *Echinochloa stagnina* improves soil structure and phytodesalinization of irrigated saline sodic Vertisols. Plant Soil 434:413–424

Ado MN, Michot D, Guero Y, Thomas Z, Walter C (2022) Monitoring and modeling of saline-sodic vertisol reclamation by *Echinochloa stagnina*. Soil Systems 6(1):4

Agostinetto D, Oliveira C, Langaro AC, Silva JDGD, Barbieri GF (2022) Differences in biochemical, physiological and molecular response mechanisms of rice, weedy rice and barnyardgrass subjected to drought. Pesquisa Agropec Trop 52:e70487

Aguirrezabal LAN, Deleens E, Tardieu F (1994) Root elongation rate is accounted for by intercepted PPFD and source-sink relations in field and laboratory-grown sunflower. Plant Cell Environ 17(4):443–450

Ahmadi MS, Haderlie LC, Wicks GA (1980) Effect of growth stage and water stress on barnyardgrass (*Echinochloa crus-galli*) control and on glyphosate absorption and translocation. Weed Sci 28(3):277–282

Al Sherif EA (2007) *Echinochloa colona* (L.) link., a promising species to cultivate salt affected soils in arid lands. Am-Euro Asian J Agric Environ Sci 2(6):767–774

Alavi S, Mazumdar SD, Taylor JR (2019) Modern convenient sorghum and millet food, beverage and animal feed products, and their technologies. In: Sorghum and millets. Elsevier, Amsterdam, pp 293–329

Antony Ceasar S, Maharajan T (2022) The role of millets in attaining united Nation's sustainable developmental goals. Plants People Planet 4(4):345–349

Aoki D, Yamaguchi H (2008) Genetic relationship between *Echinochloa crus-galli* and *Echinochloa oryzicola* accessions inferred from internal transcribed spacer and chloroplast DNA sequences. Weed Biol Manag 8(4):233–242. https://doi.org/10.1111/j.1445-6664.2008.00303.x

Aoki D, Yamaguchi H (2009) Oryza sh4 gene homologue represents homoeologous genomic copies in polyploid *Echinochloa*. Weed Biol Manag 9(3):225–233. https://doi.org/10.1111/j.1445-6664.2009.00343.x

Apse MP, Aharon GS, Snedden WA, Blumwald E (1999) Salt tolerance conferred by overexpression of a vacuolar Na+/H+ antiport in Arabidopsis. Science 285:1256–1258

Arthi N, Rajagopal B, Geethanjali S, Nirmalakumari A, Senthil N (2019) Screening of barnyard millet (*Echinochloa frumentacea*) germplasm for salinity tolerance. Electron J Plant Breed 10(2):659–666

Ashok S, Senthil A, Sritharan N, Punitha S, Divya K, Ravikesavan R (2018) Yield potential of small millets under drought condition. Madras Agric J 105(7–9):370–372

Ashraf MHPJC, Harris PJ (2013) Photosynthesis under stressful environments: an overview. Photosynthetica 51:163–190

Babu BK, Joshi A, Sood S, Agrawal PK (2017) Identification of microsatellite markers for finger millet genomics application through cross transferability of rice genomic SSR markers. Indian J Genet Plant Breed 77(01):92–98

Babu BK, Sood S, Kumar D, Joshi A, Pattanayak A, Kant L, Upadhyaya HD (2018) Cross-genera transferability of rice and finger millet genomic SSRs to barnyard millet (*Echinochloa spp.*). 3 Biotech 8(2):1–10

Bajwa AA, Jabran K, Shahid M, Ali HH, Chauhan BS (2015) Eco-biology and management of *Echinochloa crus-galli*. Crop Prot 75:151–162. https://doi.org/10.1016/j.cropro.2015.06.001

Balaji GA, Sakthivel S, Chinnusamy C, Muthuramu S (2019) Effect of nutrient management and PPFM application on growth and productivity of barnyard millet under rain-fed ecosystem. J Pharma Phytochem 8(3):3031–3033

Barbiero L, Valles V, Régeard A, Cheverry C (2001) Residual alkalinity as tracer to estimate the changes induced by forage cultivation in a non-saline irrigated sodic soil. Agric Water Manag 50(3):229–241

Benitez-Alfonso Y, Soanes BK, Zimba S, Sinanaj B, German L, Sharma V et al (2023) Enhancing climate change resilience in agricultural crops. Curr Biol 33(23):R1246–R1261

Bernstein N, Silk WK, Läuchli A (1995) Growth and development of sorghum leaves under conditions of NaCl stress: possible role of some mineral elements in growth inhibition. Planta 196:699–705

Bobkov SV (2005) Long-term regeneration in callus culture of paisa (*Echinochloa frumentacea* link.). Int Sorghum Millets Newslett 46:120–122

Bradford KJ (1994) Water stress and the water relations of seed development: a critical review. Crop Sci 34(1):1–11

Brower R (1977) Root functioning. In: Landsberg JJ, Cutting CV (eds) Environmental effects on crop physiology. Academic Press, pp 229–245

Cappelli A, Cini E (2021) Challenges and opportunities in wheat flour, pasta, bread, and bakery product production chains: a systematic review of innovations and improvement strategies to increase sustainability, productivity, and product quality. Sustain For 13(5):2608. https://doi.org/10.3390/su13052608

Cappelli A, Lupori L, Cini E (2021) Baking technology: a systematic review of machines and plants and their effect on final products, including improvement strategies. Trends Food Sci Technol 115:275–284. https://doi.org/10.1016/j.tifs.2021.06.048

Charest C, Potvin C (1993) Maternally-induced modification of progeny phenotypes in the C 4 weed *Echinochloa crus-galli*: an analysis of seed constituents and performance. Oecologia 93:383–388

Chauhan BS, Johnson DE (2009) Seed germination ecology of junglerice (*Echinochloa colona*): a major weed of rice. Weed Sci 57(3):235–240

Chauhan BS, Johnson DE (2010) Growth and reproduction of jungle rice (*Echinochloa colona*) in response to water stress. Weed Sci 58(2):132–135

Chauhan BS, Abugho SB, Amas JC, Gregorio GB (2013) Effect of salinity on growth of barnyardgrass (*Echinochloa crus-galli*), horse purslane (*Trianthema portulacastrum*), jungle rice (*Echinochloa colona*), and rice. Weed Sci 61(2):244–248

Chivenge P, Mabhaudhi T, Modi AT, Mafongoya P (2015) The potential role of neglected and underutilised crop species as future crops under water scarce conditions in sub-Saharan Africa. Int J Environ Res Public Health 12(6):5685–5711

Cho Kwang-Min CK, Lee Sang-Bok LS, Back Nam-Hyun BN, Moon Yae-Jin MY, Yang Chang-Hyu YC, Shin Pyung SP, …, Jung Nam-Jin JN (2013a) Salt accumulation capacity comparison by the parts of the plant of the forage barnyard grass under the salt stress. Conference poster, pp 530–532

Cho Kwang-Min CK, Lee Sang-Bok LS, Back Nam-Hyun BN, Moon Yae-Jin MY, Yang Chang-Hyu YC, Shin Pyung SP, …, Jung Nam-Jin JN (2013b) Seed germination of the forage barnyard millet collected for seeding under the salt stress. Conference poster, pp 533–535.

Chu-Yu Y, Dongya W, Lingfeng M (2020) The genomes of the allohexaploid *Echinochloa crus-galli* and its progenitors provide insights into polyploidization-driven adaptation. Mol Plant 13(9):1298–1310

Clayton WD, Renvoize SA (2006) Genera Graminum: grasses of the world in Kew bulletin additional series XIII, Royal Botanical Gardens Kew. University of Chicago Press, Chicago, IL

Cordeiro GM, Casu R, McIntyre CL, Manners JM, Henry RJ (2001) Microsatellite markers from sugarcane (*Saccharum spp.*) ESTs cross transferable to erianthus and sorghum. Plant Sci 160(6):1115–1123. https://doi.org/10.1016/S0168-9452(01)00365-X

Covshoff S, Szecowka M, Hughes TE, Smith-Unna R, Kelly S, Bailey KJ et al (2016) C4 photosynthesis in the rice paddy: insights from the noxious weed *Echinochloa glabrescens*. Plant Physiol 170(1):57–73

Danquah EY, Hanley SJ, Brookes RC, Aldam C, Karp A (2002) Isolation and characterization of microsatellites in *Echinochloa* (L.) Beauv. spp. Mol Ecol Notes 2:54–56

Davis KF, Gephart JA, Emery KA, Leach AM, Galloway JN, D'Odorico P (2016) Meeting future food demand with current agricultural resources. Glob Environ Chang 39:125–132. https://doi.org/10.1016/j.gloenvcha.2016.05.004

De Wet JMJ, Rao KEP, Mengesha MH, Brink DE (1983) Domestication of Sawa millet (*Echinochloa colona*). Econ Bot 37:283–291

Denton DC (1987) Food crops for waterfowl. In: Wesley DE, Leitch WG (eds) Fireside waterfowler: fundamentals of duck and goose ecology. Stackpole Books, Harrisburg, p 352

Dhanalakshmi R, Subramanian A, Thirumurugan T, Elangovan M, Kalaimagal T (2019) Genetic variability and association studies in barnyard millet germplasm under sodic soil condition. Electron J Plant Breed 10(2):430–439

Dias-Martins AM, Pessanha KLF, Pacheco S, Rodrigues JAS, Carvalho CWP (2018) Potential use of pearl millet (Pennisetum glaucum (L.) R. Br.) in Brazil: food security, processing, health benefits and nutritional products. Food Res Int 109:175–186

Divya K, Kalarani MK, Jeyakumar P, Latha KR (2019) Influence of drought at anthesis on physiological traits of barnyard millet (*Echonochloa frumentacaea* L.) genotypes. J Pharma Phytochem 8(3):3209–3212

Ejiri M, Shiono K (2019) Prevention of radial oxygen loss is associated with exodermal suberin along adventitious roots of annual wild species of *Echinochloa*. Front Plant Sci 10:254

Eliasson L (1968) Dependence of root growth on photosynthesis in Populus tremula. Physiol Plant 21:806–810

El-Shabrawi H, Kumar B, Kaul T, Reddy MK, Singla-Pareek SL, Sopory SK (2010) Redox homeostasis, antioxidant defense, and methylglyoxal detoxification as markers for salt tolerance in Pokkali rice. Protoplasma 245(1–4):85–96

Ernst WHO (1990) Ecophysiology of plants in waterlogged and flooded environments. Aquat Bot 38:73–90. https://doi.org/10.1016/0304-3770(90)90099-7

FAO and ICRISAT (1996) The world sorghum and millet economies: facts, trends and outlook. http://www.fao.org/docrep/W1808E/w1808e00.htm#Contents

FAO, IFAD, UNICEF, WFP & WHO (2020) The state of food security and nutrition in the world 2020. Transforming food systems for affordable healthy diets (FAO, 2020). http://www.fao.org/3/ca9692en/online/ca9692en.html

Farrell W (2011) Plant guide for billion-dollar grass (*Echinochloa frumentacea*). USDA-Natural Resources Conservation Service

Feng ZJ, He GH, Zheng WJ et al (2015) Foxtail millet NF-Y families: genome-wide survey and evolution analyses identified two functional genes important in abiotic stresses. Front Plant Sci 6:1–19

Florin M, Quiroz D, Bruil J (2016) Revaluing traditional plants. Farm Matt 2:6–39

Fogliatto S, Patrucco L, Milan M, Vidotto F (2021) Sensitivity to salinity at the emergence and seedling stages of barnyardgrass (*Echinochloa crus-galli*), weedy rice (*Oryza sativa*), and rice with different tolerances to ALS-inhibiting herbicides. Weed Sci 69(1):39–51

Fox TC, Green BJ, Kennedy RA, Rumpho ME (1998) Changes in hexokinase activity in *Echinochloa phyllopogon* and *Echinochloa crus-pavonis* in response to abiotic stress. Plant Physiol 118(4):1403–1409

Foyer CH, Rasool B, Davey JW, Hancock RD (2016) Cross-tolerance to biotic and abiotic stresses in plants: a focus on resistance to aphid infestation. J Exp Bot 67:2025–2037

Francis N, Ravikesavan R, Iyanar K, Raveendran M, Chitdeshwari T, Senthil A (2021) Rapid screening of small millet varieties for seedling stage drought stress tolerance. Int J Plant Soil Sci 33(23):246–252

Fukao T, Kennedy RA, Yamasue Y, Rumpho ME (2003) Genetic and biochemical analysis of anaerobically-induced enzymes during seed germination of *Echinochloa crus-galli* varieties tolerant and intolerant of anoxia. J Exp Bot 54(386):1421–1429

Fukao T, Paterson AH, Hussey MA, Yamasue Y, Kennedy RA, Rumpho ME (2004) Construction of a comparative RFLP map of *Echinochloa crus-galli* toward QTL analysis of flooding tolerance. Theor Appl Genet 108:993–1001

Gahoonia TS, Nielsen NE, Joshi PA, Jahoor A (2001) A root hairless barley mutant for elucidating genetics of root hairs and phosphorus uptake. Plant Soil 235(2):211–219. https://doi.org/10.1023/A:1011993322286

Galamay TO, Kono Y, Yamauchi A, Shimizu M (1991) Comparative root anatomy of seminal and nodal root axes of summer cereals with special reference to the development of hypodermis and cortical sclerenchyma. Jpn J Crop Sci 60(1):184–190

Gao Y, Li J, Pan X, Liu D, Napier R, Dong L (2018) Quinclorac resistance induced by the suppression of the expression of 1-aminocyclopropane-1-carboxylic acid (ACC) synthase and ACC oxidase genes in *Echinochloa crus-galli* var. zelayensis. Pestic Biochem Physiol 146:25–32

Giacomelli L, Masi A, Ripoll DR, Lee MJ, van Wijk KJ (2007) *Arabidopsis thaliana* deficient in two chloroplast ascorbate peroxidases shows accelerated light-induced necrosis when levels of cellular ascorbate are low. Plant Mol Biol 65:627–644

Guo L, Qiu J, Ye C-Y, Jin G, Lingfeng M, Zhang H et al (2017) *Echinochloa crus-galli* genome analysis provides insight into its adaptation and invasiveness as a weed. Nat Commun 8:1031

Guo L, Qiu J, Li LF, Lu B, Olsen K, Fan L (2018) Genomic clues for crop–weed interactions and evolution. Trends Plant Sci 23(12):1102–1115

Gupta P, Raghuvanshi S, Tyagi AK (2001) Assessment of the efficiency of various gene promoters via biolistics in leaf and regenerating seed callus of millets, *Eleusine coracana* and *Echinochloa crusgalli*. Plant Biotechnol 18(4):275–282

Hadebe S, Modi A, Mabhaudhi T (2017) Drought tolerance and water use of cereal crops: a focus on sorghum as a food security crop in sub-Saharan Africa. J Agron Crop Sci 203:177–191

Hakim M, Juraimi A, Ismail MR, Hanafi M, Selamat A (2013) A survey on weed diversity in coastal rice fields of Sebarang Perak in peninsular Malaysia. J Anim Plant Sci 23:534–542

Hasegawa PM, Bressan RA, Zhu JK, Bohnert HJ (2000) Plant cellular and molecular responses to high salinity. Annu Rev Plant Biol 51(1):463–499

He YL, Liu YL, Chen Q, Bian AH (2002) Thermotorerance related to antioxidation induced by salicylic acid and heat hardening in tall fescue seedlings. J Plant Physiol Mol Biol 28(2):89–95
Helalia AM, El-Amir S, Abou-Zeid ST, Zaghloul KF (1992) Bio-reclamation of saline-sodic soil by Amshot grass in northern Egypt. Soil Tillage Res 22(1–2):109–115
Hilu KW (1994) Evidence from RAPD markers in the evolution of Echinochloa millets (Poaceae). Plant Syst Evol 189:247–257
Hoshino T, Nakamura T, Seimiya Y, Kamada T, Ishikawa G, Ogasawara A (2010) Production of a fully waxy line and analysis of waxy genes in the allohexaploid crop, Japanese barnyard millet. Plant Breed 129:349–355
Ishikawa G, Seimiya Y, Saito M, Nakamura T, Hoshino T (2013) Molecular characterization of spontaneous and induced mutations in the three homoeologous waxy genes of Japanese barnyard millet [*Echinochloa esculenta* (A. Braun) H. Scholz]. Mol Breed 31:69–78
Islam F, Xie Y, Farooq MA, Wang J, Yang C, Gill RA et al (2018) Salinity reduces 2, 4-D efficacy in *Echinochloa crusgalli* by affecting redox balance, nutrient acquisition, and hormonal regulation. Protoplasma 255:785–802
Jackson MB, Fenning TM, Drew MC, Saker LR (1985) Stimulation of ethylene production and gas-space (aerenchyma) formation in adventitious roots of *Zea mays L.* by small partial pressures of oxygen. Planta 165:486–492. https://doi.org/10.1007/BF00398093
Jayakodi M, Madheswaran M, Adhimoolam K et al (2019a) Transcriptomes of Indian barnyard millet and barnyard grass reveal putative genes involved in drought adaptation and micronutrient accumulation. Acta Physiol Plant 41:66
Jayakodi M, Madheswaran M, Adhimoolam K, Perumal S, Manickam D, Kandasamy T, Yang TJ, Natesan S (2019b) Transcriptomes of Indian barnyard millet and barnyard grass reveal putative genes involved in drought adaptation and micronutrient accumulation. Acta Physiol Plant 41:66
Kaur H, Sharma S (2020) An overview of barnyard millet (*Echinochloa frumentacea*). J Pharma Phytochem 9(4):819–822
Kesh H, Devi S, Kumar N, Kumar A, Kumar A, Dhansu P, Sheoran P, Mann A (2022) Insights into physiological, biochemical and molecular responses in wheat under salt stress. In: Wheat–recent advances. Intech Open
Khanthavong P, Yabuta S, Malik AI, Hossain MA, Akagi I, Sakagami JI (2022) Combinational variation temperature and soil water response of stomata and biomass production in maize, millet, sorghum and rice. Plan Theory 11(8):1039
Khedr AH, Serag M, Alhalak O, Shaaban H (2021) Response to salt stress of two wetland grasses of forage potentialities. Rev Bras Bot 44:345–358
Kim YH, Shim IS, Kobayashi K, Usui K (1999) Relationship between Na content or K/Na ratio in shoots and salt tolerance in several gramineous plants. J. Weed Sci. Technol. 44:293–299
Kim YH, Shim IS, Kobayashi K, Usui K (2000) Accumulation of amino acids glycinebetaine by NaCl treatment and its relation to salt tolerance in three gramineous plants. J. Weed Sci. Technol. 45:96–103
Kim YH, Arihara J, Nakayama T, Nakayama N, Shimada S et al (2004) Antioxidative responses and their relation to salt tolerance in *Echinochloa oryzicola* Vasing and Setaria virdis (L.) Beauv. Plant Growth Regul 44(1):87–92
Kobayashi H, Masaoka Y, Sato S (2005) Effects of excess magnesium on the growth and mineral content of rice and *Echinochloa*. Plant Produc Sci 8(1):38–43
Kono Y, Yamauchi A, Kawamura N, Tatshumi J, Nonoyama T, Inagaki N (1987) Interspecific differences of the capacities of waterlogging and drought tolerances among summer cereals. Jpn J Crop Sci 56(1):115–129. https://doi.org/10.1626/jcs.56.115
Kreuzwieser J, Papadopoulou E, Rennenberg H (2004) Interaction of flooding with carbon metabolism of forest trees. Plant Biol 6:299–306. https://doi.org/10.1055/s-2004-817882
Kumar B, Singh AK, Bahuguna RN, Pareek A, Singla-Pareek SL (2023a) Orphan crops: a genetic treasure trove for hunting stress tolerance genes. Food Energy Secur 12(2):e436

Kumar A, Sheoran P, Mann A, Yadav D, Kumar A, Devi S, Kumar N, Dhansu P, Sharma DK (2023b) Deciphering trait associated morpho-physiological responses in pearlmillet hybrids and inbred lines under salt stress. Front Plant Sci 14:1121805

Kumari K, Muthamilarasan M, Misra G, Gupta S, Subramanian A et al (2013) Development of eSSR-markers in Setaria italica and their applicability in studying genetic diversity, cross-transferability and comparative mapping in millet and non-millet species. PLoS One 8(6):e67742. https://doi.org/10.1371/journal.pone.0067742

Laanbroek HJ (1990) Bacterial cycling of minerals that affect plant growth in waterlogged soils: a review. Aquat Bot 38:109–125. https://doi.org/10.1016/0304-3770(90)90101-P

Lamers LPM, Tomassen HBM, Roelofs JGM (1998) Sulfate-induced entrophication and phytotoxicity in freshwater wetlands. Environ Sci Technol 32:199–205. https://doi.org/10.1021/es970362f

Lata C, Gupta S, Prasad M (2013) Foxtail millet: a model crop for genetic and genomic studies in bioenergy grasses. Crit Rev Biotechnol 33(3):328–343

Lee J, Park K, Lee I, Kim C, Kown O, Park T (2014) Simple sequence repeat analysis of genetic diversity among acetyl-CoA carboxylase inhibit or resistant and susceptible Echinochloa crus-galli and E. oryzicola populations in Korea. Weed Res 55:90–100

Li X, Siddique KH (2020) Future smart food: harnessing the potential of neglected and underutilized species for zero hunger. Matern Child Nutr 16:e13008. https://doi.org/10.1111/mcn.13008

Li G, Wu SG, Yu RX, Cang T, Chen LP, Zhao XP et al (2013) Identification and expression pattern of a glutathione S-transferase in Echinochloa crus-galli. Weed Res 53:314–321

Lu AQ, Zhang FJ, Xu X, Wang XQ, Yao S (2021) Effects of salt stress on growth and physiological characteristics of *Echinochloa frumentacea* seedlings. Acta Pratacul Sin 305:84–93

Madhan Mohan M, Lakshmi Narayanan S, Ibrahim SM (2000) Chlorophyll stability index (CSI): its impact on salt tolerance in rice. Int Rice Res Notes (Philippines) 25(2):38–39

Mahelka V, Krak K, Fehrer J et al (2021) A Panicum-derived chromosomal segment captured by Hordeum a few million years ago preserves a set of stress-related genes. Plant J 105:1141–1164

Majid MA, Hamid MA, Mannujan (1989) Importance, genetic resources and breeding of small millets in Bangladesh. In: Seetharam A, Riley KW, Harinarayana G (eds) Small millets in global agriculture. Oxford & IBH Publishing Co, New Delhi, pp 71–76

Makarana G, Yadav RK, Sheoran P, Kumar R, Kumar A, Ram H, Yadav MR, Kumar D, Kumar S, Minkina T, Movsesyan HS (2023) Effects of sodic water irrigation and neutralizing amendments on physiological, biochemical, and nutritional quality traits of fodder sorghum. Agronomy 13(4):1128

Mandelbaum CI, Barbeau WE, Hilu KW (1995) Protein, calcium, and iron content of wild and cultivated species of *Echinochloa*. Plant Foods Hum Nutr 47:101–108

Mann A, Kumar A, Sanwal SK, Sharma PC (2020) Sustainable production of pulses under saline lands in India. In: Legume crops-prospects, production and uses. IntechOpen. https://doi.org/10.5772/intechopen.90304

Mann A, Lata C, Kumar N, Kumar A, Kumar A, Sheoran P (2023) Halophytes as new model plant species for salt tolerance strategies. Front Plant Sci 14:1137211. https://doi.org/10.3389/fpls.2023.1137211

Manzur ME, Grimoldi AA, Insausti P, Striker GG (2015) Radial oxygen loss and physical barriers in relation to root tissue age in species with different types of aerenchyma. Funct Plant Biol 42:9–17. https://doi.org/10.1071/FP14078

Marchezan MG, Avila LA, Schaedler CE, Amarante L, Scherner A, Schreiber F (2017) Biochemical alterations of weeds in response to stress caused by herbicides and total plant submersion. Planta Daninha 35:e017168751

Masumoto C, Miyazawa S, Ohkawa H, Fukuda T, Taniguchi Y, Murayama S, Kusano M, Saito K, Fukayama H, Miyao M (2010) Phosphoenolpyruvate carboxylase intrinsically located in the chloroplast of rice plays a crucial role in ammonium assimilation. Proc Natl Acad Sci USA 107:5226–5231

Matsuura A, Inanaga S, Sugimoto Y (2000) Growth of roots emerged from excised phytomers of three gramineous species under a low osmotic potential. Plant Produc Sci 3(1):55–60

Matsuura A, An P, Murata K, Inanaga K (2016) Effect of pre- and post-heading waterlogging on growth and grain yield of four millets. Plant Produc Sci 19(3):348–359. https://doi.org/10.1080/1343943X.2016.1146907

Matsuura A, Kato Y, Suzuki T, Murata K, An P (2022) Hypoxia tolerance of four millet species is attributable to constitutive aerenchyma formation and root hair development of adventitious roots. Plant Produc Sci 25(2):157–171

McDonald MP, Galwey NW, Colmer TD (2001) Waterlogging tolerance in the tribe Triticeae: the adventitious roots of *Critesion marinum* have a relatively high porosity and a barrier to radial oxygen loss. Plant Cell Environ 24:585–596. https://doi.org/10.1046/j.0016-8025.2001.00707.x

McDonald MP, Galwey NW, Colmer TD (2002) Similarity and diversity in adventitious root anatomy as related to root aeration among a range of wetland and dryland grass species. Plant Cell Environ 25:441–451. https://doi.org/10.1046/j.0016-8025.2001.00817.x

Mehta H, Tyagi PC, Mohapatra KP (2005) Genetic diversity in barnyard millet (*Echinochloa frumentacea Roxb.*). Indian J Genet Plant Breed 65:293–295

Michael PW (1983) Taxonomy and distribution of *Echinochloa species* with special reference to their occurrence as weeds of rice. In: Weed control in rice. International Rice Research Institute, Laguna, pp 291–306

Miškelytė D, Dikšaitytė A, Žaltauskaitė J, Januškaitienė I, Kacienė G, Sujetovienė G, …, Miliauskienė J (2017) The response of antioxidative defence system of spring barley (*Hordeum vulgare L.*) and barnyard grass (*Echinochloa crus-galli L*) under elevated CO2 and temperature. In CEST 2017: 15th international conference on environmental science and technology, 31 August-2 September 2017, Rhodes, Greece: proceedings/editor DF Lekkas. Rhodes, Greece: University of the Aegean, 2017. Rhodes, Greece: University of the Aegean, 2017

Mouko GRD, Mbemba KM, Mabika ABM, Ouamba JM (2024) Wastewater Phytopurification monitoring with Congolese planted filters, republic of Congo. Asian J Chem Sci 14(2):97–112

Murashige T, Skoog F (1962) A revised medium for rapid growth and bioassays with tobacco tissue cultures. Physiol Plant 15:473–497

Muthamilarasan M, Prasad M (2020) Small millets for enduring food security amidst pandemics. Trends Plant Sci 26:33–40

Muthamilarasan M, Prasad M (2021) Small millets for enduring food security amidst pandemics. Trends Plant Sci 26(1):33–40

Muthamilarasan M, Venkata Suresh B, Pandey G, Kumari K, Parida SK, Prasad M (2014) Development of 5123 intron-length polymorphic markers for large-scale genotyping applications in foxtail millet. DNA Res 21:41–52

Muthamilarasan M, Singh NK, Prasad M (2019) Multi-omics approaches for strategic improvement of stress tolerance in underutilized crop species: a climate change perspective. Adv Genet 103:1–38

Nah G, Im JH, Kim JW, Park HR, Yook MJ, Yang TJ et al (2015) Uncovering the differential molecular basis of adaptive diversity in three *Echinochloa* leaf transcriptomes. PLoS One 10(8):e0134419

National Institute of Food and Agriculture (2021) Retrieved from March 13, 2021. https://nifa.usda.gov/program/afri-sas

Nelson GC, Rosegrant MW, Koo J, Robertson R, Sulser T, Zhu T et al (2009) Climate change: impact on agriculture and costs of adaptation, vol 21. The International Food Policy Research Institute

Nguyen HTT, Shim IS, Kobayashi K, Usui K (2005) Effects of salt stress on ion accumulation and antioxidative enzyme activities of *Oryza sativa L.* and *Echinochloa oryzicola* Vasing. Weed Biol Manag 5(1):1–7

Norsworthy JK et al (2014) Herbicidal activity on acetolactate synthase-resistant barnyard grass (*Echinochloa crus-galli*) in Arkansas, USA. Weed Biol Manag 14:50–58

Numan M, Serba DD, Ligaba-Osena A (2021) Alternative strategies for multi-stress tolerance and yield improvement in millets. Genes 12(5):739

Nye PH (1966) The effect of the nutrient intensity and buffering power of a soil, and the absorbing power, size and root hairs of a root, on nutrient absorption by diffusion. Plant Soil 25(1):81–105. https://doi.org/10.1007/BF01347964

OECD-FAO (2011) Agricultural Outlook 2011. Retrieved online, April2012: http://www.keepeek.com/Digital-Asset-Management/oecd/agriculture-and-food/oecd-faoagricultural-outlook-2011_agr_outlook-2011-en

Opeña JL, Chauhan BS, Baltazar AM (2014) Seed germination ecology of *Echinochloa glabrescens* and its implication for Management in Rice (*Oryza sativa* L.) (J. Ali, Ed.). PLoS One 9:e92261

Padulosi S, Mal B, Bala Ravi S, Gowda J, Gowda KTK, Shanthakumar G, Yenagi N, Dutta M (2009) Food security and climate change: role of plant genetic resources of minor millets. Indian J Plant Genet Resour 22:1–16

Pan L, Guo Q, Wang J et al (2022) CYP81A68 confers metabolic resistance to ALS and ACCase-inhibiting herbicides and its epigenetic regulation in *Echinochloa crus-galli*. J Hazard Mater 428:128225

Pandey G, Mishra G, Kumari K, Gupta S, Parida SK, Chattopadhyay D, Prasad M (2013) Genome-wide development and use of microsatellite markers for large-scale genotyping applications in foxtail millet [*Setaria italica* (L.)]. DNA Res 20:197–207

Pandey P, Ramegowda V, Senthil-Kumar M (2015) Shared and unique responses of plants to multiple individual stresses and stress combinations: physiological and molecular mechanisms. Front Plant Sci 6:723

Park SJ, Yu HJ, Yook MJ, Kim DS (2016) Adaptive diversity of *Echinochloa species* to osmotic stress. Weed Turfgrass Sci 5(4):181–186

Parkash K, Ravimycin T, Santhoshkumar M (2018) Germination and growth of *Echinochloa Frumentacea L.* under various levels of NaCl stress. J Plant Stress Physiol 4:30–32

Parul KA, Chourasia S, Kumar R, Devi G, Kumar A, Mann A, Sheoran P, Sanwal SK (2023) Physiological, biochemical, and molecular responses to salt stress and seed priming approach to enhance salt tolerance in bread wheat. In: Salinity and drought tolerance in plants: physiological perspectives. Springer Nature Singapore, Singapore, pp 61–86

Passioura JB (1996) Drought and drought tolerance. Plant Growth Regul 20:79–83

Plaza-Wüthrich S, Blösch R, Rindisbacher A, Cannarozzi G, Tadele Z (2016) Gibberellin deficiency confers both lodging and drought tolerance in small cereals. Front Plant Sci 7:643

Ponnamperuma FN (1984) Effects of flooding on soils. In: Kozlowski TT (ed) Flooding and plant growth. Academic Press, New York, pp 9–45. https://doi.org/10.1016/B978-0-12-424120-6.50007-9

Poorniammal R, Prabhu S, Senthilkumar M, Anandhi K (2020) Effect of phyllosphere application of Methylobacterium on growth and yield of barnyard millet (*Echinochloa frumentacea* var. COKV 2). Int J Curr Microbiol App Sci 9(4):1860–1866

Poorniammal R et al (2021) Methylobacterium: a foliar bioinoculant for barnyard millet. Biot Res Today 3(3):197–199

Potvin C, Simon JP, Strain BR (1986) Effect of low temperature on the photosynthetic metabolism of the C 4 grass *Echinochloa crus-galli*. Oecologia 69:499–506

Prasad M (2020) Omics of neglected and underutilized crop species: one small step for NUCS, one giant leap for addressing global hunger. Nucleus 63:213–215

Rahman M, Ungar IA (1990) The effect of salinity on seed germination and seedling growth of *Echinochloa crus-galli*. Ohio J Sci 90:13–15

Rajak K, Sprae S, Rathour S, Kumari R, Nikeeta T (2018) To study the indirect plant regeneration of two cultivars in barnyard millet using different combination of plant growth regulators and compared between superior genotype for in-vitro culture. Int J Curr Microbiol Appl Sci 51:2055–2061

Rangani G, Rouse CE, Saski C et al (2022) High resistance to quinclorac in multiple-resistant *Echinochloa colona* associated with elevated stress tolerance gene expression and enriched xenobiotic detoxification pathway. Genes 13(3):515

Rani V, Joshi DC, Joshi P, Singh R, Yadav D (2023) "Millet models" for harnessing nuclear factor-Y transcription factors to engineer stress tolerance in plants: current knowledge and emerging paradigms. Planta 258(2):29

Rao AN, Johnson DE, Sivaprasad B, Ladha JK, Mortimer AM (2007) Weed management in direct-seeded rice. Adv Agron 93:153–255. https://doi.org/10.1016/s0065-2113(06)93004-1

Rawson HM, Begg JE, Woodward RG (1977) The effect of atmospheric humidity on photosynthesis, transpiration and water use efficiency of leaves of several plant species. Planta 134:5–10

Refatti JP, De Avila LA, Camargo ER, Ziska LH, Oliveira C, Salas-Perez R et al (2019) High [CO2] and temperature increase resistance to cyhalofop-butyl in multiple-resistant *Echinochloa colona*. Front Plant Sci 10:529

Renganathan VG, Vanniarajan C, Ramalingam J (2019) Genetic analysis and identification of molecular markers linked to the anthocyanin pigmentation in barnyard millet [*Echinochloa frumentacea Roxb* (Link)]. In: Proceedings of the neglected and underutilized crop species for food, nutrition, energy and environment. NIPGR, New Delhi, p 43

Renganathan VG, Vanniarajan C, Karthikeyan A, Ramalingam J (2020) Barnyard millet for food and nutritional security: current status and future research direction. Front Genet 11:500

Revathy K, Shanker SR (2023) Effect of cadmium stress on seed germination and seedling mor pho-physiological growth parameters of barnyard millet (*Echinochloa frumentacea link*). Plant Sci Today 10(1):252–260

Revoredo-Giha C, Toma L, Akaichi F, Dawson I (2022) Exploring the effects of increasing underutilized crops on consumers' diets: the case of millet in Uganda. Agric Food Econ 10:1

Rintamaki E, Kettunen R, Aro EM (1996) Differential D1 dephosphorylation in functional and photodamaged photosystem II centers. J Biol Chem 271:14870–14875

Roma-Burgos N, Rouse CE, Saski, CA, Noorai RE, Shankar V, Lawton-Rauh AL (2018) Concerted enrichment of gene networks in multiple resistant *Echinochloa colona*. In: Proceedings of the annual meeting. Weed Science Society of America, Arlington, p 387

Romanowska E, Buczyńska A, Wasilewska W, Krupnik T, Drożak A, Rogowski P et al (2017) Differences in photosynthetic responses of NADP-ME type C4 species to high light. Planta 245:641–657

Rosero A, Granda L, Berdugo-Cely JA, Šamajová O, Šamaj J, Cerkal R (2020) A dual strategy of breeding for drought tolerance and introducing drought-tolerant, underutilized crops into production systems to enhance their resilience to water deficiency. Plan Theory 9:1263

Rouse CE, Roma-Burgos N, Saski CA, Noorai RE, Shankar V (2018) Co-evolution of abiotic stress adaptation and quinclorac resistance in *Echinochloa colona*. In: Proceedings of the annual meeting. Weed Science Society of America, Arlington, p 386

Royo C, Voltas J, Romagosa I (1999) Response of four spring wheat cultivars to drought stress. Crop Sci 36:982–986

Ruiz-Santaella JP, De Prado R, Wagner J, Fischer AJ, Gerhards R (2006) Resistance mechanisms to cyhalofop-butyl in a biotype of *Echinochloa phyllopogon* (Stapf) Koss from California. J Plant Dis Prot 20:95–100

Sadeghloo A, Asghari J, Ghaderi-Far F (2013) Seed germination and seedling emergence of velvetleaf (*Abutilon theophrasti*) and Barnyardgrass (*Echinochloa crus-galli*). Planta Daninha 31:259–266

Samantaray S, Rout GR, Das P (2001) Induction, selection and characterization of Cr and Ni-tolerant cell lines of *Echinochloa colona* (L.) link in vitro. J Plant Physiol 158(10):1281–1290

Sampath TV, Razvi SM, Singh D, Bondale KV (1989) Small millets in Indian agriculture. In: Small millets in global agriculture, pp 33–44

Sanyal R et al (2021) Small millets (Nutri cereals): food for the future. Biot Res Today 3(9):793–796

Sawada H, Shim IS, Usui K, Kobayashi K, Fujihara S (2008) Adaptive mechanism of *Echinochloa crus-galli* Beauv. *Var. formosensis* Ohwi under salt stress: effect of salicylic acid on salt sensitivity. Plant Sci 174(6):583–589

Senthil A, Ashok S, Sritharan N, Punitha S, Divya K, Ravikesavan R (2018) Physiological efficiency of small millets under drought condition. Madras Agric J 105

Serba DD, Yadav RS (2016) Genomic tools in pearl millet breeding for drought tolerance: status and prospects. Front Plant Sci 7:1724

Serba DD, Yadav RS, Varshney RK, Gupta SK, Mahalingam G, Srivastava RK et al (2020) Genomic designing of pearl millet: a resilient crop for arid and semi-arid environments. In: Genomic designing of climate-smart cereal crops, pp 221–286

Serra F, Fogliatto S, Vidotto F (2018) Effect of salinity on *Echinochloa crus-galli* germination as affected by herbicide resistance. Ital J Agron 13(3):221–228

Shayghan S, Sedghi S (2013) Effects of polyethylene glicol and NaCl on seed germination do forage. J Biol Today's World 2:275–283

Singh R, Singh GS (2017) Traditional agriculture: a climate-smart approach for sustainable food production. Energy Ecol Environ 2:296–316

Singh D, Dubey A, Verma AK (2021) Assessment of biomolecules and enzymes during drought stress in *Echinochloa frumentacea* grown in Himalaya. Int J Comp Sociol 9(2):29–52

Solís-Domínguez FA, González-Chávez MC, Carrillo-González R, Rodríguez-Vázquez R (2007) Accumulation and localization of cadmium in *Echinochloa polystachya* grown within a hydroponic system. J Hazard Mater 141(3):630–636

Sood S, Khulbe R, Saini N, Gupta A, Agrawal P (2014) Interspecific hybrid between *Echinochloa esculenta* (Japanese barnyard millet) and *E. Frumentacea* (Indian barnyard millet) – a new avenue for genetic enhancement of barnyard millet. Electron J Plant Breed 5(2):248–253

Sood S, Khulbe RK, Gupta AK, Agrawal PK, Upadhyaya HD, Bhatt JC (2015) Barnyard millet–a potential food and feed crop of future. Plant Breed 134:135–147. https://doi.org/10.1111/pbr.12243

Stepien P, Johnson GN (2009) Contrasting responses of photosynthesis to salt stress in the glycophyte Arabidopsis and the halophyte Thellungiella: role of the plastid terminal oxidase as an alternative electron sink. Plant Physiol 149:1154–1165

Subramanian A, Nirmal Raj R, Elangovan M (2019) Genetic variability and multivariate analysis in sorghum under sodic soil conditions. Electron J Plant Breed 10(4):1405–1414

Subramanian A, Raj RN, Jeyaprakash P (2020) In vitro and in vivo screening of barnyard millet (*Echinochloa frumentacea* (roxb.) link) germplasm for salinity tolerance. Plant Arch 20:7389–7397

Suryawanshi V, Talke IN, Weber M, Eils R, Brors B, Clemens S et al (2016) Between-species differences in gene copy number are enriched among functions critical for adaptive evolution in *Arabidopsis halleri*. BMC Genomics 17:1034

Tabacchi M, Mantegazza R, Spada A, Ferrero A (2009) Morphological traits and molecular markers for classification of *Echinochloa species* from Italian rice fields. Weed Sci 54:1086–1093

Tadele Z (2016) Drought adaptation in millets. In: Shanker AK, Shanker C (eds) Abiotic and biotic stress in plants – recent advances and future perspectives. In Tech, Rijeka, pp 639–662

Tanesaka E, Ohno T, Yamaguchi H (2010) Species diversity of the genus *Echinochloa* (*Poaceae*), native to eastern Australia: a focus on their habitat and the threat of exotic species. J Crop Res 55:13–17. https://doi.org/10.18964/jcr.55.0_13

Taylor JR (2019) Sorghum and millets: taxonomy, history, distribution, and production. In: Sorghum and millets. AACC International Press, pp 1–21

Tomar RS (2010) Molecular markers and plant biotechnology. New India Publishing

Trivedi AK, Arya L, Verma SK, Tyagi RK, Hemantaranjan A (2017) Evaluation of barnyard millet diversity in central Himalayan region for environmental stress tolerance. J Agric Sci 155(10):1497–1507

United Nations, General Assembly (2015, October 21) Transforming our world: the 2030 agenda for sustainable development. A/RES/70/1. Sustainable development knowledge platform. Available from https://sustainabledevelopment.un.org/post2015/transformingourworld

USDA NCRS- United State Department of Agriculture Natural Resources Conservation Service. USDA NRCS Plant Materials Program (2002) Plant materials program. http://plant-materials.nrcs.usda.gov

Varshney RK, Sigmund R, Borner A, Korzun V, Stein N et al (2005) Interspecific transferability and comparative mapping of barley ESTSSR markers in wheat, rye and rice. Plant Sci 168:195–202. https://doi.org/10.1016/j.plantsci.2004.08.001

Venkata Suresh B, Muthamilarasan M, Misra G, Prasad M (2013) FmMDb: a versatile database of foxtail millet markers for millets and bioenergy grasses research. PLoS One 8:e71418

Verhage L (2021) A hitchhiker's guide to foreign genomes. Plant J 105:1139–1140

Vetriventhan M, Azevedo VC, Upadhyaya HD, Nirmalakumari A, Kane-Potaka J, Anitha S et al (2020) Genetic and genomic resources, and breeding for accelerating improvement of small millets: current status and future interventions. Nucleus 63:217–239

Visser EJW, Colmer TD, Blom CWPM, Voesenek LACJ (2000) Changes in growth, porosity, and radial oxygen loss from adventitious roots of selected mono- and dicotyledonous wetland species with contrasting types of aerenchyma. Plant Cell Environ 23:1237–1245. https://doi.org/10.1046/j.1365-3040.2000.00628.x

Wallace J, Upadhyaya H, Vetriventhan M, Buckler E, Tom Hash C, Ramu P (2015) The genetic makeup of a global barnyard millet germplasm collection. Plant Genome 8:1–7

Warren R, Price J, Graham E, Forstenhaeusler N, VanDerWal J (2018) The projected effect on insects, vertebrates, and plants of limiting global warming to 1.5 C rather than 2 C. Science 360(6390):791–795

Wasilewska-Dębowska W, Galas M, Rogowski P, Trojanowska J, Urban A, Charnyak D et al (2024) The NDH complex can be involved in the regulation of phosphorylation of LHCII proteins and tolerance of *Echinochloa crus-galli* (L.) to high light intensity. Environ Exp Bot 220:105680

Weber A (2015) Discovering new biology through RNA-Seq. Plant Physiol 169:01081

Willett W, Rockström J, Loken B, Springmann M, Lang T, Vermeulen S et al (2019) Food in the Anthropocene: the EAT–lancet commission on healthy diets from sustainable food systems. Lancet 393(10170):447–492

Williams G, Vanniarajan C, Vetriventhan M, Thiageshwari S, Anandhi K, Rajagopal B (2019) Genetic variability for seedling stage salinity tolerance in barnyard millet [Echinochloa frumentaceae (Roxb.) link]. Electron J Plant Breed 10(2):552–558

Wilson C, Read JJ (2006) Effect of mixed-salt salinity on growth and ion relations of a barnyard grass species. J Plant Nutr 29:1741–1753

Wu LM, Fang Y, Yang HN, Bai LY (2019) Effects of drought-stress on seed germination and growth physiology of quinclorac-resistant *Echinochloa crusgalli*. PLoS One 14(4):e0214480

Wu D, Shen E, Jiang B et al (2022a) Genomic insights into the evolution of *Echinochloa species* as weed and orphan crop. Nat Commun 13:689

Wu L, Yang H, Li Z, Wang L, Peng Q (2022b) Effects of salinity-stress on seed germination and growth physiology of quinclorac-resistant *Echinochloa crus-galli* (L.) Beauv. Agronomy 12(5):1193

Wu D, Jiang B, Ye CY, Timko MP, Fan L (2022c) Horizontal transfer and evolution of the biosynthetic gene cluster for benzoxazinoids in plants. Plant Commun 3(3):100320

Xie WG, Zhang XQ, Cai HW, Liu W, Peng Y (2010) Genetic diversity analysis and transferability of cereal EST-SSR markers to orchardgrass (Dactylis glomerata L). Biochem Syst Ecol 4:740

Xu W, Di C, Zhou S, Liu J, Li L, Liu F et al (2015) Rice transcriptome analysis to identify possible herbicide quinclorac detoxification genes. Front Genet 6:306

Yabuno T (1966) Biosystematic study of the genus *Echinochloa*. Jpn J Bot 19:277–323

Yabuno T (1975) The classification and geographical distribution of the genus Echinochloa. J Weed Sci Technol 20(3):97–104. https://doi.org/10.3719/weed.20.97

Yabuno T (1984) A biosystematic study on *Echinochloa oryzoides* (Ard.) Fritsch. Cytologia 49(3):673–678

Yabuno T (1987) Japanese barnyard millet (*Echinochloa utilis*, Poaceae) in Japan. Econ Bot 41(4):484–493. https://doi.org/10.1007/BF02908141

Yadav CB, Muthamilarasan M, Pandey G, Khan Y, Prasad M (2014) Development of novel microRNA-based genetic markers in foxtail millet for genotyping applications in related grass species. Mol Breed 34:2219

Yamaguchi H, Utano AYA, Yasuda K, Yano A, Soejima A (2005) A molecular phylogeny of wild and cultivated Echinochloa in East Asia inferred from non-coding region sequences of trnT-L-F. Weed Biol Manag 5:210–218. https://doi.org/10.1111/j.1445-6664.2005.00185.x

Yamasue Y (2001) Strategy of *Echinochloa oryzicola* Vasing for survival in flooded rice. Weed Biol Manag 1:28–36. https://doi.org/10.1046/j.1445-6664.2001.00008.x

Yamasue Y, Asai Y, Ueki K, Kusanagi T (1989a) Anaerobic seed germination for the habitat segregation in Echinochloa weeds. Jpn J Breed 39:159–168

Yamasue Y, Nakamura A et al (1989b) Drought resistance for the habitat segregation *in Echinochloa* weeds. Jpn J Breed 39(3):337–343

Yang X, Yu X-Y, Li YF (2013) De novo assembly and characterization of the barnyard grass (*Echinochloa crus-galli*) transcriptome using next-generation pyrosequencing. PLoS One 8:e69168

Ye CY, Lin Z, Li G, Wang YY, Qiu J, Fu F, Zhang H, Chen L, Ye S, Song W et al (2014) *Echinochloa* chloroplast genomes: insights into the evolution and taxonomic identification of two weedy species. PLoS One 9:e113657

Yu Q, Powles SB (2014) Metabolism-based herbicide resistance and cross-resistance in crop weeds: a threat to herbicide sustainability and global crop production. Plant Physiol 166:1106–1118

Yu JK, La Rota M, Kantety RV, Sorrells ME (2004) EST derived SSR markers for comparative mapping in wheat and rice. Mol Gen Genomics 271:742–751

Yuichiro M, Shinya U, Yamaguchi H (1999) Identification of polyploid groups in the genus Echinochloa by isozyme analysis. J Weed Sci Technol 44:205–217. https://doi.org/10.3719/weed.44.205

Yun S, Lee SY, Lim HJ, Shim MB, Sung JK, Kim TW (2004) Influence of salinity treatment on seed germination and polyamine synthesis in barnyard grass (*Echinochloa hispidula*). Korean J Soil Sci Fert 37(1):19–24

Yupeng JING, Chunye LIN, Peiyi ZHAO, Xiuping LI, Qiang ZHAO, Yujie LIU et al (2020) Salinized soil improvement by planting 5 plant species in Hetao irrigation area. Soils Crops 9(2):114–125

Zafar SA et al (2020) Engineering abiotic stress tolerance via CRISPR/Cas-mediated genome editing. J Exp Bot 71(2):470–479

Zare M, Mehrabi Oladi A, Sharif Zade SH (2006) Evaluation of Giberlic acid and kenitin effects on germination and development of wheat seedlings under salinity stress. J Agric Sci 12(4):855–865

Zegada-Lizarazu W, Iijima M (2005) Deep root water uptake ability and water use efficiency of pearl millet in comparison to other millet species. Plant Produc Sci 8(4):454–460

Zhang Y, Zhang Y, Zhang R, Song Y, Li G, Song Y et al (2024) Adaptability of root morphology and growth of two forage grass species in response to salt stress. Front Environ Sci 12:1406778

Zhao Z, Zhang K, Wang P, Wang L, Yin C, Tian C (2013) The effects of halophytes on salt balance in an arid irrigation district. J Food Agric Environ 11:2669–2673

Zhao C, Liu B, Piao S, Wang X, Lobell DB, Huang Y et al (2017) Temperature increase reduces global yields of major crops in four independent estimates. Proc Natl Acad Sci 114(35):9326–9331

Zhu JK (2002) Salt and drought stress signal transduction in plants. Annu Rev Plant Biol 53(1):247–273

Zhu JK, Liu J, Xiong L (1998) Genetic analysis of salt tolerance in Arabidopsis: evidence for a critical role of potassium nutrition. Plant Cell 10:1181–1191

Challenges and Opportunities of Cultivating Sandalwood (*Santalum album*) Under Abiotic Stress Conditions

20

Raj Kumar, Kamlesh Verma, Ashwani Kumar, Manish Kumar, Anita Mann, Suresh Kumar, Sneha Dobhal, and Sagar Chaudhary

Abstract

Sandalwood (*Santalum album* L.) tree is a member of the genus *Santalum*, is highly valued for its fragrant wood and essential oil. As an obligate root hemi-parasite, sandalwood depends on host species for water and essential nutrients, which are transferred through specialized vascular structures known as haustoria. The growth and survival of sandalwood are largely influenced on its ability to tolerate various abiotic stresses, including salinity, drought, temperature extremes, and nutrient deficiencies. Sandalwood exhibits moderate tolerance to salinity, allowing it to survive in salt-affected soils; however, excessive salinity can inhibit its growth and development. As sandalwood thrives in tropical climates with high temperatures, extreme heat can negatively affect its growth. Cold stress, particularly in temperate regions, can significantly limit its growth and productivity. Additionally, sandalwood is sensitive to waterlogged conditions, requiring well-drained soils for optimal growth. Effective soil and water management strategies are therefore critical for sustaining sandalwood production under abiotic stress conditions. Interestingly, moderate stress may stimulate heartwood formation, as plants often produce increased levels of metabolites in response to environmental stresses. Despite these challenges, sandalwood exhibits a remarkable degree of adaptability to abiotic stress through various physiological and biochemical mechanisms. To mitigate the effects of stress, *S. album* increases the activity of antioxidant enzymes such as peroxidase, superoxide dismutase (SOD), and catalase in response to elevated levels of reactive oxygen species (ROS). Moreover, cultivating sandalwood with a compatible and stress-

R. Kumar (✉) · K. Verma · A. Kumar · M. Kumar · A. Mann · S. Kumar
ICAR-Central Soil Salinity Research Institute, Karnal, Haryana, India

S. Dobhal · S. Chaudhary
Uttarakhand University of Horticulture and Forestry, Ranichauri, Uttarakhand, India

A. Mann et al. (eds.), *Cutting Edge Technologies for Developing Future Crop Plants*,
https://doi.org/10.1007/978-981-96-2508-6_20

tolerant host can promote satisfactory growth and biomass production under adverse environmental conditions. Co-cultivation with host plants has been shown to significantly enhance *S. album*'s survival and growth under such conditions. Overall, literature suggests significant potential for cultivating sandalwood in challenging environmental conditions, given appropriate management and host selection strategies.

Keywords
Sandalwood · Abiotic stress · Haustorium · Tree

20.1 Introduction

Modern agriculture faces numerous risks related to production, pricing, markets, finance, institutions, and social factors, many of which are exacerbated by adverse environmental impacts. Climate change and abiotic stresses pose significant threats to global food security and rural livelihoods. Research indicates that integrating trees into farming systems can mitigate yield losses under such challenging conditions (Banyal et al. 2018). Furthermore, there is an urgent need to enhance the profitability of agricultural land. One promising approach to achieving "profitable agriculture" is the cultivation of high-value crops. Sandalwood (*Santalum album*), a highly valued crop, is in high demand both domestically and internationally, with the price of dry wood ranging from Rs 2000 to Rs 6000 per kilogram.

The sandalwood tree (*Santalum album L.*) is prized for its fragrant wood and the scented oil derived from it. The species naturally occurs in the Indian states of Karnataka, Kerala, and Tamil Nadu. However, sandalwood production in India has significantly declined, dropping from 4000 metric tons (MT) in the 1950s to just 500 MT in 2007. In contrast, global demand for sandalwood is estimated to be around 5000–6000 MT of wood and 100–120 mg of oil annually. The primary challenges contributing to the depletion of sandalwood in forests include illegal harvesting, smuggling, and diseases. The high demand for fragrant sandalwood oil in both domestic and international markets has led to illegal logging and smuggling of the species from natural reserves. This has resulted in reduced species diversity, with one sandalwood species becoming extinct and three others now classified as threatened.

Sandalwood offers significant potential for cultivation, particularly within agroforestry systems, especially considering the increasing gap between demand and supply. Nonetheless, extensive planting efforts have faced challenges stemming from inadequate awareness among farmers, limited access to planting materials, governmental regulations, understanding of host-parasite interactions, region-specific cultivation techniques, and adaptability under abiotic stress conditions (Das 2021). Apart from the number of host plants, the type of soil and the distance between plants also affect the growth of sandalwood. To increase sandalwood production, it is crucial to deepen our understanding of the relationships between hosts and parasites, ensure the quality of planting materials, and develop effective cultivation and management practices, particularly in the context of abiotic stress conditions.

Arid and semiarid regions are particularly vulnerable to land degradation and abiotic stresses such as drought and salinity (Sheoran et al. 2023). These stresses negatively affect plant growth by disrupting various physiological processes (Basak et al. 2021; Dhansu et al. 2021; Kumar et al. 2023; Singh et al. 2022). Some species can tolerate stress conditions by accumulating soluble proteins and sugars, which help maintain protein stability, protect against oxidative damage, regulate osmotic balance, and support signalling and energy supply (Kesh et al. 2022; Mann et al. 2023; Sharma et al. 2024). However, many stress-tolerant tree species are less profitable, creating an urgent need to identify highly profitable trees that can thrive under an abiotic stress environments. Sandalwood presents a promising opportunity in this regard, with its potential to generate huge revenue even under stressful environments. One of its unique biological traits is its semi-parasitic root system, which highlights the importance of further research into host-parasite relationships to optimize cultivation and management practices. Currently, there is a lack of literature and studies concerning the evaluation of sandalwood's performance and the advancement of cultivation practices under abiotic stress conditions. To address this gap, it is crucial to develop strategies that assess the limitations, challenges, and opportunities for commercial sandalwood cultivation in abiotic stress environments.

20.2 Management Practices of Sandalwood Under Abiotic Stress Conditions

20.2.1 Role of Host Species in Regulating Growth Processes in Sandalwood

Sandalwood trees, belonging to the genus Santalum, are obligate root hemiparasites that rely heavily on host plants for water and nutrients. Their growth depends on establishing connections with suitable hosts, as they cannot absorb sufficient resources independently. Large-scale sandalwood cultivation has not gained traction globally, primarily due to challenges in establishing these host parasite relationships and limited knowledge about them. Without a suitable host, sandalwood's growth is stunted, often leading to yellowing or chlorosis of the leaves and other growth disorders. A suitable host significantly improves sandalwood seedling growth by providing adequate supplies of essential nutrients, such as nitrogen and calcium. Studies have shown that the presence of a host species can accelerates the growth of sandalwood seedlings through the consistent supply of these nutrients. Successful growth of sandalwood under abiotic stress conditions depends on several factors, including the type of host plant, the number of hosts, and the spacing between them and the sandalwood. A compatible, stress-tolerant host can facilitate satisfactory growth and biomass production in sandalwood under challenging environmental conditions. The nutrient intake of the host plant also plays a critical role, as variations in the host's nutrient absorption directly affect the growth of sandalwood tree. The use of multiple host species can further enhance the supply of nutrients to sandalwood, leading to better growth performance. Studies have shown that sandalwood plants with hosts form a greater number of haustorial connections, compared to those without hosts. These connections enable the transfer of essential nutrients and water from the host to the sandalwood tree, promoting its growth. Under abiotic stress, a tolerant host species can support both its own physiological processes and those of the sandalwood tree. Specific haustorial connections with different host species are crucial for the development of aromatic sandalwood heartwood. For instance, Ouyang et al. found that mature haustoria could reach the root xylem of hosts such as *Dalbergia sissoo* and *Aquilaria sinensis*. The haustoria connected to *D. sissoo* developed a mature xylem, while those connected to *A. sinensis* lacked vessel elements. Additionally, 6-year-old sandalwood trees grown alongside *Citrus aurantium*, *Punica granatum*, and *Casuarina equisetifolia* on farmland exhibited superior growth and heartwood content. Ultimately, the overall growth and performance of sandalwood trees are strongly dependent on the compatibility with their host from planting through to harvest.

Several host species have been identified as beneficial for sandalwood cultivation, each contributing differently to its growth, nutrient intake, and heartwood formation. The best host species typically exhibit high nutrient uptake, drought resistance, and adaptability to varying environmental conditions. The list of some of the best host species for cultivating sandalwood is given below.

1. ***Dalbergia sissoo***: *Dalbergia sissoo* is considered one of the most compatible hosts for sandalwood. It forms strong haustorial connections with sandalwood, enhancing water and nutrient uptake, especially nitrogen. This species has been shown to support faster growth and higher biomass production in sandalwood. It is also tolerant to salinity and drought, making it a preferred choice in areas facing water scarcity or salinity stress.
2. ***Casuarina equisetifolia***: *Casuarina equisetifolia* is another highly effective host for sandalwood cultivation. It is a nitrogen-fixing species that provides essential nutrients such as nitrogen, enhancing the growth of sandalwood. It has also been associated with improved heartwood quality and increased oil yield in mature sandalwood trees. This species is commonly used in agroforestry systems and is highly adaptable to poor soils and saline environments.
3. ***Acacia* spp.**: Various Acacia species, such as *Acacia nilotica*, *Acacia auriculiformis*, and *Acacia senegal*, are excellent host plants for sandalwood. Acacias are known for their nitrogen-fixing ability, which supports the nutrient requirements of sandalwood. These species are drought-tolerant and thrive in arid and semiarid regions, making them ideal for sandalwood cultivation in areas prone to water stress.
4. ***Azadirachta indica***: *Azadirachta indica* is one of the most effective and widely used host species for cultivating sandalwood (*Santalum album*). It is highly compatible with sandalwood due to several beneficial characteristics that support the

tree's growth and resilience, especially under stressful environmental conditions. Species offer a range of advantages as a host for sandalwood cultivation.

5. ***Citrus aurantium***: *Citrus aurantium* has been shown to support sandalwood growth effectively, particularly when grown on farmland. It provides essential nutrients and fosters better heartwood content in sandalwood trees. Citrus species are often used in mixed cropping systems, and their adaptability to different soil types makes them a viable option as host plants.
6. ***Punica granatum***: *Punica granatum* is another promising host species for sandalwood cultivation. It has been found to support good growth and enhance heartwood formation in sandalwood trees. This host is especially valuable in regions with dry climates due to its drought tolerance.
7. ***Melia dubia***: *Melia dubia* is gaining popularity as a host for sandalwood due to its fast growth rate and compatibility. It provides shade and nutrients to the sandalwood tree, supporting its overall development. This species is well-suited to a variety of soil types, including degraded soils, making it a good choice for agroforestry systems.
8. ***Sesbania grandiflora***: *Sesbania grandiflora* is a short-term nitrogen-fixing host often used in the early stages of sandalwood growth. It provides essential nutrients, particularly nitrogen, which helps young sandalwood seedlings establish themselves before transitioning to more permanent hosts.

20.2.2 Factors Affecting the Selection of Host Species

Selecting the right host species can significantly improve the growth, health, and heartwood quality of sandalwood, contributing to more successful cultivation in diverse environments. The following considerations are necessary when selecting a host species for sandalwood.

- *Nutrient availability*: The host plant should be capable of providing essential nutrients, especially nitrogen, which is crucial for sandalwood's growth and development.
- *Drought and salinity tolerance*: The hosts should withstand water stress or saline soils which is important for cultivating sandalwood in arid and semiarid regions.
- *Longevity of the host*: Since sandalwood takes several years to mature, the host species should ideally be long-lived and able to provide support throughout the growth cycle.
- *Compatibility with sandalwood*: Not all plant species are suitable hosts. The host must form effective haustorial connections with sandalwood to ensure proper nutrient and water transfer.

20.2.3 Spacing Requirements of Sandalwood and Host Species

The optimal spacing between sandalwood (*Santalum album*) and its host species depends on several factors, including soil conditions, climate, and the type of host species considered. Sandalwood can typically extend its lateral roots in the adjacent areas to a maximum of 1.5–3 metres in order to establish an association with the host plant. The minimum spacing between sandalwood trees and their hosts should be 12 × 10 feet, and the optimum spacing is 15 × 12 feet. This will help the trees to grow and form roots without competing with one another for nutrients and light. Host plants, which offer critical nutrients to the sandalwood tree, should be planted closer to the tree, often 1.5–2.5 metres apart, to allow the roots to interact and promote the parasitic nature of the tree (Fig. 20.1). The optimal spacing for sandalwood trees in a mixed-species arrangement is 3 metres, while the optimal spacing for host trees is 1.5–2 metres. Vignesh et al. observed the performance of *Santalum album* L. at five spacing treatments (2.4 × 2.4 m, 3.0 × 3.0 m, 3.6 × 3.6 m, 4.6 × 4.6 m, and 5.5 × 5.5 m), and they showed that the 3.0 × 3.0 m and 3.6 × 3.6 m spacing produced best growth in the *Santalum album* tree. An appropriate spacing allows for efficient use of both water and nutrients from soils. Under abiotic stress, the planting techniques and spacing of sandalwood could be different than those of fertile soils due to the slow growth rate of plants. In salinity and drought conditions, co-planting sandalwood with host species could increase species survival. The cultivation process can be somewhat complicated because it necessitates different host species for specific environmental conditions. Overall, for optimal growth and proper nutrient absorption, the ideal distance between sandalwood and host species should be designed based on the environmental conditions and the type of host.

Fig. 20.1 General view of the spacing between sandalwood and host species under field conditions

20.2.4 Soil and Water Management

Soil and water management are crucial for enhancing plant productivity and sustainability under biotic stress conditions, such as drought, salinity, and extreme temperatures. Sandalwood plantations require specific soil and water management practices for their optimal growth. The type of soil and the soil-forming parent material also determine sandalwood growth. Sandalwood thrives in well-drained soils with a pH range of 6.0–7.5 and requires balanced quantities of nitrogen, phosphorus, potassium, and micronutrients. Regular watering, particularly during the first 3–4 years, is essential for nutrient retention and preventing root rot and fungal diseases. Proper drainage is one of the pre-requisite requirements for sandalwood plantations, as species do not tolerate water logging. It is better to plant sandalwood in raised beds or on buds, and the height and width of the beds should be such that water does not stagnate around the tree stem. Mulching, such as organic or stone mulch, helps retain soil moisture, regulate temperature, and suppress weed growth. This could also help sandalwood plants cope with drought and low rainfall conditions. In salt-affected soil, leaching to flush salts out and the application of organic amendments and gypsum can improve the soil's physical and chemical properties. Terracing should be implemented along the slope areas, in conjunction with the cultivation of cover crops and vegetation. Strips between sandalwood rows can control soil erosion. In degraded soil conditions, the introduction of nitrogen-fixing hosts can improve soil fertility and nitrogen availability. Drip or sprinkler irrigation in water-scarce regions and water conservation techniques, such as rainwater harvesting in regions experiencing seasonal rainfall, can help in achieving optimum growth in sandalwood. These practices, along with regular soil testing and water monitoring, can ensure optimum growth and productivity in sandalwood plantations. This suggests that integrated soil and water management is essential for sustaining sandalwood production under abiotic stress conditions.

20.2.5 Heartwood Formation

Heartwood formation is a biological process that consists of vascular cambium formation, cell elongation and enlargement, cell wall deposition, programmed cell death, and heartwood formation (Ruiqin 2015). Mechanical, chemical, or biological interventions can initiate heartwood production in some tree species. In India, sandal trees begin to form heartwood at 10–12 years old. The host plant affects the formation of heartwood, and the amount and formation of oil vary with maturity (Fig. 20.2). The oil content in sandalwood is higher in the roots, followed by the stem at ground level, and gradually tapers off towards the tip of the stem (Das 2021). Genetic variability within sandalwood species can lead to significant variations in heartwood production. The host plant plays a crucial role in heartwood formation, with well-drained, nutrient-rich soils supporting faster growth and more abundant heartwood formation (Fig. 20.3). Due to their ability to fix nitrogen and improve soil fertility, legumes often serve as hosts in nutrient-deficient soils. It has

Fig. 20.2 View of the heartwood (dark brown) formation in sandalwood

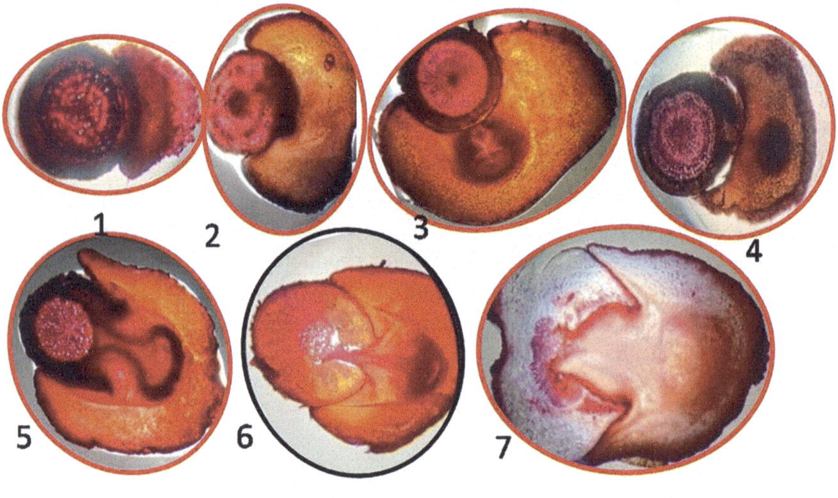

Fig. 20.3 Various stages of haustoria formation in sandalwood during the interaction process

also been observed that fertile soil promotes greater growth in sandalwood, while low to moderate fertile soil favours heartwood formation in sandalwood. Water availability is also critical; adequate moisture levels during the initial years produce good growth, and slight water stress during the later years initiates heartwood and oil formation in the sandalwood. Some researchers have suggested that slightly high temperatures and drought conditions at maturity produce a greater amount of oil-containing heartwood (Jessie et al. 2020). Similarly, a moderate level of salinity stress may also induce heartwood formation in sandalwood, as plant species tend to produce more metabolites in defense against stress. However, overwatering or waterlogging conditions can hinder heartwood development by promoting diseases such as root rot. Sandalwood grows best in warm, tropical climates with moderate humidity, with optimal heartwood formation at temperatures ranging from 12 °C to 35 °C. However, detailed and comprehensive information about the optimum growth conditions for heartwood formation in sandalwood is still lacking. Furthermore, growth rate and biomass allocation also play a role in heartwood formation. Regular pruning and thinning can stimulate heartwood formation by reducing energy directed towards excessive foliage and enhancing light penetration. Slow growth rates tend to produce more heartwood, while more resources allocated to trunk growth enhance heartwood deposition. Techniques such as scoring or girdling the tree, or the application of chemicals such as ethylene or salicylic acid, can artificially stimulate induced heartwood formation (Liu et al. 2022). The published literature showed that heartwood formation is a dynamic process that involves the interaction of environmental, genetic, and physiological factors. Moreover, effective management practices, along with favourable environmental conditions, can significantly stimulate heartwood formation and oil quality in sandalwood trees.

20.2.6 Nutrient Translocation Process Between Host and Sandalwood

The parasitic plants obtain nutrients and water from hosts through a haustorial structure. Haustorium establishes a xylem-to-xylem connection between the parasite and the host plant (Songkui et al. 2016). Hemiparasites, such as sandalwood, are also xylem feeders and obtain their resources from host xylem (Picture 3). They have high rates of transpiration, which facilitates the transfer of water, carbon, nitrogen, and mineral nutrients from the host to sandalwood (Rocha and Santhoshkumar 2022). Sandalwood has lower water potential than its hosts, sometimes by 1 MPa or more. High concentrations of soluble carbohydrates and inorganic ions in hemiparasites create a high osmotic potential gradient between host and parasite. Sandalwood relies on hosts for macro- and micronutrients such as calcium, iron, nitrogen, phosphorous, potassium, magnesium, amino acids, proline, malic acid, and sugars (Rocha et al. 2014).

Under stressful environments, sandalwood mainly relies on water and nutrients provided by their host species. For example, Kumar et al. (2024) found significant differences in nutrient content in sandalwood leaves influenced by different host species under salinity stress. The higher absorption of nutrients by *Dalbergia sissoo* and their better translocation to sandalwood via the soil-to-plant system indicate the best opportunity for cultivating sandalwood with *D. sissoo* in moderately saline soils. Similarly, Lu et al. (2014) also found higher N content in sandalwood with hosts *Dalbergia* compared to hosts *Bischofia polycarpa* and *Dracontomelon duperreranum*. Calcium plays an important role in producing various plant tissues, and a sufficient supply of calcium from the host to the sandalwood produces rapid growth and proliferous root formation. Sandalwood trees benefit more from the various nutrients that their host plants contribute when planted with a variety of hosts.

Under natural conditions, the presence of more than one type of host plant was observed to supply better nutrients for sandalwood growth. Overall, the interaction between host species and stress conditions alters the nutrient content in sandalwood leaves.

20.2.7 Identification of Germplasm

At present, the lack of quality planting material is one of the main hindrances to the large-scale adoption of sandalwood. To boost sandalwood farming and value chain industries, developing planting materials for plantation programmes is urgently required, which can increase the productivity of sandalwood plantations. Similarly, in drought and salinity conditions, identification of tolerant germplasm using tree improvement tools is required to increase sandalwood plantations' productivity. Marker-assisted selection (MAS) can be a tool to improve salt stress tolerance in sandalwood, as it is less time-consuming and more reliable. The genetic diversity-based research studies could identify sandalwood's potential germplasm for commercial cultivation in degraded and resource-poor lands.

The conventional methods of tree improvement are time-consuming and labour-intensive. Marker-assisted selection can be used to improve the sandalwood's abiotic stress tolerance. Marker-assisted selection (MAS) is the process of using morphological, biochemical, or DNA markers as indirect selection criteria for selecting important traits in plant breeding (Sushma et al. 2023). This process is used to improve the effectiveness or efficiency of selection for the traits of interest in breeding programmes. However, studies need to be carried out to assess the abiotic stress tolerance of sandalwood through marker-assisted selection and gene regulation mechanisms.

20.3 Abiotic Stress Tolerance of Sandalwood

The growth and survival of sandalwood depend largely on its capacity to withstand a range of abiotic stresses, such as salt, drought, temperature extremes, and nutrient deficiency. The detailed literature about the abiotic stress tolerance mechanisms of sandalwood is given below.

20.3.1 Salinity Stress of Sandalwood

Saline soils are prevalent in arid and semiarid regions, primarily resulting from inadequate drainage, waterlogging, and the use of poor-quality irrigation water in agriculture (Sheoran et al. 2021). In such soils, only salt-tolerant crops and trees can achieve satisfactory growth and yield. Salt tolerance in plants is a complex trait involving the interaction of numerous molecular, biochemical, and physiological processes. Under salinity stress, the reduction in water potential impairs critical biological functions such as respiration, cell membrane stability, transpiration, and the maintenance of turgor pressure. Additionally, salinity-induced oxidative stress exacerbates cellular damage through the accumulation of reactive oxygen species (ROS) and related oxidative processes (Soni et al. 2021; Parul et al. 2023). Prolonged salinization leads to the accumulation of sodium (Na^+) and chloride (Cl^-) ions in the root zone, disrupting the availability, uptake, and transport of essential nutrients within the soil-plant system (Lata et al. 2022; Aly et al. 2011). To cope with salinity stress, plants have evolved a variety of physiological and biochemical mechanisms that allow them to adapt and thrive in saline environments (Dhansu et al. 2021). These strategies enable plants to mitigate the adverse effects of high salinity, ensuring their survival and growth under challenging conditions.

Santalum album can tolerate a range of salinity levels, but the degree of tolerance depends on the host plant species (Fig. 20.4). Verma et al. (2023) have demonstrated that the salinity tolerance of sandalwood can range from 25% to 59.8% depending on the host plant. This shows that sandalwood exhibits moderate tolerance to salinity, enabling it to survive in salt-affected soils. It can tolerate salinity stress ranging from mild (2–4 dS m^{-1}) (Carter et al. 2010) to moderate (7–9 dS m^{-1}) (Verma et al. 2023). However, high salinity levels may inhibit its growth and development (Lake 2019). In such conditions, sandalwood benefits from associations with host species. Verma et al. (2023) recommended *Melia dubia, Dalbergia sissoo*, and *Azadirachta indica* as ideal hosts for optimal growth and biomass production of sandalwood under salinity stress (9 dS m^{-1}). Similarly, Sharma et al. (2024) also highlighted *Dalbergia sissoo* and *Melia dubia* as superior hosts under both water deficit and salinity stress, suggesting that these species help maintain ion homeostasis and osmotic balance under stress.

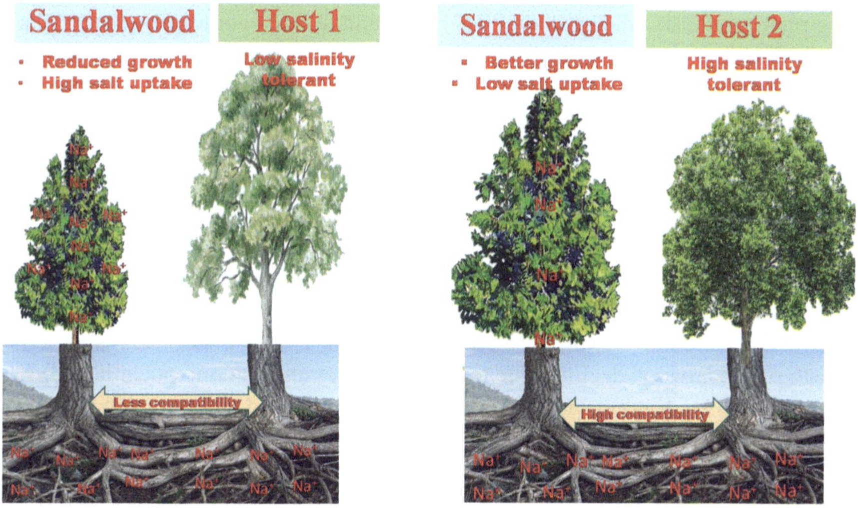

Fig. 20.4 Illustration of the variation in growth and salinity tolerance of sandalwood with two different host species

Notably, sandalwood plants without a host could not survive in saline environments and demonstrate poor growth under both water deficit and normal conditions (Sharma et al. 2024). Carter et al. (2010) observed that salt-induced soil water stress does not limit the transpiration of sandalwood trees, which respond more strongly to increasing vapour pressure deficits compared to their host trees. Photosynthesis and water-use efficiency of sandalwood are largely influenced by the response of its host plant to environmental factors (Radomiljac et al. 1999). Under salt stress, sandalwood maintains ionic balance by compartmentalizing harmful ions such as sodium (Na^+) and chloride (Cl^-) in vacuoles, which helps protect its photosynthetic and metabolic processes. Certain genotypes of sandalwood species exhibit higher salt tolerance due to their ability to exclude salt from sensitive tissues. Overall, research indicates that the growth and salinity tolerance of sandalwood are significantly influenced by the host species. Further studies are necessary to explore the mechanisms underlying salinity tolerance in sandalwood.

20.3.2 Drought Stress Tolerance of Sandalwood

Climate change is causing unpredictable weather patterns, such as more frequent droughts, especially in arid and semiarid regions. Drought stress lowers turgor pressure, which in turn slows down photosynthesis, transpiration, stomatal efficiency, and overall plant development. The accumulation of osmolytes and reactive oxygen species (ROS) during stress disrupts biological and metabolic processes (Iqbal et al. 2020; Hayat et al. 2012). Drought also triggers ROS production, leading to oxidative damage at the cellular level. This results in the degradation of chlorophyll, nucleic acids, proteins, and other vital biomolecules, ultimately reducing enzyme activity, disrupting cellular membranes, impairing stomatal conductance, and inhibiting photosynthesis (Zhang et al. 2022). Drought stress significantly affects plant gas exchange by reducing leaf growth, damaging the photosynthetic apparatus, accelerating leaf senescence, breaking down chloroplast lipids, and altering the structure of pigments and proteins. Mittler (2002) observed that ROS, including superoxide (O2−), hydrogen peroxide (H2O2), and hydroxyl radicals (-OH), can directly interact with membrane lipids, causing lipid peroxidation. Elevated ROS levels also damage proteins, nucleic acids, and cellular membranes, which severely affect the structural and functional attributes of plants (Wu et al. 2016).

As a semi-root parasitic plant, sandalwood (*Santalum album*) depends heavily on its host species, which supplies it with essential water and nutrients through specialized vascular structures known as haustoria (Li et al. 2023). Under water stress, the growth of sandalwood declines due to reduced water and osmotic potentials (Tyree and Ewers 1991; Kaur et al. 2022). However, sandalwood can maintain its water balance by either accumulating osmotically active compounds or forming direct xylem-to-xylem connections with its host (Dhaniklal 2006; Rocha et al. 2014; Světlíková 2018). In their study, Sharma et al. (2024) showed that the combination of sandalwood and *Dalbergia sissoo* demonstrates improved water status, as this host facilitates increased water absorption via direct xylem connections. Additionally, under mild to moderate drought stress, *S. album* leaves show significant increases in the activities of antioxidant enzymes such as peroxidase (POD), superoxide dismutase (SOD), and catalase (CAT), likely in response to elevated ROS levels (Bashir et al. 2019; Zhang et al. 2022). However, under severe drought conditions, the activities of these enzymes decrease, possibly due to extensive damage to the leaves. Notably, *S. album* plants grown with a host exhibit considerably higher antioxidant enzyme activity compared to those grown without a host (Zhang et al. 2022).

A key strategy for mitigating salt and water deficit stress in plants is the accumulation of proline (Zhang et al. 2022). Proline levels naturally increase under such conditions, contributing to stress tolerance by acting as an osmoprotectant and serving as a nitrogen reserve during recovery. Co-cultivation of *Santalum album* with a host plant under drought stress significantly enhances proline and soluble sugar concentrations in the leaves (Zhang et al. 2022). Under salt stress, various carbohydrates, including glucose, fructose, fructans, trehalose, and starch accumulate in the plants (Parida et al. 2004; Kerepesi and Galiba 2000). These carbohydrates play a crucial role in alleviating stress by scavenging reactive oxygen species (ROS), sequestering carbon, and providing osmoprotection. Furthermore, Mahesh et al. (2018) conducted a genome assembly and annotation of *S. album*, identifying 53 novel protein-coding genes involved in protecting the plant from abiotic stresses such as drought and salinity.

20.3.3 Sandalwood Under Temperature Stress

Sandalwood thrives in tropical climates with high temperatures, particularly in regions such as southern India and parts of Australia. However, extreme heat (above 45 °C) can affect both growth and oil quality in sandawood. While primarily a tropical species, sandalwood has some tolerance to lower temperatures, although cold stress, especially in temperate regions, can significantly limit its growth and productivity. Low temperatures, including chilling (<20 °C) and freezing (<0 °C), are among the most harmful environmental stresses for higher plants. To cope with cold stress, plants have evolved various adaptive strategies, including structural modifications and changes in gene expression, which activate a range of protective responses (Chinnusamy et al. 2007). These protective mechanisms involve increasing intracellular solutes, accumulating cryoprotectants and antioxidants, and producing antifreeze and cold-regulated (COR) proteins (Uemura et al. 2003; Chen et al. 2015). These compounds stabilize plant membranes during cold stress by maintaining lipid integrity, preserving ion balance, and neutralizing reactive oxygen species (ROS) (Yadav 2010).

Although the *Santalum album* is typically found in tropical and subtropical regions, cold damage remains a major factor limiting its cultivation in areas with lower temperatures. A survey revealed that sandalwood trees can survive temperatures above −1 °C during winter (Li 2003). Transcriptomic and physiological studies on *S. album* seedlings exposed to cold stress (4 °C for 0–48 h) showed increased accumulation of proline, malondialdehyde (MDA), soluble carbohydrates, and antioxidants. Additionally, the expression of genes related to phytohormone signal transduction increased with prolonged cold exposure (Zhang et al. 2017). Interestingly, more than 50% of the differentially expressed genes (DEGs) in the *S. album* lacked annotated homologs in the Nr database, suggesting that these may represent novel cold-responsive genes unique to sandalwood. Over the 48-h cold stress period, increases in MDA, electrolyte leakage, proline, and soluble sugar levels indicated that osmolytes may help protect plant cell membranes and enhance membrane stability during cold-induced dehydration.

Cold stress affects multiple aspects of photosynthesis, including light reactions, the Calvin cycle, and photorespiration (Friesen and Sage 2016). It can halt growth, reduce energy utilization, and cause inhibition of photosynthesis (Ruelland and Zachowski 2010). The enzymatic antioxidant system plays a critical role in defending plants against ROS, thereby improving tolerance to abiotic stress (Jithesh et al. 2006; Mittler et al. 2004). In sandalwood, increased activities of superoxide dismutase (SOD) and peroxidase (POD) have been observed, highlighting an enhanced ability to manage oxidative stress under cold conditions (Zhang et al. 2017).

20.3.4 Sandalwood Under Waterlogged Conditions

Waterlogging occurs when excess water accumulates in the soil, saturating the root zone and depriving it of oxygen. Sandalwood trees, which require well-drained soils, are particularly sensitive to waterlogged conditions and struggle to thrive in such environments. Although sandalwood is adaptable to a wide range of agro-climatic and soil conditions for natural regeneration, waterlogged areas remain a notable exception. In a 4-week nursery experiment, Fox and Millar (2001) observed no significant impact on shoot growth in seedlings of sandalwood. However, they suggested that prolonged exposure to waterlogged conditions could lead to root disintegration and eventual seedling mortality. Field studies further indicate that waterlogging hinders seed emergence, with excess moisture promoting root rot disease and impairing overall plant development.

20.4 Conclusion

Sandalwood has potential to tolerate moderate level of abiotic stresses. Planning and adopting proper cultivation practices—such as irrigation, host plant selection, and soil management—crucial for maximizing growth and oil yield in sandalwood. It offers a promising option for cultivation in poor soils and has the potential to become a valuable tree species, helping farmers generate significant revenue in challenging environments. However, the choice of host species and their ability to tolerate variable degree of stresses are key factors in the success of sandalwood under adverse conditions. Therefore, selecting suitable host plants that not only supply essential nutrients but also help mitigate harsh soil and environmental conditions is essential for optimum growth of sandalwood. Moreover, given its high economic value and growing global demand, sandalwood represents a strong investment opportunity for farmers in challenging environmental conditions. In conclusion, although abiotic stresses present considerable challenges to sandalwood cultivation, the species' inherent resilience, coupled with ongoing research and technological advancements, provides promising avenues for sustainable cultivation in a variety of environmental conditions.

References

Aly AH, Debbab A, Proksch P (2011) Fungal endophytes: unique plant inhabitants with great promises. Appl Microbiol Biotechnol 90(6):1829–1845

Banyal R, Yadav RK, Sheoran P, Bhardwaj AK, Rajkumar PK, Tolia R (2018) Managing saline soils of Indo-Gangetic plains with eucalyptus and Melia based agroforestry systems. Indian J Ecol 45(4):841–849

Basak N, Sheoran P, Sharma R, Yadav RK, Singh RK, Kumar S, Krishnamurthy T, Sharma PC (2021) Gypsum and pressmud amelioration improve soil organic carbon storage and stability in sodic agroecosystems. Land Degrad Dev 32(15):4430–4444

Bashir K, Matsui A, Rasheed S, Seki M (2019) Recent advances in the characterization of plant transcriptomes in response to drought, salinity, heat, and cold stress. F1000Research, 8

Carter JL, Smith AJ, Silberstein RP, Ali SM, Doody TM, Byrne J, Smith P (2010) Interaction between trees and groundwater in the ord river irrigation area. National Research Flagship Water for a Healthy Country, CSIRO, Australia

Chen HY et al (2015) A comparison of the low temperature transcriptomes of two tomato genotypes that differ in freezing tolerance: *Solanum lycopersicum* and *Solanum habrochaites*. BMC Plant Biol 15:132

Chinnusamy V, Zhu J, Zhu JK (2007) Cold stress regulation of gene expression in plants. Trends Plant Sci 12:444–451

Das (2021) Silviculture, growth and yield of sandalwood. https://doi.org/10.1007/978-981-16-0780-6_7

Dhaniklal G (2006) Influence of host plants on soil moisture stress on the water relations in sandal. Doctoral dissertation, M.Sc. thesis, Kerala Agricultural University, Thrissur

Dhansu P, Kulshreshtha N, Kumar R, Raja AK, Pandey SK, Goel V, Ram B (2021) Identification of drought-tolerant cocanes based on physiological traits, yield attributes and drought tolerance indices. Sugar Tech 23(4):747–761. https://doi.org/10.1007/s12355-021-00967-7

Fox JED, Millar KL (2001) Sandalwood emergence and growth in relation to salinity, waterlogging and depth of burial. Occasional Report No 5, Mulga Research Centre, Curtin University of Technology, School of Environmental Biology, Perth, Australia

Friesen PC, Sage RF (2016) Photosynthetic responses to chilling in a chilling-tolerant and chilling-sensitive Miscanthus hybrid. Plant Cell Environ. https://doi.org/10.1111/pce.12699

Hayat S, Hayat Q, Alyemeni MN, Wani AS, Pichtel J, Ahmad A (2012) Role of proline under changing environments: a review. Plant Signall Behav 7(11):1456–1466

Iqbal MS, Singh AK, Ansari MI (2020) Effect of drought stress on crop production. In: New frontiers in stress management for durable agriculture, pp 35–47

Jessie P, Tony J, Michael (2020) Enhancing the formation of heartwood in Sandalwood in Vanuatu. https://www.aciar.gov.au/sites/default/files/2020-09/final-report-FST-2016-054.pdf

Jithesh MN, Prashanth SR, Sivaprakash KR, Parida AK (2006) Antioxidative response mechanisms in halophytes: their role in stress defence. J Genet 85:237–254

Kaur G, Sanwal SK, Sehrawat N, Kumar A, Kumar N, Mann A (2022) Getting to the roots of Cicer arietinum L. (chickpea) to study the effect of salinity on morpho-physiological, biochemical and molecular traits. Saudi. J Biol Sci 29(12):103464

Kerepesi I, Galiba G (2000) Osmotic and salt stress-induced alteration in soluble carbohydrate content in wheat seedlings. Crop Sci 40(2):482–487

Kesh H, Devi S, Kumar N, Kumar A, Kumar A, Dhansu P, Sheoran P, Mann A (2022) Insights into physiological, biochemical and molecular responses in wheat under salt stress. In: Wheat–recent advances. Intech Open

Kumar A, Sheoran P, Mann A, Yadav D, Kumar A, Devi S, Kumar N, Dhansu P, Sharma DK (2023) Deciphering trait associated morpho-physiological responses in pearlmillet hybrids and inbred lines under salt stress. Front Plant Sci 14:1121805

Kumar R, Verma K, Bhardwaj A, Kumar A, Dobhal S, Kumar M, Verma RC, Yadav RK (2024) Mechanistic understanding of nutrient translocation in soil-host-parasite network of sandalwood (*Santalum album*) under salinity stress. Under review

Lake M (2019) Australian forest woods: characteristics, uses and identification. CSIRO Publishing, Australia

Lata C, Kumar A, Mann A, Soni S, Meena BL, Rani S (2022) Mineral nutrient analysis of three halophytic grasses under sodic and saline stress conditions. Ind J Agr Sci 92(9):1051–1055

Li YL (2003) Study on the introduction of sandalwood. Science Press, Beijing, pp 97–98

Li S, Lu S, Wang J, Chen Z, Zhang Y, Duan J, Liu P, Wang X, Guo J (2023) Responses of physiological, morphological and anatomical traits to abiotic stress in woody plants. Forests 14(9):1784. https://doi.org/10.3390/f14091784

Liu X, Zhang Q, Hong Z, Xu D (2022) Induction of heartwood formation in young Indian sandalwood (Santalum album L.) by gas elicitors. Front Plant Sci 28(13):961391. https://doi.org/10.3389/fpls.2022.961391

Lu JK, Xu DP, Kang LH, He XH (2014) Host-species-dependent physiological characteristics of hemiparasite *Santalum album* in association with N2-fixing and non-N_2-fixing hosts native to southern China. Tree Physiol 34(9):1006–1017

Mahesh HB, Subba P, Advani J, Shirke MD, Loganathan RM, Chandana SL, Shilpa S, Chatterjee O, Pinto SM, Prasad TS, Gowda M (2018) Multiomics driven assembly and annotation of the sandalwood (Santalum album) genome. Plant Physiol 176(4):2772–2788

Mann A, Lata C, Kumar N, Kumar A, Kumar A, Sheoran P (2023) Halophytes as new model plant species for salt tolerance strategies. Front Plant Sci 14:1137211. https://doi.org/10.3389/fpls.2023.1137211

Mittler R (2002) Oxidative stress, antioxidants and stress tolerance. Trends Plant Sci 7(9):405–410

Mittler R, Vanderauwera S, Gollery M, Van Breusegem F (2004) Reactive oxygen gene network of plants. Trends Plant Sci 9:490–498

Parida AK, Das AB, Mohanty P (2004) Investigations on the antioxidative defence responses to NaCl stress in a mangrove, Bruguiera parviflora: differential regulations of isoforms of some antioxidative enzymes. Plant Growth Regul 42:213–226

Parul KA, Chourasia S, Kumar R, Devi G, Kumar A, Mann A, Sheoran P, Sanwal SK (2023) Physiological, biochemical, and molecular responses to salt stress and seed priming approach to enhance salt tolerance in bread wheat. In: Salinity and drought tolerance in plants: physiological perspectives. Singapore, Springer Nature Singapore, pp 61–86

Radomiljac AM, McComb JA, McGrath JF (1999) Intermediate host influences on the root hemiparasite Santalum album L. biomass partitioning. For Ecol Manag 113(2):143–153

Rocha D, Santhoshkumar AV (2022) Host plant influence on haustorial growth and development of Indian sandalwood (Santalum album). In: Indian Sandalwood: a compendium, pp 229–244

Rocha D, Ashokan PK, Santhosh Kumar AV, Anoop EV, Suresh Kumar P (2014) Influence of host plant on the physiological attributes of field-grown sandal tree. J Trop For Sci 26(2):166–172

Ruelland E, Zachowski A (2010) How plants sense temperature. Environ Exp Bot 69:225–232

Ruiqin (2015) Molecular control of wood formation in trees. https://academic.oup.com/jxb/article-abstract/66/14/4119/2893369

Sharma A, Kumar A, Chahal S, Verma K, Chauhan K, Kumar R, Mann A (2024) Morpho-physiological studies of sandalwood-host interaction under individual and interactive water and salt stress. Biol Futura:1–10

Sheoran P, Basak N, Kumar A, Yadav RK, Singh R, Sharma R, Kumar S, Singh RK, Sharma PC (2021) Ameliorants and salt tolerant varieties improve rice-wheat production in soils undergoing sodification with alkali water irrigation in Indo–Gangetic Plains of India. Agric Water Manag 243:106492

Sheoran P, Kamboj P, Kumar A, Kumar A, Singh RK, Barman A, Prajapat K, Mandal S, Yousuf DJ, Narjary B, Kumar S (2023) Matching N supply for yield maximization in salt–affected wheat agri–food systems: on-farm participatory assessment and validation. Sci Total Environ 875:162573

Singh A, Kumar A, Kumar R, Sheoran P, Yadav RK, Sharma PC (2022) Multivariate analyses discern shared and contrasting eco-physiological responses to salinity stress of Ziziphus rootstocks and budded trees. S Afr J Bot 146:573–584

Songkui Y et al (2016) The haustorium, a specialized invasive organ in parasitic plants. Annu Rev Plant Biol 67(1):643–667

Soni S, Kumar A, Sehrawat N, Kumar A, Kumar N, Lata C, Mann A (2021) Effect of saline irrigation on plant water traits, photosynthesis and ionic balance in durum wheat genotypes. Saudi J Biol Sci 28(4):2510–2517

Sushma M et al (2023) Biotechnological approaches for genetic improvement of crops. https://www.researchgate.net/profile/Manoj-Tripathi-10/publication/370214283_Biotechnological_Approaches_for_Genetic_Improvement_of_Crops/links/64462e1c8ac1946c7a48af7e/Biotechnological-Approaches-for-Genetic-Improvement-of-Crops.pdf

Světlíková P (2018) Effects of abiotic factors on hemiparasitic plants. Ph.D. thesis. University of South Bohemia, Faculty of Science, School of Doctoral Studies in Biological Sciences České Budějovice Czech Republic, 86

Tyree MT, Ewers FW (1991) The hydraulic architecture of trees and other woody plants. New Phytol 119:345–360

Uemura M, Warren G, Steponkus PL (2003) Freezing sensitivity in the sfr4 mutant of Arabidopsis is due to low sugar content and is manifested by loss of osmotic responsiveness. Plant Physiol 131

Verma K, Kumar R, Kumar A, Bhardwaj AK, Verma RC (2023) Host plant regulates growth processes, ion homeostasis, and salinity tolerance of sandalwood. J Plant Growth Regul 42:4423–4435

Wu J, Kim SG, Kang KY, Kim JG, Park SR, Gupta R, Kim ST (2016) Overexpression of a pathogenesis-related protein 10 enhances biotic and abiotic stress tolerance in rice. Plant Pathol J 32(6):552

Yadav SK (2010) Cold stress tolerance mechanisms in plants. A review. Agron Sustain Dev 30:515–527

Zhang X, Teixeira da Silva JA, Niu M, Li M, He C, Zhao J, Ma G (2017) Physiological and transcriptomic analyses reveal a response mechanism to cold stress in Santalum album L. leaves. Sci Rep 7(1):42165

Zhang Q, Liu X, Xu D, Hong Z, Zhang N, Cui Z (2022) Effects of drought and host on the growth of Santalum album seedlings in pot culture. Int J Mol Sci 23(19):11241. https://doi.org/10.3390/ijms231911241

Potential of Halophytes in Greening the Barren Land and Making Use of Waste Lands

21

Pankaj, Gayatri Kumari, Monika, Sunder Singh Arya, and Sarita Devi

Abstract

The degradation of arable land resources is accelerating globally, driven by climate change factors such as soil salinization and declining water resources. This has heightened the need for strategies to transform coastal zones and degraded lands into productive agricultural areas. Salinity poses a major challenge, significantly reducing plant productivity and driving efforts to develop salt-tolerant crop varieties. One promising and cost-effective approach to rehabilitating degraded land is the use of highly productive halophytes for desalination purposes. Halophytes are plants that naturally thrive in saline environments by employing various mechanisms to either exclude or accumulate salt. These adaptive traits make them ideal candidates for cultivation on saline soils. By focusing on the domestication of halophytic cash crops through conventional breeding programs, we can enhance both their productivity and their ability to tolerate high salt levels. This approach not only helps in repurposing degraded land but also supports the development of sustainable agricultural practices in the face of increasing environmental stressors. In essence, harnessing the potential of halophytes offers a viable solution to mitigate the impacts of soil salinization and ensure the continued availability of productive agricultural land.

Keywords

Halophytes · Salinity · Abiotic stress · Productivity · Tolerance

Pankaj · G. Kumari · Monika
Department of Botany and Plant Physiology, College of Basic Science and Humanities, CCS Haryana Agricultural University, Hisar, Haryana, India

S. S. Arya · S. Devi (✉)
Department of Botany, MDU, Rohtak, Haryana, India

A. Mann et al. (eds.), *Cutting Edge Technologies for Developing Future Crop Plants*,
https://doi.org/10.1007/978-981-96-2508-6_21

21.1 Introduction

Gradual climate change results in escalated land degradation, degraded soil health and fertility and decreased crop production and yield which offers food insecurity (Mylonas et al. 2020). Excess heavy metal contaminants such as lead, cadmium, arsenic etc. and excess soluble salts such as sodium and chloride contribute to soil salinity which leads to land degradation on a global level. Around 20% of the agricultural land is affected due to salinity and still continues to increase to date (Stavi et al. 2021). Soil salinity is exacerbated by both natural processes and human activities such as excessive use of fertilizers, improper irrigation practices, weathering of rocks and other anthropogenic activities (Mastrocicco et al. 2021). On the basis of soluble salts present in the soil, salinity-affected soil can be classified into three categories: saline soils, sodic soils and saline-sodic soils. Saline soils have high concentrations of soluble salts, while sodic soils are characterized by high levels of sodium ions (Osman an Osman 2018). High salt concentration affects plant growth and physiology and finally leads to yield reduction. Salinity adversely affects the crop by disturbing ionic and osmotic balance and causing ionic toxicity, posing a challenge to the crop growth and survival of the plants and can even cause mortality in the crop.

Plants serve as effective natural systems for removing toxic heavy metals including lead, cadmium, arsenic, pesticides and organic compounds from contaminated water and soil. Certain terrestrial plants, known as halophytes, can tolerate, grow and reproduce in saline conditions with salt levels around 20 dS m^{-1} NaCl or higher (Mann et al. 2019), which is generally harmful to sensitive plants (glycophytes). Halophytes have a number of adaptations to survive under highly saline environments including ionic and osmotic balance, the presence of salt glands on their leaves and succulence mechanisms. Halophytic species present promising solutions for remediating salt-contaminated areas through bio-saline agriculture (Mann et al. 2023).

In summary, halophyte offers a potential solution for the reclamation of barren and salt-affected soil and making use of wasteland for better crop production. Different halophytic adaptations such as salt exclusion, succulence and ion homeostasis can be exploited in making use of waste land offering better crop production and improved soil fertility and health and promoting sustainable agriculture.

21.2 Classification of Halophytes

Steiner (1935) classified different halophytes into three types: succulent halophytes, non-succulent halophytes and accumulating halophytes. Succulent halophytes, such as *Salicornia herbacea* (Fig. 21.1a), can tolerate high chloride concentrations in their cell sap. Non-succulent halophytes, such as *Spartina alterniflora* (Fig. 21.1b), use salt glands to excrete excess salt. Accumulating halophytes, such as *Suaeda fruticosa* (Fig. 21.1c) and *Juncus gerardii*, lack mechanisms to expel salt and instead store it in vacuoles and cell walls, ultimately shedding their leaves.

Fig. 21.1 (**a**) Salicornia herbacea. (**b**) Spartina alterniflora. (**c**) Suaeda fruticose

Fig. 21.2 Facultative halophytes: (**a**) *Sesuvium portulacastrum*, (**b**) *Aster tripolium*, (**c**) *Plantago maritima* and habitat-indifferent halophytes: (**d**) *Chenopodium* sp., Potentilla anserine, (**e**) *Festuca rubra*, (**f**) *Agrostis stolonifera*, showing adaptability to saline and nonsaline soils

Additionally, halophytes are categorized based on their salt requirements throughout their life cycle: obligate halophytes (which are native to saline environments and require salt), facultative halophytes (which have an optimal or minimal salt requirement) and habitat-indifferent halophytes (having good salt tolerance and are not native to saline environments) (Von Sengbusch 2003). Some members of the Chenopodiaceae family are classified as obligate halophytes. Facultative halophytes, on the other hand, can thrive in both nonsaline and saline environments. These plants can best grow in nonsaline areas but also have the potential to thrive in highly saline environments. Numerous species within the Brassicaceae, Poaceae and Cyperaceae families fall into this category. Notable examples of facultative halophytes studied for their salt tolerance and valuable by-products include *Sesuvium portulacastrum* (Fig. 21.2a), *Aster tripolium portulacastrum* (Fig. 21.2b) and *Plantago maritima* (Fig. 21.2c). Habitat-indifferent halophytes typically grow in nonsaline soils but can also survive in salt-affected soils. These plants are adaptable to various soil types, such as *Chenopodium* species and *Potentilla anserine* (Fig. 21.2d). Additionally, species such as *Festuca rubra* (Fig. 21.2e), *Agrostis stolonifera* (Fig. 21.2f) and *Juncus bufonius* exhibit genetic differences depending on whether they grow in nonsaline or saline soils.

On the basis of anatomical behavior, halophytes can also be classified into mesophytes and extreme halophytes which are based on 30 species of halophytes from various saline environments, examining their anatomical behavior, strategies, relevance and ecological significance (Grigore and Toma 2010). Extreme halophytes exhibit a strong correlation between anatomical features and salinity by developing highly developed water storage tissues in aerial organs which help in high succulence, successive cambia in roots, less-developed mechanical systems and increased lignin deposition such as in *Salicornia europaea* and *Petrosimonia oppositifolia*. Some non-succulent halophytes, such as *Halimione verrucifera*, have salt-secreting hairs, while *P. triandra* and *P. oppositifolia* display foliar kranz anatomy. Mesophytes exhibit intermediate behavior between extreme glycophytes and halophytes, having endodermis, well-developed aerenchyma, storage tissues and sunken stomata such as *Scorzonera cana* and *Lepidium* species. Some amphibious halophytes, such as *Alopecurus arundinaceus* and *Carex distans*, possess bulliform cells.

21.3 Adaptation Mechanisms in Halophytes

Halophytes survive in saline environments by employing various adaptation mechanisms to tolerate high salt levels (Fig. 21.3). These mechanisms enable them to maintain growth, photosynthesis and reproduction despite the stress imposed by high salinity.

21.3.1 Salt Exclusion

Halophytes, such as hydro-halophytes and certain phreatophytes, employ root system-induced ultrafiltration mechanisms to exclude excessive salts. By creating apoplastic barriers at the roots, they enhance resistance to bypass flow, effectively preventing salts from entering the roots which reduces the accumulation of toxic ions in the shoots and other aboveground parts via the transpiration stream (Hussain et al. 2021). The root endodermis and exodermis act as separate cell barrier layers, each having specific roles associated with the two main components of the cell wall: the suberin lamella (SL) and the Casparian band (CB). All apoplastic transport is channeled into the tightly controlled symplastic system by the CB, which is made up of lignin deposits in the cell walls of the endodermis and exodermis (Lee et al. 2013). Furthermore, the whole cell wall of the exodermis and endodermis is penetrated by the SL, which is mainly composed of long-chain fatty acids. This creates a hydrophobic barrier that helps control the intake of water and ions. Through the transpiration stream, this mechanism reduces the buildup of harmful ions in the

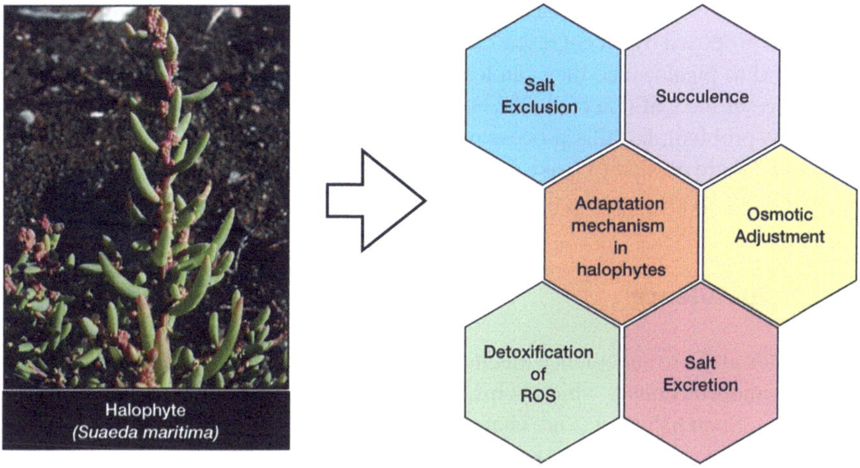

Fig. 21.3 Mechanism of halophytes to adapt under saline environments

shoots (Wang et al. 2020). A few studies that focused primarily on halophytes have shown that apoplastic barriers are necessary for species such as *A. marina* (Cheng et al. 2020), *Avicennia officinalis* (Krishnamurthy et al. 2014) and *S. salsa* (Liu et al. 2020) to adapt to salt. On the other hand, compared to the Casparian bands, not much is known about the function of suberin lamella in plant salt tolerance. Thus, additional investigation is required to assess the role of suberin lamella in saline ion exclusion.

21.3.2 Salt Excretion

Recreto-halophytes have significantly higher adaptability as compared to non-halophytes because they secrete much more salt ions onto their leaf surfaces in order to maintain cellular ion homeostasis (Lu et al. 2021). This process is made possible by specialized epidermal structures called salt glands e.g., several species from the *Limonium* genus, such as *L. gmelinii, L. bicolor, L. aureum*, *L. sinuatum* and *L. otolepis*, which enhance salt tolerance by increasing salt secretion through their salt glands (Mi et al. 2021). Similarly, species from the *Tamarix* genus exhibit salt tolerance by increasing salt gland density and secretion rate (Wei et al. 2020). Despite the structural variations and cell count differences in salt glands across halophyte species, they share common features: (i) they are encased by a thick cuticle (ii) they contain numerous plasmodesmata connection between cells, a large number of fully developed mitochondria, a large number of tiny vesicles in the cytoplasm and (iii) their cells do not have chloroplasts (Devi et al. 2016). Many cations (K^+, Na^+, Ca^{2+}, Mn^{2+}, Mg^{2+}, Zn^{2+} and Fe^{2+}) and anions (Br^-, Cl^-, I^-, PO_4^{3-}, SO_4^{2-} and NO^{3-}) are secreted by salt glands, and their contents mostly rely on the surrounding conditions (Lu et al. 2021). Although the precise mechanisms by which salt is secreted from these glands remain unclear, a number of possible mechanisms have been proposed by recent research. It has been demonstrated that salt glands can be harmful to plants since they can lead to leaf dryness by forming salt crystals on the surface of the leaf (Lu et al. 2021). The synthesis of suitable solutes can help to offset this problem, but this process requires a lot of energy around 30–109 molecules of ATP are needed to make one molecule of any compatible solute which will eventually affect crop productivity (Roy et al. 2014).

21.3.3 Succulence

Halophytes also use succulence mechanisms to maintain ion homeostasis and accumulate osmoprotectants, which helps to preserve cell turgor pressure under saline conditions. Such halophytes are characterized by waxy thick cuticles on leaves and stem, larger mesophyll cells, reduced intercellular space, increased tissue water content and high turgor pressure (Nikalje et al. 2017). Succulent leaves contain more mitochondria, which supply extra energy needed for salt sequestration and compartmentalization (Nikalje et al. 2017). Many halophytes, such as *Salsola drummondii, Achras sapota* and *Sarcocornia fruticose*, exhibit efficient succulent mechanisms, allowing them to tolerate high salt levels by accumulating salts in their leaves and stems while maintaining enhanced photosynthetic efficiency (Elnaggar et al. 2020). However, much remains to be understood about the detailed mechanisms of succulence. Therefore, current research should focus on: (i) understanding the complex structural-functional coordination that regulated physiological processes in succulent plants, (ii) utilizing transcriptomics and gene-editing techniques to gain insights into the functional biology of succulents and (iii) applying genetic engineering to introduce succulent traits into non-succulent plants to improve their resilience to salinity-induced environmental stress.

21.3.4 Osmotic Adjustment

Halophytic plants adapted to grow in highly saline environments play a crucial role in the remediation of saline soils. One of the key mechanisms they employ for survival and growth under such harsh conditions is osmotic adjustment. This physiological process enables halophytes to maintain cell turgor pressure and water uptake despite high external salt concentrations, making them valuable for improving saline soils.

Two main strategies by which plants maintain their osmotic balance are: (i) increased uptake of inorganic ions and organic molecules and (ii) maintenance of turgor pressure by the synthesis of organic osmolytes within plant cells by maintenance of Na^+/K^+ ratio in the cells and accumulating excess ions in the vacuoles along with other compatible organic solutes such as proline, glycine betaine content and sugars in the cytoplasm, which maintain osmotic balance under saline environment (Munns et al. 2019). Typically, ROS-activated nonselective cation channels

(NSCCs) and guard cell outward rectifying K^+ (GORK) channels are the two main channels via which salinity-induced K^+ efflux from the cell occurs (Zhao et al. 2020).

21.3.5 Detoxification of ROS

Plants rely on efficient antioxidant defense mechanisms to protect against oxidative stress caused by the overaccumulation of ROS under high salinity conditions (Castro et al. 2021). To mitigate ROS-induced damage, plants have developed a robust antioxidant defense system, which includes nonenzymatic antioxidants such as flavonoids, AsA, GSH and tocopherols, as well as enzymatic antioxidants such as SOD, CAT, APX, GR and GST. Improved salt tolerance in various halophytes is linked to their ability to maintain ROS homeostasis through the effective activation of their antioxidant systems under salt stress (Mann et al. 2018, 2024). For instance, in the plant *H. salicornicum*, enhanced SOD activity, elevated GSH levels and higher AsA/DHA and GSH/GSSG ratios are crucial for scavenging ROS, thereby aiding in oxidative stress resistance and enhancing salt tolerance (Panda et al. 2019).

Halophytic plants adjust the levels of endogenous phytohormones to regulate various defense mechanisms, including antioxidant defenses in response to salinity stress. The accumulation of phytohormones in halophytes varies depending on the species and the type of salt stress. For example, the accumulation of JA increases in *S. persica* but decreases in *H. salicornicum* (Panda et al. 2020) and *A. tripolium* (Wiszniewska et al. 2021) under salinity. SA levels remain relatively unchanged in *H. salicornicum* (Panda et al. 2020) and *S. persica* Kumari and Parida 2018), while they decrease significantly in *A. tripolium* (Wiszniewska et al. 2021).

Pathways for the synthesis of JA include the linolenic acid oxidation pathway. Hence, JA levels in plants under stressed conditions can be directly correlated with the amount of ROS present in the cell and the rate of ROS-induced lipid peroxidation. On the other hand, the level of ABA, which is known as stress hormone and has role in stomatal opening and closing to control transpiration rate in plants, significantly increases in *A. tripolium, Lycium humile, H. salicornicum* and *S. persica* in response to salinity stress ((Wiszniewska et al. 2021; Palchetti et al. 2021; Panda et al. 2020; Kumari and Parida 2018).

21.4 Halophytes in Soil Salinity Management

Common methods for reclaiming salt-affected areas include leaching and the application of chemical and organic amendments. Deeper soil layers receive the salt particles that are moved from the surface by leaching. The effectiveness of leaching is limited due to the lack of water as it can adversely affect microbial activity, total N and OC content, soil stability and overall fertility (Shaygan and Baumgartl 2022). The chemical method for reclaiming saline soil involves ion exchange utilizing $CaCO_3$ or other chemical alternatives. However, the limited solubility, quality concerns and high costs restrict the extensive application of these chemicals in soil reclamation. Notably, the application of organic amendments at optimal rates (not exceeding 50 T ha^{-1}) can markedly enhance the dissolution of $CaCO_3$. Additionally, these amendments improve soil physical properties such as structure, permeability and water holding capacity, as well as chemical properties including pH and cation exchange capacity, thereby promoting plant growth and microbial activity (Wu et al. 2021). However, the advantages of using organic amendments are constrained by the potential release of various contaminants, including heavy metals, phenolic compounds, ethylene, ammonia, excess salts and organic acids. These limitations are prompting farmers to explore alternative strategies that are environmentally sustainable, economically viable and visually appealing (Agarwal et al. 2022).

Phytoremediation, also referred to as vegetative bioremediation or biological reclamation, represents a promising alternative to conventional methods. This technique entails the cultivation of salt-tolerant plant species to extract salt ions from the soil, thereby rehabilitating soils degraded by salinity (Khan et al. 2023). Its increasing popularity can be attributed to several advantages. Phytoremediation is (i) cost-efficient, as it obviates the need for chemical purchases, (ii) improves nutrient availability in the soil, consequently enhancing crop yields, (iii) more uniformly removes excess salts, even from deeper soil strata, compared to gypsum and (iv) is environmentally sustainable, sequestering carbon in both above- and below-ground biomass (Al Zohbi 2023). Despite these advantages, the mechanistic understanding of phytoremediation of saline soils using halophytes remains limited. The intricate interactions involved pose challenges to fully comprehending the plants' overall mode of action in the saline soil remediation process and their beneficial effects on soil properties (Roe and MacFarlane 2022).

Halophytic plants can aid in salt removal by absorbing salt ions through their root systems and storing them in aboveground parts. According to Liang and Shi (2021), *Suaeda salsa* plants can accumulate around 27 milligrams of salt per gram of dry weight in their aboveground biomass. Similarly, Shaygan et al. (2018a, b) reported that the shoots of *Tecticornia pergranulata* have the capacity to accumulate 98 grams of salt per kilogram of dry weight. Additionally, certain research indicates that planting halophytes can mitigate salt ion accumulation in soils by improving the soil's physical properties. Deep-rooted halophytic plants serve as natural tillage agents, a phenomenon referred to as biological drilling. This process has been demonstrated to stabilize soil structure by reducing soil bulk density, thereby enhancing soil porosity and improving hydraulic characteristics. These modifications facilitate the leaching of sodium ions (Na^+) and their substitution with other cations while also promoting greater efficiency in nutrient utilization (Silva et al. 2016).

In accordance with these findings, intercropping *Suaeda salsa* with cotton (*Gossypium hirsutum*) resulted in significant reductions in soil salinity and bulk density, coupled with increases in soil porosity, soil organic carbon content, root growth, total aboveground biomass and cotton yield compared to conventional monoculture systems (Liang and Shi 2021). Additionally, *Atriplex nummularia* enhanced hydraulic conductivity and decreased bulk density, promoting a greater presence of macropores and higher rates of water infiltration under salinity conditions of 40.86 dS m^{-1} (Silva et al. 2016). Ashraf et al. (2010) also documented that the cultivation of halophytes lowered bulk density and increased soil porosity and hydraulic conductivity, thereby enhancing the removal of salt ions from saline-alkali soils.

Root respiration and the breakdown of organic matter are essential for decreasing Na^+ levels in saline-sodic soils by elevating CO_2 levels in the soil atmosphere. When irrigation water and soil moisture interact with CO_2, they produce H_2CO_3. As H_2CO_3 dissociates and nitrogen fixation occurs in halophyte roots, protons (H^+) are released, leading to the dissolution of calcite ($CaCO_3$) into Ca^{2+}, CO_2 and H_2O (Qadir et al. 2005).

The released Ca^{2+} helps displace Na^+ from the cation exchange sites of soil colloids. The displaced Na^+ is then removed through two main pathways: (i) uptake by halophytic plant roots and (ii) leaching facilitated by flooding irrigation prior to sowing, ultimately reducing soil sodicity and salinity (Qadir et al. 2005). However, relying on halophytes as a desalination tool is not universally reliable. Its success depends on environmental variables such as precipitation and temperature, as well as agricultural practices such as irrigated versus nonirrigated methods and the potential for salts to be re-circulated into the soil through plant residues (Shabala 2013). Therefore, comprehensive evaluations of different halophytes' capabilities, including understanding their mechanisms of action, are crucial before employing them for desalinating salt-affected regions. Moreover, further research is necessary to assess the efficacy of plant-mediated phytoremediation. If proven effective, this approach could offer a promising strategy for rehabilitating salinity-affected areas, especially in soils lacking calcium carbonate and under conditions where leaching is not viable.

21.4.1 Selection of Halophyte Species

The selection of species for utilizing halophytes in greening barren lands and using wastelands is crucial for ecological restoration. Halophytes can thrive in saline conditions where traditional crops fail due to their unique adaptability to harsh environments. Species such as *Salicornia bigelovii*, *Atriplex nummularia* and *Suaeda maritima* have demonstrated high salt tolerance and vigorous growth in saline soils. These plants help in soil stabilization and reclamation and contribute toward biodiversity by providing habitats for wildlife. Halophytes used for phytoremediation can be further utilized for biofuel production or as fodder for animals, thereby transforming wastelands into productive ecosystems (Slama et al. 2015). Thus, the strategic selection and cultivation of halophytes can play a pivotal role in sustainable land management and environmental conservation.

Barren land types such as deserts, salt flats, volcanic barrens, alpine barrens etc. are characterized by extreme aridity, salinity or rocky terrain leading to sparse vegetation and rocky or stony lands bearing minimal soil cover and limited plant growth. These areas typically support only specialized or adapted plant and animal life due to its harsh environmental conditions. These are often characterized by high salinity and low fertility and, therefore typically unsuitable for conventional agriculture (Li et al. 2020). Halophytes are widely distributed in various saline habitats such as coastal areas, post-industrial wastelands and marshlands (Hsouna et al. 2022). Table 21.1 presents notable halophyte species found in major barren regions of the world, including the Sahara Desert, Arabian Desert, Thar Desert and Great Victoria Desert. The use of native halophytes in landscape design has been explored in multiple African and Middle Eastern countries to enhance ecological sustainability (Sağlam and Önder 2018). Cultivation of halophytes on marginal and degraded lands using saline water offers a valuable resource for commercial-scale biofuel production (Abideen et al. 2023). Anatomical adaptations and ionic homeostasis in aquatic halophytes such as *Cyperus laevigatus* contribute to its resilience in highly saline environments (Mumtaz et al. 2021). The potential of halophytes to grow on saline-alkali lands in arid areas has also been highlighted. This emphasizes the

Table 21.1 Halophyte species of major barren areas of the world (Sahara Desert, Arabian Desert, Thar Desert and Great Victoria Desert)

Desert	Family	Species	Type of halophyte	Morphology
Sahara	Amaranthaceae	*Salicornia senegalensis*	Succulent halophyte	Succulent, jointed stems, small scale-like leaves
	Tamaricaceae	*Tamarix aphylla*	Shrub/tree halophyte	Small tree or large shrub, slender branches, scale-like leaves
	Plumbaginaceae	*Limonium axillare*	Herbaceous halophyte	Rosette of basal leaves, branched flowering stems, papery flowers
Arabian	Amaranthaceae	*Haloxylon persicum*	Shrub halophyte	Woody shrub, needle-like leaves, small inconspicuous flowers
	Chenopodiaceae	*Salsola imbricata*	Shrub halophyte	Dense, bushy shrub, small scale-like leaves, spiny branches
	Tamaricaceae	*Tamarix nilotica*	Shrub/tree halophyte	Shrub or small tree, feathery branches, tiny leaves, pink flowers
Thar	Amaranthaceae	*Salicornia brachiata*	Succulent halophyte	Succulent, jointed stems, small scale-like leaves
	Chenopodiaceae	*Salsola baryosma*	Shrub halophyte	Bushy shrub, small fleshy leaves, spiny tips
	Zygophyllaceae	*Zygophyllum simplex*	Herbaceous halophyte	Low growing, fleshy leaves, small yellow flowers
Great Victoria	Chenopodiaceae	*Atriplex nummularia*	Shrub halophyte	Spreading shrub, silver-gray leaves, small inconspicuous flowers
	Amaranthaceae	*Tecticornia arbuscula*	Succulent halophyte	Succulent, jointed stems, small scale-like leaves
	Plumbaginaceae	*Limonium austral*	Herbaceous halophyte	Rosette of basal leaves, branched flowering stems, papery flowers

importance of microbial diversity and community related to these plants in these areas (Gao et al. 2022).

Utilization of halophytes is essential for developing saline-alkali lands prone to desertification and ecological vulnerability (Duan et al. 2022). Halophytes have also been considered for forage potential, with specific taxa such as Poaceae and Fabaceae being identified as key species for breeding programs (Del Rosario Marinoni et al. 2019). The adaptation of halophytes to different habitats underscores their versatility in easing pressure on cropping systems, restoring degraded lands and conferring stress tolerance through gene transfer (González 2020). The prime potential of halophytes should be harnessed focusing on their application in the replanting and ecological recovery of barren and saline tracts. This includes utilizing halophytes to provide biomass for renewable energy, mitigating climate change, sequestering CO_2 and aiding in biological reclamation. Halophytes are economically important due to their potential use as sources of food, fodder, medicine, chemicals and seaweeds. These applications can provide viable livelihood opportunities for marginal farmers in barren or saline-affected regions.

In conclusion, the selection of halophytes for barren land reclamation holds significant promise for improving soil quality, restoring degraded lands, and enhancing ecological sustainability. By leveraging the unique characteristics of halophytes, such as their salt tolerance and adaptability to harsh environments, barren lands can be transformed into productive and biodiverse ecosystems.

21.4.2 Economic Aspect of Halophytes

Halophytes offer a range of economic benefits by serving as a source of food, forage and biofuel (Liu and Wang 2021). Their economic value extends to their role in greening efforts, coastal protection and combating climate change-related issues (Hameed et al. 2024). Halophytes have also been recognized for their nutritional and medicinal properties (Ahmed et al. 2024). The cultivation of halophytes presents a promising opportunity to increase productivity on salt-affected lands, which constitute a substantial portion of irrigated lands globally (Goehring et al. 2019). Prolonged cultivation of halophytes has been shown to enhance soil microbial diversity after the third year of cultivation (Zhao et al. 2024). Furthermore, halophytes can be utilized for various purposes such as biochar production, biogas generation and bio-oil extraction, making them valuable resources for sustainable agriculture (Wang et al. 2023). Harnessing the economic and ecological benefits of halophytes can aid in overcoming challenges related to limited freshwater resources by utilizing saline water for irrigation and restoring salt-affected lands (Bhatt et al. 2023). In conclusion, the economic potential of halophytes is vast and multifaceted, ranging from agriculture and bioenergy production to environmental management and ecosystem restoration (Table 21.2).

Table 21.2 Halophytes, their edible portion and their economic importance

Plant name	Edible portion	Economic importance
Salicornia europaea	Stems	Used in gourmet salads, rich in vitamins and minerals, potential biofuel source
Atriplex hortensis	Leaves	Consumed as a leafy vegetable, with high nutritional value, used in traditional medicine
Suaeda maritima	Leaves, young shoots	Edible raw or cooked, used in salads and soups, potential in coastal agriculture
Salicornia bigelovii	Seeds, stems	Potential oilseed crop, used in gourmet cuisine and biofuel production
Crithmum maritimum	Leaves, stems	Known as rock samphire, used in salads and pickles, aromatic and medicinal properties
Salsola soda	Leaves, stems	Traditional Italian vegetable (agretti), high in vitamins and minerals, culinary uses
Tecticornia arbuscula	Stems	Edible stems, potential for saline agriculture, used in traditional dishes
Mesembryanthemum crystallinum	Leaves, stems	Edible succulent, used in salads, rich in antioxidants, potential in cosmetics
Batis maritima	Leaves, young shoots	Can be used as an oil crop due to its high oil content and also used as herbal medicine for thyroid disorders

21.4.3 Halophytes as Biodiesel Source

Biodiesel is produced through the transesterification of fats and oils into fatty acid methyl esters. It is biodegradable, nontoxic and produces lower levels of carbon monoxide and unburned hydrocarbons. Traditionally, biodiesel is derived from crops such as soybean, canola and palm oil, which requires significant amounts of freshwater and fertile land. In contrast, halophytes offer a sustainable alternative because they thrive under barren lands and saline water resources. These plants are valuable for biodiesel production due to their ability to accumulate high levels of lipids, particularly triacylglycerols (TAGs) rich in saturated fatty acids (Duarte et al. 2019). Furthermore, the phytochemical composition of halophytes presents opportunities for biorefining, where biofuel production can be coupled with the extraction of value-added chemicals (Cybulska et al. 2014). Halophytes have also demonstrated potential in marine integrated multi-trophic aquaculture systems (Custódio et al. 2017). *Salicornia bigelovii* is a halophyte that has garnered significant attention due to its high oil content and ability to grow in seawater. It is native to coastal areas and produces seeds with up to 30% oil content, making it a viable candidate for biodiesel production (Katel et al. 2023). *Spartina alterniflora*, commonly known as smooth cordgrass, which is well-adapted to salt marshes and estuarine environments is another potential candidate (Konisky 2003). It exhibits rapid growth and significant biomass production although its oil yield is lower than *Salicornia bigelovii*. *Atriplex nummularia*, *Suaeda maritima*, *Haloxylon salicornicum* and *Tamarix* spp. represent a diverse group of plants that can be utilized for biodiesel production

(Joshi et al. 2019). By integrating these halophytes into biodiesel production systems, we can reduce reliance on freshwater resources and mitigate the environmental impact of biofuel production.

21.4.4 Halophytes as a Forage Source

Halophytes offer a promising alternative as potential forage crops, especially in arid and semiarid regions where conventional forage crops may struggle to grow (Li et al. 2021). High palatability, digestibility and good nutritional value indicated by high protein content and low fiber, ash and oxalate contents signify high fodder quality (Shaer 2006). Alkali grass (*Puccinellia distans*) is a halophyte having high protein content and good palatability for livestock, making it a valuable option for forage (Zhang et al. 2020). *Atriplex amnicola* also has been found to have high nutritive value compared to traditional forage crops as quality animal feed (Eissa et al. 2016). It is crucial to consider the palatability of halophytic forages when evaluating their suitability for different animal species. The palatability of halophytic species as forage crops for different animal species is a critical aspect to consider when evaluating their potential use in animal feed production. Some halophytic species may face limitations due to their salt content and the presence of antinutritional compounds (Koyro and Lieth 2011). Certain halophytic grass species, such as Sacaton grass (*Sporobolus airoides*) and Seashore paspalum (*Paspalum vaginatum*), are well-suited for cultivation in marginal lands that are unsuitable for conventional forage crops. These can maintain acceptable forage quality and palatability, even when exposed to high salinity levels. Other promising halophyte species for forage use are *Nitraria billardieri*, *Cynodon dactylo*, *Medicago sativa* and Kikuyu grass. They have also shown potential for use in sequential reuse systems, where saline-sodic drainage waters are used for irrigation (Grattan et al. 2002). While some halophytes offer high-quality forage with good palatability, factors such as salt content and antinutritional compounds can influence their acceptability by animals. Further research into the palatability of specific halophytic species is necessary to determine their suitability for different livestock and ensure optimal animal nutrition.

21.4.5 Halophytes for Landscaping

Because of how much emphasis is given to their appearance, ornamentals are typically irrigated with quality water because salinity typically has an adverse effect on both yield and quality (Shillo et al. 2002). Halophytes can also be utilized for their potential ornamental use in landscaping. The use of halophytes as ornamental plants offers an eco-friendly option for landscaping high-salinity areas. Along with esthetic value, they also contribute to the restoration of degraded areas (Shaygan et al. 2018a, b). Ornamental plants grown under saline conditions show a reduction of the photosynthetic rate due to stomatal limitations leading to the breakdown of accessory and photosynthetic pigments. Halophytic vegetation influences erosional and depositional patterns and impacts marsh survival (Geng et al. 2021). High biomass production capabilities of halophytes can be harnessed for mass-scale propagation and domestication strategies along with landscaping (Bhat and Hakeem 2020). Furthermore, the use of native halophytes in landscape design has been explored in various regions, such as Central Anatolia and Turkey, where specific halophytic plant species have been identified for landscape design purposes (Sağlam and Önder 2018).

21.5 Challenges Faced in Halophyte Cultivation

While halophytes are valued for their nutritional content, they also can accumulate heavy metals and other antinutritional factors such as nitrates, oxalates, phenols, tannins and saponins. Some halophytes exhibit a high level of heavy metal accumulation, with external salt content influencing heavy metal speciation and absorption rates (Lutts and Lefèvre 2015). This occurs due to the ability of certain halophytes to act as hyperaccumulators of elements such as cadmium, zinc, lead and copper. These accumulated ions can be excreted through salt glands or trichomes on the leaves in a process known as phytoexcretion. Another significant obstacle is the lack of understanding regarding consumer acceptance of halophyte products. It is essential for commercialization and encouraging farmers to invest in halophyte cultivation (Hamed et al. 2021). In conclusion, while halophytes offer significant nutritional benefits and have several applications, including food, forage, biofuel production and landscaping, they can also accumulate undesired factors such as heavy metals and toxic elements which can be toxic to herbivores. Understanding halophytes' dual nature in accumulating beneficial and undesired components is crucial for their sustainable utilization in various fields.

21.6 Halophytes for Future Agriculture

According to the current scenario, yield production from major crops is not enough to meet increasing global food demand which is expected to increase by at least 70 percent. By introducing specific genes from halophytes related to ion transporters, antioxidants and defense proteins into crop plants, it is possible to improve their salt tolerance and ensure sustainable agriculture in salinity-prone areas (Mann et al. 2021; Rahman et al. 2021). Recent estimations suggest that an additional 500 million acres of new croplands are required to feed the increasing global population over the next 30 years. However, only half of the proposed land area is available for farming expansion as most of this land is currently covered with forests and bushes to preserve the ecosystem. Additionally, freshwater resources are depleted, and the current agricultural irrigation practices require fresh water. To address this challenge, the use of seawater or brackish water for both inland and coastal sandy soils is required. Increasing the practice of biosaline agriculture is a legitimate option as empty barren areas are available to practice saline agriculture. The utilization of halophytes in agriculture not only aids in improving crop yields but also presents environmental advantages. Halophytes can also contribute to bioremediation efforts by treating effluents such as aquaculture waste, increasing biodiversity and facilitating efficient CO_2 sequestration through the expansion of aquatic biomass farms (Basto et al. 2024). Industrial wastelands can also be utilized as these plants exhibit efficient biochemical tolerance mechanisms against heavy metals. Additionally, halophytes are considered important alternative cash crops for sustainable saline production systems, thriving in conditions where conventional crops struggle, such as salt-affected soils and saline irrigation water (Custódio et al. 2021). This integration of halophytes into agricultural practices holds significant promise for sustainable and resilient farming systems. Halophytes offer improved crop productivity in saline environments along with phytoremediation, biodiversity enhancement and carbon sequestration which marks them as valuable assets in addressing food security and environmental challenges.

References

Abideen Z, Ansari R, Hasnain M, Flowers TJ, Koyro H-W, El-Keblawy A, Abouleish M, Khan MA (2023) Potential use of saline resources for biofuel production using halophytes and marine algae: prospects and pitfalls. Front Plant Sci 14. https://doi.org/10.3389/fpls.2023.1026063

Agarwal H, Kashyap VH, Mishra A, Bordoloi S, Singh PK, Joshi NC (2022) Biochar-based fertilizers and their applications in plant growth promotion and protection. 3 Biotech 12(6):136. https://doi.org/10.1007/s13205-022-03195-2

Ahmed F, El-Saied A-B, Din RSE, Swaify ZE (2024) Phytochemical screening and anticancer activities of some terrestrial and aquatic plants growing in saline habitats. Al-Azhar J Agric Res 48:89–106. https://doi.org/10.21608/ajar.2024.220188.1184

Al Zohbi G (2023) Review on sustainable technologies for liquid waste treatment and management. In: Waste recovery and management. CRC Press, pp 47–68

Ashraf MY, Ashraf M, Mahmood K, Akhter J, Hussain F, Arshad M (2010) Phytoremediation of saline soils for sustainable agricultural productivity. In: Plant adaptation and phytoremediation, pp 335–355. https://doi.org/10.1007/978-90-481-9370-7-15

Basto A, Peixoto D, Machado M, Costas B, Murta D, Valente LMP (2024) Physiological response of european sea bass (*Dicentrarchus labrax*) juveniles to an acute stress challenge: the impact of partial and total dietary fishmeal replacement by an insect meal. J Mar Sci Eng 12:815. https://doi.org/10.3390/jmse12050815

Bhat RA, Hakeem KR (2020) Biomass production of various halophytes. In: Handbook of halophytes: from molecules to ecosystems towards biosaline agriculture. Springer eBooks, pp 1–13

Bhatt A, Bhat NR, Suleiman MK, Al-Mansour H (2023) Prioritization of potential native plants from arabian peninsula based on economic and ecological values: implication for restoration. Sustain For 15:6139. https://doi.org/10.3390/su15076139

Castro B, Citterico M, Kimura S, Stevens DM, Wrzaczek M, Coaker G (2021) Stress-induced reactive oxygen species compartmentalization, perception and signalling. Nature plants 7(4):403–412

Cheng H, Inyang A, Li CD, Fei J, Zhou YW, Wang YS (2020) Salt tolerance and exclusion in the mangrove plant *Avicennia marina* in relation to root apoplastic barriers. Ecotoxicology 29:676–683. https://doi.org/10.1007/s10646-020-02203-6

Custódio M, Villasante S, Cremades J, Calado R, Lillebø AI (2017) Unravelling the potential of halophytes for marine integrated multi-trophic aquaculture (IMTA)— a perspective on performance, opportunities and challenges. Aquac Environ Interact 9:445–460. https://doi.org/10.3354/aei00244

Custódio V, Gonin M, Stabl G, Bakhoum N, Oliveira MM, Gutjahr C, Castrillo G (2021) Sculpting the soil microbiota. Plant J 109:508–522. https://doi.org/10.1111/tpj.15568

Cybulska I, Brudecki G, Alassali A, Thomsen M, Brown J (2014) Phytochemical composition of some common coastal halophytes of The United Arab Emirates. Emir J Food Agric 26:1046. https://doi.org/10.9755/ejfa.v26i12.19104

Del Rosario ML, Zabala JM, Taleisnik EL, Schrauf GE, Richard GA, Tomas PA, Giavedoni JA, Pensiero JF (2019) Wild halophytic species as forage sources: Key aspects for plant breeding. Grass Forage Sci 74:321–344. https://doi.org/10.1111/gfs.12410

Devi S, Nandwal AS, Angrish R, Arya SS, Kumar N, Sharma SK (2016) Phytoremediation potential of some halophytic species for soil salinity. Int J Phytoremediation 18:693–696

Duan Q, Zhu Z, Wang B, Chen M (2022) Recent progress on the salt tolerance mechanisms and application of Tamarisk. Int J Mol Sci 23:3325. https://doi.org/10.3390/ijms23063325

Duarte B, Matos AR, Marques JC, Caçador I (2019) Lipids in halophytes: stress physiology relevance and potential future applications. In: Halophytes and climate change: Adaptive mechanisms and potential uses. CABI eBooks, pp 359–371

Eissa MA, Ahmed EM, Reichman SM (2016) Production of the forage halophyte *Atriplex amnicolain* metal-contaminated soils. Soil Use Manag 32:350–356. https://doi.org/10.1111/sum.12275

Elnaggar A, Mosa KA, El-Keblawy A, Tammam A, El-Naggar M (2020) Physiological and biochemical insights for salt stress tolerance in the habitat-indifferent halophyte *Salsola drummondii* during the vegetative stage. Botany 98:1–17. https://doi.org/10.1139/cjb-2019-0160

Gao L, Huang Y, Liu Y, Mohamed OAA, Fan X, Wang L, Li L, Ma J (2022) Bacterial community structure and potential microbial coexistence mechanism associated with three halophytes adapting to the extremely hypersaline environment. Microorganisms 10:1124. https://doi.org/10.3390/microorganisms10061124

Geng L, D'Alpaos A, Sgarabotto A, Gong Z, Lanzoni S (2021) Intertwined eco-morphodynamic evolution of salt marshes and emerging tidal channel networks. Water Resour Res 57. https://doi.org/10.1029/2021wr030840

Goehring N, Verburg N, Saito N, Jeong N, Meki N (2019) Improving modeling of quinoa growth under saline conditions using the Enhanced Agricultural Policy Environmental eXtender model. Agronomy 9:592. https://doi.org/10.3390/agronomy9100592

González MB (2020) Adaptation of halophytes to different habitats. In: Seed dormancy germination. IntechOpen eBooks

Grattan SR, Zeng L, Shannon MC, Roberts SR (2002) Rice is more sensitive to salinity than previously thought. Calif Agric 56:189–198. https://doi.org/10.3733/ca.v056n06p189

Grigore MN, Toma C (2010) Halophytes: an integrative anatomical study. Springer

Hamed KB, Castagna A, Ranieri A, García-Caparrós P, Santin M, Hernandez JA, Espin GB (2021) Halophyte based Mediterranean agriculture in the contexts of food insecurity and global climate change. Environ Exp Bot 191:104601. https://doi.org/10.1016/j.envexpbot.2021.104601

Hameed A, Hussain S, Rasheed A, Ahmed MZ, Abbas S (2024) Exploring the potentials of halophytes in addressing Climate Change-Related issues: a synthesis of their biological, environmental, and socioeconomic aspects. WORLD 5:36–57. https://doi.org/10.3390/world5010003

Hsouna AB, Michalak M, Kukula-Koch W, Saad RB, Romdhane WB, Zeljković SĆ, Mnif W (2022) Evaluation of halophyte biopotential as an unused natural resource: the case of *Lobularia maritima*. Biomol Ther 12:1583. https://doi.org/10.3390/biom12111583

Hussain S, Khalid MF, Sohail M, Anjum MA, Ejaz S, Nafees M, Zakir I, Ahmad M, Ali S, Ahmad S (2021) Role of transporters in accumulating salt ions by halophytes. In: Approaches in remediation of inorganic pollutants, pp 11–40. https://doi.org/10.1007/978-981-15-6221-1_2

Joshi A, Kanthaliya B, Arora J (2019) Halophytes of Thar desert: potential source of nutrition and feedstuff. Int J Bioassays 8:5674–5683

Katel S, Yadav SPS, Oli S, Adhikari R, Shreeya S (2023) Exploring the potential of Salicornia: A halophyte's impact on Agriculture, the Environment, and the Economy. Peruvian J Agron 7:220–238. https://doi.org/10.21704/pja.v7i3.1991

Khan S, Masoodi TH, Pala NA, Murtaza S, Mugloo JA, Sofi PA, Kumar A (2023) Phytoremediation prospects for restoration of contamination in the natural ecosystems. Water 15(8):1498. https://doi.org/10.3390/w15081498

Konisky RA (2003) Spatial modeling and visualization of habitat response to hydrologic restoration in New England salt marshes. Doctoral Dissertation, University of New Hampshire, Durham

Koyro H, Lieth M (2011) Halophytic crops: a resource for the future to reduce the water crisis? Emir J Food Agric 23:01–16. https://doi.org/10.9755/ejfa.v23i1.5308

Krishnamurthy P, Jyothi-Prakash PA, Qin L, He J, Lin Q, Loh CS, Kumar PP (2014) Role of root hydrophobic barriers in salt exclusion of a mangrove plant *Avicennia officinalis*. Plant Cell Environ 37:1656–1671. https://doi.org/10.1111/pce.12272

Kumari A, Parida A (2018) Metabolomics and network analysis reveal the potential metabolites and biological pathways involved in salinity tolerance of the halophyte *Salvadora persica*. Environ Exp Bot 148:85–99. https://doi.org/10.1016/j.envexpbot.2017.12.021

Lee Y, Rubio MC, Alassimone J, Geldner NA (2013) Mechanism for localized lignin deposition in the endodermis. Cell 153:402–412. https://doi.org/10.1016/j.cell.2013.02.045

Li J, Yuan F, Liu Y, Zhang M, Liu Y, Zhao Y, Wang B, Chen M (2020) Exogenous melatonin enhances salt secretion from salt glands by upregulating the expression of ion transporter and vesicle transport genes in *Limonium bicolor*. BMC Plant Biol 20:1–11. https://doi.org/10.1186/s12870-020-02703-x

Li J, Ali EF, Majrashi A, Eissa MA, Ibrahim OHM (2021) Compost enhances forage yield and quality of River Saltbush in arid conditions. Agriculture 11:595. https://doi.org/10.3390/agriculture11070595

Liang J, Shi W (2021) Cotton/halophytes intercropping decreases salt accumulation and improves soil physicochemical properties and crop productivity in saline-alkali soils under mulched drip irrigation: a three-year field experiment. Field Crop Res 262:108027. https://doi.org/10.1016/j.fcr.2020.108027

Liu L, Wang B (2021) Protection of halophytes and their uses for cultivation of Saline-Alkali soil in China. Biology 10:353. https://doi.org/10.3390/biology10050353

Liu Q, Yang Y, Zhang X, Liu R, Song J (2020) Chloride allocation in the euhalophyte *Suaeda salsa* from different habitats in field and controlled saline conditions. Aquat Bot 167:103292. https://doi.org/10.1016/j.aquabot.2020.103292

Lu C, Yuan F, Guo J, Han G, Wang C, Chen M, Wang B (2021) Current understanding of role of vesicular transport in salt secretion by salt glands in recretohalophytes. Int J Mol Sci 22:2203. https://doi.org/10.3390/ijms22042203

Lutts S, Lefèvre I (2015) How can we take advantage of halophyte properties to cope with heavy metal toxicity in salt-affected areas? Ann Bot 115:509–528. https://doi.org/10.1093/aob/mcu264

Mann A, Kumar N, Lata C, Kumar A, Kumar A (2018) Halophytes-miracle plants to survive in extreme saline environment. J Agric Res 3(5):000176

Mann A, Kumar N, Lata C, Kumar A, Kumar A, Meena BL (2019) Functional annotation of differentially expressed genes under salt stress in *Dichanthium annulatum*. Plant Physiol Rep 24:104–111. https://doi.org/10.1007/s40502-019-0434-8

Mann A, Kumar N, Kumar A, Lata C, Kumar A, Meena BL, Mishra D, Grover M, Gaba S, Parameswaran C, Mantri N (2021) de novo transcriptomic profiling of differentially expressed genes in grass halophyte *Urochondra setulosa* under high salinity. Sci Rep 11:5548. https://doi.org/10.1038/s41598-021-85220-7

Mann A, Lata C, Kumar N, Kumar A, Kumar A, Sheoran P (2023) Halophytes as new model plant species for salt tolerance strategies. Front Plant Sci 14:1137211. https://doi.org/10.3389/fpls.2023.1137211

Mann A, Kumar N, Lata C, Kumar A, Meena BL, Kumar A (2024) Physiological and differential gene expression reveals a trade-off between antioxidant capacity and salt tolerance in *Urochondra setulosa* and *Dichanthium annulatum*. Plant Growth Regul 102:555–570

Mastrocicco M, Gervasio MP, Busico G, Colombani N (2021) Natural and anthropogenic factors driving groundwater resources salinization for agriculture use in the Campania plains (Southern Italy). Sci Total Environ 758:144033. https://doi.org/10.1016/j.scitotenv.2020.144033

Mi P, Yuan F, Guo J, Han G, Wang B (2021) Salt glands play a pivotal role in the salt resistance of four recretohalophyte *Limonium* Mill. species. Plant Biol 23:1063. https://doi.org/10.1111/plb.13284

Mumtaz S, Saleem MH, Hameed M, Batool F, Parveen A, Amjad SF, Mahmood A, Arfan M, Ahmed S, Yasmin H, Alsahli AA, Alyemeni MN (2021) Anatomical adaptations and ionic homeostasis in aquatic halophyte *Cyperus laevigatus* L. Under high salinities. Saudi J Biol Sci 28(5):2655–2666. https://doi.org/10.1016/j.sjbs.2021.03.002

Munns R, Passioura JB, Colmer TD, Byrt CS (2019) Osmotic adjustment and energy limitations to plant growth in saline soil. New Phytol 225:1091–1096. https://doi.org/10.1111/nph.15862

Mylonas I, Stavrakoudis D, Katsantonis D, Korpetis E (2020) Better farming practices to combat climate change. In: Climate change and food security with emphasis on wheat. Academic Press, pp 1–29. https://doi.org/10.1016/B978-0-12-819527-7.00001-7

Nikalje GC, Srivastava AK, Pandey GK, Suprasanna P (2017) Halophytes in biosaline agriculture: mechanism, utilization, and value addition. Land Degrad Dev 29:1081–1095. https://doi.org/10.1002/ldr.2819

Osman KT, Osman KT (2018) Saline and sodic soils. In: Management of soil problems, pp 255–298. https://doi.org/10.1007/978-3-319-75527-4_10

Palchetti MV, Reginato M, Llanes A, Hornbacher J, Papenbrock J, Barboza GE, Luna V, Cantero JJ (2021) New insights into the salt tolerance of the extreme halophytic species *Lycium humile* (Lycieae, Solanaceae). Plant Physiol Biochem 163:166–177. https://doi.org/10.1016/j.plaphy.2021.03.054

Panda A, Rangani J, Parida AK (2019) Cross talk between ROS homeostasis and antioxidative machinery contributes to salt tolerance of the xero-halophyte *Haloxylon salicornicum*. Environ Exp Bot 166:103799. https://doi.org/10.1016/j.envexpbot.2019.103799

Panda A, Rangani J, Parida AK (2020) Unraveling salt responsive metabolites and metabolic pathways using non-targeted metabolomics approach and elucidation of salt tolerance mechanisms in the xero-halophyte *Haloxylon salicornicum*. Plant Physiol Biochem 15:284–296. https://doi.org/10.1016/j.plaphy.2020.11.012

Qadir M, Noble AD, Oster JD, Schubert S, Ghafoor AJSU (2005) Driving forces for sodium removal during phytoremediation of calcareous sodic and saline–sodic soils: a review. Soil Use Manag 21(2):173–180. https://doi.org/10.1111/j.1475-2743.2005.tb00122.x

Rahman MM, Mostofa MG, Keya SS, Siddiqui MN, Ansary MM, Das AK, Rahman MA, Tran LS (2021) Adaptive mechanisms of halophytes and their potential in improving salinity tolerance in plants. Int J Mol Sci 22:10733. https://doi.org/10.3390/ijms221910733

Roe RA, MacFarlane GR (2022) The potential of saltmarsh halophytes for phytoremediation of metals and persistent organic pollutants: an Australian perspective. Mar Pollut Bull 180:113811. https://doi.org/10.1016/j.marpolbul.2022.113811

Roy SJ, Negrão S, Tester M (2014) Salt resistant crop plants. Curr Opin Biotechnol 26:115–124. https://doi.org/10.1016/j.copbio.2013.12.004

Sağlam C, Önder S (2018) The use of native halophytes in landscape design in the central Anatolia, Turkey. Türk Tarım Gıda Bilim Ve Teknoloji Dergisi 6:1718–1726. https://doi.org/10.24925/turjaf.v6i12.1718-1726.1954

Shabala S (2013) Learning from halophytes: physiological basis and strategies to improve abiotic stress tolerance in crops. Ann Bot 112(7):1209–1221. https://doi.org/10.1093/aob/mct205

Shaer HME (2006) Halophytes as cash crops for animal feeds in arid and semi-arid regions. In: Biosaline agriculture and salinity tolerance in plants. Birkhäuser Basel eBooks, pp 117–128

Shaygan M, Baumgartl T (2022) Reclamation of salt-affected land: a review. Soil systems 6(3):61. https://doi.org/10.3390/soilsystems6030061

Shaygan M, Mulligan D, Baumgartl T (2018a) The potential of three halophytes (*Tecticornia pergranulata, Sclerolaena longicuspis,* and *Frankenia serpyllifolia*) for the rehabilitation of brine-affected soils. Land Degrad Dev 29:2002–2014. https://doi.org/10.1002/ldr.2954

Shaygan M, Mulligan D, Baumgartl T (2018b) The potential of three halophytes (*Tecticornia pergranulata, Sclerolaena longicuspis*, and *Frankenia serpyllifolia*) for the rehabilitation of brine-affected soils. Land Degrad Dev 29(6):2002–2014. https://doi.org/10.1002/ldr.2954

Shillo R, Ding M, Pasternak D, Zaccai M (2002) Cultivation of cut flower and bulb species with saline water. Sci Hortic 92:41–54. https://doi.org/10.1016/s0304-4238(01)00276-x

Silva YJ, Silva YJ, Freire MB, Lopes EA, Santos MA (2016) *Atriplex nummularia* Lindl. as alternative for improving salt-affected soils conditions in semiarid environments: a field experiment. Chil J Agric Res 76(3):343–348. https://doi.org/10.4067/S0718-58392016000300012

Slama I, Abdelly C, Bouchereau A, Flowers T, Savouré A (2015) Diversity, distribution and roles of osmoprotective compounds accumulated in halophytes under abiotic stress. Ann Bot 115:433–447. https://doi.org/10.1093/aob/mcu239

Stavi I, Thevs N, Priori S (2021) Soil salinity and sodicity in drylands: a review of causes, effects, monitoring, and restoration measures. Front Environ Sci 9:712831. https://doi.org/10.3389/fenvs.2021.712831

Steiner M (1935) On the ecology of the salt marshes of the northeastern United States of North America. Jahrb F Wiss Botonay 81:94–202

Von Sengbusch P (2003) Halophyte Botanik online. University of Hamburg

Wang P, Wang CM, Gao L, Cui YN, Yang HL, De Silva NDG, Ma Q, Bao AK, Flowers TJ, Rowland O (2020) Aliphatic suberin confers salt tolerance to Arabidopsis by limiting Na^+ influx, K^+ efflux and water backflow. Plant Soil 448:603–620. https://doi.org/10.1007/s11104-020-04464-w

Wang Y, Lin Q, Liu Z, Liu K, Wang X, Shang J (2023) Salt-affected marginal lands: a solution for biochar production. Biochar 5. https://doi.org/10.1007/s42773-023-00219-9

Wei X, Yan X, Yang Z, Han G, Wang L, Yuan F, Wang B (2020) Salt glands of recretohalophyte *Tamarix* under salinity: their evolution and adaptation. Ecol Evol 10:9384–9395. https://doi.org/10.1002/ece3.6625

Wiszniewska A, Koźmińska A, Hanus-Fajerska E, Dziurka K, Dziurka M (2021) Insight into phytohormonal modulation of defense mechanisms to salt excess in a halophyte and a glycophyte from Asteraceae family. Plant Soil 463:55–76. https://doi.org/10.1007/s11104-021-04889-x

Wu L, Zhang S, Ma R, Chen M, Wei W, Ding X (2021) Carbon sequestration under different organic amendments in saline-alkaline soils. Catena 196:104882. https://doi.org/10.1016/j.catena.2020.104882

Zhang W, Liu J, Zhang Y, Qiu J, Li Y, Zheng B, Hu F, Dai S, Huang X (2020) A high-quality genome sequence of alkaligrass provides insights into halophyte stress tolerance. Sci China Life Sci 63:1269–1282. https://doi.org/10.1007/s11427-020-1662-x

Zhao C, Zhang H, Song C, Zhu JK, Shabala S (2020) Mechanisms of plant responses and adaptation to soil salinity. Innovations 1. https://doi.org/10.1016/j.xinn.2020.100017

Zhao S, Liu X, Banerjee S, Hartmann M, Peng B, Elvers R, Zhao Z-Y, Zhou N, Liu J-J, Wang B, Tian C-Y, Jiang J, Lian T-X (2024) Continuous planting of euhalophyte *Suaeda salsa* enhances microbial diversity and multifunctionality of saline soil. Appl Environ Microbiol 90:e0235523. https://doi.org/10.1128/aem.02355-23

The manufacturer's authorised representative in the EU is Springer Nature Customer Service Centre GmbH, Europaplatz 3, 69115 Heidelberg, Germany. If you have any concerns regarding our products, please contact ProductSafety@springernature.com

Printed and bound by CPI Group (UK) Ltd, Croydon, CR0 4YY
01/10/2026
02225941-0015